A Laboratory Manual and Study Guide for

ANATOMY and PHYSIOLOGY

Third Edition

by **Kenneth G. Neal**

and **Barbara H. Kalbus**

Long Beach City College
Long Beach, California

Burgess Publishing Company • Minneapolis, Minnesota

Printed in the United States of America
ISBN 0-8087-1423-6

10 9 8 7 6 5 4 3

Preface

This manual was designed to be used in an introductory course in anatomy and physiology. It contains sufficient experiments so that it can be used in either a one or two semester course. It is suitable for use by students in the health technologies, physical education, and home economics, as well as by the general student.

The manual has been designed to be used by students to develop laboratory skills, to become familiar with the methods of science, and to learn the significant concepts of anatomy and physiology. The student is provided with opportunities to apply that which has been learned in lectures in experiments which are both interesting and informative.

This new edition includes many additional illustrations (almost all of the illustrations have been redrawn to improve clarity of detail) of both the human and the fetal pig. Also new to this edition are directions for dissection of the musculature and central nervous system of the fetal pig. Many of the physiology experiments were rewritten to incorporate in the main body of the manual the directions for the kymograph (which were formerly in an appendix), and to include directions for two different models of the physiograph. The directions for the physiograph are supported with illustrations showing the actual connections and settings on the physiograph. This both simplifies and clarifies the experiments for the student. Most of the many supportive illustrations for physiological experiments are entirely new.

The general organization of the manual is unchanged from the previous edition, since we feel that the following features facilitate the study of anatomy and physiology:

1. Each exercise or experiment includes a precisely stated purpose (objective), the materials needed, a discussion (when appropriate), and a procedure with detailed instructions. The numerous illustrations, diagrams, and tables clarify the experimental directions. As a general rule, little or no additional orientation is required. This saves valuable time in preparing and performing laboratory activities for both the student and the instructor.
2. The Results and Questions section of each exercise is designed to be removed from the

manual for recording data during an experiment, for answering questions, and for submitting the material for evaluation.

3. Experiments of a physiological nature are supported (and well-balanced) with anatomical studies involving the dissection of the fetal pig, sheep heart, sheep brain, sheep eye, sheep kidney, and the frog. Original drawings are included with the instructions for dissection, which greatly facilitate the identification of structures in the specimen.
4. Each major unit includes a relatively comprehensive self-test that covers lecture, laboratory, and practical applications. Keys are at the end of each test. Most of the illustrations can also serve as self-tests with keys for each illustration.

The flexibility of the manual allows for the demonstration of key concepts with or without assorted pieces of apparatus. For example, living frog experiments may be conducted for the purpose of demonstrating the principles of nerve, muscle, and circulatory physiology without using any elaborate equipment. However, experiments involving the use of physiological apparatus are also included, and even if these are not performed, they provide a reference for the student which may improve his understanding of the nature of the experiments described in lecture and the text.

The fetal pig is emphasized in the manual, and it may be desirable to limit all dissection to this single, inexpensive animal. In addition, detailed instructions for the dissection of several organs of the sheep that are included may be readily adapted to beef or other mammalian structures. A limited dissection of the frog is also included.

Dissection may be done in the first few laboratories and then attention may be concentrated on physiological experiments or, as suggested by the location of the exercises in the manual, dissection may be performed throughout the course. The latter is the method recommended by the authors, wherein a portion of most laboratory periods are used for dissection. In many cases, this may amount to just 15 or 20 minutes. Furthermore, we suggest that students should be allowed to continue dissections outside of the regular laboratory time.

All of the experiments and exercises in this manual have been tested under laboratory conditions over the past several years by the authors.

Kenneth G. Neal
Barbara H. Kalbus

Spring 1976

Acknowledgments

We would like to acknowledge our deep appreciation for the long hours spent by Robert Bush in preparing most of the illustrations in this manual; to Judith Ramos, Ph.D., for her preparation of the illustrations and write-up for the dissection of the muscular system of the fetal pig; to Linda Neal for her hours spent in preparing label lines, modification and correction of illustrations, and proofreading much of the manuscript.

All illustrations used in this manual are the copyrighted property of E.L.O.T. Publishing Company (P.O. Box 8294, Long Beach, CA 90808). Our thanks to E.L.O.T. Publishing Company for granting us permission to use the illustrations in this publication.

We would be remiss if we failed to mention the patience of our families, who aided us in a number of tangible ways in the preparation of this manual, and who bore with us during the long hours spent trying to meet deadlines.

Many of our students have participated in helping us improve the quality of the manual by providing us with many functional suggestions.

The Authors

Contents

Preface iii

Acknowledgments iv

Orientation and Suggestions for Students xvii

1 Introduction to Anatomy 1

Exercise 1-A. Organization of the Body as a Whole 1
Exercise 1-B. Anatomic Terminology 2
Exercise 1-C. The External Characteristics of the Fetal Pig 6

Self-Test—Introduction to Anatomy 9
Results and Questions for Chapter 1 11

2 Basic Microscopy 13

Exercise 2-A. Structure of the Microscope 13
Exercise 2-B. Microscope Drawings 15
Exercise 2-C. Orientation of Images Viewed through the Compound Microscope 16
Exercise 2-D. Thread Slide 17

Self-Test—Microscopy 17
Results and Questions for Chapter 2 19

3 Cells, Mitosis, and Tissues 21

Exercise 3-A. The Structure and Functions of Cells 21
Exercise 3-B. Mitosis 24
Exercise 3-C. Epithelial Tissue 24
Exercise 3-D. Connective Tissue 27
Exercise 3-E. Muscular Tissue 31
Exercise 3-F. Nerve Tissue 33
Exercise 3-G. The Skin 34

Self-Test—Cells, Mitosis, Tissues, and Skin 36
Results and Questions for Chapter 3 41

4 The Skeletal System 47

Exercise 4-A. Composition and Structure of Bone 47
Exercise 4-B. The Appendicular Skeleton 48
Exercise 4-C. The Axial Skeleton 58
Exercise 4-D. The Fetal Skeleton 69
Exercise 4-E. Joints 69
Exercise 4-F. The Skeletal System of the Fetal Pig 70
Self-Test—Skeletal System 71
Results and Questions for Chapter 4 75

5 Physical Transport of Materials 79

Exercise 5-A. Filtration 79
Exercise 5-B. Diffusion 80
Exercise 5-C. Dialysis 81
Exercise 5-D. Plasmolysis 82
Exercise 5-E. Osmosis 83
Exercise 5-F. Hemolysis and Crenation 83
Exercise 5-G. Brownian Movement 84
Self-Test—Physical Transport of Materials 84
Results and Questions for Chapter 5 87

6 The Muscular System 91

Exercise 6-A. The Gross Anatomy of the Major Superficial Human Muscles 91
Exercise 6-B. Dissection of the Muscular System of the Fetal Pig 91
Self-Test—Muscular System 101
Results and Questions for Chapter 6 105

7 Introduction to Stimulating and Recording Apparatus 107

Exercise 7-A. Faradic Current 107
Exercise 7-B. Tetanizing Stimulation 109
Exercise 7-C. Use of the Kymograph and Ink Recording Apparatus 109
Exercise 7-D. The Signal Magnet 112
Exercise 7-E. The Tuning Fork 113
Exercise 7-F. Use of the Physiograph 115
Results and Questions for Chapter 7 133

8 Muscle Physiology 141

Exercise 8-A. Strength of Stimulus and Height of Contraction 141
Exercise 8-B. Genesis of Tetanus 148
Exercise 8-C. Duration of a Single Muscle Twitch 150

Exercise 8-D. Muscle Fatigue 151
Exercise 8-E. Physiology of Human Muscle 152
Self-Test—Muscle Physiology 153
Results and Questions for Chapter 8 157

9 **The Digestive System 165**

Exercise 9-A. Anatomy of the Human Digestive System 165
Exercise 9-B. Dissection of the Digestive System of the Fetal Pig 168
Exercise 9-C. The Action of Salivary Amylase 174
Exercise 9-D. Protein Digestion 176
Exercise 9-E. The Effect of Rennin on Milk 177
Exercise 9-F. Pancreatic Digestion of Fat 177
Exercise 9-G. Bile 178
Self-Test—Digestive System 178
Results and Questions for Chapter 9 185

10 **Metabolism 191**

Exercise 10-A. Basal Metabolic Rate 191
Exercise 10-B. Diurnal Variation in Body Temperature 192
Exercise 10-C. Effect of Exercise on Temperature 193
Exercise 10-D. Mapping of Sweat Glands 193
Self-Test—Metabolism 194
Results and Questions for Chapter 10 197

11 **The Respiratory System 203**

Exercise 11-A. Anatomy of the Human Respiratory System 203
Exercise 11-B. Dissection of the Respiratory System of the Fetal Pig 206
Exercise 11-C. Chest Measurements during Respiration 208
Exercise 11-D. Respiratory Volumes 209
Exercise 11-E. Respiratory Movements 210
Exercise 11-F. The Effect of Carbon Dioxide on the Respiratory Center 216
Exercise 11-G. Temperature and Composition of Exhaled Air 216
Self-Test—Respiratory System 217
Results and Questions for Chapter 11 221

12 **The Circulatory System 229**

Exercise 12-A. Blood Tests 229
Exercise 12-B. Structure of an Artery, Vein, and a Lymph Vessel Valve 232
Exercise 12-C. Gross Anatomy of the Human Circulatory System 233
Exercise 12-D. The Dissection of the Sheep Heart 233
Exercise 12-E. The Dissection of the Circulatory System of the Fetal Pig 240
Exercise 12-F. Physiology of the Frog Heart 244
Exercise 12-G. Microcirculation in the Fish or Frog 254
Exercise 12-H. Heart Sounds and Pulse Rate 255
Exercise 12-I. Blood Pressure 258
Exercise 12-J. Electrocardiography with the Physiograph 264

Self-Test—Circulatory System 270
Results and Questions for Chapter 12 277

13 The Nervous System 297

Exercise 13-A. Anatomy of the Nervous System 297
Exercise 13-B. The Dissection of the Sheep Brain 298
Exercise 13-C. Dissection of the Brain and Spinal Cord of the Fetal Pig 302
Exercise 13-D. Reflex Action in the Normal, Spinal, and Double-Pithed Frog 303
Exercise 13-E. Demonstration of Individual Human Reflexes 304
Exercise 13-F. Reaction Time 305
Self-Test—Nervous System 306
Results and Questions for Chapter 13 311

14 Sense Organs 317

Introduction 317
Exercise 14-A. Gross Anatomy of the Human Eye 317
Exercise 14-B. Dissection of the Sheep Eye 318
Exercise 14-C. Accommodation 321
Exercise 14-D. Measuring the Near Point of Accommodation 321
Exercise 14-E. Snellen Chart Test for Visual Acuity 322
Exercise 14-F. Test for Astigmatism 322
Exercise 14-G. Color Blindness 323
Exercise 14-H. The Blind Spot 323
Exercise 14-I. Anatomy of the Ear 323
Exercise 14-J. Bone Conduction of Sound Waves 324
Exercise 14-K. Localization of Sounds 327
Exercise 14-L. Localization 327
Exercise 14-M. Cutaneous Sensations 328
Exercise 14-N. Pressure Sensation 328
Exercise 14-O. Receptor Adaptation 329
Exercise 14-P. Referred Pain 329
Exercise 14-Q. Location of Taste Buds 329
Self-Test—Sense Organs 330
Results and Questions for Chapter 14 335

15 The Urinary System 345

Exercise 15-A. Anatomy of the Human Urinary System 345
Exercise 15-B. Dissection of the Urinary System of the Fetal Pig 346
Exercise 15-C. Urinalysis 349
Self-Test—The Urinary System 352
Results and Questions for Chapter 15 355

16 The Endocrine System 361

Exercise 16-A. Endocrine Glands 361
Exercise 16-B. Action of Hormones 362
Self-Test—The Endocrine System 364
Results and Questions for Chapter 16 367

17 The Reproductive System 371

Exercise 17-A. The Human Male Reproductive System 371
Exercise 17-B. The Human Female Reproductive System 375
Exercise 17-C. The Endometrial (Menstrual) Cycle 375
Exercise 17-D. The Placenta 381
Exercise 17-E. Dissection of the Reproductive System of the Fetal Pig 382
Exercise 17-F. Embryology 385

Self-Test—The Reproductive System 387
Results and Questions for Chapter 17 391

18 Genetics 397

Exercise 18-A. Directions for Working Genetics Problems 397
Exercise 18-B. Genetics Problems 403
Exercise 18-C. Variability in Man 404

Self-Test—Genetics 406
Results and Questions for Chapter 18 409

Appendix 1. Formulas Used in Preparation of Solutions 413

Appendix 2. Weights and Measures 415

Appendix 3. Respiration and Metabolism 417

3-A. Dubois Body Surface Chart 417
3-B. Predicted Vital Capacity—Males 418
3-C. Predicted Vital Capacity—Females 419

Appendix 4. Definitions for Genetics 421

Appendix 5. The Determination of pH 425

List of Figures

1-1 Body Cavities 2
1-2 Regions of the Body 3
1-3 Planes of the Body 5
2-1 The Microscope 14
3-1 Structure of Cell (observed with the light microscope) 21
3-2 Generalized Cell (as viewed with an electron microscope) 22
3-3 *Ameba proteus* 23
3-4 Mitosis (as observed in whitefish blastula) 25
3-5 Simple Squamous Epithelium (surface view) 26
3-6 Simple Columnar Epithelium (cross section) 26
3-7 Ciliated Columnar Epithelium (cross section) 26
3-8 Simple Cuboidal Epithelium (cross section) 26
3-9 Stratified Squamous Epithelium (cross section) 26
3-10 Areolar Connective Tissue 28
3-11 Adipose Tissue 28
3-12 Hyaline Cartilage 29
3-13 Cross Section of Diaphysis (magnified) 30
3-14 Longitudinal Section through Compact Bone (430X) 30
3-15 Cross Section through Two Haversian Systems 30
3-16 Human Blood Cells 32
3-17 Skeletal Muscle Fiber (l.s.) 32
3-18 Smooth (visceral) Muscle Fibers (teased) 33
3-19 Cardiac Muscle Tissue (l.s.) 33
3-20 Motor Neuron (smear) 34
3-21 Myelinated Nerve Fiber (l.s.) 34
3-22 The Skin (cross section) 35
4-1 Longitudinal Section through Femur 48
4-2 Right Clavicle (cranial aspect), Superior Surface 49
4-3 Posterior View of the Right Scapula 49
4-4 Right Humerus (anterior view) 50
4-5 Right Humerus (posterior view) 50
4-6 Right Radius (anterior view) 51
4-7 Right Ulna (anterior view) 51
4-8 Dorsal Surface of Right Hand 52

4-9 Lateral View of the Right Os Coxae 53
4-10 The Female Pelvis 54
4-11 Right Femur (anterior view) 55
4-12 Right Femur (posterior view) 55
4-13 Right Fibula (anterior view) 56
4-14 Right Tibia (anterior view) 56
4-15 Bones of the Right Foot (medial view) 57
4-16 Bones of the Right Foot (dorsal view) 57
4-17 The Skull (lateral view) 58
4-18 Anterior View of the Skull 59
4-19 Sagittal Section of the Skull 60
4-20 Superior View of the Ethmoid 61
4-21 Anterior View of the Sphenoid 61
4-22 Superior View of the Sphenoid 62
4-23 Inferior View of the Occipital 62
4-24 Left Palatine 63
4-25 Hyoid 63
4-26 Nasal Bones, outer surface 63
4-27 Lacrimal (lateral aspect) 63
4-28 Vomer 63
4-29 Right Inferior Nasal Concha (lateral surface) 63
4-30 Zygomatic (Malar) 63
4-31 Auditory Ossicles 63
4-32 The Mandible 64
4-33 Lateral View of the Right Maxilla 64
4-34 The Right Temporal (lateral aspect) 65
4-35 The Right Temporal (basal aspect) 65
4-36 The Atlas (first cervical vertebra) 66
4-37 The Axis (second cervical vertebra) 66
4-38 Seventh Cervical Vertebra 66
4-39 Typical Thoracic Vertebra 67
4-40 Typical Lumbar Vertebra 67
4-41 The Sacrum and Coccyx (ventral aspect) 67
4-42 The Bony Thorax 68
4-43 Fetal Skull (superior surface) 69
4-44 Fetal Skull (lateral view) 69
4-45 Anterior View of Knee Joint 70
4-46 Lateral View of Knee Joint (sagittal section) 70
4-47 Anterior View of Shoulder Joint 71
5-1 Setup for Filtration Experiment 80
5-2 *Elodea* Cell 82
5-3 Apparatus for Osmosis Demonstration 83
6-1 Anterior View of the Muscles 92
6-2 Posterior View of the Muscles 93
6-3 Lateral View of Major Muscles of Head and Neck 94
6-4 Anterior View of Major Muscles of Head and Neck 94
6-5 Lateral View of Superficial Muscles 96
6-6 Ventral View of Deep Muscles 98
6-7 Ventral View of Superficial and Deep Muscles 100
7-1 Model 330 Induction Stimulator 108
7-2 Movement of the Kymograph Drum 109
7-3 The Signal Magnet 112
7-4 Diagram for Tuning Fork Setup 114
7-5 Physiograph Recording Channel 115

7-6A Transducer—The Photoelectric Pulse Pickup to be used with DMP-4A and Physiograph Four-A 116
7-6B Transducer—Photoelectric Pulse Pickup to be used with DMP-4B 116
7-6C Transducer—Bellows Pneumograph to be used with either the DMP-4A or DMP-4B Models 117
7-6D Preamplifier—Impedance Pneumograph to be used with DMP-4A 117
7-6E Preamplifier—Hi-Gain Preamplifier to be used with DMP-4A 117
7-6F Preamplifier—Electrosphygmograph to be used with DMP-4A 117
7-6G Coupler—Electrosphygmograph Coupler to be used with DMP-4B 118
7-6H Coupler & Channel Amplifier—Impedance Pneumograph Coupler to be used with DMP-4B 118
7-6I Coupler & Channel Amplifier—Hi-Gain Coupler to be used with DMP-4B 118
7-7 Physiograph Four-A with Recording Channels and Plug-in Modules 119
7-8 The Desk Model Physiograph DMP-4B 120
7-9 Paper Loading for the Physiograph Four-A 121
7-10 Paper Loading for the DMP-4B 122
7-11 Pen Assembly 122
7-12 Inking Assemblies 123
7-13 Channel Amplifier on the Physiograph Four-A and DMP-4A 124
7-14 Channel Amplifier on the DMP-4B 125
7-15 Stimulator 127
7-16 Setup for Stimulation of Tongue 128
7-17 Photoelectric Pulse Pickup with Physiograph Four-A and DMP-4A 129
7-18 Recording of Pulse Wave 130
7-19 Photoelectric Pulse Pickup with the DMP-4B 131
8-1 Decapitation of the Frog 142
8-2 Muscle Lever with Ink Adapter 143
8-3 Faradic Stimulation of Muscle 143
8-4 Setup for Frog Muscle Stimulation with Physiograph Four-A 146
8-5 Setup for Frog Muscle Stimulation with DMP-4B 147
8-6 Stimulator Model 340 149
8-7 Duration of a Muscle Twitch 151
8-8 The Ergograph 153
9-1 Digestive Organs 166
9-2 Longitudinal Section through a Molar 167
9-3 The Deciduous Teeth 167
9-4 The Permanent Teeth 167
9-5 Diagram Showing Initial Incisions for Fetal Pig Dissection 169
9-6 Dissection of the Salivary Glands 170
9-7 The Oral Cavity 171
9-8 Superficial View of Digestive Organs of the Fetal Pig 172
9-9 Digestive Organs of the Fetal Pig (with liver and spleen pulled back and small intestine removed on the right side) 173
9-10 Human Abdominal Blood Vessels and Ducts 175
11-1 Larynx (anterior view) 204
11-2 Larynx (posterior view) 204
11-3 Sagittal Section through the Head 204
11-4 Respiratory Tract 205
11-5 Superficial View of Thoracic Cavity with Neck Dissected 207
11-6 Dissection of the Organs of the Thoracic Cavity of the Fetal Pig 208
11-7 Diagram of Pneumograph Setup 211
11-8 Diagram Showing Position of Bellows Pneumograph 212
11-9 Tracing of Two Normal Breaths at Paper Speed 2.5 cm/sec 213
11-10 Impedance Pneumograph 214

11-11 Impedance Pneumograph, with Physiograph DMP-4B 215
12-1 Digestive Organs and Portal Circulation 234
12-2 Major Arteries of the Body 235
12-3 Major Veins of the Body 236
12-4 Blood Vessels of the Neck and Interior of the Heart 237
12-5 Ventral View of the Sheep Heart 238
12-6 Right Side of Sheep Heart 239
12-7 Veins of the Thorax and Neck 241
12-8 Major Arteries of the Fetal Pig (heart pulled to the right) 243
12-9 Coronal Section through Fetal Pig Heart 244
12-10 Pithing the Frog 245
12-11 Ventral View of the Internal Organs of the Frog 247
12-12 Ventral View (frog heart) 248
12-13 Longitudinal Section (frog heart) 248
12-14 Dorsal View (frog heart) 248
12-15 Diagram of Setup for Stimulation of Frog Heart using Kymograph 249
12-16 Heart Lever 250
12-17 Setup for the Stimulation of the Frog Heart with the DMP-4B 251
12-18 Ligature in Atrioventricular Groove 253
12-19 Photoelectric Pulse Pickup with DMP-4B 256
12-20 Record of Pulse with Photoelectric Pulse Pickup 257
12-21 Setup for the Electrosphygmograph 260
12-22 Impedance Pneumograph, with Physiograph DMP-4B 262
12-23 Setup for Electrosphygmograph with the DMP-4B 263
12-24 Tracing of Blood Pressure with Electrosphygmograph 264
12-25 Normal Electrocardiogram Showing Three Beats of the Heart 265
12-26 Setup for Simultaneously Recording Respiration and Heart Action with the Physiograph Four-A 267
12-27 Equipment Setup for Simultaneously Recording EKG and Respiration with the Physiograph DMP-4B 268
12-28 Setup for Recording Electrocardiogram 269
12-29 Tracing of EKG and Respiration 270
13-1 Sagittal Section of the Brain 299
13-2 Inferior Surface of the Brain 299
13-3 Dorsal View of Sheep Brain (with cerebellum separated from cerebral hemispheres) 300
13-4 Ventral View of Sheep Brain 301
13-5 Sagittal Section through Sheep Brain 302
13-6 Diagram of Setup for Measuring Reaction Time 305
13-7 Cross Section of Spinal Cord 312
14-1 Midsagittal Section through the Eye 319
14-2 Anterior View of Eye Showing Lacrimal Apparatus 319
14-3 Extrinsic Muscles of the Right Eye 320
14-4 Osseous Labyrinth 324
14-5 The Ear 325
14-6 Axial Section through Entire Cochlea Showing Canals 326
14-7 Radial Section through One Coil of the Cochlea 326
15-1 Longitudinal Section through the Kidney 346
15-2 Renal Blood Supply 346
15-3 The Nephron 347
15-4 Renal Corpuscle 347
15-5 Ventral View of Urinary System (with intestine pulled to the right) of the Male Fetal Pig 348
16-1 Location of Endocrine Glands 363
17-1 Cross Section through a Seminiferous Tubule 372

17-2 Cross Section through a Seminiferous Tubule (enlarged) 372
17-3 Cross Section through the Epididymis 372
17-4 Vertical Section through the Testis and Adjacent Structures 373
17-5 Parasagittal Section through Male Pelvis 374
17-6 Comparison of Spermatogenesis and Oogenesis 376
17-7 Diagrammatic Section through Human Ovary 377
17-8 Coronal Section through the Female Reproductive Organs 377
17-9 Median Sagittal Section through Female Reproductive Organs 378
17-10 Female External Genital Organs (perineal view) 379
17-11 Sagittal Section through the Breast 379
17-12 Section of Placenta Showing Scheme of Circulation 381
17-13 Reproductive Organs of the Female Fetal Pig (with intestine and urethra pulled to the right and right ovary turned to left)
17-14 Ventral View of Urinary System and Reproductive Organs of the Male Fetal Pig (with intestine pushed to the right) 384
17-15 Stages of Starfish Embryology 385
17-16 Twenty-six Day Old Embryo 386
17-17 Thirty-four Day Old Embryo 386
17-18 Forty-three Day Old Embryo 386
17-19 Fifty-six Day Old Embryo 386
17-20 Full Term Human Fetus 387
18-1 The Structure of DNA 398

Orientation and Suggestions for Students

A. Introduction

The laboratory provides an opportunity to study at close range materials related to the topics described in lecture. A number of the features of the manual were suggestions of students.

The laboratory experience can be the most rewarding part of the entire course, if a student approaches it with the appropriate attitude—namely, with the desire for knowledge, an interest in the development of skills and techniques, and a willingness to subject himself to the discipline of the laboratory. The latter is especially important in the development of habits essential to good scientific work. Such discipline and training are required in virtually all branches of science, and may be life-saving in many areas. The development of a scientific attitude is the result of thinking and acting in a scientific manner—carelessness, sloppiness, tardiness, preconceived ideas, and inefficiency may be accepted in some areas of human endeavor, but not in science.

During a laboratory, one of the most significant questions a student can ask is: "What is the purpose of this activity or experiment?" In this manual, where the purpose (objective) of each experiment (or activity) is indicated, the student should let the stated objectives permeate his thinking as he engages in the activities prescribed.

Critically observe all specimens or materials which you are asked to study. Your own observations should provide answers to most of the questions that arise during the laboratory. *Learn to think independently.* The instructor will offer helpful suggestions, but he should not be expected to be a source of ready answers to each of your questions. No one can learn for you.

Do not interfere with the work of others. Work is to be done on an individual basis unless you are specifically instructed otherwise. It is often necessary to do certain experiments with a laboratory partner. This is a cooperative effort and the work should be shared on an equal basis. Furthermore, both partners should understand the experiment and share the results. However, the write-ups should be the results of individual effort. Ordinarily, there is no objection to the discussion of laboratory work with other students.

B. Laboratory Supplies

In the spaces provided, check the supplies your instructor indicates are to be provided by you.

1. ____ 3H pencil
2. ____ 4H pencil
3. ____ red pencil
4. ____ blue pencil
5. ____ box of colored pencils
6. ____ manila folder
7. ____ 6 microscope slides
8. ____ 6 cover glasses
9. ____ lens paper
10. ____ dissecting kit
11. ____ biology filler paper
12. ____ grease (wax) pencil
13. ____ black, felt-tip pen
14. ____ 6 inch ruler
15. ____ eraser (pink pearl)
16. ____ tape (Magic transparent)
17. ________________________

C. Assignments

All, or most, of the assignments will be given in advance of the laboratory. The student is expected to read the introductions and objectives, as well as to preview procedures, **before** the date of the assigned laboratory. In addition, a brief orientation will be given at the beginning of the laboratory.

D. General Laboratory Rules

There are reasons for every rule imposed upon the student. If these are not readily apparent to you, **ask your instructor for an interpretation.** Since the laboratory environment and situation vary somewhat from school to school, and since methods employed may also vary from instructor to instructor, some of the following rules may not apply. On the other hand, your instructor may invoke still additional rules.

1. Be on time. Instructions are given only at the beginning of laboratory. It is your responsibility to be present to receive them.
2. Bring your textbook, laboratory manual, and laboratory supplies to the laboratory.
3. Handle all equipment loaned to you with care. Broken or damaged equipment will be charged against the liable student or students.
4. Check with the instructor concerning the policy for laboratory "breaks."
5. Do not leave the laboratory early without permission of the instructor.
6. Visitors are not allowed except by special permission.
7. All work to be handed in should be arranged in sequence, and placed in a manila folder. Write your name, class, day and hour class meets, and (if one is assigned) your section number on the index portion of the manila folder.
8. Before leaving, check to see if you have done the following:
   a. Cleaned the laboratory desk.
   b. Turned out the desk lamp.
   c. Locked up the microscope and/or other supplies.
   d. Cleaned, dried, and returned all equipment consigned to your care.
   e. Returned microscope slides to their proper location.
9. Keep the sinks in your area clean. Do not put solid matter, such as dissectable materials, in the sinks. Waste baskets are for solids, sinks for liquids.
10. Don't waste time at the beginning of the laboratory period. Whenever possible, begin work as soon as you enter the room.
11. Always keep in mind the meaning of the work you are doing.
12. Acknowledge the fact that extra time on laboratory work is often necessary.
13. Don't hesitate to consult your instructor, if you have any problems with the course.

E. Care of Preserved Specimens

1. Fetal Pig

The fetal pig is packaged in a plastic bag containing a liquid preservative. The tip of the bag should be cut open with scissors, and the fluid poured into the sink. At the conclusion of any laboratory involving the dissection of the pig, two or three thicknesses of paper towels moistened with tap water should be used to completely cover the specimen. The pig should be replaced in the plastic bag, the opening secured with a rubber band, and the bag then placed in the storage cabinet. This procedure will ordinarily keep the specimen in good condition throughout the course.

2. Organs (e.g., sheep eye, heart, kidney, brain, etc.)

Each specimen should be wrapped in a double thickness of moistened paper towels and placed in the plastic bag with the fetal pig, or in individual plastic bags.

An alternate method is to place the organs in a 5% solution of formaldehyde. This solution is sufficiently concentrated to maintain the preserved state of the organs for the period of time they are being used.

F. Written Work

Each unit (exercise) in this manual contains a section entitled Results and Questions. To facilitate the recording of data during the performance of an experiment or exercise, remove the Results and Questions section from the manual. Data and answers to questions should always be written neatly and legibly, and all kymograph records should be pasted or taped in the appropriate spaces. Use either number 2 pencils or ink (blue or black only) in recording your data and writing answers to the questions. When completed, write your name on each page and place your assignments in a labeled manila folder (Section D, 7.). Do not clip or staple the pages together.

G. Laboratory Tests

As one aspect of evaluation of work done in the laboratory, your instructor may give laboratory tests. The frequency of such tests will be announced by the instructor. The following methods are often employed in testing in the laboratory:

1. Practical

This type of test may include actual specimens (e.g., bones), microscope slides, drawings, and questions. There are several ways in which such a test may be given. One method commonly used involves the establishment of from 30 to 40 test stations. One or more specimens, a microscope with a slide in focus, a diagram or drawing with structures to be identified, or a card containing written questions are located at each station. The student is provided with an answer sheet and stands at a specified station. At the signal to begin, the student answers the questions (usually from three to five) asked at that station. The time alloted for this task varies from 45 seconds to 1-1/2 minutes. When time is called, the student moves to the next station, and so forth, until the test is completed.

2. Orals

In this type of test, the student is asked to bring a specimen he has dissected to the instructor, who then asks the student to identify selected structures. The student may also be asked to give functions of the structures identified. In some cases, models may be used for giving oral tests.

3. Written Tests

These may take a variety of forms (short answer, problem solving, identification, essay, etc.). These tests are designed to determine the student's understanding of the experiments and exercises performed in the laboratory.

H. Vocabulary

Your ability to communicate and to comprehend and to reason and interpret depends upon the acquisition of a functional vocabulary. This is especially true of any science course, wherein the vocabulary practically constitutes a "foreign language." It is therefore important for you to develop an understanding of all unfamiliar words with which you come in contact in the process of your reading and investigations. Define all new words in terms you understand. Make a conscious effort to use the new words until, by constant reinforcement, they are firmly entrenched in your mind.

I. Rules Governing Chemicals

1. Treat all chemicals with respect.

2. Avoid contaminating chemicals within their containers.
3. Keep all containers of chemicals clean and dry.
4. Use only pipettes (medicine droppers) already in the bottles.
5. Avoid placing reagent bottles or other caustic agents on the tops of decks.
6. Pour chemicals from bottles on the side away from the label (to avoid damaging the label).
7. Return all containers of chemicals to the appropriate location. Do not remove the chemicals from the area, unless otherwise directed.

J. Illustrations as Self-Tests

Many of the drawings in this manual are designed so that they may be used as self-tests. Prestudy of similar illustrations in the textbook, references, wall charts, as well as the prestudy of models and specimens, is one way to prepare the student for the self-tests. The student may then try to identify the parts indicated on an illustration (writing the answers on a separate piece of paper). The answers may then be checked with the keys found in the chapter.

K. Self-Tests

At the end of each chapter is a true-false self-test. Each test item includes a key word, words, or phrase in **boldface** type. If the test item is false, the **boldface** portion should be corrected to make the statement true. In general, the self-tests cover material found in texts, covered in lecture, and applied in the laboratory. Keys are located at the end of each self-test.

1. Suggestions for Using the Self-Tests

a. Take them as a **pretest** before studying the material covered.
   (1) Write your answers on a separate sheet of paper, but do not correct them at this time.
   (2) This procedure will provide you with clues as to what to look for in your reading of the text, what to listen for in lecture, and what to watch for in the laboratory.
b. After you have had an opportunity to study the material covered, take the self-test again (**post-test**).
   (1) Correct your pretest and post-test by using the key at the end of the test.
   (2) Compare the results of the post-test with those you achieved on the pretest to determine how much you actually learned.
c. Other suggestions:
   (1) Always try to determine the reason why a given statement is or is not true. Merely knowing the answer to the statement may provide very little depth.
   (2) Any words in a statement that are not understood should be defined from a suitable reference.
   (3) Try to correlate the concepts indicated in the statements with the material presented in lectures, discussions, the text, and the laboratory.

2. Limitations of Self-Tests

a. They may be a useful tool, but they are not comprehensive.
b. Since courses vary in their emphasis of certain concepts, this must be taken into consideration when the self-tests are used.
c. Tests usually sample bits of a student's knowledge—never the whole. Study-type tests, such as the self-tests in this manual, merely provide a basis for learning subject matter and showing trends in achievement.
d. Analysis of questions missed on a self-test may often imply that the student needs to study the entire block of knowledge from which the question was derived.
e. If not used as suggested (e.g., as a supplement to other learning devices), the student may fail to attain the depth and insights essential for this course.

1

Introduction to Anatomy

Exercise 1-A Organization of the Body as a Whole

Objectives

To locate and identify the cavities and anatomical regions of the body, and to identify the organs within the cavities and those underlying the regions.

Materials

Dissectable torso; anatomic charts; and reference books.

Discussion

The human body is made up of **cells**. Cells that have similar specializations and are united in the performance of a particular function form a **tissue**. Tissues, in turn, make up an **organ**, which may be defined as several tissues grouped into a structural unit having a special function. A set or series of organs united in a common function is referred to as a **system**. While it is possible to distinguish individual parts (e.g., a cell, a tissue, or an organ) in the human organism, these parts function as an integrated whole.

In doing the following exercises, and throughout your study of anatomy and physiology, keep the following question constantly before you: how does this part of the human body fit in with the concept of the body as an integrated whole?

Part 1. Cavities of the Body

Procedure

1. Using your text and reference books for orientation, locate the following cavities of the body on the dissectable torso:

   a. cranial
   b. nasal
   c. buccal (oral)
   d. thoracic
   e. abdominal
   f. pelvic
   g. ventral
   h. dorsal

2. Under each label on Figure 1-1, indicate the major structures found in the cranial, thoracic, abdominal, pelvic, and vertebral cavities. Use anatomic charts, your text, and references to aid you in this activity.

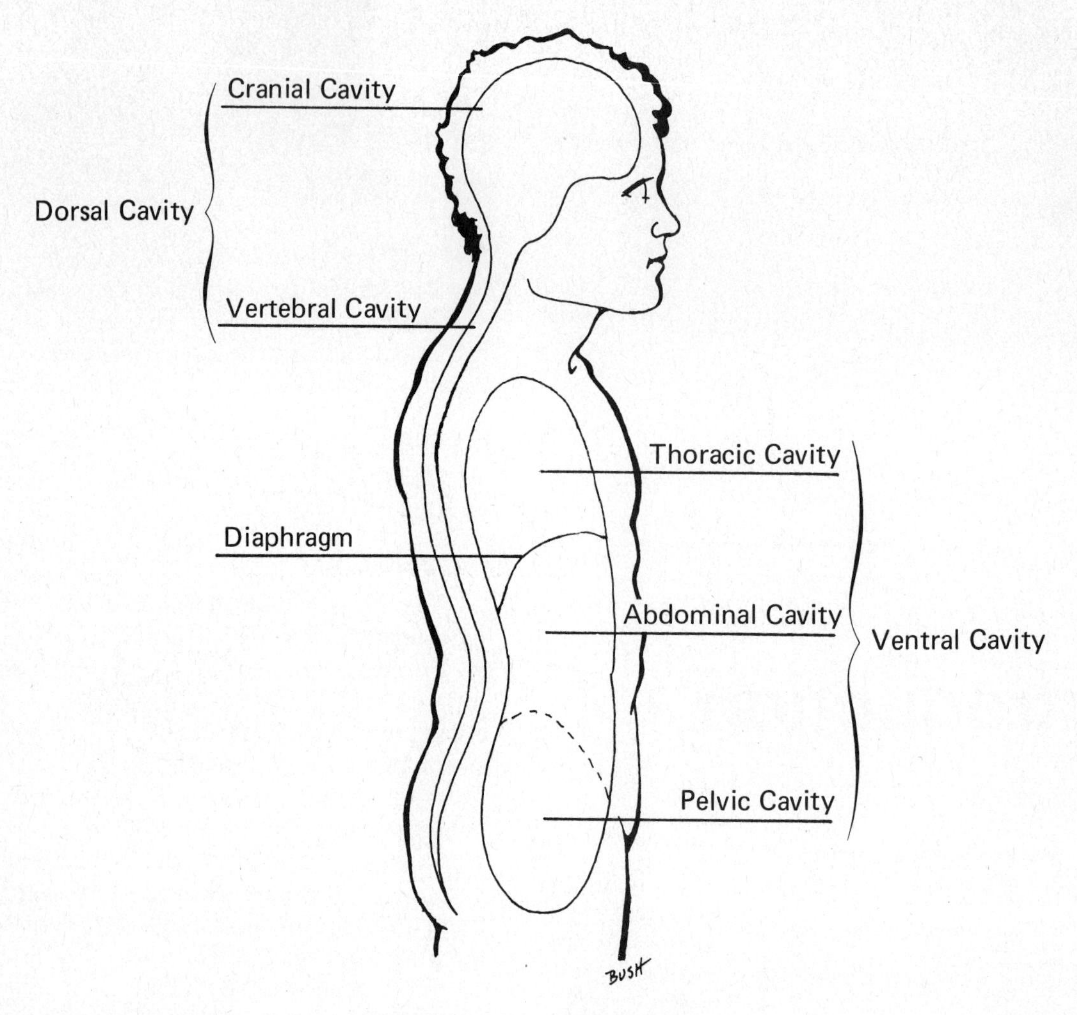

Figure 1-1. Body Cavities

Part 2. Regions of the Body

Procedure

1. Locate the following nine regions of the body on the dissectable torso and on your own body: epigastric, umbilical, hypogastric (pubic), right and left hypochondriac, right and left lumbar (lateral), and the right and left inguinal (iliac). Use your text and other references for orientation.
2. Use Figure 1-2 as a self-test for the nine body regions. See Section J, p. xx (Illustrations as Self-Tests) for a suggested approach. Under each label, write the major structure or structures underlying the region.

Exercise 1-B
Anatomic Terminology

Objective

To develop a functional knowledge of anatomic terminology.

Materials

Dissectable torso, anatomic charts, and reference books.

Discussion

Terminology used in locating various structures of the human body is vital to any precision in the study of anatomy and physiology.

KEY TO FIGURE 1-2:

1. right hypochondriac
2. right lumbar
3. right inguinal
4. left hypochondriac
5. epigastric
6. umbilical
7. left lumbar
8. left inguinal
9. hypogastric

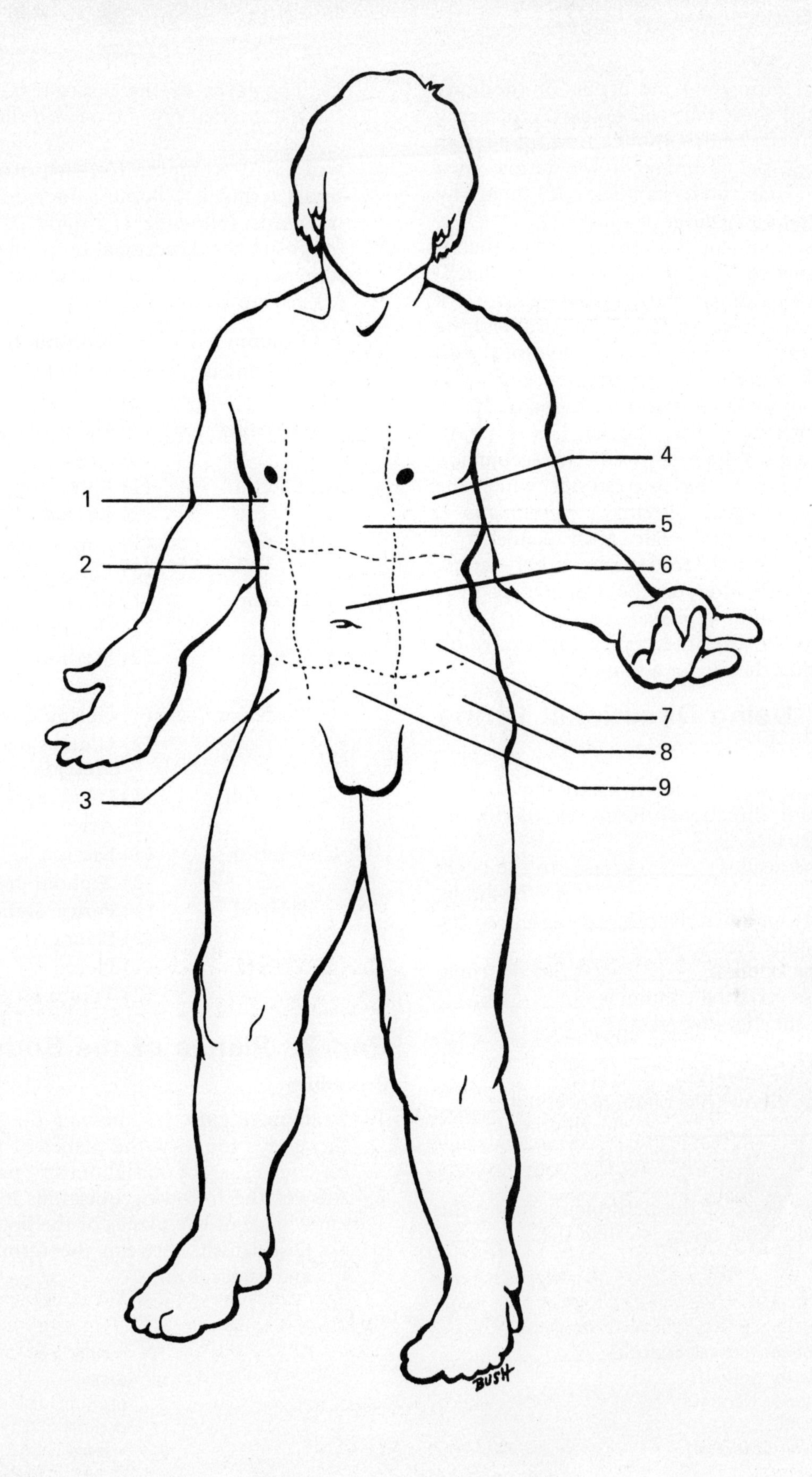

Figure 1-2. Regions of the Body

Directional terms and the planes of the body are essential tools that will be used repeatedly throughout the course. Build a firm foundation by learning the meaning of the terms, then reinforce your understanding of them by applying them whenever possible.

In lower animals the **ventral** surface (belly) faces downward while the **dorsal** surface (back) faces upward. Since man assumes an upright position, the ventral surface is forward, and the term ventral is, in humans, synonymous with the word **anterior.** Furthermore, dorsal is synonymous with **posterior** in humans. In an animal, such as a dog, the head is anterior, while in humans the head is **superior (cephalic).** The tail of the dog is posterior while the tailbone (coccygeal vertebrae) in humans is **inferior** (or caudal). The term **caudad** (or caudal) may be used to describe either the tail portion of the dog or the lower part of a human. The term **cephalad** (or cephalic) may be used to describe the cranial (or head) portion of either the dog or a human.

Part 1. Using Directional Terms

Procedure

1. Self-test: without referring to the key below, use directional terms to fill in the blank spaces.
   a. The head is ____________ to the neck.
   b. The foot is ____________ to the ankle.
   c. The umbilicus is ____________ to the lumbar vertebrae.
   d. The fibula is ____________ to the tibia.
   e. The vertebral column is ____________ to the digestive tract.
   f. The ulna is on the ____________ side of the forearm.
   g. The elbow (olecranon process) is ____________ to the shoulder.
   h. The kneecap (patella) is ____________ to the ankle.
   i. The layer of the peritoneum lining the abdominal cavity is called the ________ layer.
   j. The layer of the peritoneum covering the internal organs is called the ____________ layer.
2. Write two sentences for each of the directional terms in Column I. Include the word or words following (1) under Column II (opposite the directional term) in your first sentence, and the word or words following (2) in your second sentence.

Column I	Column II
a. Peripheral	(1) Body hair (2) Epidermis
b. Anterior	(1) Mammary glands (2) Nose
c. Lateral	(1) Ears (2) Radius
d. Distal	(1) Fingernails (2) Toes
e. Proximal	(1) Elbow (2) Head of femur
f. Medial	(1) Sternum (2) Ulna
g. Posterior	(1) Occipital bone (2) Gluteus maximus muscle
h. Superior	(1) Diaphragm (2) Atlas
i. Inferior	(1) Sacrum (2) Xiphoid process
j. Parietal	(1) Pericardium (2) Pleura
k. Visceral	(1) Pleura (2) Tunica serosa

Part 2. Planes of the Body

Procedure

1. Examine Figure 1-3 showing the planes of the body. Identify the planes of the body on yourself and your laboratory partner.
2. Answer the following questions involving a knowledge of the planes of the body:
   a. Distinguish between the terms sagittal and midsagittal.

KEY TO SELF-TEST

a. superior (cranial, cephalic)
b. inferior (distal)
c. anterior (ventral)
d. lateral
e. posterior (dorsal)
f. medial
g. distal
h. proximal (superior)
i. parietal
j. visceral

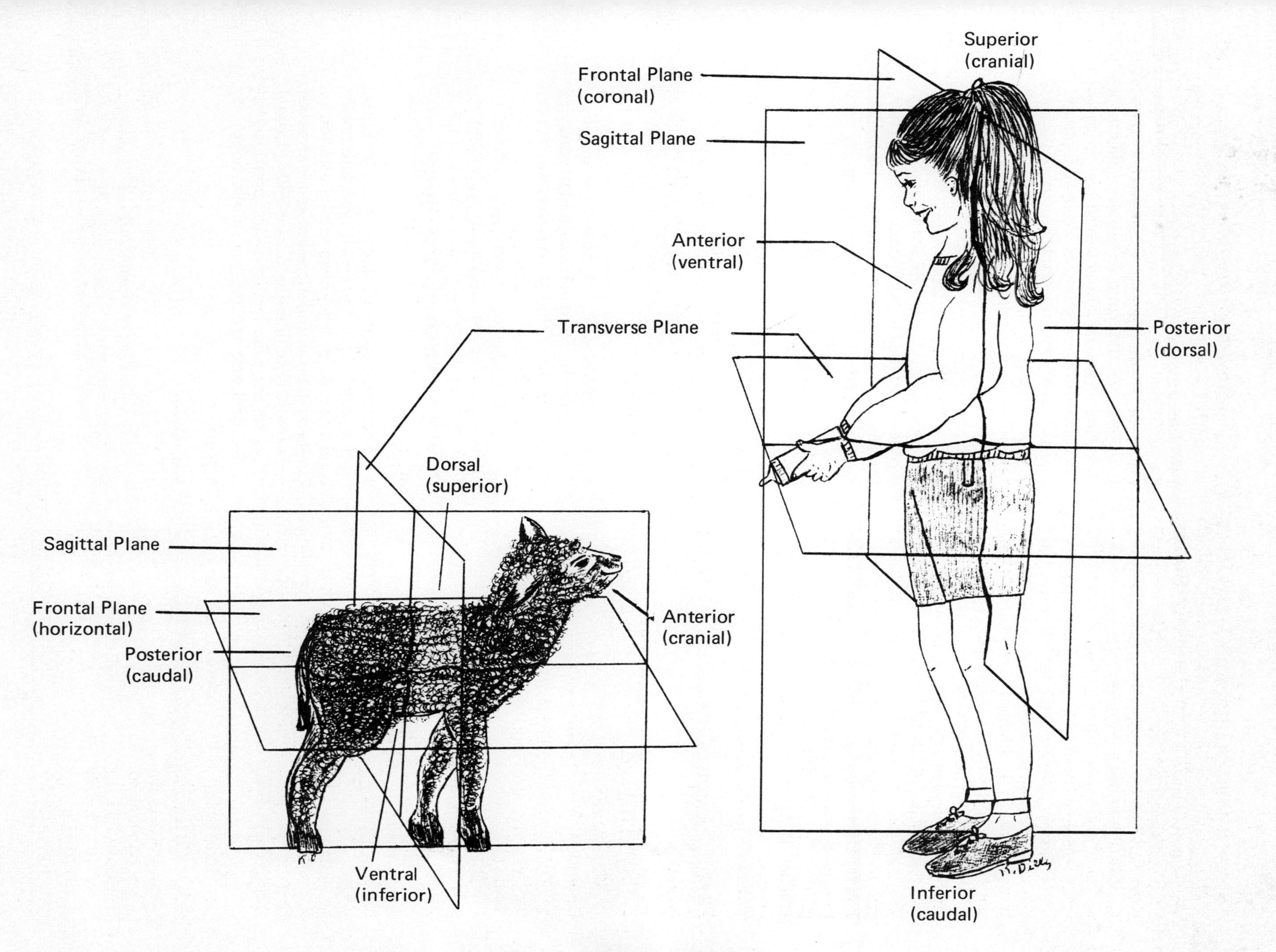

Figure 1-3. Planes of the Body

b. Dividing the body to expose the coronal plane through the thorax would expose what structures? (Use the dissectable torso and anatomic charts to aid you in answering this question.)
c. Is the transverse plane limited to one specific area of the body or is this a general term that can be applied to many parts of the body? Explain your answer giving examples.
d. Define the following terms and give examples of how each might be used:
(1) anterior aspect
(2) posterior aspect
(3) lateral aspect

Exercise 1-C. The External Characteristics of the Fetal Pig

Objective

To become familiar with the external characteristics of the fetal pig.

Materials

Fetal pig; dissecting board or tray; and string.

Discussion

The specimens of pigs used in this laboratory exercise are fetal pigs. These animals are readily available, since farmers find it profitable to breed any female which they plan to sell. Thus pig fetuses are by-products of the slaughterhouses. The period of gestation is 112-115 days, and there are, on the average, about seven to eight offspring in a litter. At birth the pigs vary from 12 to 14 inches in length. The approximate age of the fetus can be determined by measuring the length of the body from the tip of the snout to the rump (not including the tail). The following are approximate body length-to-age relationships:

Body length (mm)	Approximate age (days)
11 (0.44 in)	21
17 (0.68 in)	35
28 (1.12 in)	49
40 (1.6 in)	56
220 (8.8 in)	100
300 (12.0 in)	Full Term

As a laboratory animal the fetal pig has a number of advantages. They are relatively inexpensive so that usually a maximum of two students can be assigned to an animal. Since they are small, they do not require much storage space. The animals are mammals and, therefore, their structures are similar to those of humans. In addition to relatively mature organs, there are also fetal structures present that are directly comparable to those of human beings. These include the umbilical cord and the circulatory structures which are specialized for fetal circulation.

As the fetal pig is dissected and studied throughout the course, the structures identified should be compared with those of the human. Dissection is not merely "cutting" the animal, but a systematic technique of bringing into view structures which, in their normal position, cannot readily be seen. Follow instructions exactly. Do not cut or remove any structure unless directed to do so. Always separate structures **carefully**, especially blood vessels, by dissecting away connective tissue.

You may find that the substances used to preserve the specimens are irritating to your skin. If so, apply "Pro-tek" (or some similar substance) to your hands before dissecting, or wear thin rubber or plastic gloves. Remove as much of the preservative from your specimen as possible by frequent washings with tap water. Keep your fingers away from your eyes during dissection.

At the conclusion of each laboratory period, clean up the working area thoroughly. Wrap the pig in wet paper towels and replace in the plastic bag provided. Do not leave any solid material in the sink. Clean and dry the laboratory table.

The terms **right** and **left** always refer to the pig's right and left. In a quadruped, **anterior** or **cranial** refers to the head end; **posterior** or **caudal** to the tail end; **dorsal** or **superior** to the back; **ventral** or **inferior** to the belly. **Lateral** refers to the side; **medial** to the position of a structure nearer the midline of the body (see Fig. 1-3).

Procedure

1. Examine the pig for body hair, although this is usually not conspicuous at this time.
2. Note the **epitrichium** which is a layer of embryonic skin that is visibly peeling. This is lost as the hair develops. It may be removed by rinsing the pig in tap water.

This fetal skin will plug the sink and care should be taken to prevent this.

3. On the head locate the following structures.
   a. The **mouth**, bounded by upper and lower **jaws** and soft lips, is sometimes partially open, revealing a soft **tongue**. The front end of the head is prolonged into a **snout**. The snout is used for rooting around in the soil for roots, insects, and other materials used by the pig for food.
   b. Observe the two **nostrils** (**external nares**) at the end of the snout.
   c. The **eyes** (usually closed) are covered by upper and lower **eyelids** fringed with **eyelashes**. Make an incision extending forward from the anterior corner of the eye and pull the upper eyelids apart. The **nictitating membrane** should be visible in the medial corner of the eye. This membrane can move across the eyeball to help keep it clean. Is this structure present in the human eye?
   d. The opening into the ear is called the **external acoustic (auditory) meatus** and the flattened flap of skin is called the **pinna**. The pinna and the external acoustic meatus make up the **external ear** in the pig as well as in the human.
4. Note that the short **neck** joins the thorax in front of the first pair of legs. There is usually an incision in the right lateral part of the neck where blood is withdrawn and colored latex injected.
5. The **trunk** can be divided approximately into two general regions, consisting of an anterior **thorax** and a posterior portion, the **abdomen**.
   a. Note that the front limbs are attached to the thorax. The ribs making up the thorax are soft at this stage of development because they are made up of cartilage.
   b. Locate the nipples which are present in both sexes. These form a double row of small **teats** or **mammary papillae** on the ventral surface of the abdomen. The number and location of mammary glands vary in different species, but are characteristic of all mammals.
   c. Observe the **umbilical cord** near the center of the ventral surface of the abdomen. Make a fresh cut across the end of the cord. Three large openings should now be visible. The largest of these is the **umbilical vein** which carries blood from the placenta to the fetal pig. This vessel may contain blue latex. The other two, smaller and with thick walls, are the **umbilical arteries**, which may contain red latex. These vessels carry blood from the fetus to the placenta. Between or near the umbilical arteries a small, hard core of tissue, the **allantoic stalk**, can be felt. All the structures present in the cord are embedded in a gelatinous connective tissue.
   d. Locate the **anus** just ventral to the tail. This is the posterior opening of the digestive tract.
   e. Determine the sex of your specimen. In the female, the external **urogenital opening** (with a small **genital papilla** projecting from it) is ventral to the anus. This is the common orifice of the urinary and reproductive tracts. In the male the external **urogenital** opening is a very small hole just posterior to the umbilical cord at the tip of the **penis**. If your specimen is a male, note the two **scrotal sacs** below and ventral to the anus. The **penis** lies under the skin, passing from the urogenital opening posteriorly between the hind legs. Feel the penis through the skin. Later this will be dissected out. Each student is expected to identify the sexual organs in both sexes. Compare your fetal pig with that of the opposite sex.
6. Note that there are only four toes or **digits** on each limb as compared to five in humans.
7. Examine the legs and note that they have the same general structure as that of humans and other animals, although they are somewhat modified.
   a. Examine the posterior surface of one of the hind legs and note the large protuberance about two inches above the toes. This is comparable to the human heel, and the region from it to the toes corresponds to the human foot. Since the pig walks on the tips of the toes, the ankle and most of the foot are above the ground.
   b. Locate the **wrist** and **elbow** of the forelimb and the **knee** and **ankle** of the hindlimb.

8. Tie a 12-18 inch length of cord around each ankle. These will be used to tie the pig to either a dissecting board or pan and should be left permanently attached to the pig.
9. Tie a label to the tail of the animal with a piece of string that is long enough to allow the label to project out of the plastic bag. The label should have printed on it, with pencil (ink runs), your name, your partner's name, the course, section number, and time of meeting.

Self-Test—Introduction to Anatomy

Directions

The following statements are either true or false. If a test item is false, the portion in bold type should be altered to make the statement true. When taking the self-test, it is suggested that the answers be written on a separate sheet of paper. This simplifies using these tests for review where it is usually undesirable to have the answers exposed. Answers may be checked by referring to the key at the end of the test. For more extensive suggestions concerning the use of self-tests, their functions and limitations, see Section K (Self-Tests) under Orientation and Suggestions for Students.

1. The **midsagittal** plane divides the body into a top and bottom half.
2. **Sagittal and coronal** planes divide the body into upper and lower parts.
3. The knee is **proximal** to the foot.
4. The part of the nervous system found within the **dorsal cavity** is called the central nervous system.
5. In relation to the mandible, the maxilla is **inferior.**
6. **Proximal** is a term meaning the part located farthest from the trunk.
7. **Visceral** is applied to organs located within the body cavities.
8. A cut through the body, or body structures, dividing it (or them) into front and back portions, is called a **transverse cut.**
9. The hands are the **proximal** portion of the upper extremities.
10. The spinal cord and brain are found within the **ventral** cavity.
11. When an individual stands in anatomic position the radius is **mesial** in position.
12. A part of the body located toward the outside of the body is said to be **peripheral.**
13. In the anatomical position, the arms are at the side with the palms of the hands **supinated.**
14. With reference to the esophagus, the stomach is **inferior.**
15. Two terms that apply to the back part of humans are **posterior and ventral.**
16. The large intestine is located within the **abdominal and pelvic cavities.**
17. **Homeostasis** is a term which means relative uniformity of cellular environment.
18. The spleen is found within the **mediastinum.**
19. The lungs are located in the **thoracic cavity.**
20. The liver is in the **abdominal cavity.**

21. The type of mirror-image symmetry that is shown by a human being is known as **radial** symmetry.

22. The surface of a pig's body that forms its back is known as the **ventral** surface.

23. In man, the back of the body may be called **either the dorsal or the posterior** surface.

24. The **anterior** end of a pig's body is the one to which the head is attached.

25. When you lie on your ventral surface you are said to be in a **supine** position.

26. Fingerprints are made by pressing the **palmar** surfaces of the fingers against an object.

27. Any part of the body which is away from the midline is said to be **mesial** in location.

28. The **abdomen** may be divided into four sections called quadrants.

29. The part of an appendage that lies farthest from the midline of the body is the **caudal** portion.

30. The crest of the hip bone is **dorsal** to the vertebrae.

31. The pain of appendicitis is frequently felt in the **lower left** quadrant of the abdomen.

32. The female gonads are located within the **vertebral cavity.**

33. The **buccal cavity** opens into the oropharynx.

KEY

1. transverse
2. transverse
5. superior
6. distal
8. frontal (coronal)
9. distal
10. dorsal
11. lateral
15. posterior and dorsal
18. abdominal cavity
21. bilateral
22. dorsal
25. prone
27. lateral
29. distal
30. lateral
31. lower right
32. pelvic cavity

Name ____________________

Results and Questions for Chapter 1

Exercise 1-A
Organization of the Body as a Whole

Part 1. Cavities of the Body:

In the blanks provided indicate the specific body cavity (or cavities) in which each of the following is located (dorsal and ventral are not answers):

1. stomach ____________________
2. spinal cord ____________________
3. urinary bladder ____________________
4. uterus ____________________
5. Fallopian tubes ____________________
6. large intestine ____________________
7. prostate gland ____________________
8. tongue ____________________
9. pancreas ____________________
10. aortic arch ____________________
11. brain ____________________
12. lungs ____________________
13. heart ____________________
14. kidneys ____________________
15. teeth ____________________
16. ovaries ____________________
17. appendix ____________________
18. small intestine ____________________
19. adrenal glands ____________________
20. spleen ____________________

Part 2. Regions of the Body:

1. Through what region of the body would a surgeon make an incision to perform an appendectomy? ____________________
2. Through what region of the body would an incision be made to perform surgery on the right kidney? ____________________
3. What is the name given to several tissues grouped into one structure performing one or more general functions? ____________________
4. Which organ system has the function of conducting air? ____________________
5. Which organ system eliminates body wastes? ____________________
6. Which organ system synthesizes hormones? ____________________
7. To which organ system do sweat glands belong? ____________________
8. Of which organ system are leukocytes a part? ____________________
9. Of which organ system are lymph nodes a part? ____________________
10. Of which organ system is gastric juice a secretion? ____________________

Exercise 1-B
Anatomic Terminology

Match the following anatomic terms used in describing the human body. Select the best answer for each blank. Use each term only once.

____ a. two terms used for the back side of the body
____

____ b. term meaning farther away from a point of reference, such as the trunk

____ c. toward the midline of the body

____ d. layer of a membrane lining the walls of a cavity

____ e. layer of a membrane covering the organs within a cavity

____ f. toward the tail end of the body

____ g. sole of the foot

____ h. away from the center of the body; toward the outside of the body

____ i. term describing the location of a structure located in the middle of another structure such as the arm

____ j. two terms used for the front side of the body
____

____ k. palm of the hand

____ l. term describing the location of the little toe with respect to the big toe

1. anterior
2. caudal
3. cranial
4. distal
5. dorsal
6. dorsum
7. inferior
8. lateral
9. medial
10. median
11. parietal
12. peripheral
13. plantar
14. posterior
15. proximal
16. superior
17. ventral
18. visceral
19. volar

2

Basic Microscopy

Exercise 2-A Structure of the Microscope

Objectives

To become familiar with the functions of the parts of the microscope and to become acquainted with the focusing procedure.

Materials

Microscope and microscope lamp.

Procedure

1. Examine the drawing of the microscope (Fig. 2-1). Compare this drawing with your microscope and learn the names and functions of all parts indicated by the instructor.
2. Use of the microscope:
   a. When it is necessary to lift the microscope, hold the **arm** of the microscope with one hand, and add support at the **base** with the other hand.
   b. Place the microscope on the deck with the arm facing you.
   c. Since tilting results in poor light regulation and unwanted movement in fresh preparations, do not tilt this instrument unless instructed to do so.
   d. Always begin by cleaning the **ocular** (eyepiece), **objectives, slide,** and **mirror** (when present), with **lens paper**.
   e. To avoid eyestrain and subsequent headaches, it is best to look through the microscope with both eyes open. Practice makes this a simple procedure.
   f. Focusing procedure:
      (1) Turn the **revolving nosepiece** until the **low power objective** (10X) clicks into position over the hole in the **stage** of the microscope. Always begin with the low power objective.
      (2) Open the **iris diaphragm** approximately halfway or turn the disc diaphragm to the III setting.
      (3) If your microscope is equipped with a mirror, arrange the microscope lamp and concave side of the mirror so that a maximum amount of light will be reflected into the condenser. This adjustment is made while you are looking through the eyepiece (ocular).

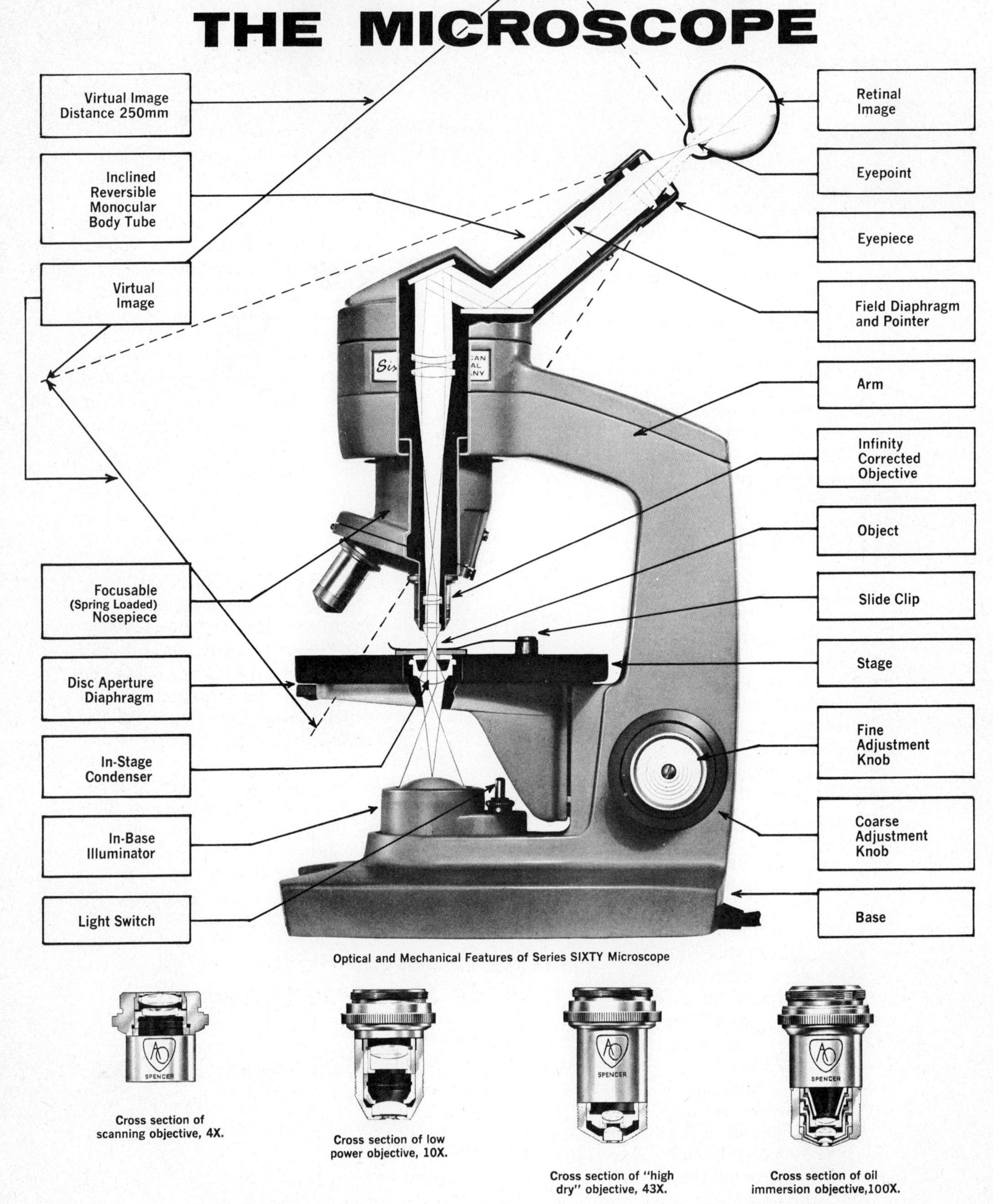

Optical and Mechanical Features of Series SIXTY Microscope

Cross section of scanning objective, 4X.

Cross section of low power objective, 10X.

Cross section of "high dry" objective, 43X.

Cross section of oil immersion objective,100X.

Figure 2-1. The Microscope (Courtesy of American Optical Corporation, Scientific Instrument Division, Buffalo, N.Y. 14215)

To regulate light intensity adjust the iris or disc diaphragm (not the mirror) when the slide is in focus.

If your microscope is equipped with an in-base illuminator, turn on the illuminator.

(4) Use the coarse adjustment knob to raise the nosepiece for easier access to the stage and then place a slide on the stage of the microscope. Position the specimen to be observed directly above the center of the condenser.

(5) Usually a stop is present on the microscope which will prevent the low power objective from striking the slide. Determine if this is true of your instrument by lowering the 10X objective as close to the slide as possible without touching it, or until the tube can be lowered no farther.

(6) If the microscope is equipped with an autofocus, when the low power objective is as close to the slide as possible the specimen can be brought into sharp focus by slowly rotating the **fine adjustment knob** either down or up. If the microscope is not equipped with an autofocus, look through the eyepiece and turn the coarse adjustment to **slowly raise** the objective from the slide until the specimen comes into focus. The image should then be brought into sharp focus with the fine adjustment knob. This knob should also be used for observing depth in a specimen.

(7) Adjust the amount of light to the optimum when the slide is in focus. If there is too much light, so that the specimen cannot be seen clearly against the glare, rotate the disc diaphragm to a smaller opening. If a condenser with an iris diaphragm is being used, close down the diaphragm. When using a mirror, if the field of view is not uniformly illuminated, adjust the mirror or the position of the illuminator so that the specimen field is evenly illuminated.

(8) Move the slide to find the specific area you desire to view in more detail. Since the quality of slides and individual portions of the slides may vary, always select a slide (or area of the slide) that seems typical.

(9) Once the object is in sharp focus and centered with the 10X objective, rotate the revolving nosepiece to the **high power objective** without changing the position of the coarse adjustment knob. (The microscope should be **parfocal**. This means that when the high power objective is brought into position in place of the low power objective, the object will be in focus, or almost in focus, provided you make certain that the material to be viewed is centered and in sharp focus before turning from low power.)

(10) Adjust the fine adjustment knob to bring the specimen into sharp focus with the high power objective. **Never focus downward with the coarse adjustment knob when using high power**. Each time a different objective is used, the disc or iris diaphragm setting must be changed.

(11) Unless otherwise directed, drawings should be made using the high power objective.

(12) Always return the slide to the correct position in the slide box after use.

Exercise 2-B Microscope Drawings

Objectives

To become familiar with basic microscope technique and the procedure to be used for drawings.

Materials

Plain white paper (or Biology Filler paper); 3H pencil; practice slides.

Discussion

Drawings are a valuable means of recording and checking the results of your observations. A drawing should indicate the accuracy and detail of all observations made. Properly completed drawings provide a basis for reviewing material you have observed under the microscope. They also provide evidence to the instructor that you have seen the significant structures.

The practice slides used in this exercise may consist of assorted tissues on damaged slides. It is not the purpose of this exercise to identify the various tissues; the student should, however, use these slides to learn how to use the microscope and how to make a drawing.

Procedure

1. Select one or more of the practice slides and, using the procedure indicated in Exercise 2-A, practice focusing (on both low and high power) until you are thoroughly familiar with the technique.
2. On one of the practice slides, locate two or three cells and draw them in accordance with the following specifications. (This is the procedure to be used on any drawings requested, unless instructions to the contrary are given.)
   a. Make all drawings on Biology Filler (or plain white) paper.
   b. In the upper right-hand corner of the page, print your name. Under your name, print the title of the course and the section number of your laboratory.
   c. In the upper left-hand corner, print the exercise number and, under this, the date the work was done.
   d. Centered at the top of the page, print the title of the exercise.
   e. As a general rule, do not include drawings of more than one exercise on a page. Use only one side of a page.
   f. Drawings must be large and neat, with the size of the cells in correct proportion to the drawing as a whole. When making the drawing, use firm continuous lines—not several indistinct, discontinuous lines. Individual drawings should not be boxed off in squares, rectangles, or circles.
   g. The drawing should be approximately centered on the page (or slightly to the left) with the labels to the right, except where excessive crowding makes this impractical.
   h. Draw only that which is actually observed, but always look for a typical view of the structure being observed.
   i. An appropriate subtitle, which identifies the specific drawing, should be placed under each drawing. The subtitle should include the following: magnification and type of section, such as cross section (x.s. or c.s.), longitudinal section (l.s.), tangential section (t.s.), whole mount, fresh mount, or smear.
   j. Print all labels on lines **drawn with a ruler** parallel to the top and bottom margins of the page. None of the label lines should cross each other. The label line should be touching the edge or should be within the structure indicated. Arrows should not be used.
   k. Use brackets to enclose related labels. For example, the parts of the nucleus should be labeled separately, and then bracketed with the general title "nucleus."
   l. Be accurate in your spelling (look up the word if you are in doubt).

Exercise 2-C Orientation of Images Viewed through the Compound Microscope

Objective

To determine orientation of images viewed through the compound microscope.

Materials

Microscope; letter "e" slide; and newspaper slide.

Part 1. Letter "e" Slide

Procedure

1. Obtain a prepared slide containing the printed letter "e" under the cover glass. Note that you can read the letter with the naked eye.
2. Place the slide on the stage of your microscope and observe with the low power objective.

3. Draw the letter "e" as you observed it with the naked eye (Drawing A) and as seen through the microscope (Drawing B).

(Drawing A) Drawing B

4. Move the slide so that the letter appears in the upper left corner of the field. Now, center the letter while looking through the microscope. In which directions did you move the slide?____________________

Part 2. Newspaper Slide

Procedure

1. Obtain a newspaper slide (approximately 1 inch x 3 inches piece with printing on its surface).
2. Draw an asymmetrical letter as it normally appears on the paper slide (Drawing A).
3. Orient the slide under the low power objective of the microscope so that the asymmetrical letter is in the same relative position as when observed in No. 2.
4. Draw the letter as it appears under the microscope (Drawing B).

(Drawing A) (Drawing B)

5. Move the slide so that the letter appears in the upper left corner of the field. Now, center the letter while looking through the microscope. In which directions did you move the slide?____________________

Exercise 2-D
Thread Slide

Objective

To develop technique in determining depth in microscopic specimens.

Materials

Microscope and thread slide.

Discussion

This exercise requires careful technique. It demonstrates the fact that objects viewed through a microscope are three-dimensional, and that detail exists at varying levels which can only be observed by careful manipulation of the fine adjustment knob.

Procedure

1. Obtain a prepared slide containing three colored, crossed threads.
2. Begin with the slide slightly out of focus **below** the three threads.
3. Using the fine adjustment knob, turn the body of the microscope upward until the bottom thread comes into focus, then the middle thread, then the upper thread. Repeat this procedure until you have determined the correct sequence of the threads. (Use the low power objective, then check the results with the high power objective.)
4. Record your observations in the spaces provided below.
   Slide Letter __________
   a. Color of thread on bottom __________
   b. Color of the middle thread __________
   c. Color of thread on top __________
5. Have the instructor check your results.
   Approved __________

Self-Test—Microscopy

Directions: See Chapter 1, Self-Test, p. 9.

1. The **nosepiece** of the microscope is a device for holding two or more readily interchangeable objectives.

2. **Parfocal** refers to objectives or lenses on a nosepiece that focus at the same position.

3. The **iris diaphragm** controls the amount of light passing through the object.

4. A **larger** portion of a specimen is seen with a high power objective than with a low power objective.

5. More detail of the specimen can be observed using a **high power objective** than with a **low power objective.**

6. The ability of a lens to distinguish the fine detail in the structure of a specimen is known as its **magnification** power.

7. A microscope that includes both objective and eyepiece lenses is called a **compound** microscope.

8. If an object is observed with a microscope having a 5X eyepiece and a 20X objective, it is magnified **50** times.

9. When using the high power objective, the **coarse adjustment knob** is used for focusing.

10. A micron is **1000 times** smaller than a millimeter.

11. A cell is 100 microns long. It would require approximately **1000** such cells to stretch an inch when the cells are placed end to end.

12. The **higher** the power of the objective, the easier it is to locate the specimen to be observed.

13. Light rays passing from the atmosphere into water bend. This phenomenon is known as **refraction.**

14. By using a lens of the proper shape, light rays can be made to converge to a point known as the **focus**, from which point they go on to form an image beyond.

15. An electron microscope converges **light rays** to a focus by means of electromagnets or an electrostatic field.

16. A possible abbreviation for a longitudinal section is **t.s.**

17. Selective regulation of light intensity **is essential** to good microscopy.

18. The diameter of an erythrocyte is about 8 microns. Therefore, **100** erythrocytes would be visible if lined up across a microscope field 400 microns wide.

KEY

4. smaller
6. resolving
8. 100
9. fine adjustment knob
11. 250
12. lower
15. electron beams
16. l.s.
18. 50

Name ____________________

Results and Questions for Chapter 2

Exercise 2-A. Structure of the Microscope

1. What is the function of stage clips? ____________________

2. What is the function of the iris diaphragm? ____________________

3. Why should the coarse adjustment knob never be used to focus downward when high power is used? ____________________

4. A word that refers to objectives or lenses that focus at the same position is

____

5. The magnification obtained with the microscope is the product of the magnification indicated on the ocular (usually 10X) times that on the objective (10X for low power). Determine the magnification obtained with the low power objective.

____

6. What magnification is obtained with the high power objective? ____________________

7. If the magnification of the ocular (eyepiece) is 15X and that of the objective is 20X, what is the total magnification obtained? ____________________

8. If the dimensions of a cell are 0.2 mm by 0.5 mm, what are the dimensions of this cell in microns? ____________________

9. The diameter of the low power field in a microscope is 1.5 mm. Express this diameter in microns. ____________________

10. Size of cells may be calculated from the formula.

$$\frac{\text{diameter of field}}{\text{no. of cells visible}} = \text{size of cell}$$

If 10 cells are visible across the width of the microscope field (diameter 0.6 mm), and 4 visible down the length of the microscope field, what are the dimensions of one cell in millimeters?

____

11. What are the dimensions of one cell (in No. 10) in microns? ____________________

12. What is the area of one cell (in No. 10) in square microns? ____________________

Exercise 2-C. Orientation of Images Viewed through the Compound Microscope

1. Draw the letter p as it would appear through the microscope. ____________________

2. In Part 1, No. 4 or in Part 2, No. 5, in which directions did you move the slide in order to center the letter? ____________________

Exercise 2-D. Thread Slide

1. In this exercise which gave the best results, low or high power objectives?

____

2. Relate this exercise to the procedure necessary when observing the parts of a cell.

3

Cells, Mitosis, and Tissues

Exercise 3-A The Structure and Functions of Cells

Objectives

To become familiar with the structure of cells; to learn selected functions of cells and their parts; and to observe phagocytosis.

Materials

Living amebae (*Ameba proteus* or *Chaos chaos*); live **ciliates** *Tetrahymena pyriformis* (or other species); slides (plain and depression); and cover glasses.

Part 1. Structures of a cell observed with a Light Microscope compared to those observed with an Electron Microscope

Discussion

Cells are highly diversified, both structurally and functionally, but all living cells have certain things in common. This part of the exercise provides an introduction to structures and functions that almost all cells have in common.

Procedure

1. Except for the nucleus, few of the structures of a cell are visible with the light microscope. Figure 3-1 illustrates cellular

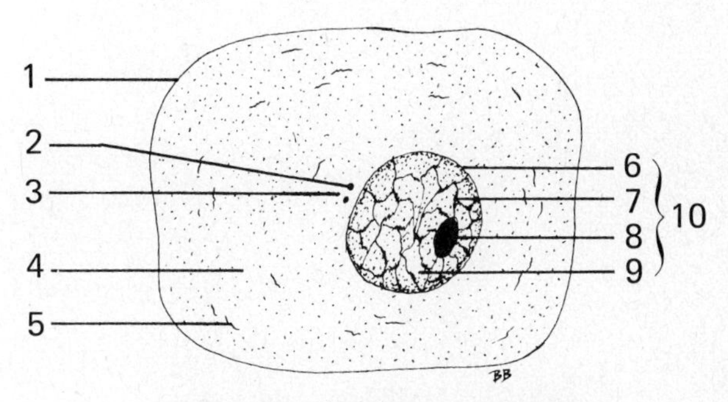

Figure 3-1. Structure of Cell (observed with the light microscope)

KEY TO FIGURE 3-1

1. plasma membrane
2. centriole
3. centrosphere
4. cytoplasm
5. mitochondrion
6. nuclear membrane
7. chromatin
8. nucleolus
9. nucleoplasm
10. nucleus

KEY TO FIGURE 3-2

1. Golgi apparatus
2. mitochondria
3. lipid droplets
4. plasma membrane
5. lysosome
6. centrosphere
7. centrioles
8. nuclear membrane
9. vacuole
10. nucleus (nucleoplasm)
11. endoplasmic reticulum
12. nucleoli
13. ribosomes
14. glycogen inclusions
15. pinocytic vesicle
16. cytoplasm

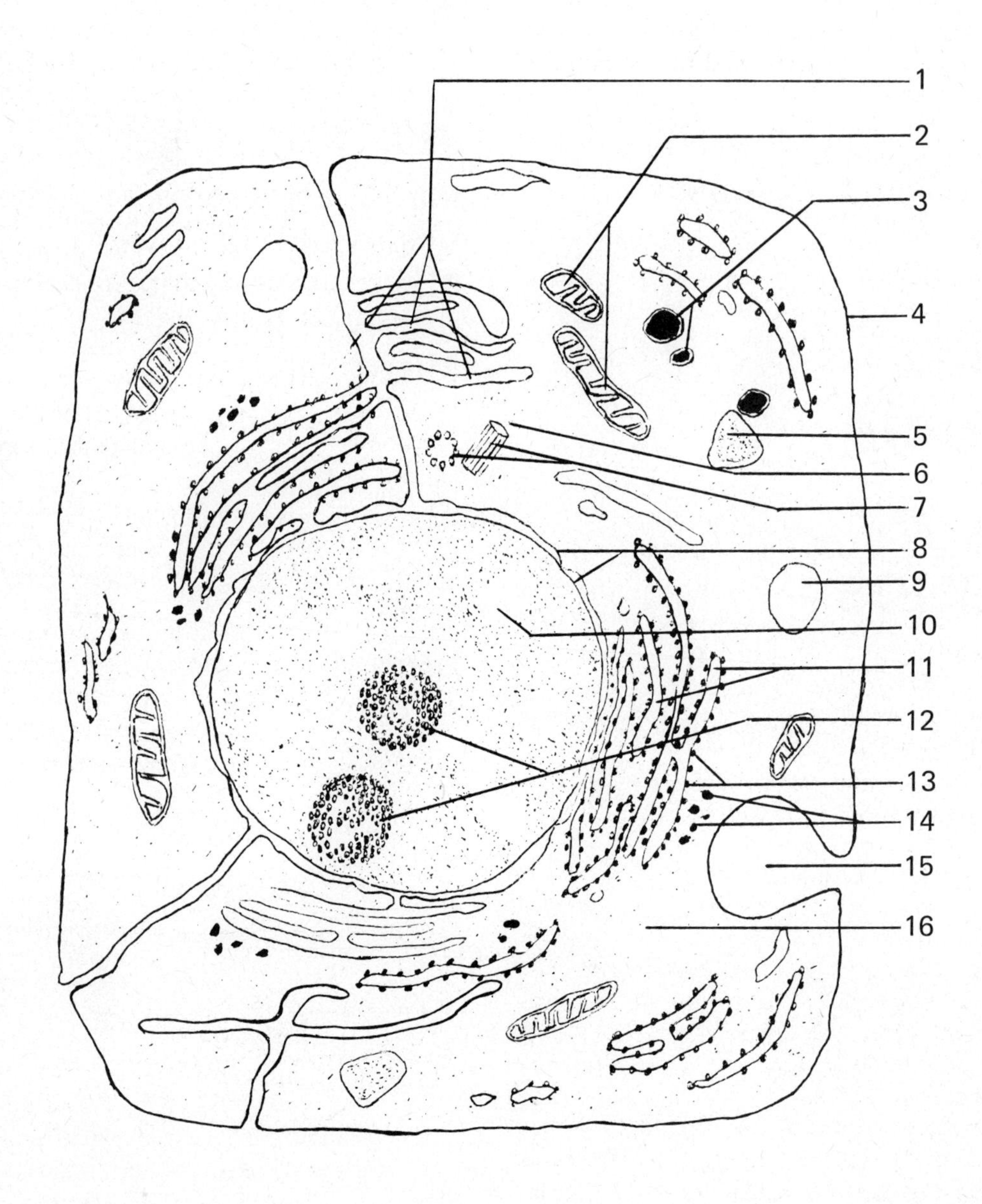

Figure 3-2. Generalized Cell (as viewed with an electron microscope)

structures visible in a properly stained slide. Most of these structures may also be observed in the cells making up tissues. Figure 3-1 may be used as a self-test (see Section J, p. xx for a suggested approach).

2. Figure 3-2 illustrates the structure of a generalized cell as viewed with an electron microscope. Figure 3-2 may be used as a self-test.
3. Using your text and/or other references, determine the functions of the structures indicated in Figures 3-1 and 3-2 and record them in the places provided under Results and Questions.

Part 2
Observation of an Ameba

Discussion

Amebae are unicellular animals. There are many similarities between these organisms and white blood cells (leukocytes) in humans. Since they move quite slowly, amebae are ideal for observing the properties of protoplasm and the structures of a living cell.

Procedure

1. Prepare a clean depression (well) slide and cover glass. If a depression slide is not available use a plain slide. Be certain that the slides and cover glasses used have not had formaldehyde or other preservatives on them at any time, since amebae are very sensitive to such chemicals and it is almost impossible to remove these contaminants by merely washing the slides.
2. Obtain a drop of culture containing amebae from your instructor.
3. If a depression slide is not used, be very cautious in placing the cover glass over the culture drop. If you drop it, the amebae may be crushed. The best procedure is to touch an edge of the cover glass to the drop and then allow the other edge to descend gradually, while supporting it with the tip of a pencil or probe.
4. If a depression slide is available, the culture drop may be placed directly into the well (depression) and the cover glass applied, or the **hanging-drop** method may be used. If the latter method is used, ring the depression on the slide with petroleum jelly; then place a drop of culture on the cover glass and invert this over the depression.
5. Avoid continued exposure of the amebae to the heat emitted by the microscope lamp, as this will induce cyst formation (the organisms "ball up"). Encystment may be avoided by using only room lights.
6. Make your initial observations with the naked eye while holding the slide over a dark background. The amebae can be seen as tiny irregular masses of transparent material. They tend to glisten as light passes through them.
7. Next, examine the cell under low power. The amebae are difficult to find, since they are nearly transparent. Therefore, it is necessary to reduce the light source considerably by reducing the size of the opening of the iris diaphragm. In the dim light they can be readily observed due to the motion of the internal granules. There should be from three to ten amebae on your slide.
8. Do not ask for more material until you have searched the slide for at least five minutes. If you are then convinced your slide contains no organisms, call the instructor.
9. In Figure 3-3 some of the structures of an ameba are identified. See how many of these you can locate on the living specimen.
10. Observe the method of locomotion employed by the ameba. The fingerlike structures that seem to flow out of the cell are pseudopods (false feet). Movement by the formation of pseudopodia is called ameboid movement.

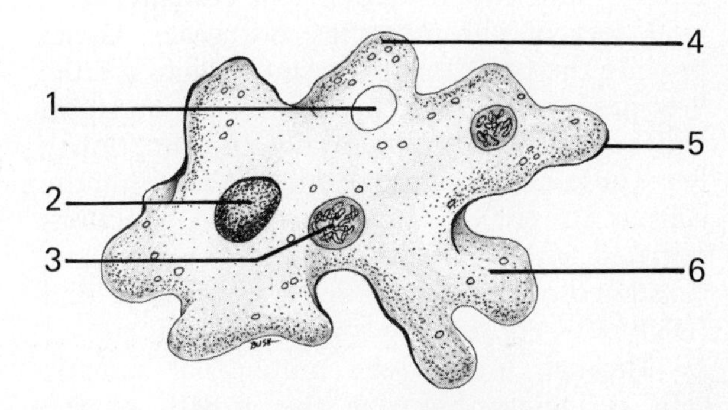

Figure 3-3. *Ameba proteus*

KEY TO FIGURE 3-3

1. contractile vacuole
2. nucleus
3. food vacuole
4. withdrawing pseudopod
5. plasma membrane
6. endoplasm

Part 3. Phagocytosis

Discussion

Phagocytosis is the ingestion and digestion of organisms and particles by cells such as amebae (unicellular organisms) and leukocytes (white blood cells). *Tetrahymena pyriformis* is a small protozoan which is readily ingested by amebae that have been starved for 24-48 hours.

Procedure

1. Obtain a drop of culture containing starved *Chaos chaos* or *Ameba proteus* cells and place it in a depression slide.
2. Using the low power objective of the microscope, locate an ameba; then add a drop of *Tetrahymena pyriformis* culture to the slide while keeping the ameba under observation.
3. Observe the formation of the so-called "food cup" as a *Tetrahymena* is engulfed.

Exercise 3-B. Mitosis

Objective

To identify the major phases of mitosis and to understand the significance of mitosis.

Materials

Microscope; prepared slides of whitefish blastula; charts; and references.

Discussion

Mitosis is a systematic sequence of events resulting in the production of two daughter cells from a single initiating or mother cell. Each of the two daughter cells contains identical sets of chromosomes and genes. Genes provide the basis for inheritance. They are the "recipes" which determine what an organism will become (both structurally and functionally). These determinations take place within the cells as a result of many biochemical reactions initiated either directly or indirectly by the substance of the genes—deoxyribonucleic acid (DNA).

Through mitosis the number of somatic cells is increased during the growth of the body. Mitosis is also the means by which worn-out cells of the body are replaced.

The blastula, which is an early stage of embryological development, shows all of the various phases of animal mitosis. In this exercise the whitefish blastula is used because these slides are readily available and the phases of mitosis that occur are the same as those in humans.

Procedure

1. Select a prepared slide of the whitefish blastula and focus under the low power objective.
2. Now, using the high power objective, explore the slide and locate the following stages of mitosis: interphase, prophase, metaphase, anaphase, telophase, and daughter cells.
3. As each stage of mitosis is located, compare it with Figure 3-4. Note the characteristics that identify each stage of mitosis.
4. Figure 3-4 may be used as a self-test.

Exercise 3-C Epithelial Tissue

Objective

To learn the characteristics of selected epithelial tissues.

Materials

Microscope; prepared slides of simple columnar, simple cuboidal, and stratified squamous epithelial tissues; 3H pencil; drawing paper; 1% methylene blue; and toothpicks.

Discussion

Epithelial tissues are specialized to protect, absorb, and secrete. They cover and line various surfaces. Secretory epithelium usually exists in the form of glands, which are groups of epithelial cells that have grown down into the underlying tissue from a surface epithelial membrane.

Preliminary study should precede examination of tissues under the microscope. The drawings of epithelial tissues (Figs. 3-5, 6, 7, 8, and 9) have been made from prepared slides similar to the ones used in the laboratory. Use these, and other reference material, as a guide. In studying tissues, think in terms of the characteristic appearance of the cells making up the tissues, as well as the locations and functions of each tissue.

KEY TO FIGURE 3-4

A. interphase
B. early prophase
C. prophase
D. metaphase
E. metaphase cell cut through equator
F. anaphase
G. early telophase
H. late telophase
I. daughter cells

1. nucleus
2. nuclear membrane
3. cytoplasm
4. plasma membrane
5. astral rays
6. centrosphere
7. chromatin
8. chromosomes
9. chromosomes at equator
10. spindle fibers
11. cleavage furrow

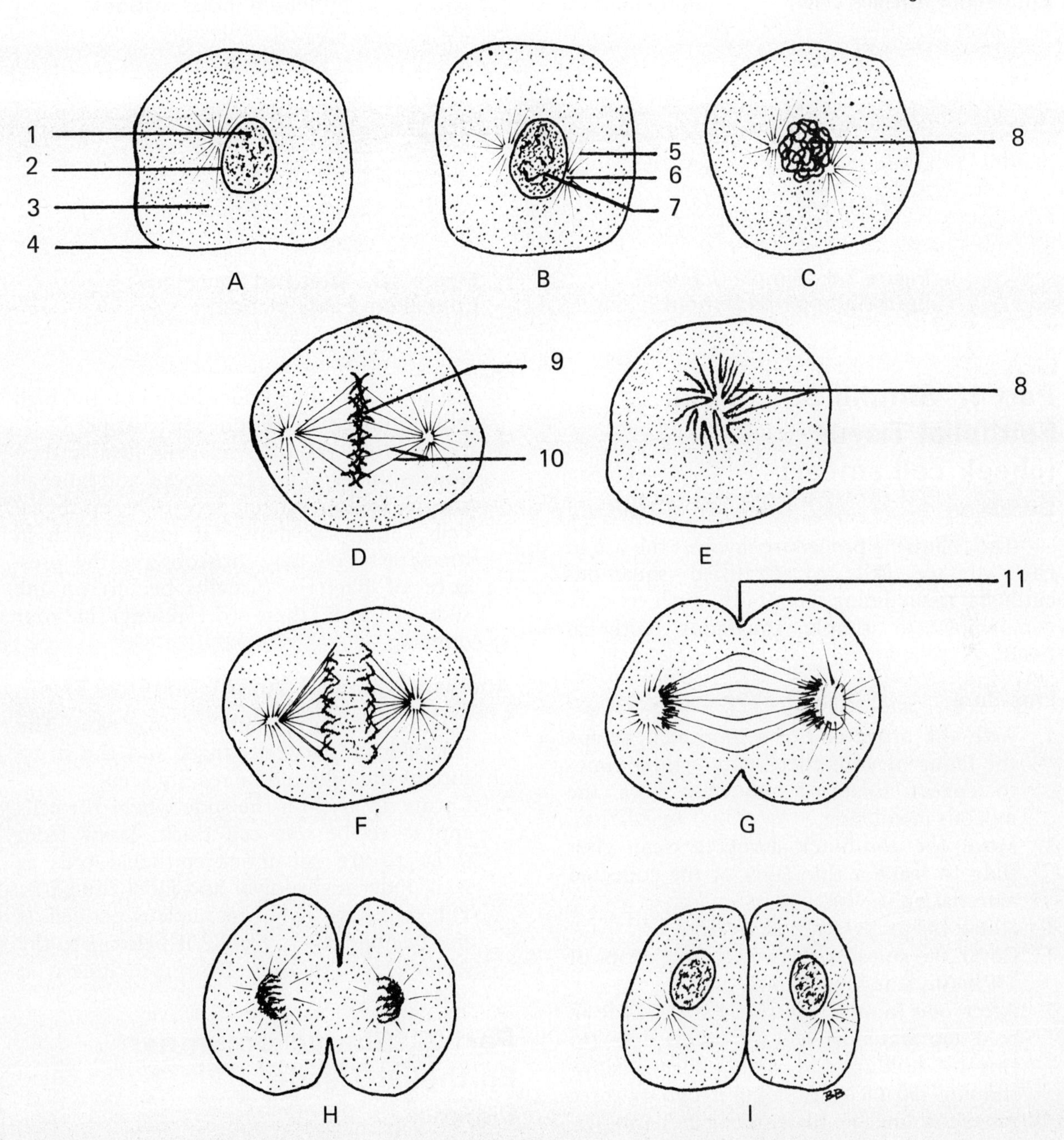

Figure 3-4. Mitosis (as observed in whitefish blastula)

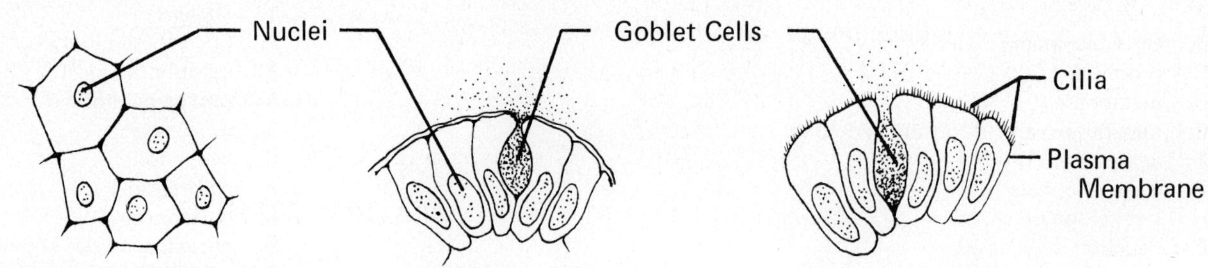

Figure 3-5. Simple Squamous Epithelium (surface view)

Figure 3-6. Simple Columnar Epithelium (cross section)

Figure 3-7. Ciliated Columnar Epithelium (cross section)

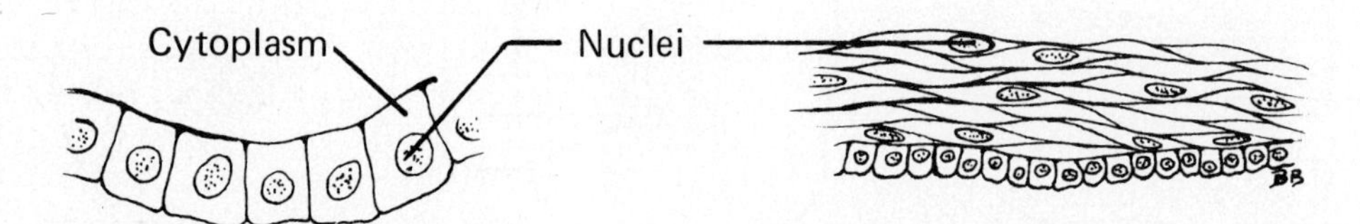

Figure 3-8. Simple Cuboidal Epithelium (cross section)

Figure 3-9. Stratified Squamous Epithelium (cross section)

Part 1. Simple Squamous Epithelial Tissue (cheek cell smear)

Discussion

The following procedure involves the use of the surface cells of stratified squamous epithelial tissue lining the mouth. Such cells are comparable to simple squamous epithelial tissue.

Procedure

1. With the broad end of a toothpick, scrape the lining of your cheek two or three times to collect some of the cells from the mucous membrane.
2. Move the toothpick across a clean glass slide to leave a thin layer of the collected material on it.
3. Allow the preparation to air dry.
4. Cover the smear with two to three drops of 1% methylene blue stain.
5. After **one minute** gently rinse the slide in cold tap water to remove excess dye. Do this by holding the slide under a slow-running tap or by flooding it two or three times by means of a medicine dropper. Gently blot (do not rub) the slide dry with a paper towel.
6. Examine the slide under both low and high power objectives of your microscope.
7. **Draw** from three to six cells just as they appear under your microscope and label all the parts you can identify. The epithelial cells should be drawn at least 1 inch in diameter. You may also observe the presence of bacteria (usually bacilli) on the slide. Include these (if present) in your drawing.

Alternate Procedure—Amphibian (Frog) Skin

1. Select a prepared slide of frog skin. The frog skin is two cells thick and the structure of the cells can be readily seen.
2. Locate an area on the slide where the cells appear to be one cell thick. **Draw** from three to six squamous epithelial cells as seen under high power and label the parts. The cells have only one nucleus per cell. If another nucleus is visible, it belongs to the second layer of cells. Do not include it in your drawing.

Part 2. Simple Columnar Epithelial Tissue

Discussion

Simple columnar epithelium makes up the outer layer of the villi in the wall of the small intestine. The villi, in cross-sectional view, have the appearance of fingerlike projections. **Ciliated columnar epithelium** is a modified type of epithelial tissue which contains cilia on the surface of each cell. Goblet cells are unicellular glands which secrete mucus; these are found in both simple and ciliated columnar epithelium.

Procedure

1. Obtain a slide entitled "simple columnar epithelium, intestine."
2. Locate the villi on low power and select an area where individual columnar cells may be seen clearly. Then turn to high power and observe the details of the cells.
3. Draw from three to six cells (at least 1/2 inch wide) and label all the parts you can identify. Include a goblet cell.
4. Your drawing should show clearly the shape, size, and location of the nucleus, the parts of the nucleus, the brush border, and any additional structures you can identify.

Part 3. Simple Cuboidal Epithelial Tissue

Discussion

The cells of this tissue are almost cube shaped. This tissue is found in many glands.

Procedure

1. Select a prepared slide and locate the cuboidal epithelial tissue under the microscope.
2. Draw from three to six cuboidal cells as observed under the high power objective and label the parts.

Part 4. Stratified Squamous Epithelial Tissue

Discussion

As the name implies, this tissue is layered (in strata). The deepest cells are almost cube-shaped, but they become more and more flattened as the surface is approached. At the surface the cells are completely flat (similar to simple squamous epithelium). The tissue varies from a few to many layers thick, depending on its location.

Procedure

1. Locate the tissue on low power and move the slide until a relatively thin area is found. Select an area which shows some of the detail of individual cells.
2. Using the high power objective draw a section, at least four cells wide, from the deepest layer of the tissue to the surface.

Exercise 3-D Connective Tissue

Objective

To learn the characteristics of selected connective tissues.

Materials

Microscope; prepared slides of areolar connective tissue, adipose tissue, hyaline cartilage, bone, and blood; 3H pencil; and drawing paper.

Discussion

Connective tissues are specialized for providing support and holding other tissues together. They are characterized by a much larger content of intercellular substances (the matrix) than other tissues.

Part 1. Areolar (Loose) Connective Tissue

Discussion

This tissue is distributed quite extensively throughout the body. It consists of white **collagenous** fibers, yellow **elastic** fibers, assorted tissue cells, and a semi-fluid to fluid ground substance. The fibers and the fluid (tissue fluid) constitute the matrix. Before viewing a slide of areolar tissue, examine Figure 3-10 and other reference material for orientation.

Procedure

1. Obtain a slide of areolar connective tissue.
2. Locate the tissue on low power; then turn to high power to distinguish the different types of fibers and cells.
3. To distinguish between the two kinds of fibers, look for differences in thickness. The white collagenous fibers occur in parallel bundles and thus resemble slightly wider bands. The yellow elastic fibers are single, thin strands which may branch.
4. **Fibroblasts** will be found among the fibers in the matrix. The cytoplasm in these cells is difficult to see without careful light regulation, although the nuclei are rela-

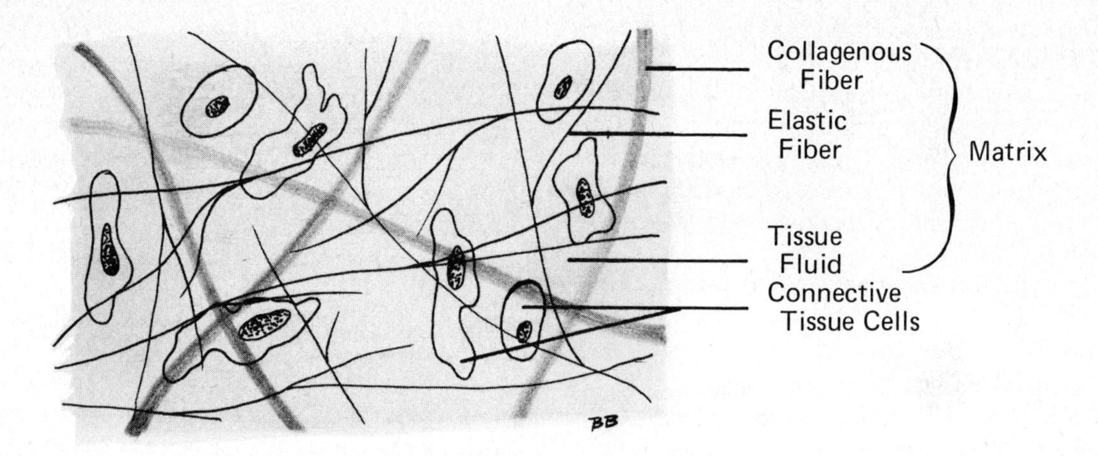

Figure 3-10. Areolar Connective Tissue

tively easy to observe. The elongated or star-shaped bodies of the fibroblasts send out sharp processes. Histiocytes or macrophages are usually elongated, spindle-shaped cells.

5. Draw a section of the slide including the following: matrix, collagenous fibers, elastic fibers, and connective tissue cells. Label all structures drawn.

Part 2. Adipose Tissue

Discussion

Fat cells are scattered singly or in groups in areolar tissue. A mature fat cell contains one large drop of fat. The cytoplasm is reduced to a thin membrane surrounding the drop of fat. When fat cells form in large numbers and crowd out other cells, areolar tissue is transformed into adipose tissue.

Examine Figure 3-11 and other reference material before viewing a prepared slide of adipose tissue.

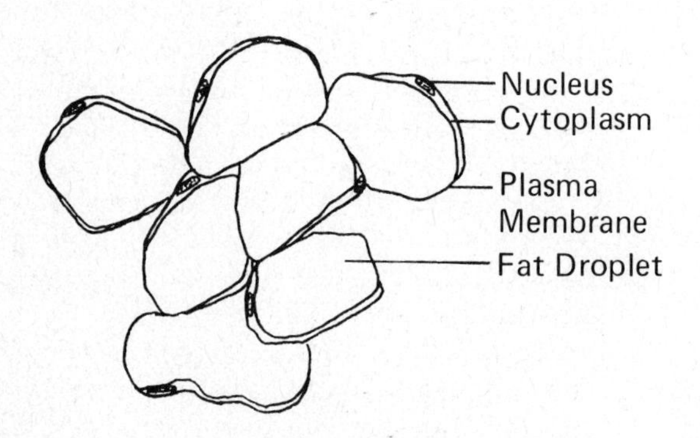

Figure 3-11. Adipose Tissue

Procedure

1. Obtain a slide containing adipose tissue.
2. Locate the tissue on low power, then examine it under the high power objective. (If the tissue has been stained, the cells will appear black if osmic acid was used, or orange if Sudan III was used.)
3. Draw from three to six cells and label all of the parts.

Part 3. Hyaline Cartilage

Discussion

Hyaline cartilage forms the tracheal rings, part of the larynx, the covering of articular surfaces of bones, etc. Each **chondrocyte** (cartilage cell) is enclosed within a **lacuna.** The chondrocyte completely fills this depression during life. There may be as many as four cells in a lacuna in mature cartilage. The cells are nourished by the **perichondrium**, a fibrous membrane covering the cartilage.

Fibrous cartilage (white fibrocartilage) differs from hyaline cartilage in that it has collagenous (white) fibers within the matrix. This cartilage forms the symphysis pubis and the intervertebral discs.

Elastic cartilage (yellow fibrocartilage) differs from hyaline cartilage in that it has elastic fibers in the matrix. This tissue may also have a limited amount of collagenous fibers within the matrix. It is found in the external ear, the eustachian tube, parts of the larynx, etc.

Examine Figure 3-12 of hyaline cartilage before viewing the prepared slide under the microscope.

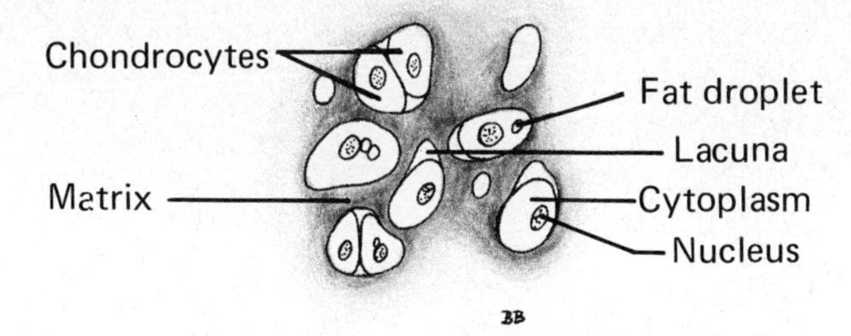

Figure 3-12. Hyaline Cartilage

Procedure

1. Obtain a slide of hyaline cartilage.
2. Locate the tissue first on low power, then observe it on high power.
3. On the basis of your observations, draw and label a segment of this tissue which includes the matrix, lacunae, and chondrocytes. You should be able to observe nuclei and other structures in the cells.
4. Which tissues can be observed on either side of the strip of cartilage?
5. Demonstrations:
   a. Examine the slide under the microscope showing fibrous cartilage, noting the characteristics indicated in the preceding discussion.
   b. Look at the slide under the microscope showing elastic cartilage. Note the characteristics indicated in the preceding discussion.

Part 4. Osseous (Bony) Tissue

Discussion

The shaft (diaphysis) of a typical long bone consists of a cortex of **compact** (dense) osseous tissue, with an inner lining of **cancellous** (spongy) bony tissue. Spongy bone consists of plates and bars forming a network which is well adapted for providing mechanical support. It lacks haversian canals and related structures that are found in compact bone.

Most of the mass of a bone is made up of layers (lamellae) of calcified **interstitial substance** (or bone matrix). The lamellae are arranged differently in compact than in spongy bone tissue. **Lacunae** (cavities) completely filled with **osteocytes** (bone cells) are found in the interstitial substance. Minute canals called **canaliculi** arise from very fine apertures in the walls of these lacunae and penetrate the matrix in all directions, branching profusely to form a network in which frequent anastomoses occur.

Within the compact bone of the diaphysis of any long bone are numerous **haversian canals.** These are cylindrical, vertically branching and anastomosing canals which may contain one or more blood vessels (usually two capillaries). They communicate with the external surface of the bone and the bone marrow cavity by means of the horizontal **Volkmann's canals.**

Before viewing the slide of ground, compact bone, examine the following three figures showing relationships between gross bone and haversian systems. Figure 3-13 shows a small portion of the diaphysis of a bone and its marrow. It has been magnified to demonstrate both longitudinal and cross-sectional aspects of haversian systems. Figure 3-14 provides a microscopic view of a longitudinal section of haversian canals. Figure 3-15 shows a cross section of two haversian systems as viewed under the microscope.

Procedure

1. Obtain a slide entitled "Ground Bone" or "Bone, Human."
2. Locate the tissue on low power, then draw while observing under high power. Since the tissue is quite thick, with the cover glass raised, use caution on high power to avoid breaking the slide.
3. Diagram one complete haversian system. Include a wedge from the center to the edge of the system, drawn in detail. Include the following structures:
   a. **Haversian canal**, the large opening in the center. This may appear empty, may contain remnants of blood vessels, or it may appear dark due to the grinding process.
   b. **Lacunae**, small depressions arranged more or less symmetrically around the haversian canal. Osteocytes are not present in these lacunae due to the grinding process. Observe the position and shape of the lacunae.
   c. **Canaliculi**, tiny canals that lead to and from the lacunae and haversian canal. Observe the position, shape, and length of the canaliculi. The canaliculi in the outer ring loop back into the haversian system.
   d. **Matrix**, the ground substance between the lacunae and canaliculi.
   e. **Lamellae**, the circular layers of bone laid down by the osteocytes. Each lamella consists of a ring of lacunae and matrix.

KEY TO FIGURE 3-13

1. periosteum
2. periosteal vein
3. periosteal artery
4. haversian system
5. haversian canal
6. Volkmann's canal
7. medullary artery
8. marrow cavity

KEY TO FIGURE 3-14

1. haversian canal
2. Volkmann's canal
3. lacuna

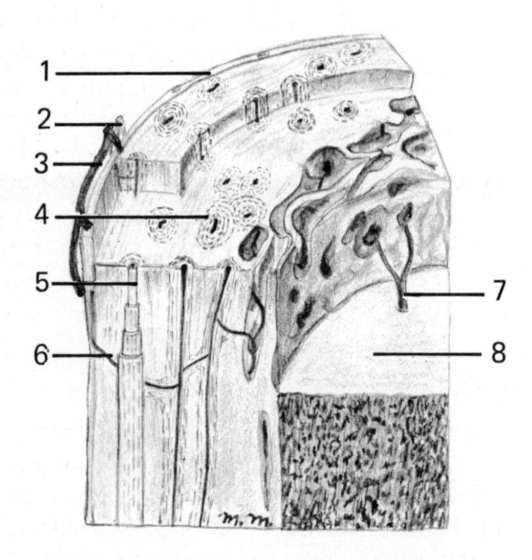

Figure 3-13. Cross Section of Diaphysis (magnified)

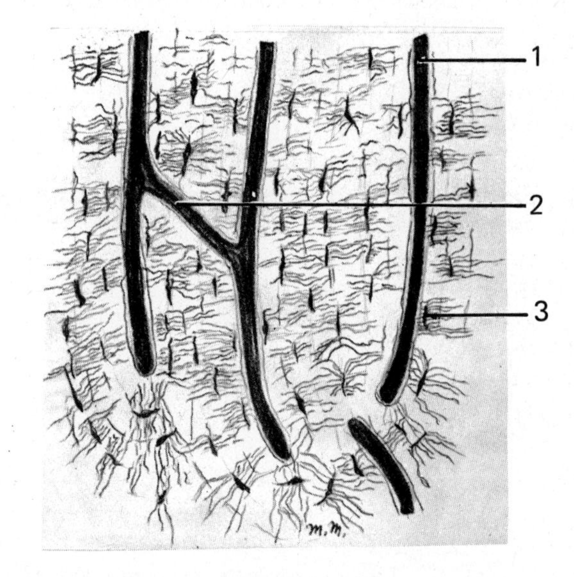

Figure 3-14. Longitudinal Section through Compact Bone (430X)

KEY TO FIGURE 3-15

1. matrix
2. haversian canal
3. canaliculus
4. lacuna containing osteocyte
5. interstitial lamella

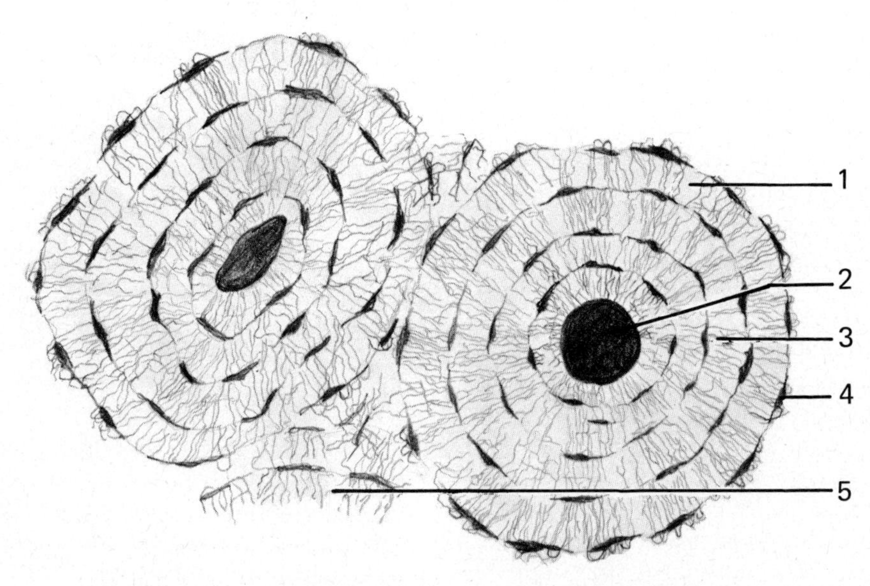

Figure 3-15. Cross Section through Two Haversian Systems

Part 5. Blood

Discussion

The slide to be examined consists of a thin smear of blood stained with either Wright's stain or Giemsa's strain. With these dyes, granules in the cytoplasm of various kinds of leukocytes are differentially stained. The nuclei of the leukocytes are stained blue and the erythrocytes are stained a pale red color.

The following outline provides some of the identifying characteristics of the cells that may be observed on the prepared slide:

A. **Erythrocytes** (red blood cells): The great majority of the cells you will observe will be of this type. They are small, **bioconcave** disks. Many vertebrates have nucleated red cells (e.g., amphibians such as frogs and salamanders); however the erythrocytes of mammals are **anucleate**.

B. **Leukocytes** (white blood cells with nuclei):
   1. **Lymphocytes** (20-25% of all leukocytes). These are found in large numbers in lymph nodes and lymph as well as in the circulating blood. They have a large, single nucleus which is approximately spherical in shape and a narrow band of cytoplasm.
   2. **Monocytes** (3-8% of all leukocytes). These cells consist of a single non-lobed nucleus. The nucleus is frequently curved in the form of a crescent. The cytoplasm is granular and more abundant than in lymphocytes (i.e., microlymphocytes).
   3. **Granulocytes**. The cytoplasm of these cells contains numerous granules, which may vary in density and coarseness depending on the type. They are distinguished on the basis of the color they stain with various dyes. These differentiating stains are included in Wright's stain.
      a. **Neutrophils** (65-75% of all leukocytes). They are also called polymorphonuclear leukocytes because their nuclei consist of several (usually three to five) lobes. Upon careful observation (high power) it will be noted that the lobes are interconnected. The granules in the cytoplasm are relatively fine-grained, stain with a neutral dye, and appear faint red in color.
      b. **Eosinophils** (2-5% of all leukocytes). The dense granules of these cells, seen in the cytoplasm, are stained red by the eosin in Wright's stain. The nucleus is usually bilobed, but this may be difficult to observe due to the dense granular cytoplasm.
      c. **Basophils** (½-1% of all leukocytes). These cells contain such coarse granules that they all but blot out the nucleus. The granules take a basic dye and appear to be blue or purple in color.
   4. **Thrombocytes** (Platelets). The platelets are about half the size of an erythrocyte. (They are derived from the fragmentation of giant megakaryocytes (which are up to 100 microns in diameter) found in the marrow.

Procedure

1. Obtain a slide of a human blood smear.
2. For orientation, refer to the drawings of blood cells (Fig. 3-16) and other references.
3. Note that the great majority of cells on the slide are erythrocytes. Draw two or three of the cells as they appear in different positions.
4. Identify and draw a lymphocyte.
5. Identify and draw a neutrophil.
6. Optional: If time permits, identify and draw an eosinophil, a basophil, a monocyte, and a group of thrombocytes.

Exercise 3-E Muscular Tissue

Objective

To distinguish the characteristics of the three kinds of muscle tissue.

Materials

Microscope; slides of skeletal, smooth, and cardiac muscle; drawing paper, and a 3H pencil.

Part 1. Skeletal (Striated) Muscle

Discussion

Each skeletal muscle cell is a multinucleate fiber containing many bands called striations. The longest striated muscle fibers are estimated to be about 40,000 microns.

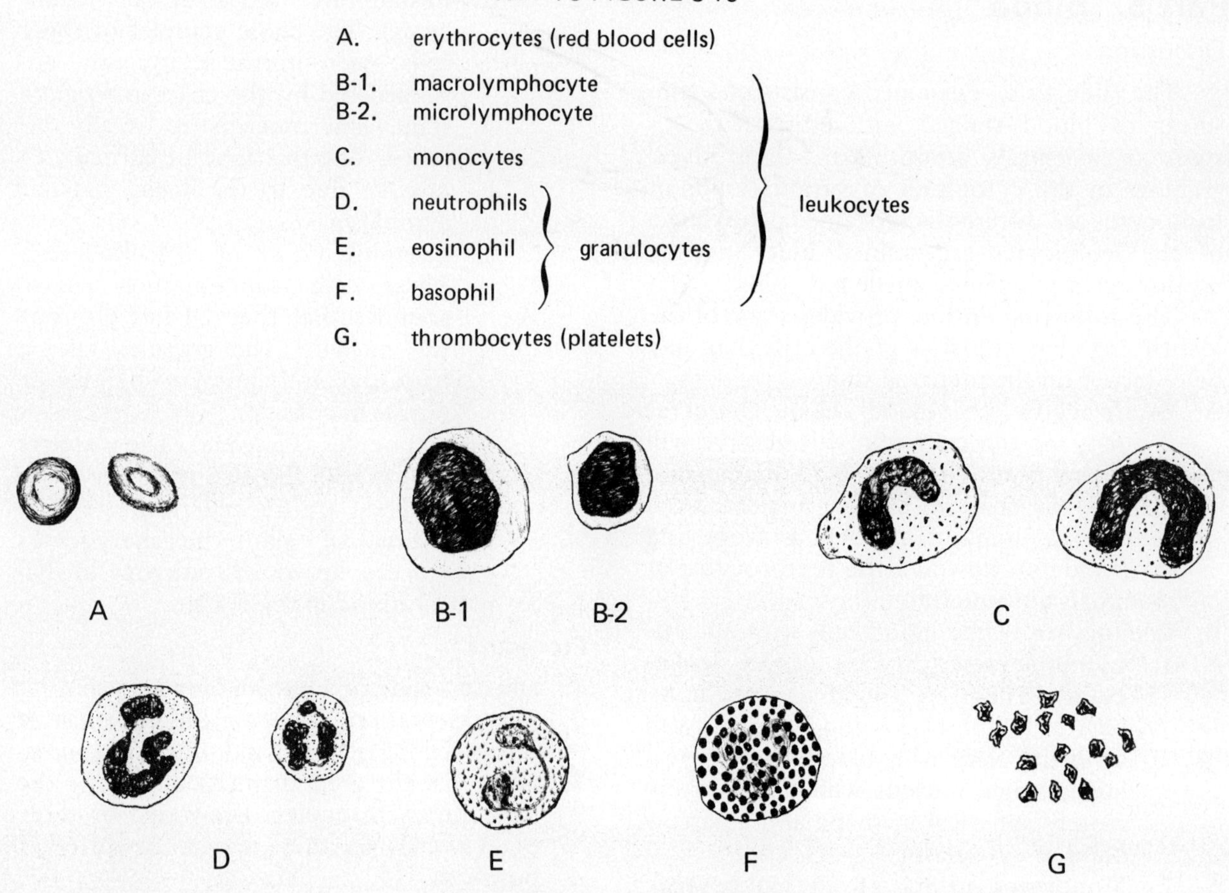

Figure 3-16. Human Blood Cells

Procedure

1. Obtain a slide showing teased skeletal muscle tissue.
2. Compare a fiber, as seen under high power, with Figure 3-17.
3. Draw one muscle fiber as seen in longitudinal section.
4. Label sarcolemma, sarcoplasm, and striations. Include nuclei if they are visible. Note their size (relative to the striations) and their position.

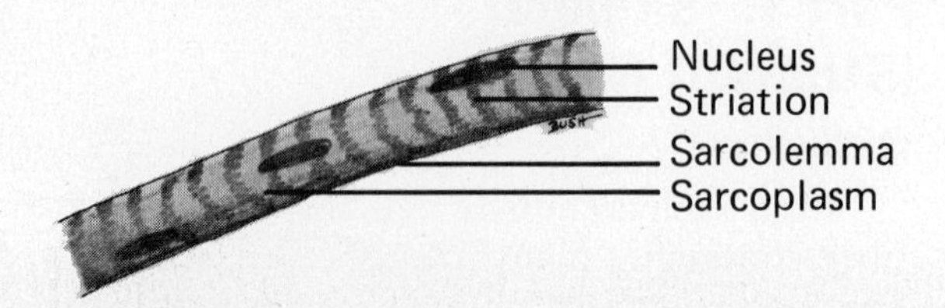

Figure 3-17. Skeletal Muscle Fiber (l.s.)

Part 2. Visceral (Smooth) Muscle

Discussion

The cells of this tissue are up to 500 microns in length; each spindle-shaped cell has tapered ends.

Procedure

1. Obtain a slide of smooth (visceral) muscle tissue.
2. Compare the cells observed with Figure 3-18.
3. Draw two or three representative cells and label all parts identified.

Part 3. Cardiac Muscle

Discussion

Somewhat fainter striations are observed in this tissue than in skeletal muscle tissue. Also,

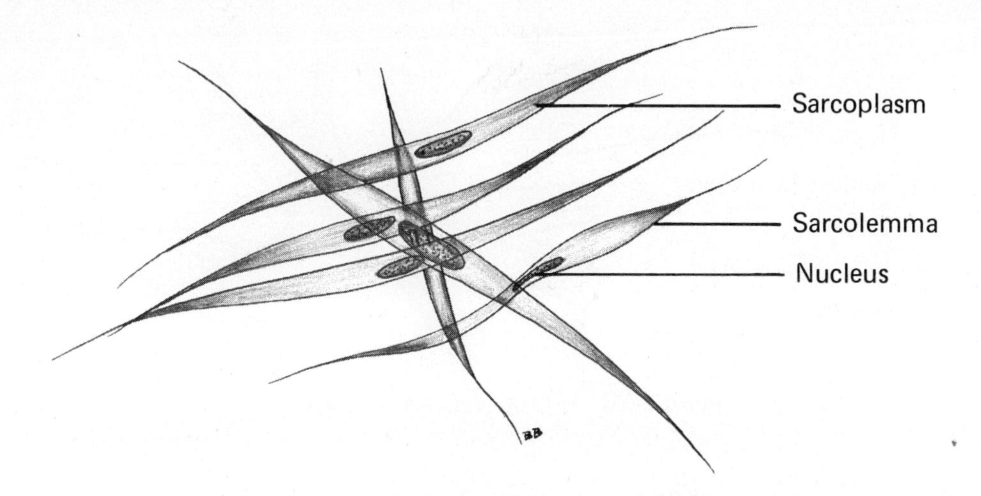

Figure 3-18. Smooth (visceral) Muscle Fibers (teased)

this tissue contains unique dark bands called **intercalated discs**. The electron microscope has revealed that these discs are places where two cell membranes abut at the ends of adjacent cardiac fibers. The cardiac fibers form a continuum by repeatedly branching.

Procedure

1. Obtain a slide of cardiac muscle tissue.
2. Compare this tissue, as observed under high power, with Figure 3-19.
3. Draw and label three or four fibers showing the branching nature of the tissue, striations, and intercalated discs.

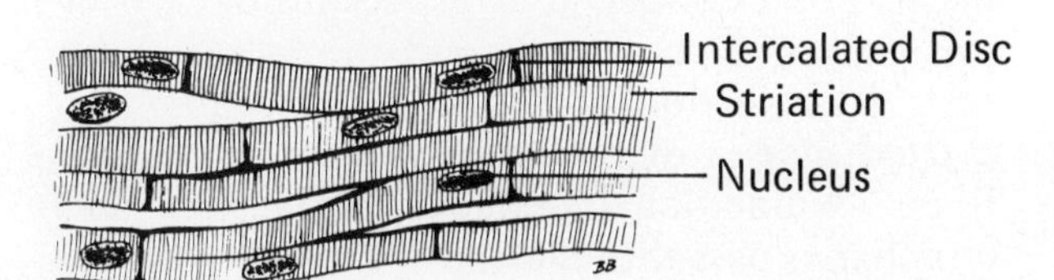

Figure 3-19. Cardiac Muscle Tissue (l.s.)

Exercise 3-F Nerve Tissue

Objectives

To observe the characteristics of a motor neuron and a myelinated nerve fiber.

Materials

Microscope; slides of motor neurons and myelinated nerve fibers; drawing paper; and a 3H pencil.

Part 1. Motor Neurons

Discussion

The slide to be examined was prepared by smearing and straining the ventral horn of the spinal cord. The cell bodies (**cytons**) of motor neurons in a spinal cord smear are among the largest in the cord. In the process of preparing the slide, the axon and dendrites of these large cells were broken. On this slide, neuroglia cells are visible between the separated neurons.

Procedure

1. Obtain a slide entitled "Nerve cells, ox spinal cord."
2. Locate one neuron and compare it with the motor neuron illustrated in Figure 3-20.
3. Draw the neuron under low power. Include in your drawing (and label) the cyton, axon, dendrites, and nucleus with nucleolus and any other visible parts.

Part 2. Myelinated (Medullated) Nerve Fiber

Discussion

Myelinated nerve fibers are located in the central nervous system and in peripheral nerves. Their name is derived from the fact that such fibers are surrounded by a pearly white, lipid substance called the myelin sheath. In the central nervous system the myelin sheath is formed by glial cells, and fibers containing it make up the white matter (gray matter is composed mainly of cell bodies and fibers which usually lack myelin, hence, it is gray in

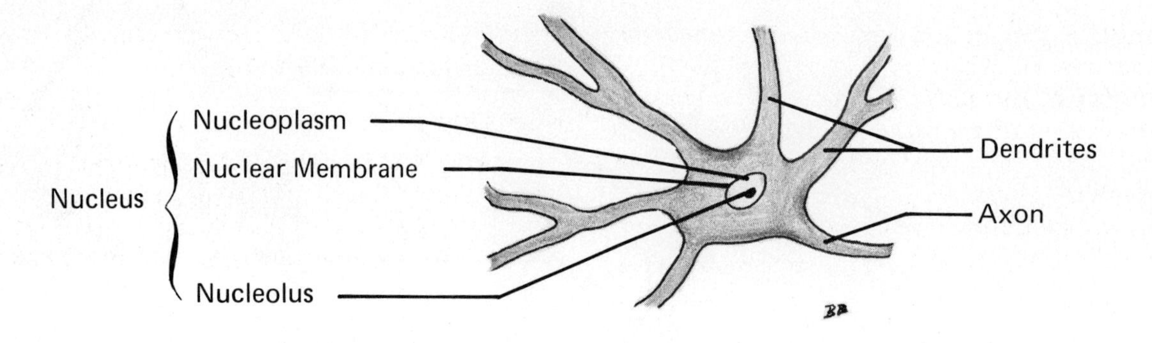

Figure 3-20. Motor Neuron (Smear)

color). In the peripheral nerves, axons of motor neurons are surrounded by sheaths of myelin which are formed by neurilemmal cells. All myelin sheaths are interrupted at regular intervals at points referred to as nodes of Ranvier. All nerve fibers outside the central nervous system have the thin membrane called the neurilemma, whether they are myelinated or not. The flat cells forming the neurilemma are called cells of Schwann. There is usually only one of these cells between two successive nodes of Ranvier.

Procedure

1. Obtain a slide containing teased myelinated nerve fibers.
2. Locate a fiber that appears to be typical and compare it with Figure 3-21.
3. Using the high power objective, draw a segment of the fiber which includes two nodes of Ranvier, neurilemma, nucleus of Schwann cell, myelin sheath, and axis cylinder.

Exercise 3-G The Skin

Objective

To become familiar with the structure of the skin.

Materials

Microscope; slide of human skin; model of skin; references; and anatomical charts.

Discussion

Skin consists of two principal layers: the **epidermis** (cuticle), and the **dermis** (corium or true skin). The epidermis contains an outer layer of cells which are keratinized (protoplasm replaced by nonliving protein called keratin). These cells are flattened and scalelike and are constantly being lost by the body. This layer of cells is known as the **stratum corneum** or horny layer; the cells are replaced as a result of the mitotic divisions of the lower layer of the epidermis, which is known as the **stratum germinativum**.

The dermis (corium), like the epidermis, is variable in its thickness in different parts of the body. For example, it is especially thick over the palms of the hands and the soles of the feet. This portion of the skin is composed of matted masses of connective tissue and elastic fibers. Numerous blood vessels, nerves, and lymphatics pass through the dermis.

Appendages of the skin include hair, nails, **sudoriferous** (sweat) glands, and **sebaceous** (oil) glands.

The subdermal tissue (subcutaneous layer) contains fat cells, connective tissue, blood vessels, lymphatics, and nerves. These tissues

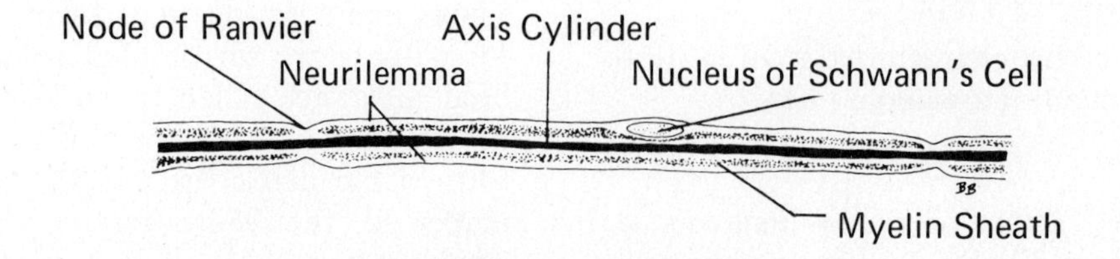

Figure 3-21. Myelinated Nerve Fiber (l.s.)

provide a connection between the skin and deeper tissues. Where the connecting fibers are loose (as in the neck) the skin can be moved quite easily. Where the skin is attached more firmly (as in the palms and soles), only limited movement is possible.

The specialized receptors located within the skin (for receiving such stimuli as cold, heat, touch, pain, etc.) are discussed under the Introduction in Chapter 12.

Procedure

1. Using Figure 3-22 as a guide, examine the structures on the model of human skin.
2. Obtain a slide of human skin. Using references, anatomical charts, the skin model,

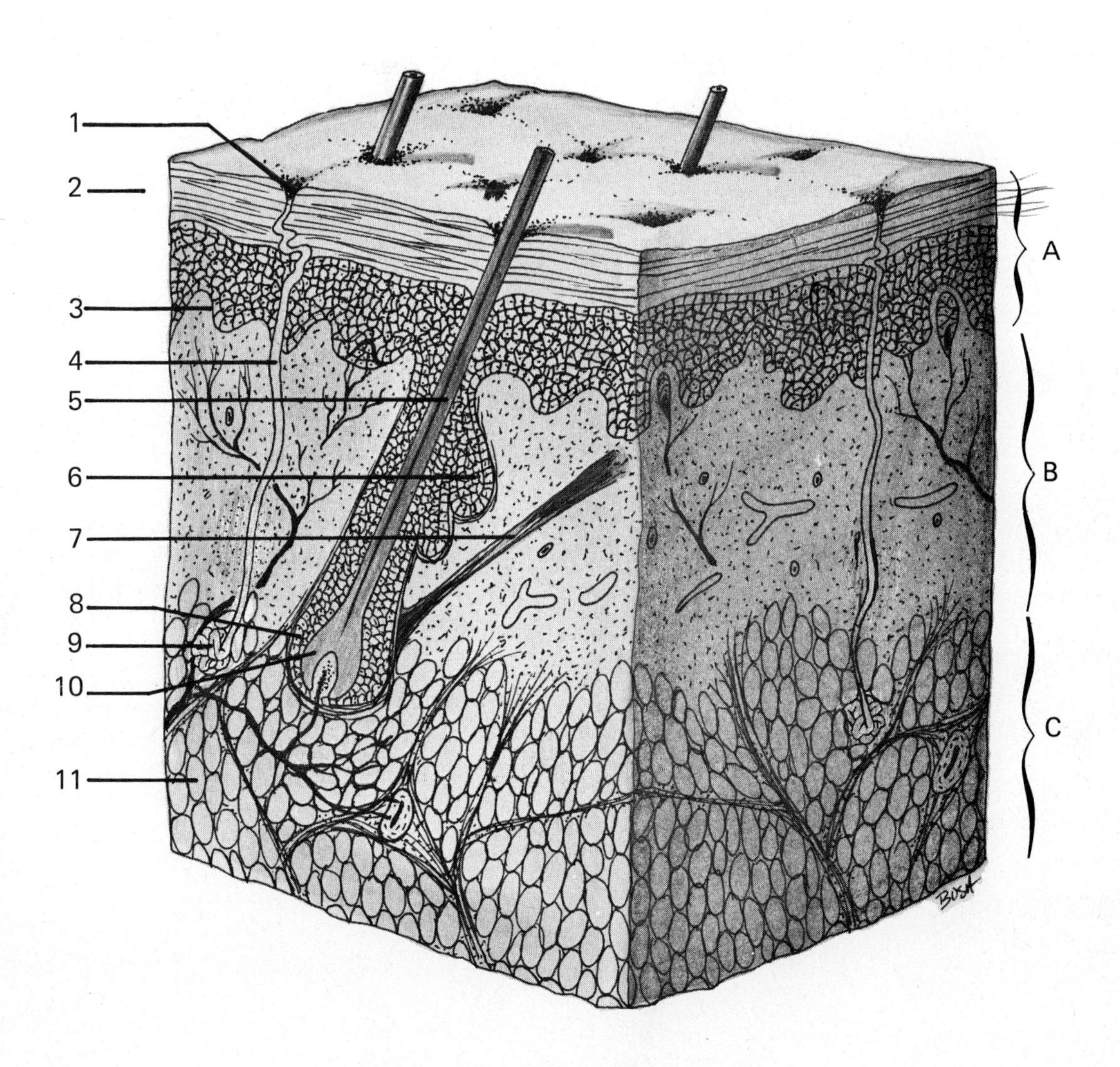

Figure 3-22. The Skin (cross section)

KEY TO FIGURE 3-22

A. epidermis (stratified squamous epithelium)
B. dermis (true skin)
C. subdermis (subcutaneous tissue)
1. pore of sweat gland
2. stratum corneum (horny layer)
3. stratum germinativum
4. duct of sweat gland
5. hair shaft
6. sebaceous (oil) gland
7. arrector pili muscle
8. hair follicle
9. sudoriferous (sweat) gland
10. bulb of hair (with papilla)
11. adipose tissue

and Figure 3-22, explore the slide for the following structures (check off each structure identified).

a. ____ adipose tissue

b. ____ arrector pili muscle

c. ____ blood vessel

d. ____ bulb of hair

e. ____ dermis

f. ____ duct of sweat gland

g. ____ epidermis

h. ____ hair shaft

i. ____ nerve fiber

j. ____ pore of sweat gland

k. ____ sebaceous gland

l. ____ stratified squamous epithelium

m. ____ stratum germinativum

n. ____ stratum corneum

o. ____ subdermis (subcutaneous layer)

p. ____ sudoriferous gland

Self-Test—Cells, Mitosis, Tissues, and Skin

Directions: See Chapter 1, Self-Test, p. 9.

1. A group of cells of similar structure constitutes **an organ.**

2. Epithelial tissue **covers and lines.**

3. Nervous tissue **supports and binds.**

4. Epithelial tissue can be distinguished from connective tissue by the **sparseness of intercellular materials.**

5. Goblet cells are unicellular glands found in **certain connective tissues.**

6. Cartilage is a variety of connective tissue which contains **osteocytes** in lacunae.

7. All connective tissue arises from **mesenchyme.**

8. One location of **elastic cartilage** is within the external ear.

9. Histiocytes, also known as fixed macrophages, become **phagocytic** in case of infection.

10. Skin is **a tissue.**

11. A single layer of flat cells which line the blood vessels is classified as **mesothelium.**

12. **Simple columnar epithelium** lines the stomach and intestines.

13. Stratified squamous epithelium is found, primarily, **covering the outside of the body.**

14. **Pseudostratified ciliated columnar epithelium** is found lining much of the respiratory tract.

15. Cilia are found on the surface of **all epithelial cells.**

16. Goblet cells produce **mucus.**

17. **Mucous membranes** line the alimentary, respiratory, and genitourinary tract.

18. **Serous membranes** line body cavities and cover organs which lie within the cavities.

19. **Haversian canals** are minute channels in the matrix of bone tissue extending to lacunae.

20. The haversian system is a nutritive arrangement of cells in **cancellous** bone.

21. The periosteum is a **fibrous membrane** covering bone.

22. The cell is both the **structural and functional** unit of the body.

23. By definition, living cells contain **only organic compounds.**

24. **All** cells have cell walls.

25. **All** cells have cell membranes.

26. The skin contains stratified squamous epithelium.

27. The **epidermis** of the skin contains the secreting portion of the sweat glands.

28. A **corn** is an inward thickening of the epidermis, which arises as a result of pressure in a local area.

29. The skin contains **striated** muscle fibers.

30. A hair is a product of **epidermal** cells.

31. The dermis contains **fibrous connective tissue.**

32. The upper surface of the epidermis is the **stratum granulosum.**

33. Sensitivity of the skin is determined by sense organs lying in the **epidermis.**

34. The most important means of protection by the skin is due to its **unbroken surface which prevents entrance of bacteria.**

35. Freckles are patches of **melanin.**

36. **Hemoglobin** is the dark brown pigment that gives color to hair, eyes, and the skin.

37. "Freckle removers" are only temporarily effective, because **they bleach only the top layer of skin,** and the pigment producing freckles is formed in the deep layers of the skin.

38. A suntan "fades" in winter, because **normally pigmented cells in the deep skin layers** are pushed upward to replace the highly pigmented cells on the surface as they are worn away.

39. Some injections with hypodermic needles hurt more than others because **pain receptors** may be hit.

40. A danger in squeezing pimples and boils is that **it may force pus into the blood stream.**

41. **Contractions of arrector pili muscles** push hair up and pull the skin down, causing "goose-pimples."

42. The **epidermis** layer of the skin is the only layer lacking a direct blood supply.

43. As an excretory structure the skin eliminates **protein metabolites.**

44. Acne is an inflammation of **sudoriferous** glands.

45. A **bunion** is an inflammation of the capsular ligament surrounding the joint of the great toe, and it may become a bony enlargement.

46. The **epidermis** is the thickest layer of skin.

47. The epidermis consists **only of dead cells.**

48. **Sebaceous glands** lubricate the skin and keep it from cracking.

49. Wrinkles are the result of **creases formed where the skin is habitually folded.**

50. The skin has friction ridges over its **entire** surface.

51. **Melanoblasts** are specialized cells in the lower part of the epidermis which produce melanin.

52. When chromosomes line up in the center of the spindle during mitosis, it is called **anaphase.**

53. The nuclear membrane disappears during the **metaphase** of mitosis.

54. The chromosomes consist of a pair of chromatids held together by a centromere in the **prophase** of mitosis.

55. During metaphase, the centromeres **divide** and the chromatids start to **separate from each other.**

56. During mitosis, chromosomes appear to be short rods due to **the spiral arrangement of each chromosome.**

57. In the **anaphase** of mitosis, there are individual chromosomes, each with a single centromere.

58. The **nucleus** of a cell contains the Golgi apparatus.

59. Mitochondria are found in **both the cytoplasm and the nucleus of a cell.**

60. Energy is released as the result of enzymatic reactions occuring in **ribosomes.**

61. The Golgi apparatus is apparently concerned with **secretion.**

62. Tendons consist primarily of **elastic connective tissue.**

63. **Vacuoles** may serve as a storage place for foods.

64. Deoxyribonucleic acid is found primarily in **cytoplasm.**

65. Secretion in the body **always** involves some form of epithelial tissue.

66. **Mucous membranes** line passageways which open to the outside, whereas **serous membranes** line closed cavities.

67. **Osteoblasts** form the hard intercellular matrix of bone.

68. **Connective tissue** forms the surface and **epithelial tissue** the underlying layer of mucous, cutaneous, and serous membranes.

KEY

1. a tissue
3. conducts
5. columnar epithelium
6. chondrocytes
10. an organ
11. endothelium
15. ciliated columnar cells
19. canaliculi
20. compact (dense)
23. both organic and inorganic compounds
24. plant
27. dermis
29. smooth (visceral)
32. stratum corneum
33. dermis
36. melanin
43. mainly water and a little salt
44. sebaceous
46. dermis
47. of living and dead
50. mostly hands and feet—not pronounced elsewhere
52. metaphase
53. prophase
58. cytoplasm
59. only in the cytoplasm
60. mitochondria
62. white fibrous connective tissue
64. the nucleus
68. epithelial tissue; connective tissue

Name ______________________

Results and Questions for Chapter 3

Exercise 3-A. The Structure and Functions of Cells

Part 1. Structure of a Cell Observed under the Light Microscope Compared to that Observed with an Electron Microscope

1. Give the primary function of each of the following structures:

   a. Vacuole ______________________

   b. Nucleus ______________________

   c. Chromosomes ______________________

   d. Centrioles ______________________

   e. Mitchondria ______________________

   f. Nucleolus ______________________

   g. Inclusions ______________________

   h. Endoplasmic reticulum ______________________

   i. Ribosomes ______________________

   j. Cytoplasmic membrane ______________________

   k. Lysosomes ______________________

   l. Pinocytic vacuole ______________________

   m. Golgi apparatus ______________________

2. Which of the structures listed under No. 1 can only be observed with an electron microscope? (Write the letter preceding the structure in answering this question.) ______________________

Part 2. Observation of an Ameba

1. Briefly describe the movement of the ameba. ______________________

   ____

2. To what cells in the human body is the ameba comparable? ______________________

3. What properties of protoplasm were you able to discern as a result of your observations of the ameba? ______________________

   ____

Part 3. Phagocytosis

1. How does the ameba react when contact is made with a particle? ______________________

   ____

2. Explain the significance of phagocytosis as it occurs in humans. ______________________

   ____

3. To what other type of engulfment by cells is phagocytosis comparable? ______________________

____

Exercise 3-B. Mitosis

1. What is the significance of mitosis? ______________________

____

2. In the spaces provided, indicate the phase of mitosis described in each of the following statements:

   a. Nuclear membrane present; two centrioles in a single centrosphere; chromosomes not visible. ______________________

   b. Chromatids completely separate into individual chromosomes, and begin moving to opposite poles. ______________________

   c. The cleavage furrow has just made its appearance. ______________________

   d. The chromosomes are oriented along the equator. ______________________

   e. Centrioles begin moving to opposite poles. ______________________

   f. Nuclear membrane disappears; distinct chromosomes first observed. ______________________

   ____

3. Where does mitosis occur in the human body? ______________________

4. What cells in the human body do not undergo mitosis? ______________________

Exercise 3-C. Epithelial Tissue

1. For each of the tissues listed below give (a) two locations; (b) two functions; and (c) two identifying characteristics:

Simple squamous: (a) ____________ ____________

(b) ____________ ____________

(c) ____________ ____________

Simple columnar: (a) ____________ ____________

(b) ____________ ____________

(c) ____________ ____________

Ciliated columnar: (a) ____________ ____________

(b) ____________ ____________

(c) ____________ ____________

Name ______________

Simple cuboidal: (a) ______________ ______________

(b) ______________ ______________

(c) ______________ ______________

Stratified squamous: (a) ______________ ______________

(b) ______________ ______________

(c) ______________ ______________

2. Which of the preceding tissues contain blood vessels? ______________

Exercise 3-D. Connective Tissue

Part 1. Areolar (Loose) Connective Tissue

1. Where is areolar connective tissue located? ______________
2. What is the function of collagenous fibers? ______________
3. What is the function of elastic fibers? ______________
4. Give two functions of areolar connective tissue. ______________

____

Part 2. Adipose Tissue

1. List three functions of adipose tissue. ______________

____

2. What is the matrix of adipose tissue? ______________

Part 3. Hyaline Cartilage

1. List four locations of hyaline cartilage. ______________

____

2. What is the function of hyaline cartilage? ______________

____

3. How does hyaline cartilage differ from fibrous and elastic cartilage? ______________

____

Part 4. Osseous (Bony) Tissue

1. What is the function of the haversian canal? ______________
2. What is the function of a canaliculus? ______________
3. By what cells is the bony matrix deposited? ______________

4. The preceding cells are located in depressions called ____________

Part 5. Blood

1. What is the major function of erythrocytes? ____________
2. What is the major function of leukocytes? ____________
3. What is the major function of thrombocytes? ____________
4. Where are erythrocytes manufactured in adults? ____________
5. Where are erythrocytes produced before birth? ____________
6. Where are leukocytes produced? ____________
7. What kind of blood cell is anucleate? ____________
8. In a normal human, what is the rarest kind of leukocyte? ____________
9. In a normal human, what is the most common kind of leukocyte? ____________

Exercise 3-E. Muscular Tissue

1. Where is skeletal muscle tissue located in the body? ____________
2. Why is skeletal muscle tissue also called voluntary muscle? ____________

   ____
3. Why is visceral muscle tissue also called smooth muscle? ____________
4. Where is visceral muscle tissue located? ____________
5. What does cardiac muscle tissue have in common with skeletal muscle? ____________

   ____
6. What does cardiac muscle tissue have in common with visceral muscle? ____________

   ____
7. Give an identifying characteristic of cardiac muscle tissue. ____________

   ____

Exercise 3-F. Nerve Tissue

1. What is the function of myelin? ____________
2. What is the function of the neurilemma? ____________
3. In what parts of the nervous system is myelin found? ____________
4. What general subdivision of the nervous system lacks a neurilemma? ____________

   ____

Name ______________________

Exercise 3-G. The Skin

1. Where, within the skin, is melanin located? ______________________

2. What is the function of melanin? ______________________

3. What causes a suntan? ______________________

4. What is the function of the arrector pili muscle? ______________________

5. What region of the skin is made up of stratified squamous epithelium? ______________________

____

6. What is the function of a sudoriferous gland? ______________________

7. What is the function of a sebaceous gland? ______________________

8. Is a hair alive or dead? __________ How is it nourished? ______________________

____

9. What is the function of the stratum germinativum? ______________________

____

10. In what region of the skin are Pacinian corpuscles located? ______________________

11. In what region of the skin are Meissner's corpuscles located? ______________________

12. What are freckles? ______________________

13. What is a melanoma? ______________________

4

The Skeletal System

Exercise 4-A Composition and Structure of Bone

Objective

To study the composition and structure of a typical long bone.

Materials

Sectioned femur; bone soaked in dilute acetic acid; and baked bone.

Discussion

The skeletal system includes the bones and the joints. The study of the bones is termed osteology; that of the joints arthrology. The functions of the skeletal system include the following: protection of internal organs such as the heart and lungs, support of the body, attachment for muscles, storage of calcium and phosphorus, and manufacture of blood cells (hemopoiesis).

The dried bones used in the laboratory have been cleansed of all membranes and muscle tissue. In a fresh bone, a membrane called the **periosteum** surrounds the **diaphysis** (shaft) of a long bone. The periosteum serves for growth in circumference, for repair, and for nutrition of the bone. The periosteal blood vessels penetrate into the bone by way of Volkmann's canals, carrying blood rich in food and oxygen. A similar membrane, called the **endosteum**, lines the marrow cavity (medullary cavity). In children this membrane contains cells called **osteoclasts** which destroy bone, enlarging the marrow cavity. The **articular cartilage**, composed of hyaline cartilage, covers the **epiphyses** (ends) of the long bones. This serves to prevent friction at joints. In a sectioned bone of an adult, an **epiphyseal line** can be seen at each end of the bone at the junction between the epiphysis and the diaphysis. This is the remnant of the epiphyseal cartilage, which serves for growth in length of the bone.

In a fresh bone, the shaft of the bone is filled with yellow bone marrow, composed primarily of adipose tissue. The ends of some of the long bones, flat bones, bodies of vertebrae, cranial bones, the sternum, and the ribs contain red bone marrow. This type of marrow manufactures erythrocytes, granulocytes (leukocytes containing granules), and thrombocytes.

Procedure

1. Examine the sectioned femur. Locate the following structures on this bone:
   a. Compact bone. Note the thick layer of compact bone in the shaft and the thin layer covering the epiphyses.
   b. Cancellous (spongy) bone. This type is contained in the epiphyses; a thin layer of spongy bone lines the marrow cavity.
   c. Marrow cavity.
   d. Epiphyseal line.
2. Use Figure 4-1, a longitudinal section through the femur, as a self-test. See Section J, p. xx (Illustrations as Self-Tests) for suggestions on using such figures as self-tests.
3. Examine the bone that has been soaked in dilute acetic acid for several days. The acid dissolves the minerals in the bone, leaving only organic materials. Examine the consistency of the bone. Try to bend the bone.
4. Examine the bone that has been baked at high temperature for eight hours. The heat removes the organic materials in the bone, leaving the minerals. Note the color of the bone. Gently press on the bone. Determine whether this bone can be bent.
5. Review the microscopic structure of bone in Exercise 3-D, Part 4. Study Figures 3-13, 14, and 15 (in Exercise 3-D) to determine how food is transported to the osteocytes and to the marrow cavity.

Exercise 4-B The Appendicular Skeleton

Objectives

To identify the bones of the appendicular skeleton and the major processes on these bones.

Materials

Articulated and disarticulated skeletons; text; and reference books.

Discussion

The skeleton in a typical adult contains 206 bones, although this number can vary. The skeleton is divided into two parts, the appen-

KEY TO FIGURE IV-1

1. periosteum
2. endosteum
3. compact bone
4. cancellous (spongy) bone
5. epiphyseal line
6. yellow bone marrow
7. marrow cavity
8. epiphysis
9. diaphysis

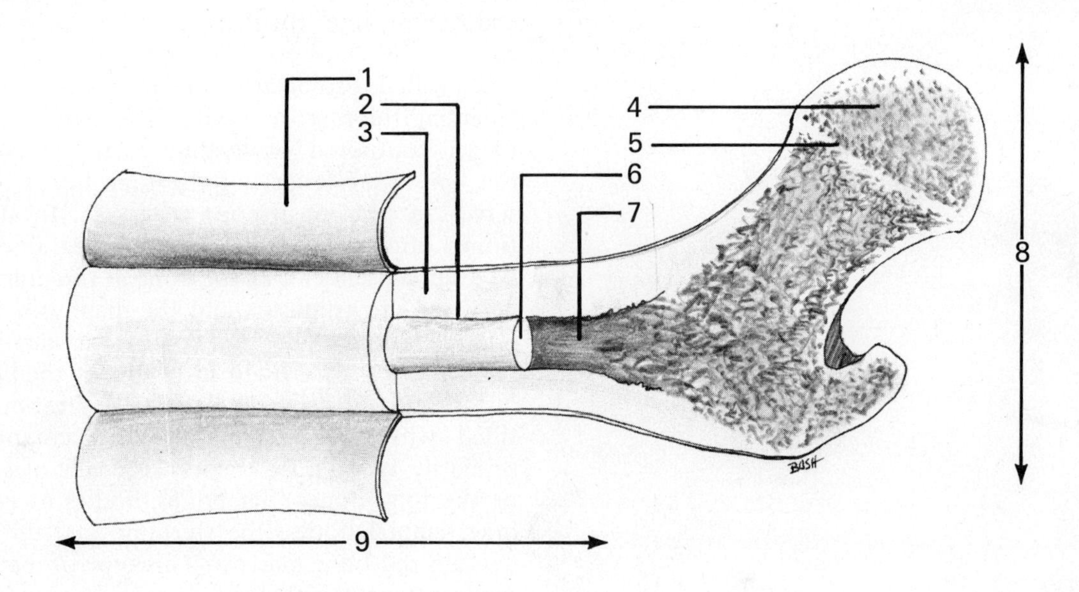

Figure 4-1. Longitudinal Section through Femur

dicular skeleton and the axial skeleton. The appendicular skeleton, composed of 126 bones, includes bones of the appendages and the girdles (shoulder and pelvic), which connect the appendages to the axial skeleton.

Procedure

1. Locate each of the illustrated bones on an articulated skeleton.
2. Compare the parts of the bone indicated in each figure with those in your text and other references, and with those of the bones of a disarticulated skeleton. Determine the function of each part indicated by using references.
3. Each of the illustrations may be used as self-tests.

KEY TO FIGURE 4-2

1. acromial extremity (lateral end)
2. conoid tubercle
3. sternal extremity (medial end)

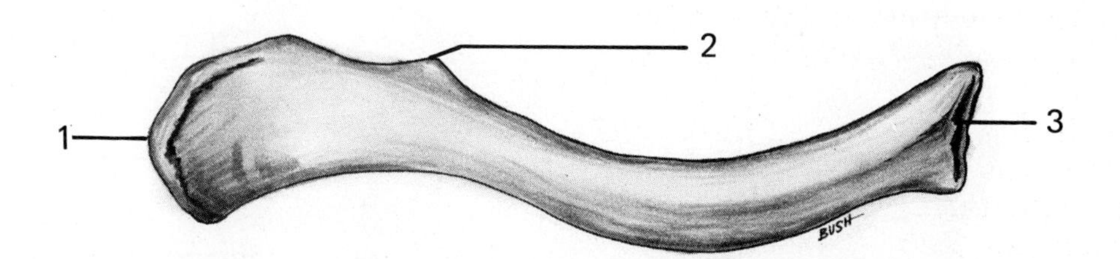

Figure 4-2. Right Clavicle (cranial aspect), Superior Surface

KEY TO FIGURE 4-3

1. medial (superior) angle
2. superior border
3. supraspinous fossa
4. scapular notch
5. spine
6. infraspinous fossa
7. vertebral (medial) border
8. coracoid process
9. acromion process
10. glenoid cavity
11. lateral angle
12. axillary (lateral) border
13. inferior angle

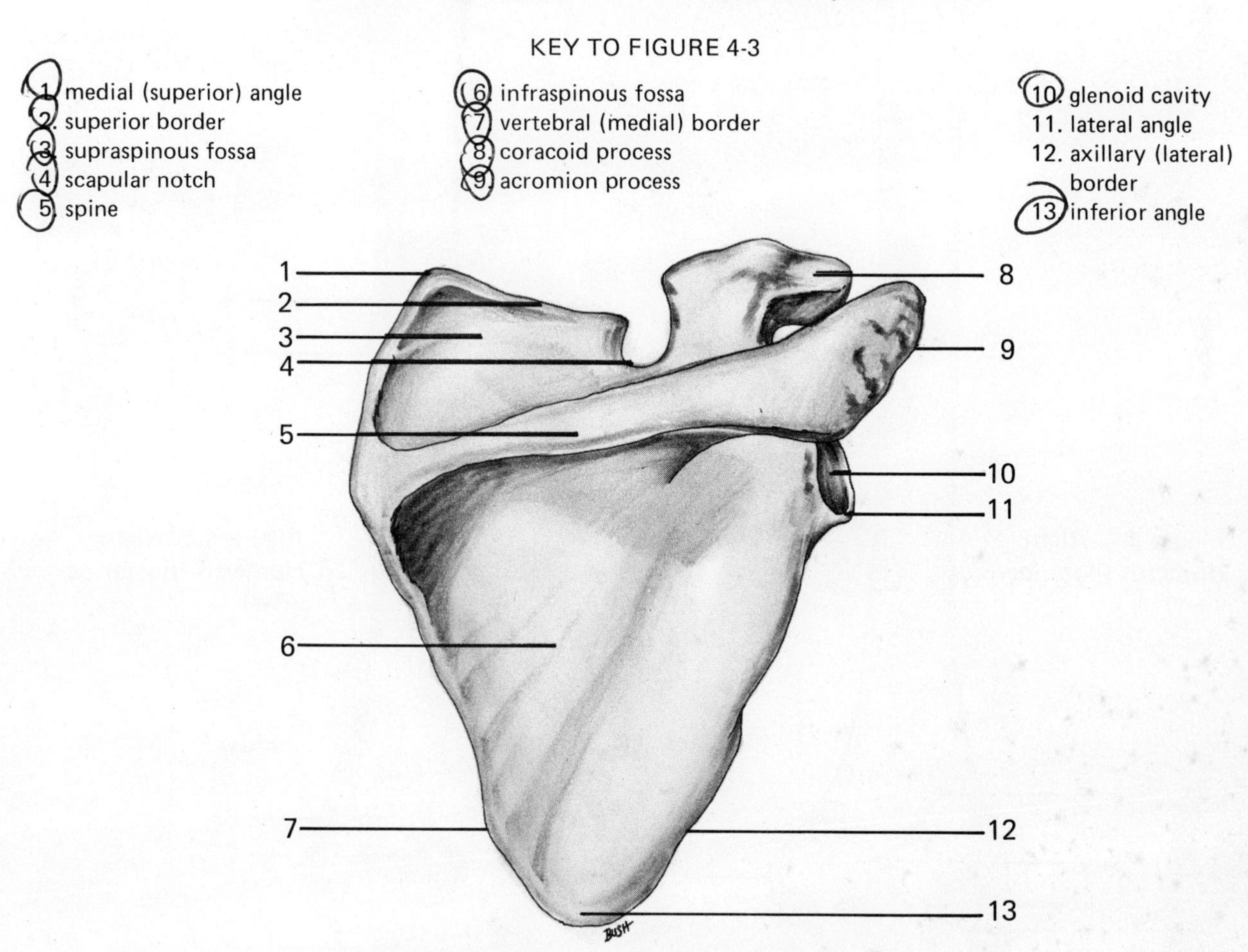

Figure 4-3. Posterior View of the Right Scapula

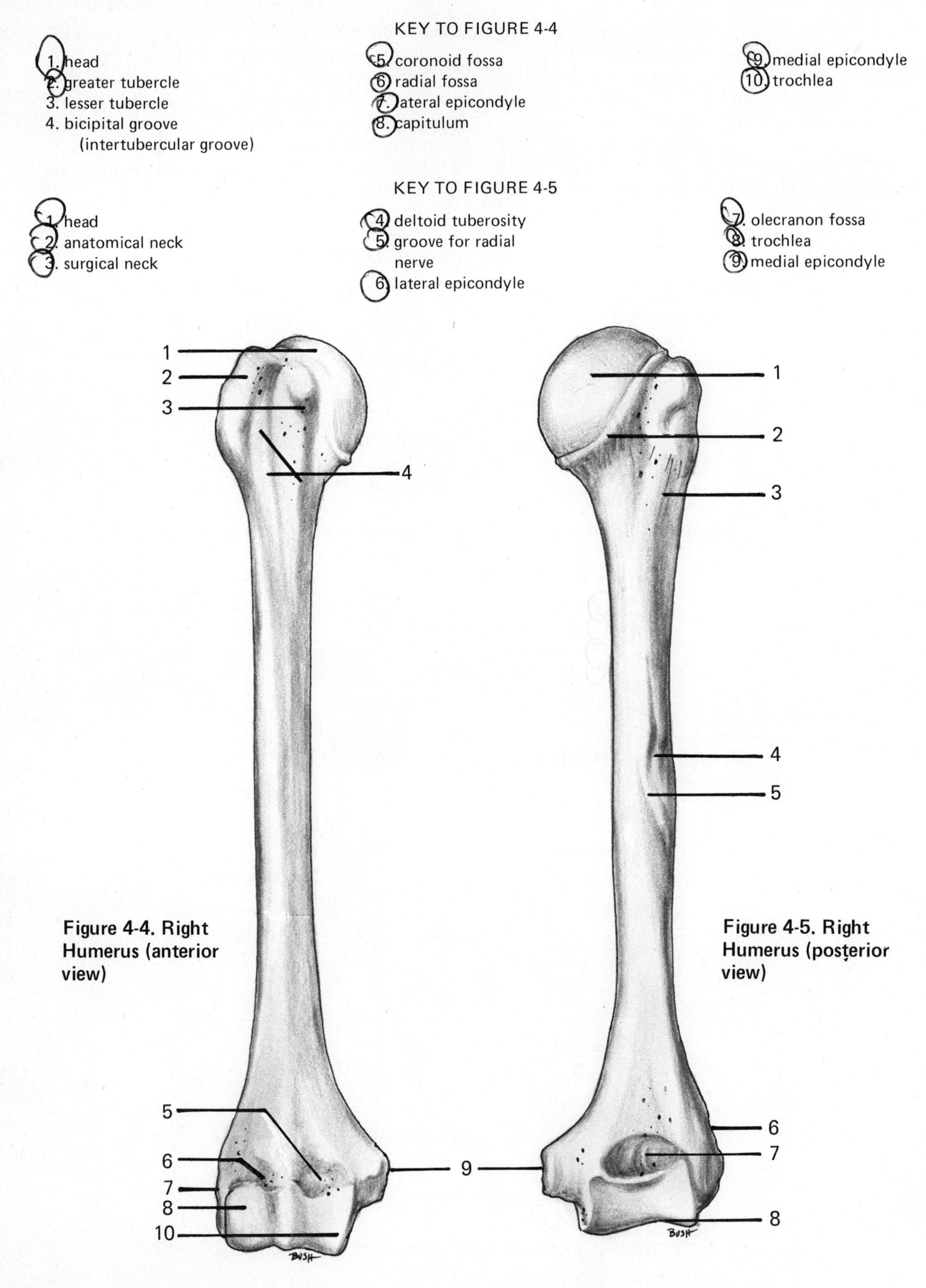

Figure 4-4. Right Humerus (anterior view)

Figure 4-5. Right Humerus (posterior view)

KEY TO FIGURE 4-6

1. head
2. neck
3. radial tuberosity
4. styloid process
5. ulnar notch

KEY TO FIGURE 4-7

1. olecranon process
2. semilunar notch
3. coronoid process
4. radial notch
5. head
6. styloid process

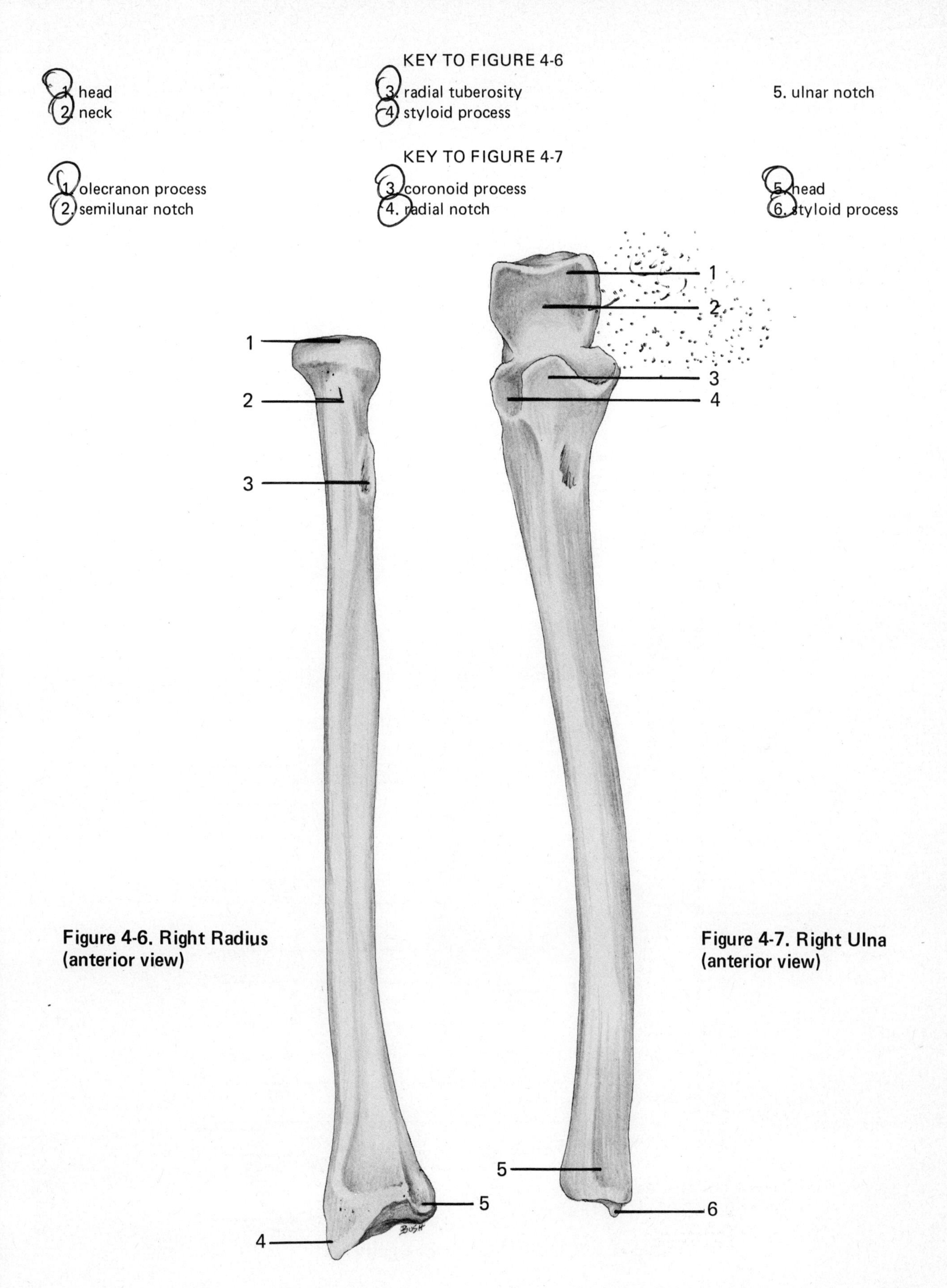

Figure 4-6. Right Radius (anterior view)

Figure 4-7. Right Ulna (anterior view)

KEY TO FIGURE 4-8

1. third phalanx
2. second phalanx
3. first phalanx
4. phalanges
5. fifth metacarpal
6. capitate
7. hamate
8. triangular (triquetrum)
9. pisiform
10. lunate (semilunar)
11. navicular (scaphoid)
12. lesser multangular (trapezoid)
13. greater multangular (trapezium)
14. carpals
15. metacarpals

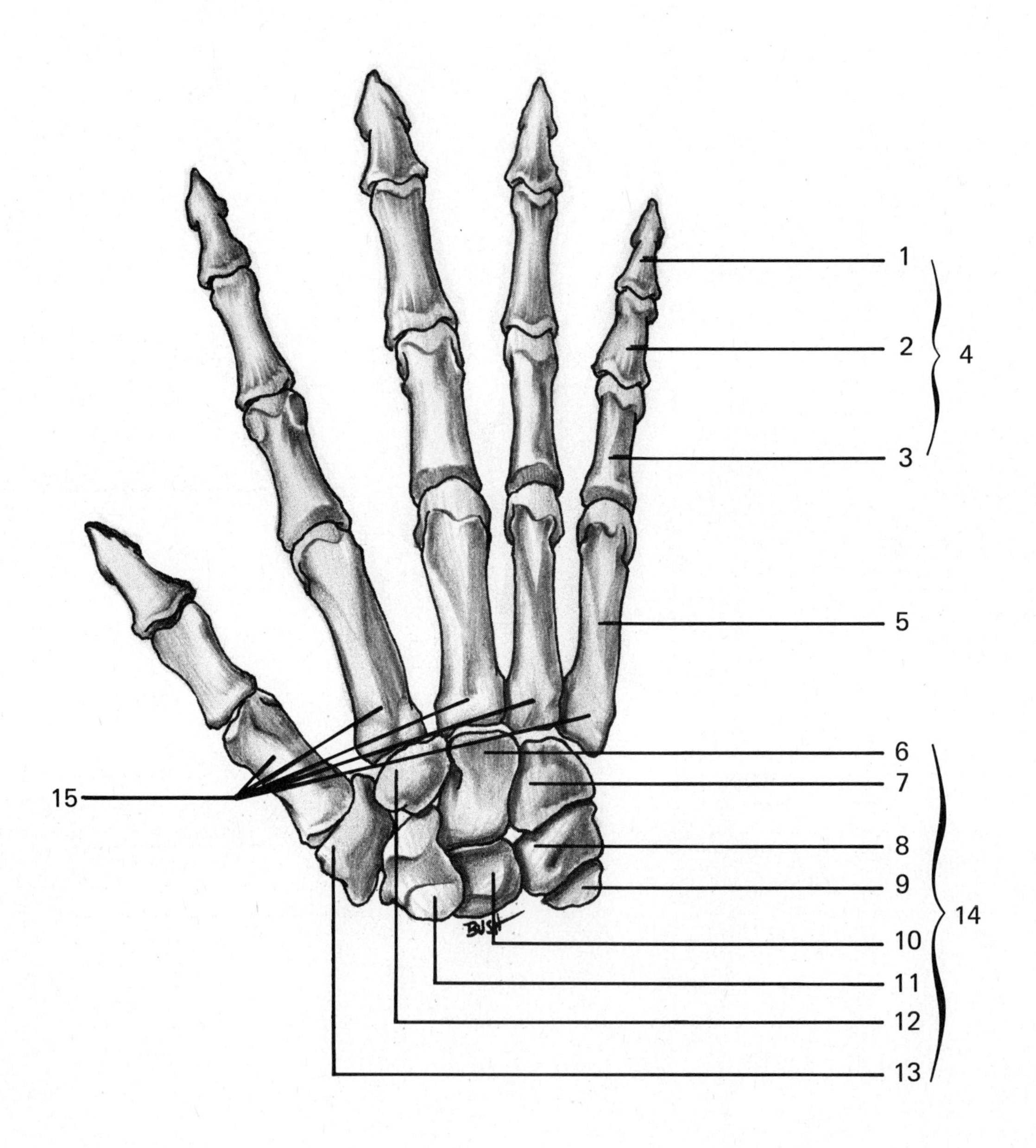

Figure 4-8. Dorsal Surface of Right Hand

KEY TO FIGURE 4-9

1. ilium
2. posterior superior iliac spine
3. posterior inferior iliac spine
4. greater sciatic notch
5. ischial spine
6. lesser sciatic notch
7. ischium
8. ischial tuberosity
9. iliac crest
10. anterior superior iliac spine
11. anterior inferior iliac spine
12. acetabulum
13. crest of pubis
14. pubis
15. obturator foramen

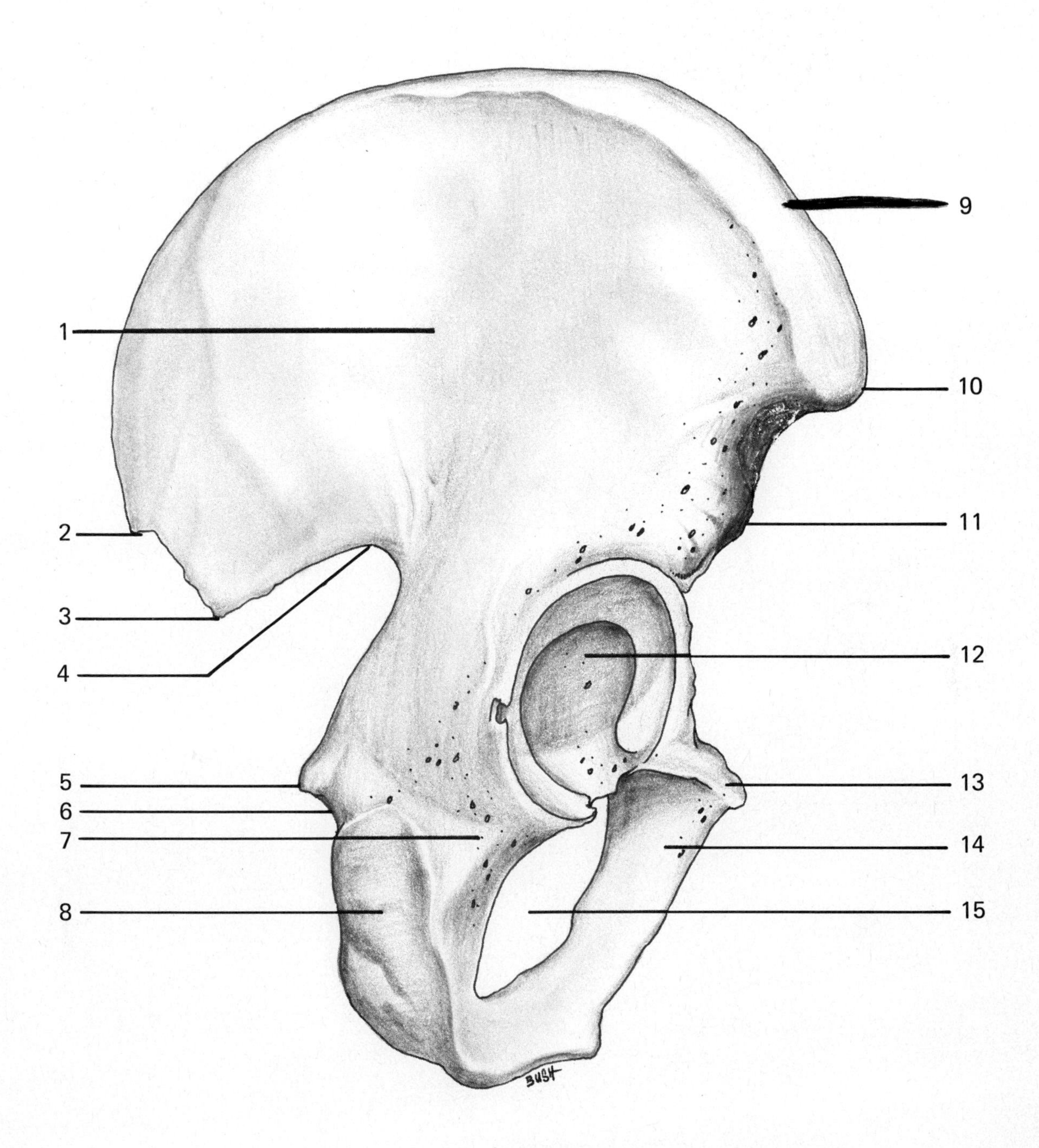

Figure 4-9. Lateral View of the Right Os Coxae

KEY TO FIGURE 4-10

1. sacral promontory
2. iliac fossa
3. articular (auricular) surface on ilium
4. iliac crest
5. anterior sacral foramen
6. sacrum
7. arcuate (iliopectineal line)
8. anterior superior iliac spine
9. coccyx
10. brim of pelvis
11. ischial spine
12. pelvic inlet
13. acetabulum
14. superior ramus of pubis
15. pubic symphysis
16. obturator foramen
17. pubic arch

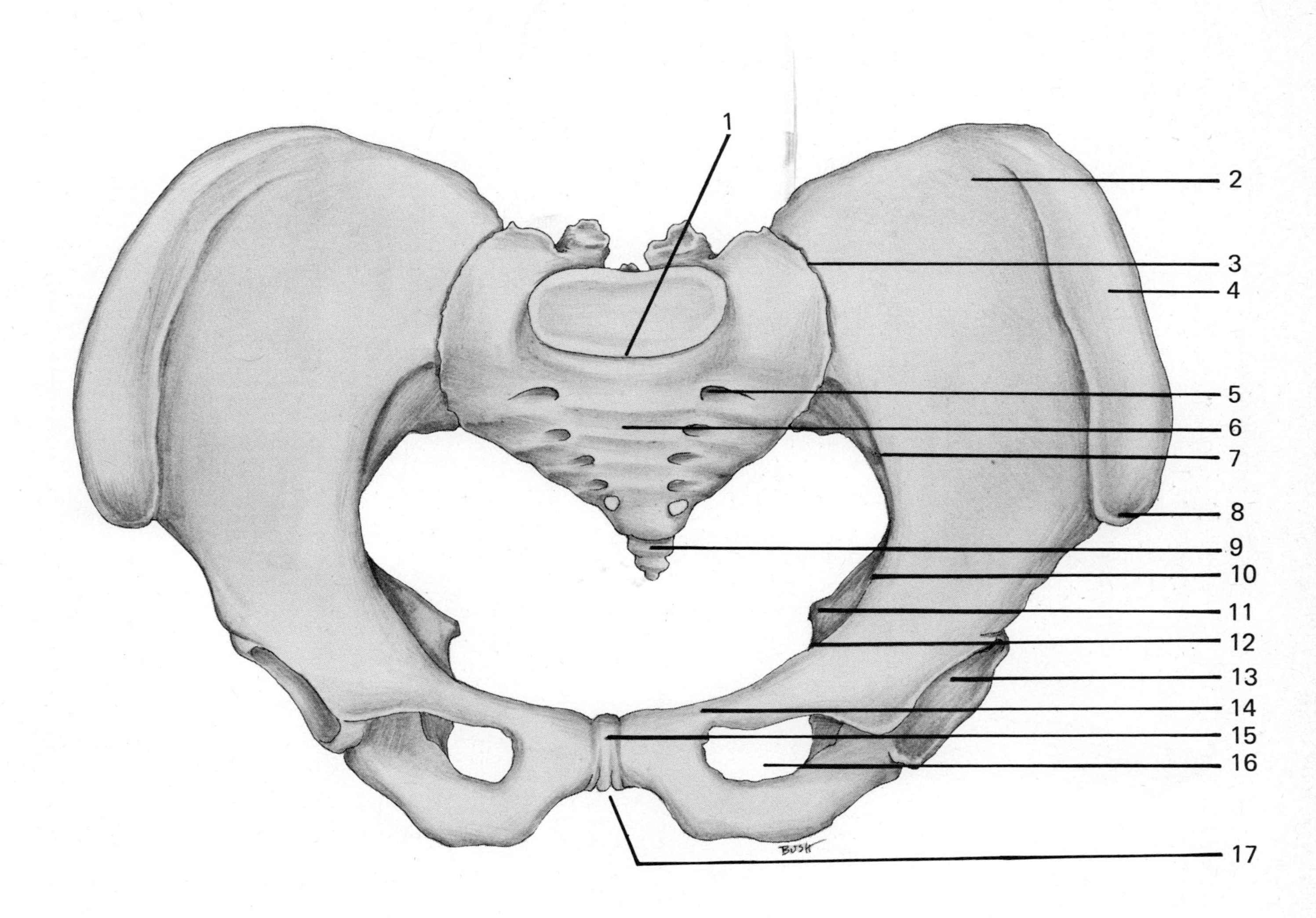

Figure 4-10. The Female Pelvis

KEY TO FIGURE 4-11

1. greater trochanter
2. lesser trochanter
3. lateral epicondyle
4. patellar surface
5. head
6. neck
7. medial epicondyle

KEY TO FIGURE 4-12

1. head
2. greater trochanter
3. neck
4. lesser trochanter
5. linea aspera
6. intercondyloid fossa
7. lateral condyle
8. medial condyle

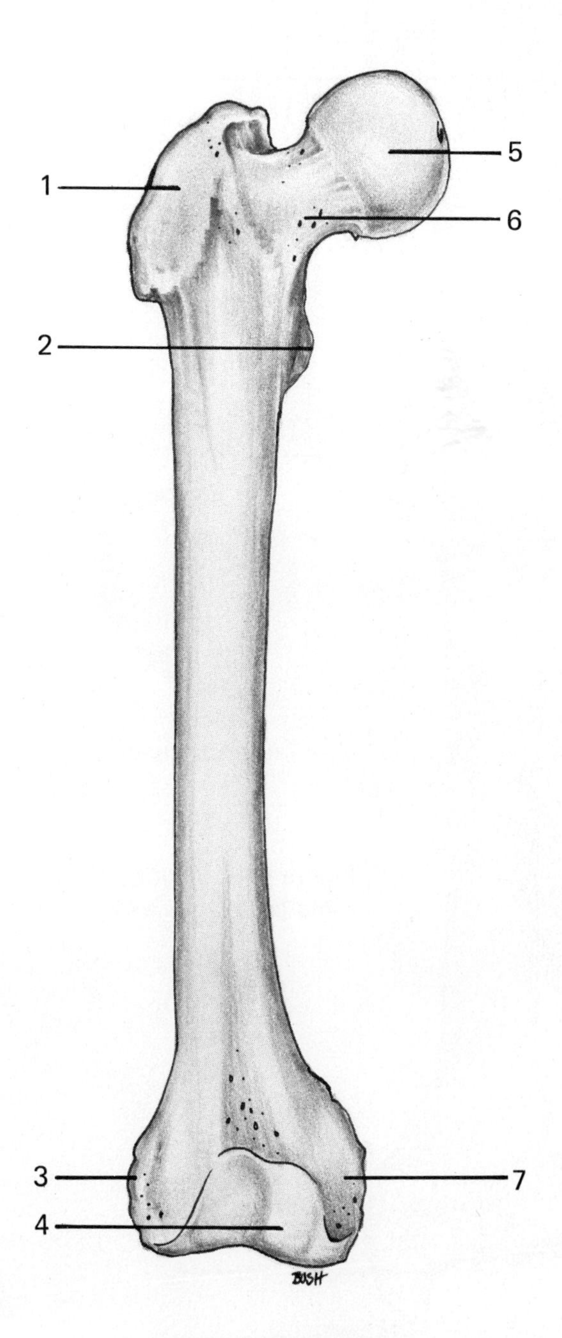

Figure 4-11. Right Femur (anterior view)

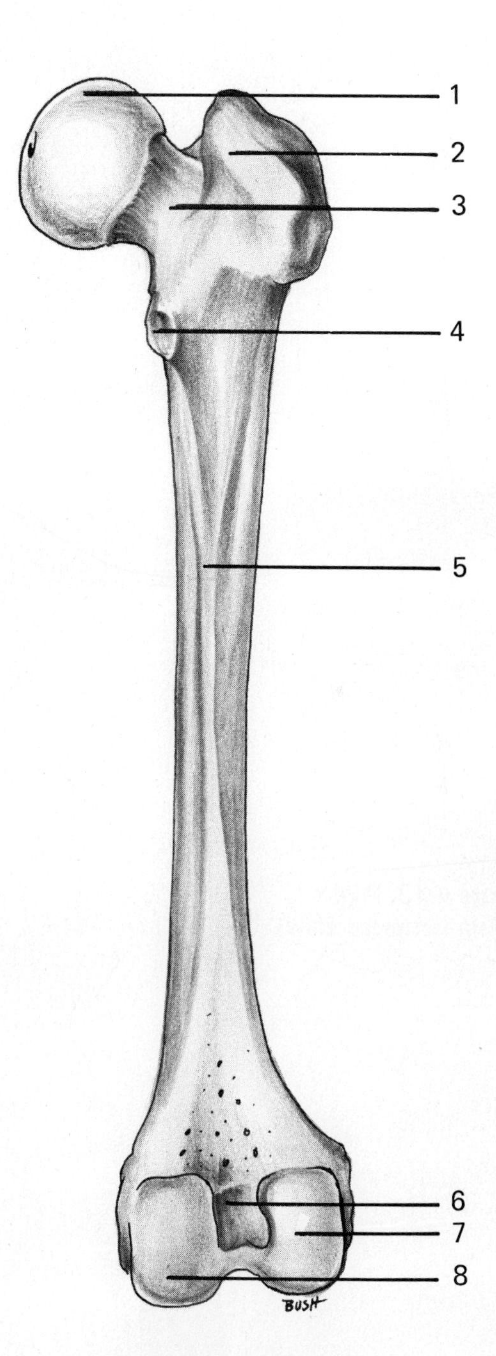

Figure 4-12. Right Femur (posterior view)

KEY TO FIGURE 4-13

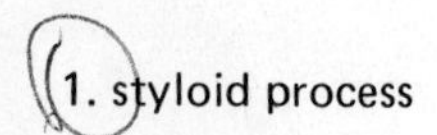

1. styloid process
2. lateral malleolus

KEY TO FIGURE 4-14

1. intercondyloid eminence
2. medial condyle
3. lateral condyle
4. tibial tuberosity
5. crest
6. medial malleolus

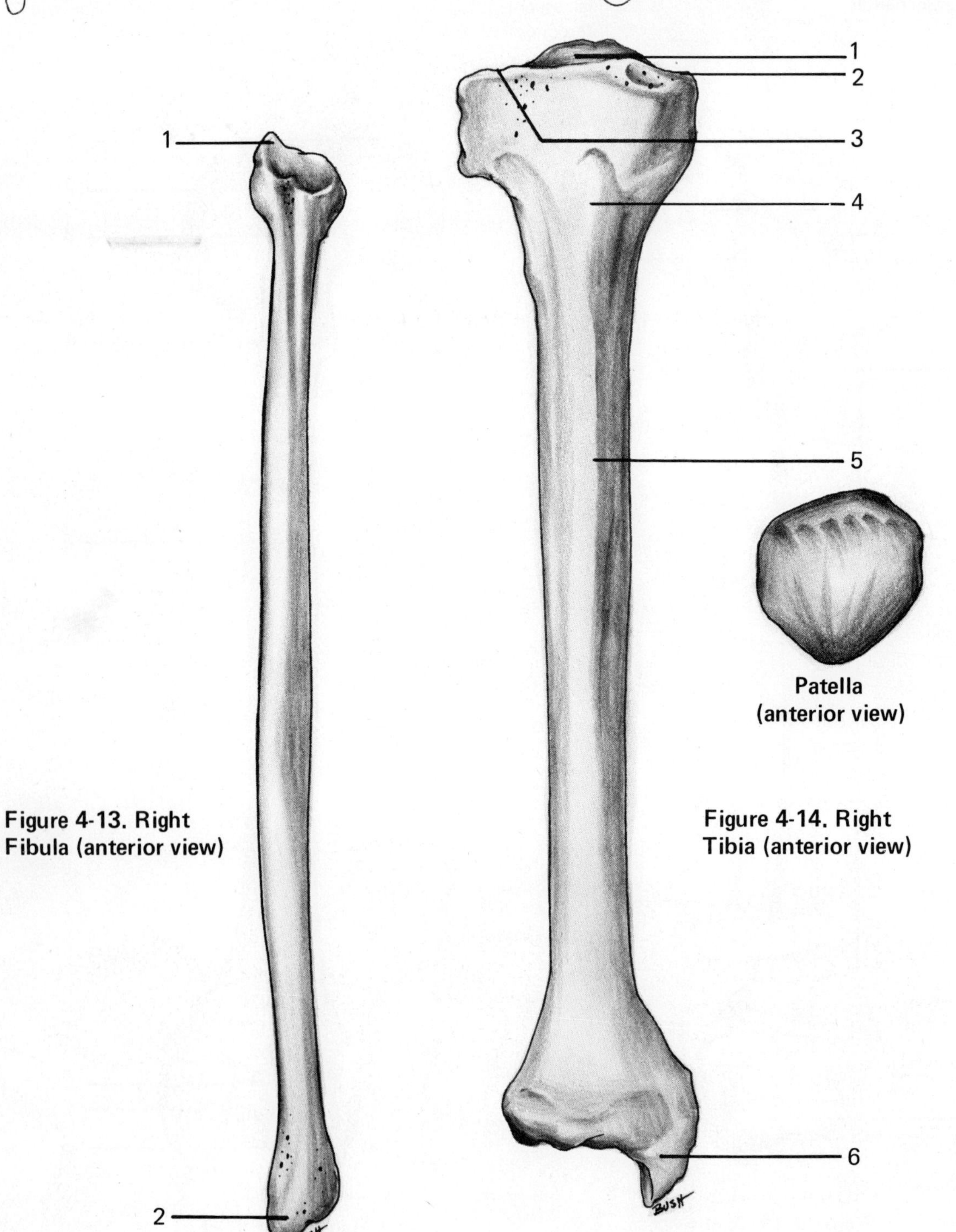

Figure 4-13. Right Fibula (anterior view)

Figure 4-14. Right Tibia (anterior view)

KEY TO FIGURE 4-15

1. third phalanx
2. second phalanx
3. first phalanx
4. phalanges
5. first metatarsal
6. first (medial) cuneiform
7. navicular
8. talus
9. calcaneus

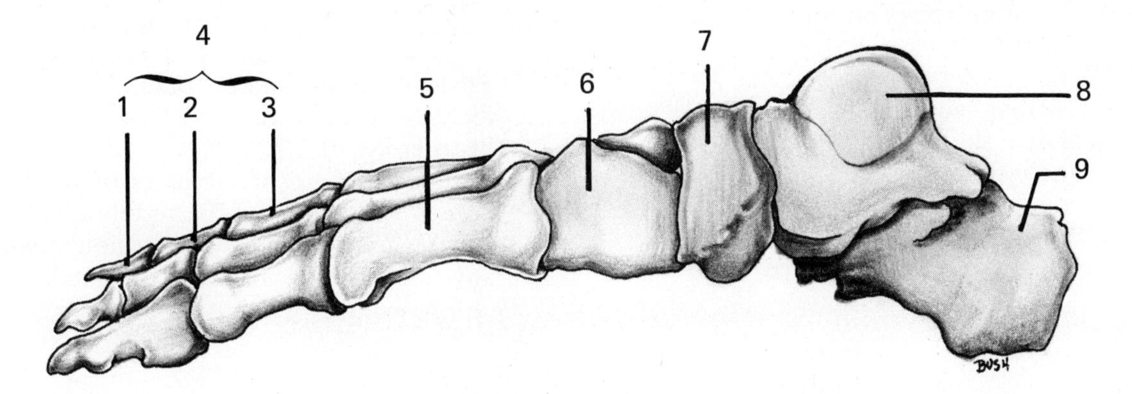

Figure 4-15. Bones of the Right Foot (medial view)

KEY TO FIGURE 4-16

1. third phalanx
2. second phalanx
3. first phalanx
4. phalanges
5. fifth metatarsal
6. metatarsals (metatarsus)
7. third (lateral) cuneiform
8. second (intermediate) cuneiform
9. first (medial) cuneiform
10. cuboid
11. navicular
12. talus - lets you flex your foot
13. calcaneus - heel
14. tarsals

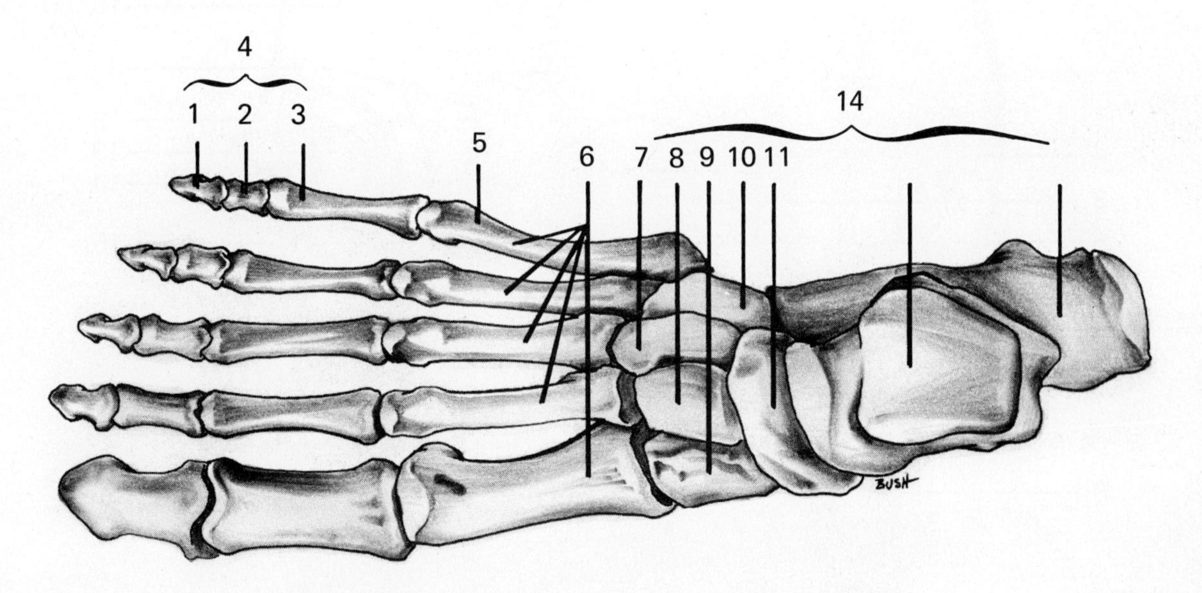

Figure 4-16. Bones of the Right Foot (dorsal surface)

Exercise 4-C
The Axial Skeleton

Objectives

To identify the bones of the axial skeleton and the major processes on these bones.

Materials

Articulated and disarticulated skeletons; disarticulated skull; skull sectioned to show sinuses and interior of skull.

Discussion

The axial skeleton, composed of 80 bones, includes the bones of the skull, vertebral column, and thorax.

Procedure

1. Follow Steps 1-5 of the procedure for Exercise 4-B.
2. Practice identifying the bones of the disarticulated skull until you can distinguish each of the separate bones. Verify your identification of the facial bones with the instructor.
3. Determine the characteristics of the vertebrae in each region of the vertebral column. Be prepared to identify the region in which any isolated vertebra may be located.

KEY TO FIGURE 4-17

1. parietal bone
2. lambdoidal suture
3. squamosal suture
4. temporal bone
5. wormian bone
6. occipital bone
7. mastoid process
8. styloid process
9. coronal suture
10. frontal bone
11. sphenoid bone
12. nasal bone
13. lacrimal bone
14. ethmoid bone
15. zygomatic bone
16. maxilla
17. zygomatic arch
18. external acoustic (auditory) meatus
19. mandible

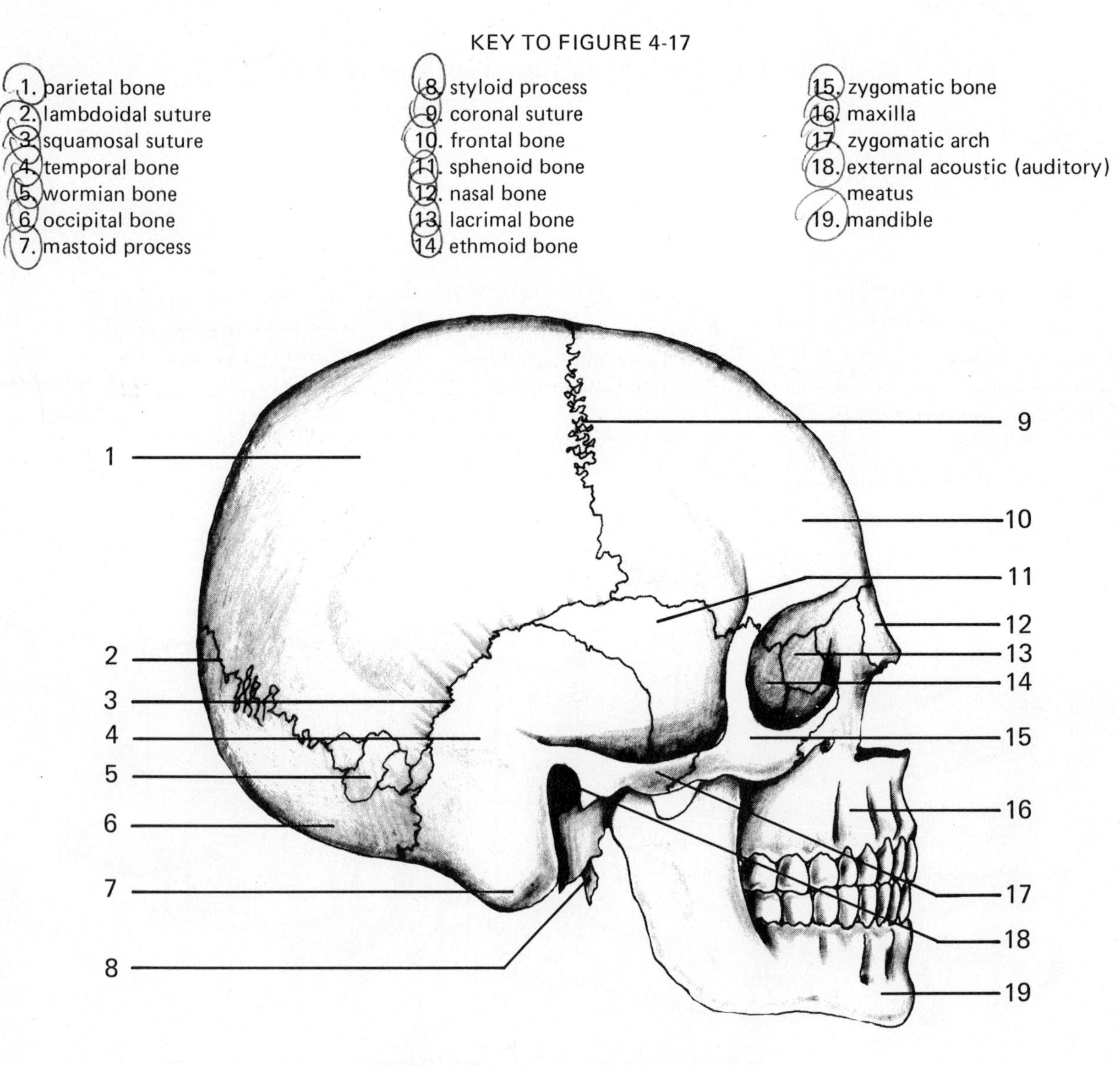

Figure 4-17. The skull (lateral view)

KEY TO FIGURE 4-18

1. sagittal suture
2. supraorbital notch
3. greater wing of sphenoid
4. temporal bone
5. optic foramen
6. superior orbital fissure
7. sphenoid (orbital surface)
8. inferior orbital fissure
9. middle nasal concha
10. infraorbital foramen
11. inferior nasal concha
12. vomer
13. mandible
14. coronal suture
15. parietal bone
16. frontal bone
17. superciliary arch
18. glabella
19. supraorbital margin
20. nasal bone
21. ethmoid
22. lacrimal bone
23. zygomatic bone
24. perpendicular plate of ethmoid
25. maxilla
26. mental foramen

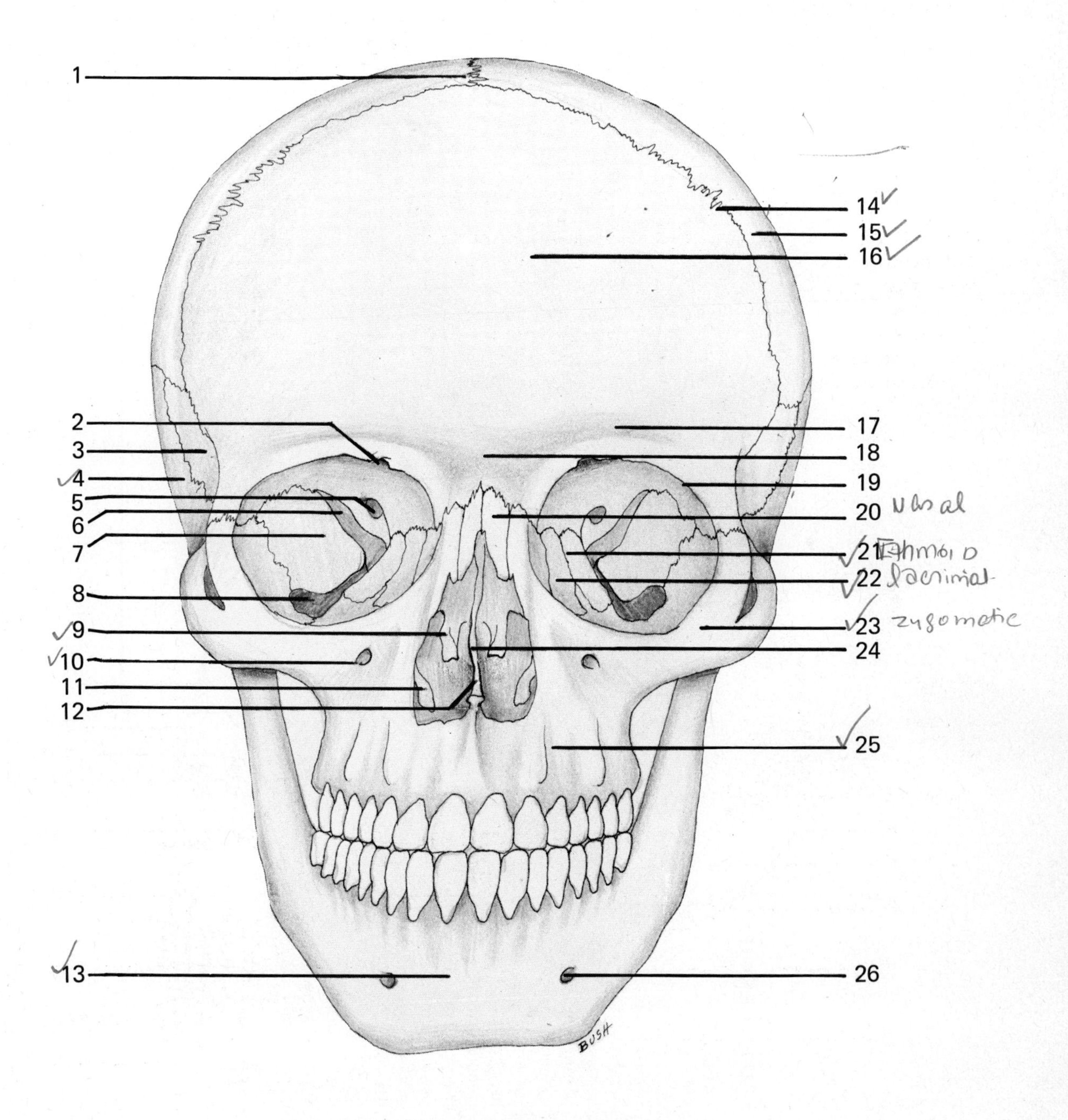

Figure 4-18. Anterior View of the Skull

KEY TO FIGURE 4-19

1. frontal bone
2. sella turcica
3. optic foramen
4. frontal sinus
5. nasal bone
6. perpendicular plate of ethmoid
7. sphenoidal sinus
8. vomer
9. hard palate
10. alveolar process
11. pterygoid process
12. parietal bone
13. internal acoustic meatus
14. temporal bone
15. jugular foramen
16. occipital bone
17. hypoglossal canal
18. occipital condyle
19. styloid process

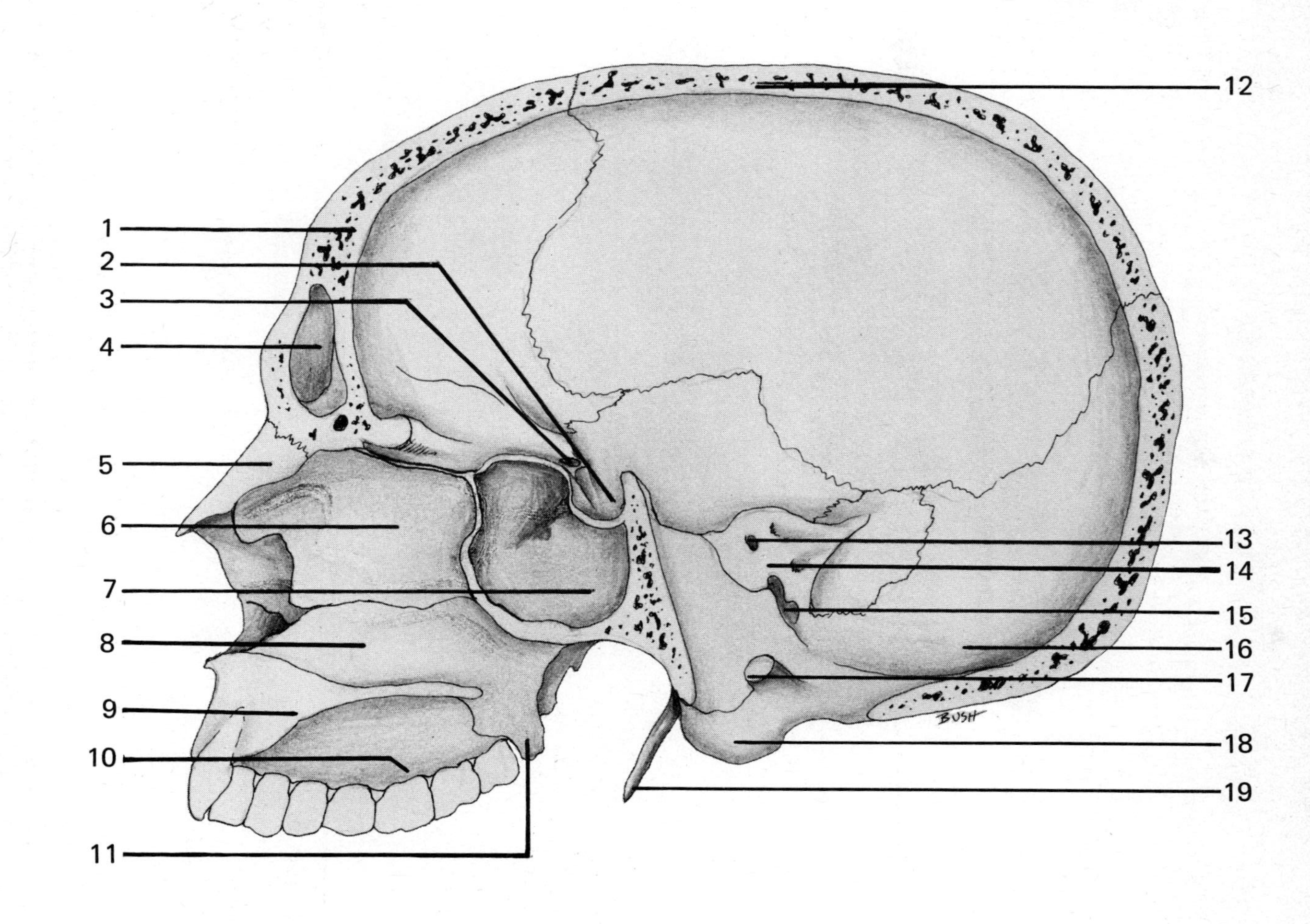

Figure 4-19. Sagittal Section of the Skull

KEY TO FIGURE 4-20

1. perpendicular plate
2. crista galli
3. cribriform plate
4. ethmoidal air cells
5. lamina papyracea (lateral masses)

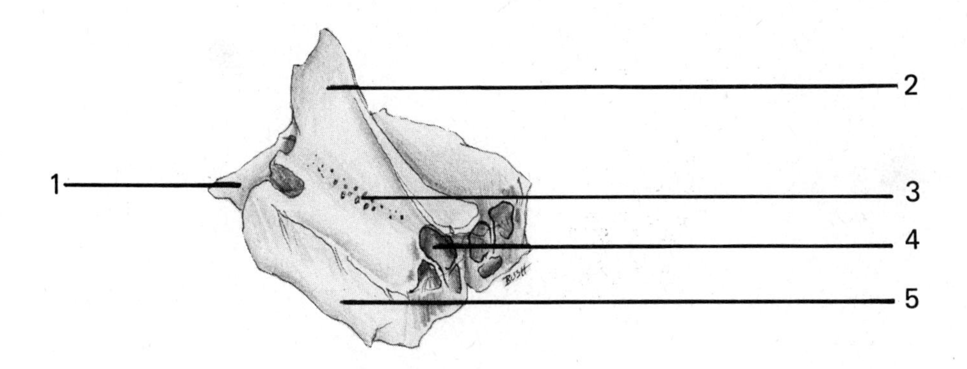

Figure 4-20. Superior View of the Ethmoid

KEY TO FIGURE 4-21

1. optic foramen
2. sphenoid sinus
3. body
4. sella turcica
5. lesser wing
6. inferior orbital fissure
7. greater wing
8. foramen rotundum
9. lateral pterygoid process
10. medial pterygoid process

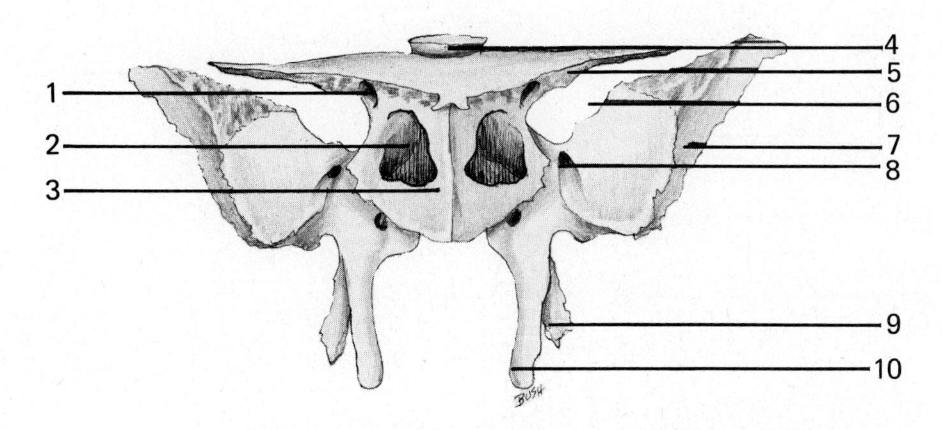

Figure 4-21. Anterior View of the Sphenoid

KEY TO FIGURE 4-22

1. greater wing
2. superior orbital fissure
3. anterior clinoid process
4. foramen rotundum
5. posterior clinoid process
6. foramen ovale
7. foramen spinosum
8. lesser wing
9. optic foramen
10. optic groove
11. sella turcica
12. dorsum sellae

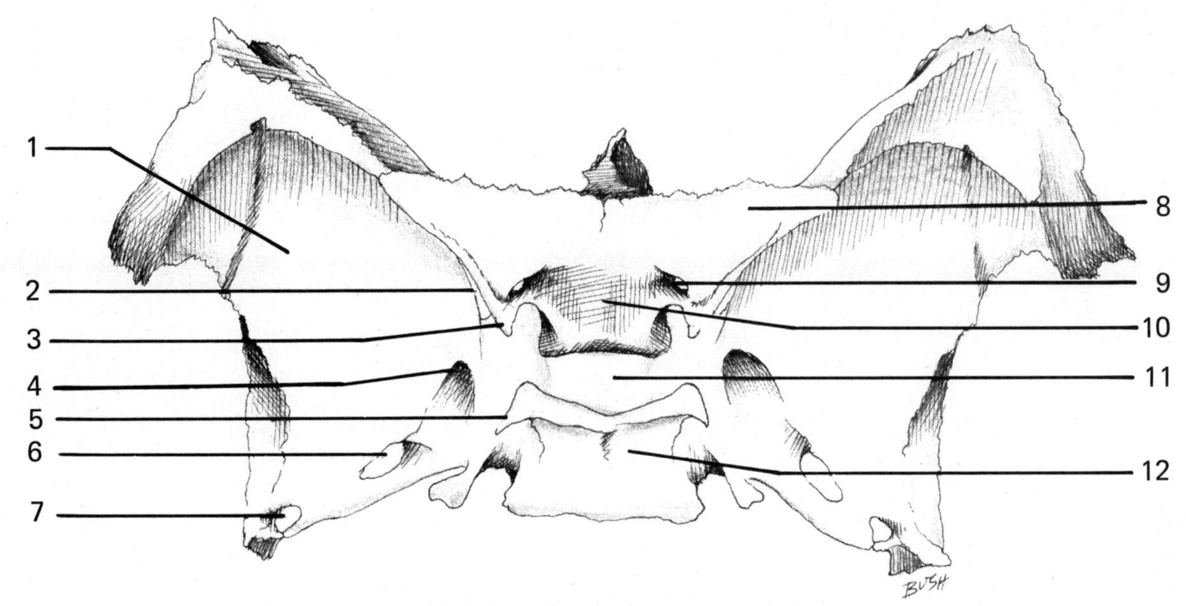

Figure 4-22. Superior View of the Sphenoid

KEY TO FIGURE 4-23

1. superior nuchal line
2. condyloid canal
3. occipital condyle
4. hypoglossal canal
5. external occipital protuberance
6. external occipital crest (median nuchal line)
7. inferior nuchal line
8. foramen magnum
9. jugular notch
10. basilar part

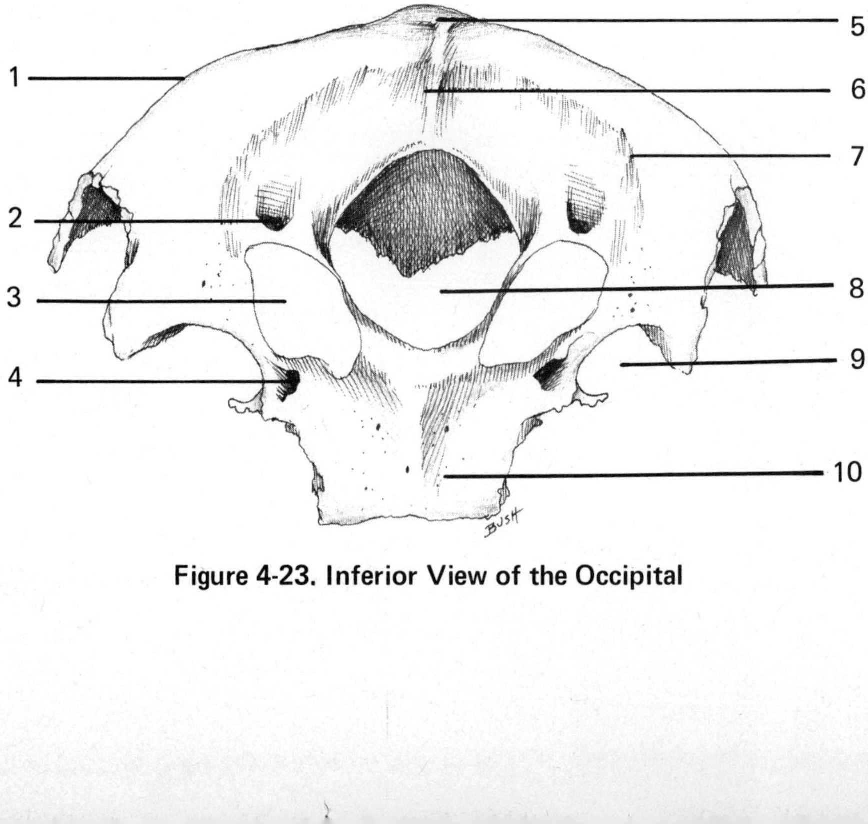

Figure 4-23. Inferior View of the Occipital

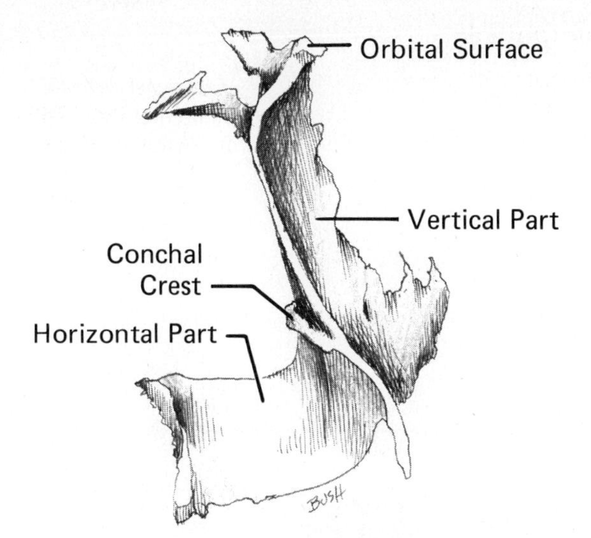

Figure 4-24. Left Palatine

Figure 4-28. Vomer

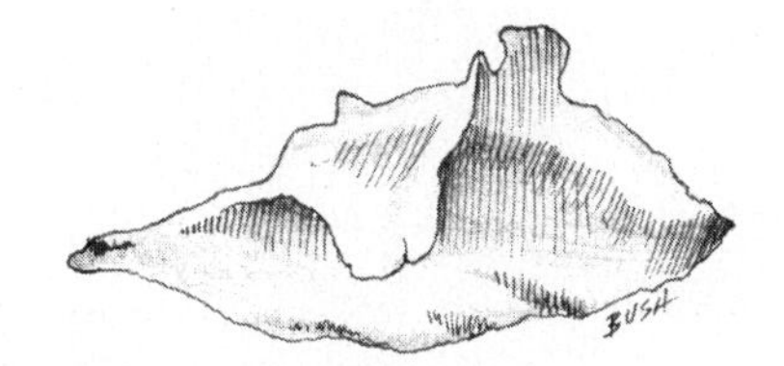

Figure 4-29. Right Inferior Nasal Concha (lateral surface)

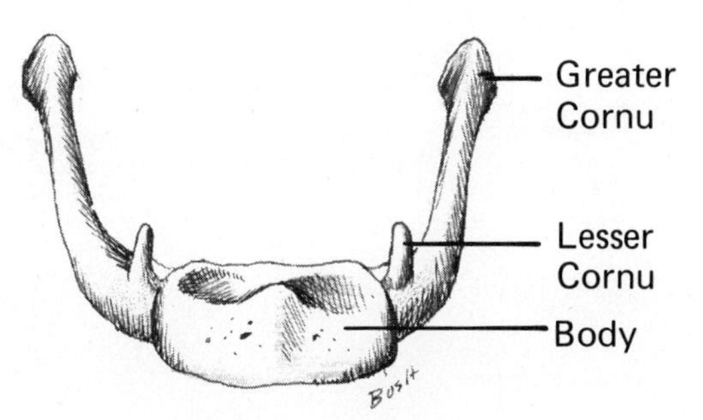

Figure 4-25. Hyoid

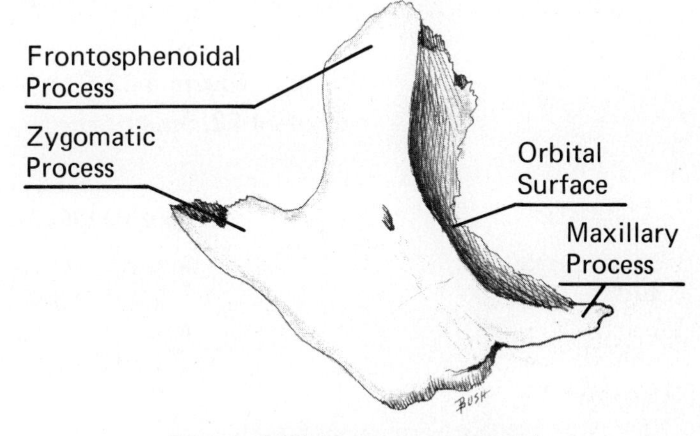

Figure 4-30. Zygomatic (Malar)

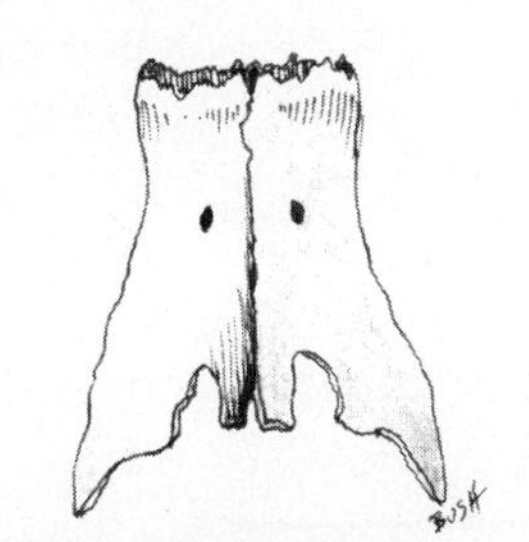

Figure 4-26. Nasal Bones (outer surface)

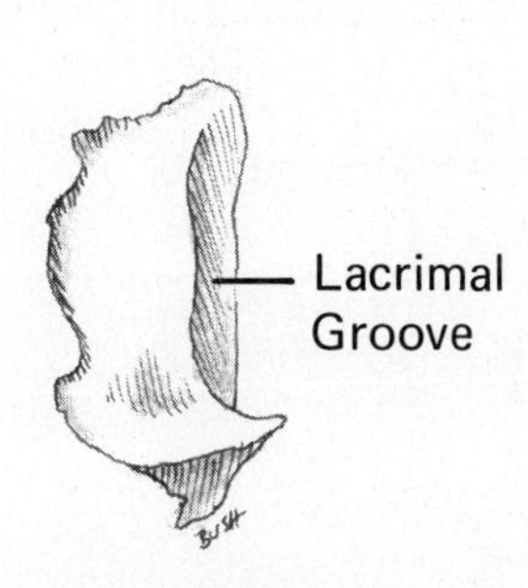

Figure 4-27. Lacrimal, (lateral aspect)

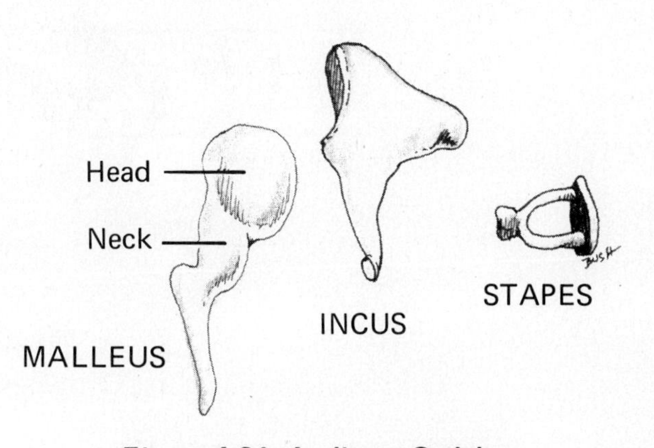

Figure 4-31. Auditory Ossicles

KEY TO FIGURE 4-32

1. coronoid process
2. mandibular notch
3. condyloid process
4. ramus
5. body
6. mandibular foramen
7. alveolar process
8. mental foramen
9. mental protuberance

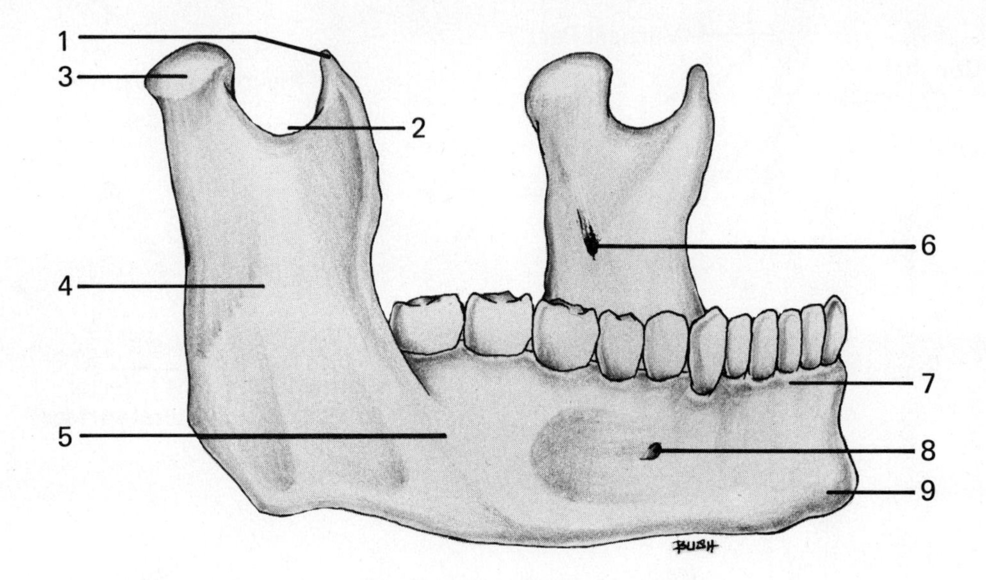

Figure 4-32. The Mandible

KEY TO FIGURE 4-33

1. orbital surface
2. infraorbital groove
3. zygomatic process
4. frontal process
5. infraorbital foramen
6. anterior nasal spine
7. alveolar process

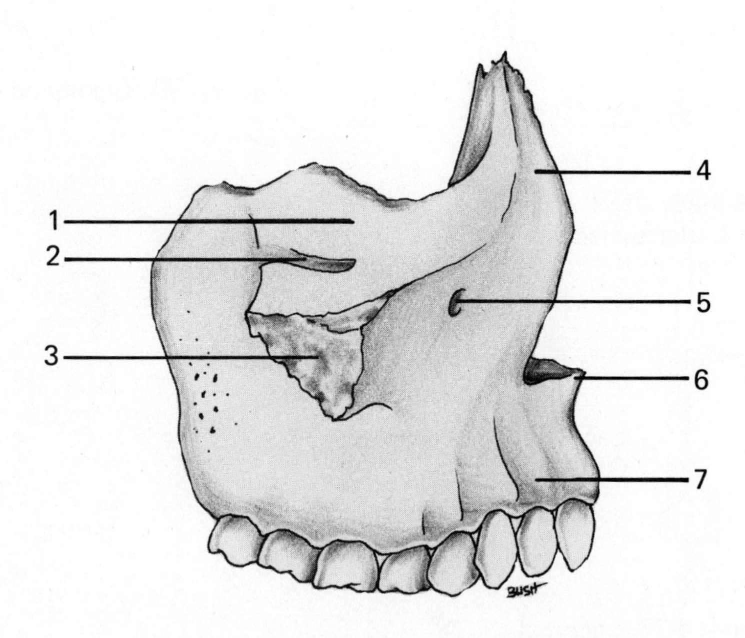

Figure 4-33. Lateral View of the Right Maxilla

KEY TO FIGURE 4-34

1. squama
2. external acoustic meatus
3. mastoid process
4. zygomatic process
5. mandibular fossa
6. styloid process

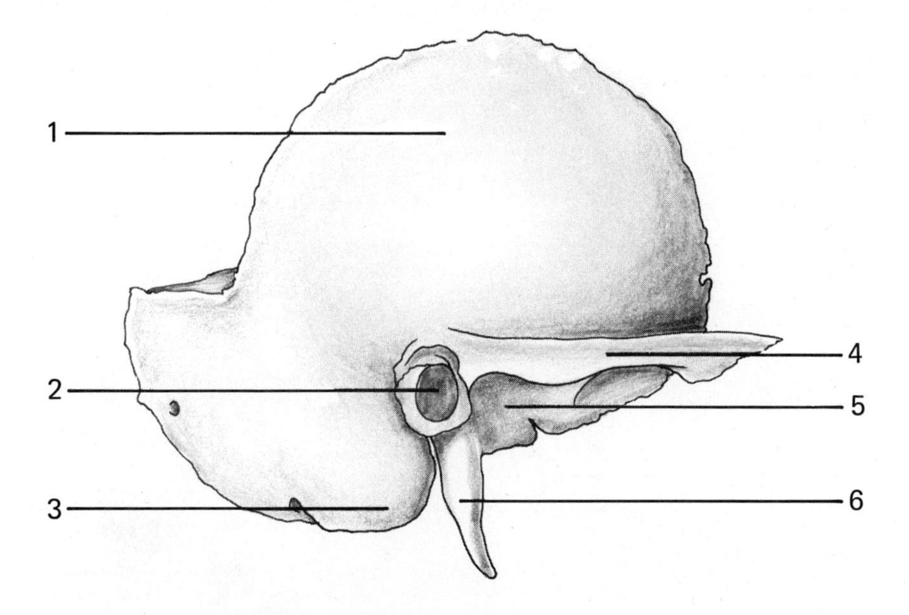

Figure 4-34. The Right Temporal (lateral aspect)

KEY TO FIGURE 4-35

1. zygomatic process
2. styloid process
3. external acoustic meatus
4. mastoid process
5. carotid canal
6. petrous portion
7. carotid canal

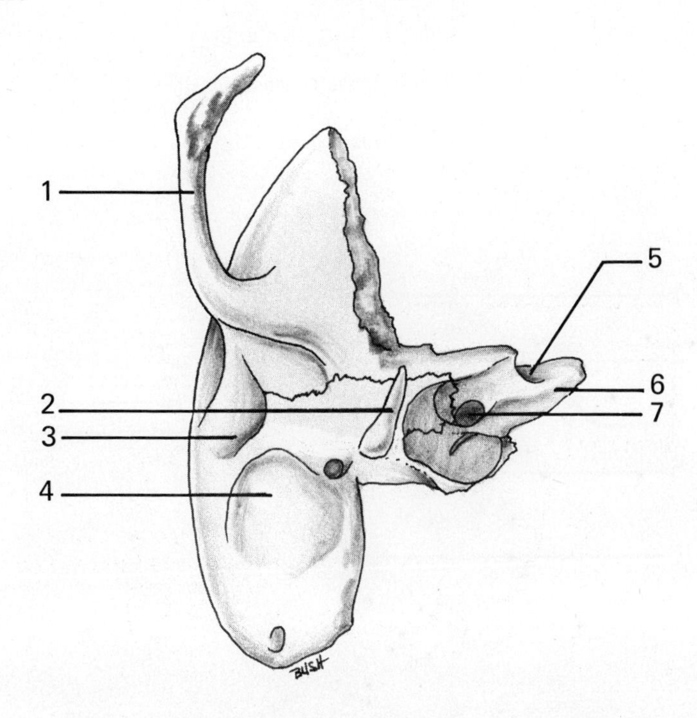

Figure 4-35. The Right Temporal (basal aspect)

KEY TO FIGURE 4-36

1. anterior arch
2. superior articular surface
3. transverse foramen
4. posterior arch
5. transverse process
6. groove for vertebral artery

KEY TO FIGURE 4-37

1. body
2. dens (odontoid process)
3. superior articular surface
4. transverse foramen
5. transverse process
6. lamina
7. bifid spinous process

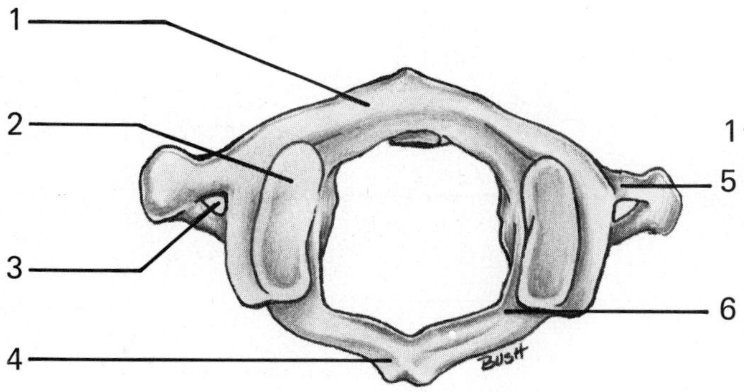

Figure 4-36. The Atlas (first cervical vertebra)

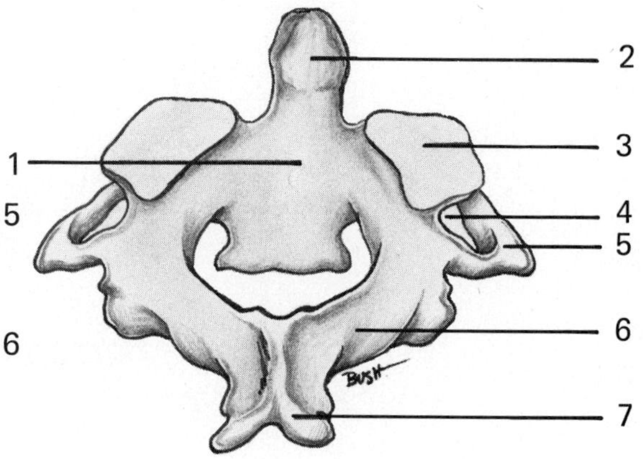

Figure 4-37. The Axis (second cervical vertebra)

KEY TO FIGURE 4-38

1. body
2. transverse foramen
3. pedicle
4. transverse process
5. lamina
6. spinous process
7. superior articular process
8. vertebral (spinal) foramen

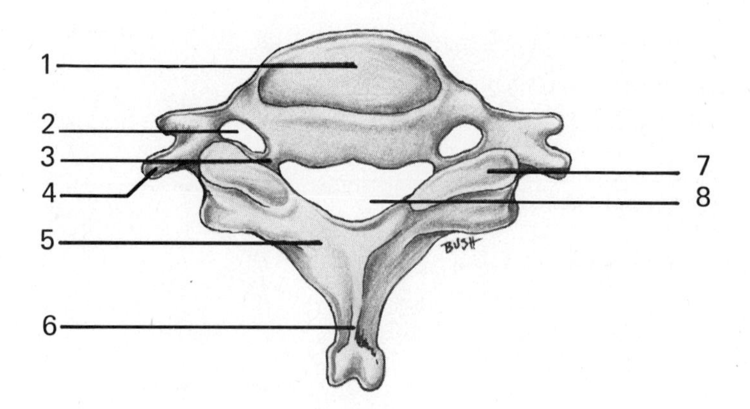

Figure 4-38. Seventh Cervical Vertebra

KEY TO FIGURE 4-39

1. costal demifacet
2. body
3. vertebral notch
4. superior articular process
5. facet for rib
6. transverse process
7. inferior articular process
8. spinous process

KEY TO FIGURE 4-40

1. spinous process
2. vertebral (spinal) foramen
3. inferior articular process
4. lamina
5. transverse process
6. superior articular process
7. pedicle
8. body (centrum)

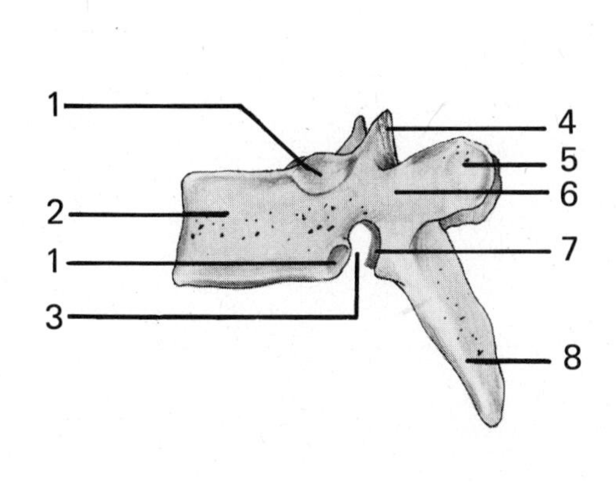

Figure 4-39. Typical Thoracic Vertebra

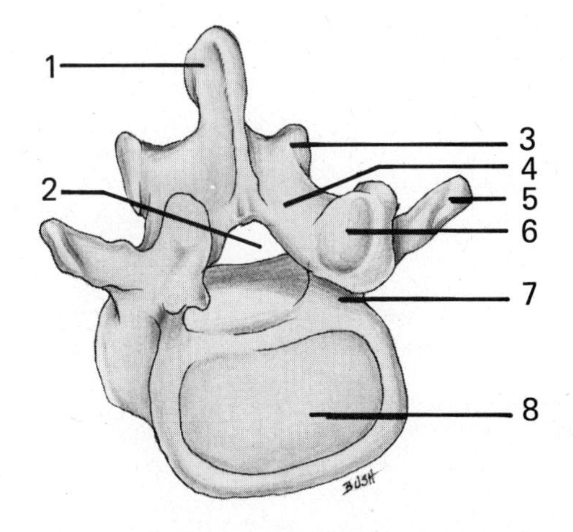

Figure 4-40. Typical Lumbar Vertebra

KEY TO FIGURE 4-41

1. ala
2. lateral mass
3. pelvic surface
4. articular process
5. articular surface
6. sacral promontory
7. anterior sacral foramen
8. body of third sacral vertebra (pelvic surface)
9. coccyx

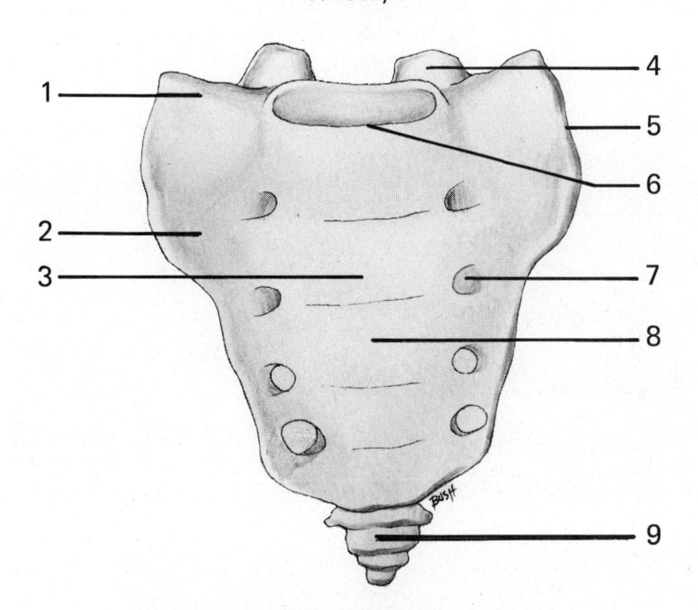

Figure 4-41. The Sacrum and Coccyx (ventral aspect)

KEY TO FIGURE 4-42

1. clavicular notch
2. suprasternal notch (jugular notch)
3. sternal angle
4. costal notch
5. manubrium
6. costal cartilage
7. rib
8. body (gladiolus)
9. xiphoid (ensiform) process
10. intervertebral disc
11. 12th thoracic vertebra
12. floating rib
13. lumbar vertebra

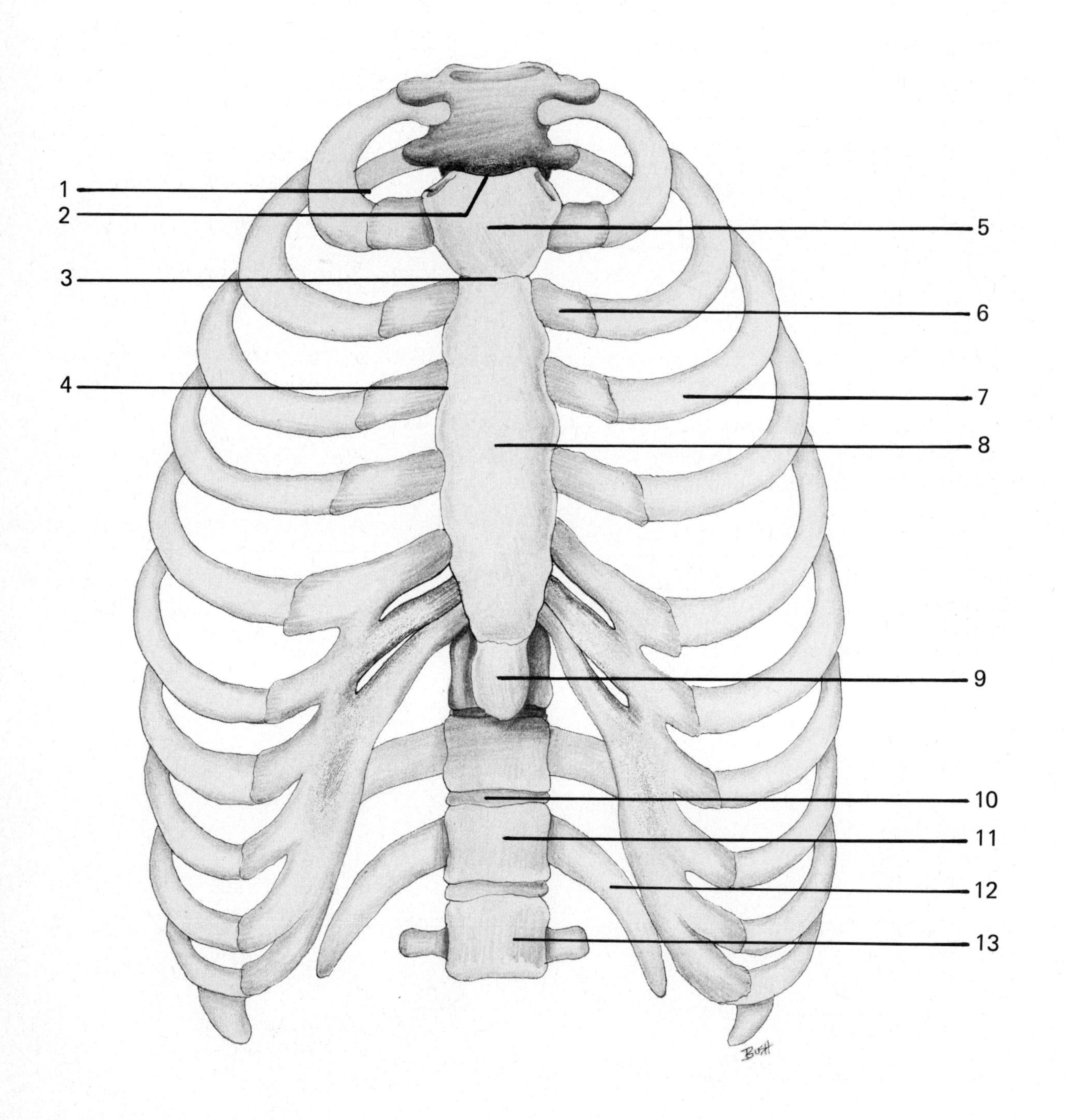

Figure 4-42. The Bony Thorax

Exercise 4-D The Fetal Skeleton

Objective

To compare the fetal skeleton with the adult skeleton.

Materials

Skeletons of an adult and a fetus.

Discussion

The number of separate bones in the fetus and child is greater than the 206 found in the typical adult skeleton. This is due to the fact that many bones in the adult are composed of several parts fused together; these parts are separate bones in the fetus and child. For example, in a child, the three regions of the os coxae bone (the ilium, ischium, and pubis) form separate bones; in the adult these have united to make one hip bone.

Procedure

1. Use Figures 4-43 and 4-44, two views of the skull at birth, as self-tests.
2. Examine the fetal and adult skeletons, and observe any differences in degree of ossification. Give special attention to the structures listed below, in the fetal skeleton. Compare the bone or process with the corresponding structure in the adult. Record your observations in the chart under Results and Questions.
   a. frontal bone
   b. sternum
   c. patella
   d. os coxae
   e. spinous processes of vertebrae
   f. carpals
   g. proximal epiphysis of humerus
3. Determine the sex of the fetal skeleton.

Exercise 4-E. Joints

Objective

To determine the structure of a typical joint.

Materials

Beef knee joint; x-ray viewing box; and x-rays of bones and joints.

KEY TO FIGURE 4-43

1. frontal bone
2. frontal suture
3. anterior (frontal) fontanel
4. sagittal suture
5. parietal bone
6. posterior (occipital) fontanel
7. lambdoidal suture
8. occipital bone

KEY TO FIGURE 4-44

1. coronal suture
2. lambdoidal suture
3. anterolateral fontanel
4. squamosal suture
5. posterolateral fontanel

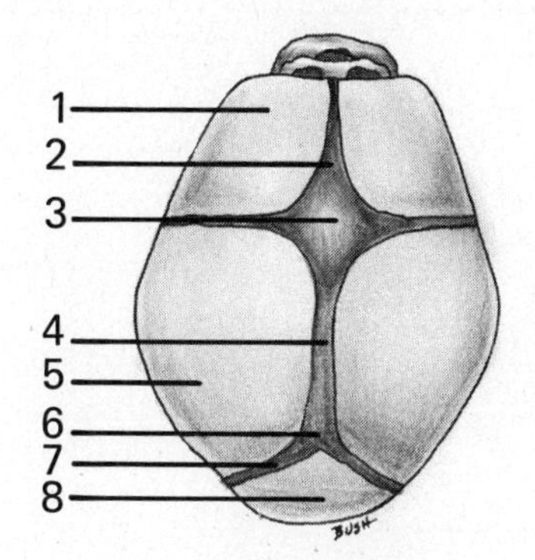

Figure 4-43. Fetal Skull (superior surface)

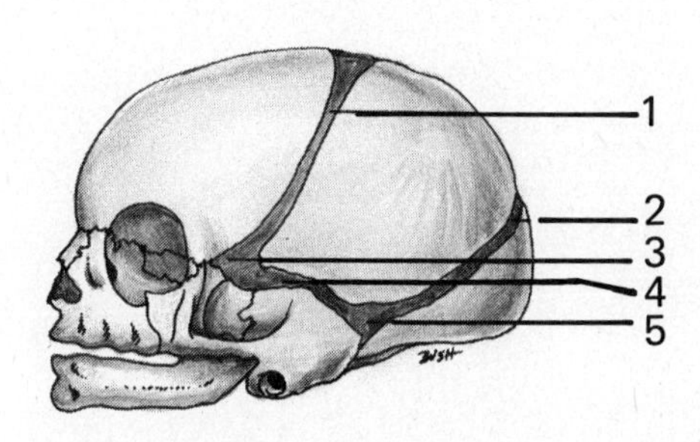

Figure 4-44. Fetal Skull (lateral view)

Discussion

A joint is the site of articulation between two bones. Joints may be divided into three classes, depending on the amount of movement between the articulating bones: synarthroses (immovable joints), amphiarthroses (slightly movable joints), and diarthroses (freely movable joints). Joints may also be divided into five classes, based on the structure of the joint (synostoses, syndesmoses, synchondroses, symphyses, and synovial joints). Most of the common joints in the body are freely movable synovial joints.

Procedure

1. The instructor will demonstrate a series of x-ray films. This will give you an opportunity to study the bones and joints as they are commonly seen in the hospital or doctor's office.
2. Study Figures 4-45 and 4-46; then locate the following structures on a prepared section through a beef knee joint.
   a. synovial membrane
   b. articular cartilage
   c. synovial fluid (joint cavity)
   d. ligaments
   e. tendons
   f. patella

Exercise 4-F The Skeletal System of the Fetal Pig

Due to the immaturity of the fetal pig and the ready availability of human bones (or other adult mammal skeletons), the skeleton of the fetal pig is not included.

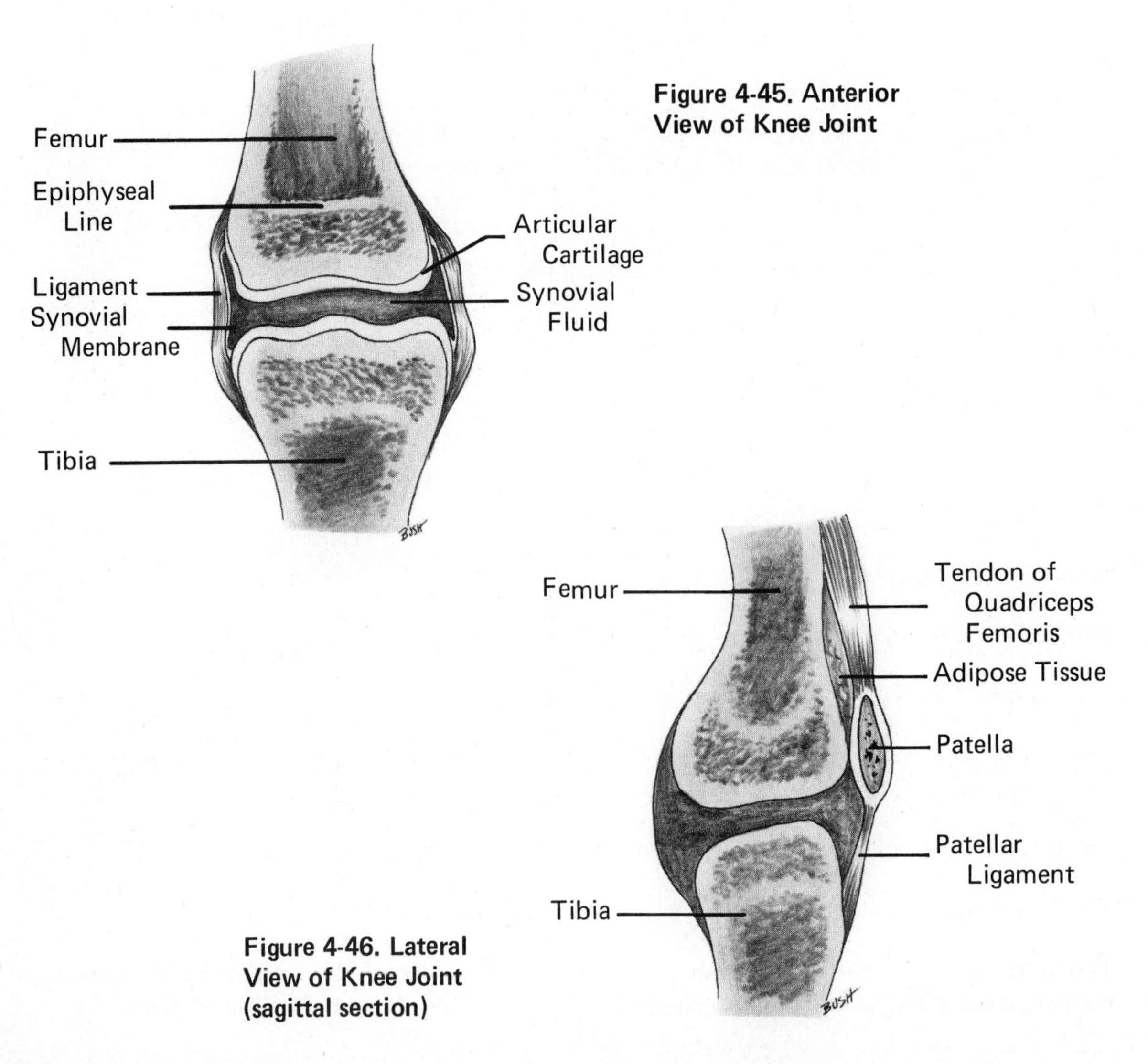

Figure 4-45. Anterior View of Knee Joint

Figure 4-46. Lateral View of Knee Joint (sagittal section)

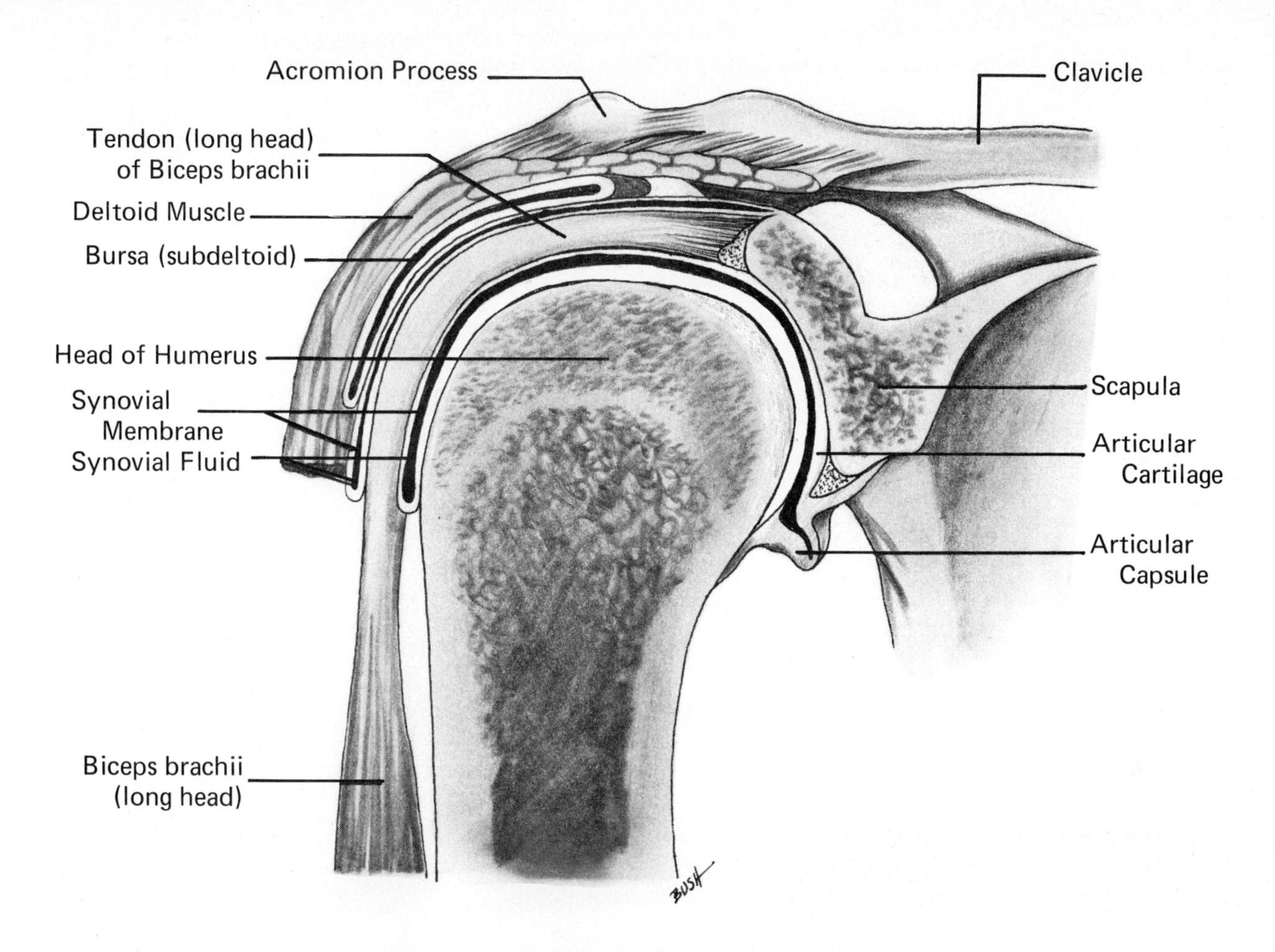

Figure 4-47. Anterior View of Shoulder Joint

Self-Test—Skeletal System

Directions: See Chapter 1, Self-Tests, p. 9.

1. The **malleus** is attached to the tympanic membrane.

2. Bones **lack** blood vessels.

3. The **periosteum** is the membrane responsible for nourishing the bone.

4. A **greenstick** fracture is one wherein the bone breaks the skin.

5. The **pelvic girdle** is actually several bones joined together for the protection of organs in the **pelvic** region.

6. The **floating** ribs provide protection for the heart.

7. The toes contain **a greater** number of bones than the fingers.

8. The bones of the appendicular skeleton and those of the thorax are formed by the process of **endochondral ossification**.

9. Ossification begins at the **epiphyses** and extends to the **diaphysis.**

10. Spongy bone is found in **flat bones** as well as in **long bones.**

11. The two types of bone formation involve the same processes, except that in endochondral ossification there is an initial period of **destruction of cartilage.**

12. The function of the **trochlea** is to articulate with the ulna.

13. Haversian systems are responsible for nutrition in **compact** bony tissue.

14. Haversian systems are characteristic of **spongy** bone.

15. Red bone marrow is found in the **medullary cavity** of some long bones of the adult.

16. The constricted portion of a long bone is called the **shaft.**

17. **Epiphyses** of bones tend to be **rounded** into condyles, trochanters, and heads.

18. The acetabulum is the socket in which the head of the **humerus** articulates.

19. The long bones of the extremities continue to grow during childhood at the **epiphyseal cartilage.**

20. Most of the cranial and facial bones develop by **intramembranous ossification.**

21. Hemopoiesis may occur in the adult in the **ribs and sternum.**

22. The lining of the medullary cavity is called the **periosteum.**

23. The primary curves of the vertebral column are the thoracic and **lumbar.**

24. Bending the elbow is an example of **flexion.**

25. Styloid processes are found upon the radius, ulna, and temporal, **but not** on the femur.

26. The fusion time of the anterior fontanel of the baby's skull is approximately **18 months.**

27. The **longitudinal arch** extends from the heel to the base of the toes.

28. Touching the little finger with the thumb is called **opposition.**

29. The iliopectineal line is found in the **skull.**

30. The **sella turcica** contains the pituitary gland.

31. The **male** pelvis has a circular pelvic inlet.

32. The **female** pelvis includes wide, flared ilia.

33. Spinal nerves are transmitted by way of **intervertebral foramina.**

34. The spinal cord is located in the **neural arches.**

35. The nasal septum is comprised of the **ethmoid** and **vomer** bones, as well as hyaline cartilage.

36. The air sinuses which open directly into the nasal cavity **do not** include the mastoid sinus.

37. The **secondary curves** of the vertebral column are the cervical and lumbar.

38. Movement of the ankle to turn the sole outward is called **eversion.**

39. When the long strands formed by mesenchymal cells become invested with bone matrix they are called **trabeculae.**

40. The condyles of the occipital bone and the depressions of the **atlas** allow for nodding movements of the head.

41. In turning the head from side to side the **atlas** rotates around a process of the axis called the **dens.**

42. A **foramen** is a tube-shaped passageway.

43. A hollow or depression is called a **cavity.**

44. A small, rounded projection is a **tuberosity.**

45. The viscosity of synovial fluid **is independent of** changes in temperature.

46. **Synchondroses** are freely moveable joints with joint cavities.

47. **Circumduction** combines several types of joint movement.

48. **Both** prominent bursae and articular capsules may be present at ball and socket and hinge joints.

49. A **symphysis** joint occurs between bodies of vertebrae.

50. The elbow is an example of a **ball and socket** joint.

51. Swinging the arms around at the shoulder is an example of **circumduction.**

52. The position of the foot when standing on the toes is known as **plantar flexion.**

53. Nodding the head involves **both flexion and extension.**

54. Moving the arms straight out from the sides is called **adduction.**

55. Bending the trunk backward is an example of **extension.**

56. Moving the thigh forward, as in marching, is an example of **flexion.**

57. Spreading the fingers apart is **adduction.**

58. Turning the palm of the hand up is **pronation.**

59. Shaking the head is **rotation.**

60. A major function of the bony rib cage is to protect the soft organs within the **abdomen.**

61. A **sulcus** is a groove or furrow.

62. Lateral curvature of the spine is called **kyphosis.**

63. A **sprain** is a joint injury with stretching or tearing of the ligaments.

64. Wearing high-heeled shoes is a common cause of **lordosis** in young women.

65. Moving a bone backward is known as **retraction.**

66. A patient was diagnosed as having a greenstick fracture, left femur. The difference between this and an ordinary fracture is that this is **a clean break.**

67. Judging from the diagnosis the patient is a child, because children have more organic matter in their bones than adults, so the bones are more likely to **break clean.**

68. The x-rays showed obvious bands of cartilage near the epiphyses, which meant that this patient's bones **were still growing.**

69. The **periosteum** is important during healing of a fracture because it nourishes the bone and promotes growth and repair.

70. Milk products provide **calcium** which is essential in building strong bones.

71. A depressed skull fracture is a bone **below** normal level that may press on the brain.

72. A simple skull fracture is usually a **shattered** bone.

73. A **craniotomy** was performed on a patient. This is an opening in the cranium to elevate the depressed bone fragment.

74. A **deviated nasal septum,** which means the partition between the nasal cavities is bent to one side, may obstruct breathing.

75. An increase in the thoracic curvature is called **lordosis or swayback.**

76. In a sternal puncture **yellow** marrow is obtained from the sternum.

77. A baby is born who has a spina bifida. This congenital defect is caused by a failure of the **vertebral laminae** to unite.

78. Spina bifida is most likely to occur in the **cervical** and **thoracic** region.

KEY

2. have
4. compound
6. true
7. the same
9. diaphysis; epiphyses
14. compact
15. proximal epiphyses
16. neck
18. femur
22. endosteum
23. pelvic (sacral)
29. pelvis
31. female
34. vertebral foramina
42. meatus
43. fossa
44. tubercle
45. varies with
46. diarthroses
50. hinge
54. abduction
57. abduction
58. supination
60. thorax
62. scoliosis
66. not a clean break
67. splinter
72. a crack in the
75. kyphosis or hunchback
76. red
78. lumbar and sacral

Name ____________________

Results and Questions for Chapter 4

Exercise 4-A. Composition and Structure of Bone

1. What is the scientific term used for the shaft of a long bone?____________________

2. From which part of a long bone were the slides of bone used in Exercise 3-D, Part 4, obtained?

____

3. What is the function of the organic material in bone?____________________

____

4. What is the function of the inorganic material in bone?____________________

____

5. Name the cells that build bone.____________________

6. Where are the bone-building cells located in bone?____________________

7. Describe the appearance of the bone soaked in acid.____________________

____

8. Describe the appearance of the bone subjected to intense heat.____________________

____

9. Name the most abundant inorganic salts in bone.____________________

10. Define a "greenstick" fracture.____________________

11. In what age group does a "greenstick" fracture occur most commonly?________Explain why:

____

12. What is the function of each of the following:

   a. Haversian canals____________________

   b. Canaliculi____________________

   c. Volkmann's canals____________________

   d. Periosteal blood vessels____________________

   e. Nutrient foramen____________________

Exercise 4-B. The Appendicular Skeleton

1. Name the bones comprising the shoulder girdle.____________________

2. Name the bones comprising the pelvic girdle.____________________

3. Name the bones comprising the bony pelvis. ______

4. How does the female pelvis differ from the male pelvis? ______

______

______

______

5. Define Colle's fracture. ______

6. What is a common cause of Colle's fracture? ______

7. Why is the clavicle particularly likely to fracture if a person falls on his shoulder?

______

8. What is the common name of each of the following?

a. Clavicle ______ d. Phalanges ______

b. Scapula ______ e. Femur ______

c. Os coxae ______ f. Calcaneus ______

9. How many carpals are there in the entire adult skeleton? ______

10. How many tarsals are there in the entire adult skeleton? ______

Exercise 4-C. The Axial Skeleton

1. Define:

a. Lordosis ______

b. Kyphosis ______

c. Scoliosis ______

2. What is another name for the first cervical vertebra? ______

3. What is another name for the second cervical vertebra? ______

4. What is meant by a "slipped disc"? ______

5. What is the cause of spina bifida? ______

6. Name the bones making up each of the following structures:

a. Nasal septum ______

b. Hard palate ______

Name ______________________

c. Orbit ______________________

____

7. List the bones possessing paranasal sinuses. ______________________

____

8. Into what do the paranasal sinuses drain? ______________________

9. List the five regions of the vertebral column and the distinguishing characteristics of each region.

Region	*No. of Vertebrae*	*Distinguishing Characteristics*

10. What bones does the lambdoidal suture unite? ______________________

11. Name the bones forming the bony thorax. ______________________

12. What are wormian bones? ______________________

Exercise 4-D. The Fetal Skeleton

1. Compare the fetal and adult skeletons by filling in the following table.

	Description of Bone in Fetal Skeleton
Frontal bone	
Sternum	
Patella	
Os coxae	
Spinous processes of vertebrae	
Carpals	
Proximal epiphysis of humerus	

2. What is a fontanel? ______

3. How many bones are there in the typical fetal skeleton? ______

4. What was the sex of the fetal skeleton? ______

5. What characteristics of the fetal skeleton enabled you to determine its sex? ______

______

Exercise 4-E. The Joints

1. Fill in the following table concerning the joints.

Name of Joint	*Bones Comprising Joint*	*Types of Movement Possible at Joint*
Hip joint		
Knee joint		
Elbow joint		
Wrist joint		
Interphalangeal joint		

2. How does the hip joint differ from the shoulder joint? Discuss both the structure of the joint and the range of movement. ______

______

______

3. What is the function of the synovial fluid? ______

4. What is a meniscus? ______

5. What is the function of a ligament? ______

6. What is the function of the patella? ______

7. What is a sesamoid bone? ______

8. Define the ligamenta flava. ______

9. What type of joint (classified as to structure **and** as to amount of movement possible) is found between

   a. the bodies of the vertebrae? ______

   b. the articular process of the vertebrae? ______

   c. the os coxae bone and the sacrum? ______

5

Physical Transport of Materials

Exercise 5-A Filtration

Objectives

To compare filterable materials with non-filterable materials; to demonstrate the importance of pressure induced by the weight of a fluid column.

Materials

Funnel; filter paper; test tube; beaker; Lugol's solution; stand; clamp; medicine dropper; and a mixture of charcoal, copper sulfate, and boiled starch in water.

Discussion

Filtration is concerned with the passage of substances in solution through a membrane. If filter paper, serving as the membrane, is placed in a funnel which is then filled with fluid, the fluid will be seen to drip through rapidly when the funnel is full, but as the height of the fluid decreases, the individual drops appear less frequently.

The difference in rate of drop formation depends upon the surface area allowed for filtration and the hydrostatic pressure (i.e., the weight of the fluid column). In this experiment it is gravity which "pulls" the particles of the substance through the pores of the membrane (filter).

Additional force could be exerted by other means. If, for example, the upper part of the filter were contained in a tubular system that was closed and also possessed a pump, an increase in pressure could be exerted upon the filter. In the circulatory system of the body, both gravity and pumping contribute to the hydrostatic pressure exerted at capillary membranes (the filter). A prime example of filtration in the human body occurs in the kidneys.

In the following experiment, copper sulfate is used as an example of a crystalloid, starch as a colloid, and charcoal as an example of particles forming a suspension.

Procedure

1. Fold a piece of filter paper as directed, and place it in a funnel. Add a few drops of water to fix the paper in position.
2. Place the funnel over a beaker as illustrated in Figure 5-1.

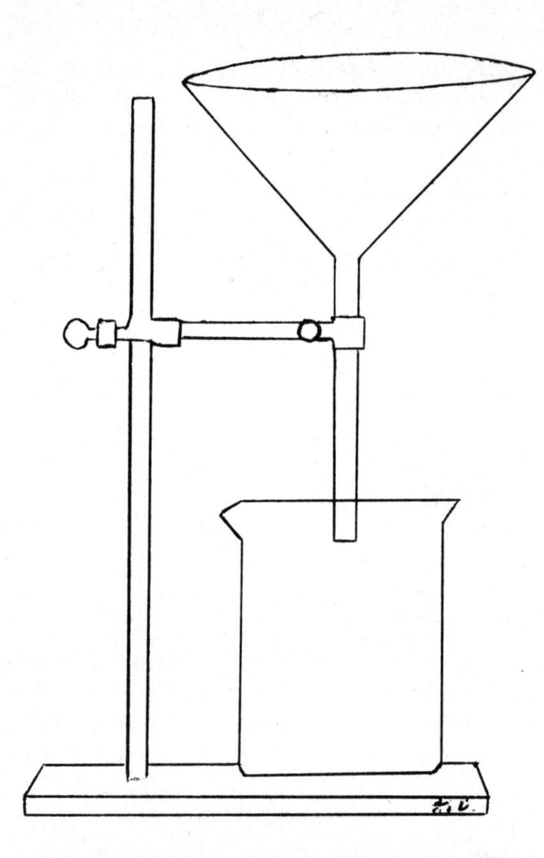

Figure 5-1. Setup for Filtration Experiment

3. Shake the mixture of powdered wood charcoal (black), copper sulfate (blue), starch (white), and water, and pour into the funnel until it almost reaches the top edge of the filter paper.
4. To observe the difference in rate of filtration as the height of the solution in the funnel decreases, count the number of drops passing through the filter in the first 10 seconds. (Make an estimate if an exact count is not possible.) 30-46
5. When the fluid level is reduced by about one-third, count the number of drops passing through the filter for a period of 10 seconds. 15
6. Repeat this procedure when the level is reduced by about two-thirds. 23
   Record these results under Results and Questions.
7. Continue the filtration until the funnel is empty. Observe which substances passed through the filter paper by their color. Check the filter paper to determine whether any colored particles were not filterable.
8. To determine whether starch passed through the filter paper, place approximately 2 cc of the filtrate in a test tube; add a few drops of Lugol's solution. A pale blue to blue-black color is a positive indication that starch is present.

Exercise 5-B Diffusion

Objective

To become familiar with the process of diffusion.

Discussion

Diffusion is the dispersion of substances as a result of the movement of their ions or molecules. Under appropriate conditions, gas particles may diffuse into another gas, or liquid, or a solid; a liquid may diffuse into a gas, another liquid, or a solid; and a solid may diffuse into a gas, a liquid, or another solid. Diffusion of solute particles occurs from a region of higher concentration of the particles to a region of lower concentration of the particles. When equilibrium is reached, diffusion occurs equally in all directions. Since the living cells of organisms are not actively involved in the process, diffusion is considered to be a passive transport mechanism.

Part 1. Diffusion of a Liquid into a Gas

Objective

To demonstrate the diffusion rate of ether into the atmosphere.

Materials

Ethyl ether and a petri dish.

Procedure

1. Close all of the doors and windows in the classroom.
2. Ether will now be poured into a petri dish at the front of the room.
3. Obtain the following data, and record them under Results and Questions:
   a. The exact time the ether was poured into the petri dish.
   b. The seconds required for the first student to detect the odor.
   c. The seconds required for students in the first row to detect the odor. (Each student will raise his hand as soon as the odor is detected.)

d. The seconds required for students in each subsequent row to detect the odor.
e. The percentage of students detecting the odor at the conclusion of the experiment.

4. The experiment will be conducted for a minimum of five minutes.

Part 2. Diffusion of a Solid in a Liquid

Objective

To demonstrate the diffusion of potassium permanganate in water.

Materials

A crystal of potassium permanganate and a beaker of water.

Procedure

1. Drop a single crystal of potassium permanganate into a beaker of water.
2. Observe this preparation at the beginning of the experiment and at intervals during the laboratory period. The container should not be touched or disturbed in any way. Record your observations under Results and Questions.

Part 3. Diffusion of a Solid in Agar

Objective

To demonstrate the relationship between molecular weight and speed of diffusion.

Materials

Petri dishes containing 2% agar; crystals of potassium permanganate and methyl orange.

Discussion

Agar molecules form a colloid when mixed with water. The molecules of the crystals used in this experiment diffuse through water channels in the agar.

The molecular weight of potassium permanganate is 158; that of methyl orange is 327.

Procedure

1. Place a single crystal of potassium permanganate and a single crystal of methyl orange about two inches apart on a film of agar in a petri dish.
2. After one hour, observe the size of the colored ring around each crystal. Record your observations under Results and Questions.
3. At the completion of all observations, scrape out the agar into a paper towel, wrap securely, and place in the waste paper basket. Clean and dry the petri dish.

Exercise 5-C Dialysis

Objective

To demonstrate the types of substances that are dialyzable.

Materials

Dialysis shell; prepared solution containing sodium chloride, glucose, albumin, and starch in distilled water; distilled water; Benedict's solution; concentrated nitric acid; Lugol's solution; 1% silver nitrate solution ($AgNO_3$); beaker; test tubes; and a bunsen burner.

Discussion

Since some qualitative tests are to be made in this experiment, it is important that certain precautions be taken to prevent contamination. Because of the presence of chlorides in tapwater, only distilled water should be used throughout this exercise in preparing solutions and in washing test tubes or other glassware.

Procedure

1. Rinse all equipment (beaker, dialysis shell, and one test tube) with a small amount of distilled water.
2. Place a prepared solution containing starch, sodium chloride, glucose, and albumin in a dialysis shell. Be very careful not to spill any liquid down the outside of the membrane. Do not touch the membrane on the part that is to be placed in the water (i.e., below the planned waterline) because of the salt contained in perspiration.
3. Place the dialysis shell in a beaker of distilled water.
4. After one hour perform the following tests on the water in the beaker to see which, if any, of the materials placed in the dialysis shell have diffused through it.
   a. Sodium chloride: Place about 2 cc of the beaker water into a clean test tube

and add several drops of silver nitrate ($AgNO_3$). The formation of a white cloudy precipitate indicates the presence of sodium chloride.

b. Starch: Place about 2 cc of the beaker water into a test tube. Add a drop of Lugol's solution. A blue or blue-black color is positive for starch.

c. Glucose: Place about 5 cc of Benedict's solution into a test tube and add to it 5 cc of beaker water. Boil for two minutes. Formation of a green, yellow, orange, or red color or precipitate in the test tube is positive for glucose. If the test is negative, repeat with a larger amount of beaker water.

d. Albumin: Place 2 cc of beaker water in a test tube. Tilt the test tube and carefully add several drops of concentrated nitric acid (HNO_3) so that it moves down the inner wall of the tube. The acid will cause a white coagulum to appear if the test is positive.

Exercise 5-D Plasmolysis

Objective

To demonstrate osmotic relationships through the process of plasmolysis.

Materials

Elodea; slide; cover glass; and 5% sodium chloride (NaCl) solution.

Discussion

Elodea is a water plant with numerous relatively small leaves which are only two cells thick. An entire leaf may thus be observed under the microscope and some of the characteristics of living cells can be studied. In contrast to animal cells, plant cells include (outside the plasma membrane) a permeable cell wall consisting primarily of cellulose (a polysaccharide) which provides them with a varying degree of rigidity. In green plants, such as *Elodea,* specialized structures called chloroplasts (absent in animal cells) may be observed. These are involved in the process of photosynthesis (manufacture of food using light as the source of energy).

Procedure

1. Remove a healthy looking leaf from a sprig of *Elodea* found in the container at the front of the room, and mount it on a clean slide. Add one or two drops of water to the leaf from the container and cover with a cover glass.
2. Locate a cell that is typical of the majority. The green bodies are chloroplasts. See Figure 5-2, which may be used as a self-test, to

KEY TO FIGURE 5-2

1. nucleolus
2. nuclear membrane
3. nucleoplasm
4. chromatin
5. nucleus
6. cell wall
7. plasma membrane
8. cytoplasm
9. vacuole
10. chloroplast

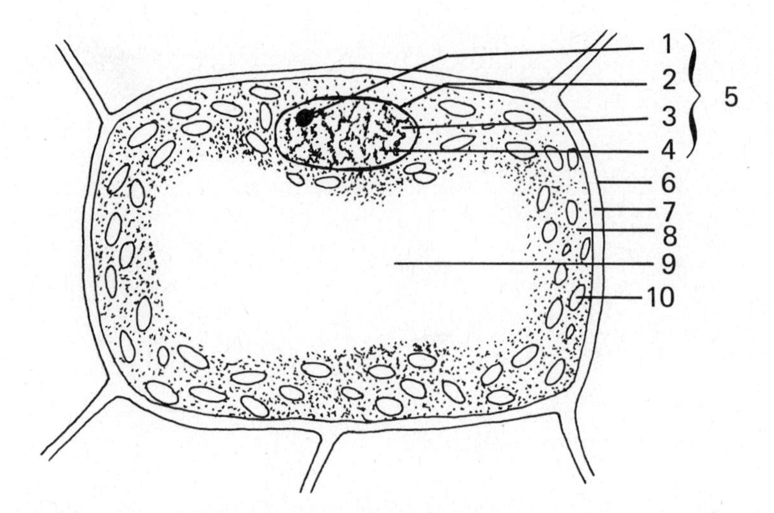

Figure 5-2. ***Elodea*** **Cell**

aid in identification of the structures in the cell. (Optional—draw a typical cell.)

3. With the cell in focus, remove the water from under the cover glass by holding a piece of paper towel or blotting paper at its edge. When most of the water has been removed, introduce a 5% solution of NaCl by holding a pipette (medicine dropper) at the edge of the cover glass, and gently squeeze out the fluid (which will move under the cover glass). Continue this procedure until the leaf is completely surrounded by the solution.
4. Immediately observe the cell (or cells) you were looking at previously. Continue these observations until distinct changes occur.
5. Draw one or two cells in which a noticeable change has occurred.
6. Remove the leaf and soak it in distilled water for 5 minutes; then remount it. Examine the leaf to observe whether plasmolysis is reversible.

Exercise 5-E Osmosis

Objective

To demonstrate the occurrence of osmosis by means of an osmometer.

Materials

Osmometer containing a 10% NaCl solution stained with methylene blue and a beaker of water.

Procedure

1. Examine the osmometer in the front of the room. Compare this setup with that illustrated in Figure 5-3.
2. Note the level of the fluid in the tube at the beginning of the laboratory period and again near the end of the laboratory.

Exercise 5-F Hemolysis and Crenation

Objectives

To demonstrate hemolysis and crenation.

Materials

Test tubes; slides; cover glasses; 0.9% (or 0.15 M) NaCl/distilled water; 3% NaCl; and blood (human or ox).

Discussion

The term hemolysis is used to indicate a special case of lysis in erythrocytes wherein the hemoglobin is lost from the cell. A 0.9 % (or 0.15 M) solution of NaCl forms an isotonic solution for erythrocytes. Hemolysis may occur when such cells are placed in a sufficiently hypotonic solution. Hemolysis also occurs when the erythrocytes (RBC's) are placed in a solvent (e.g., ether) that destroys the plasma membrane, or in cases where solute particles can penetrate the plasma membrane.

Blood becomes a transparent cherry red color (instead of a dull opaque color) when the hemoglobin becomes uniformly dissolved in the surrounding liquid. Thus, a change from opaque blood to transparent blood may be regarded as an indication of the occurrence of hemolysis.

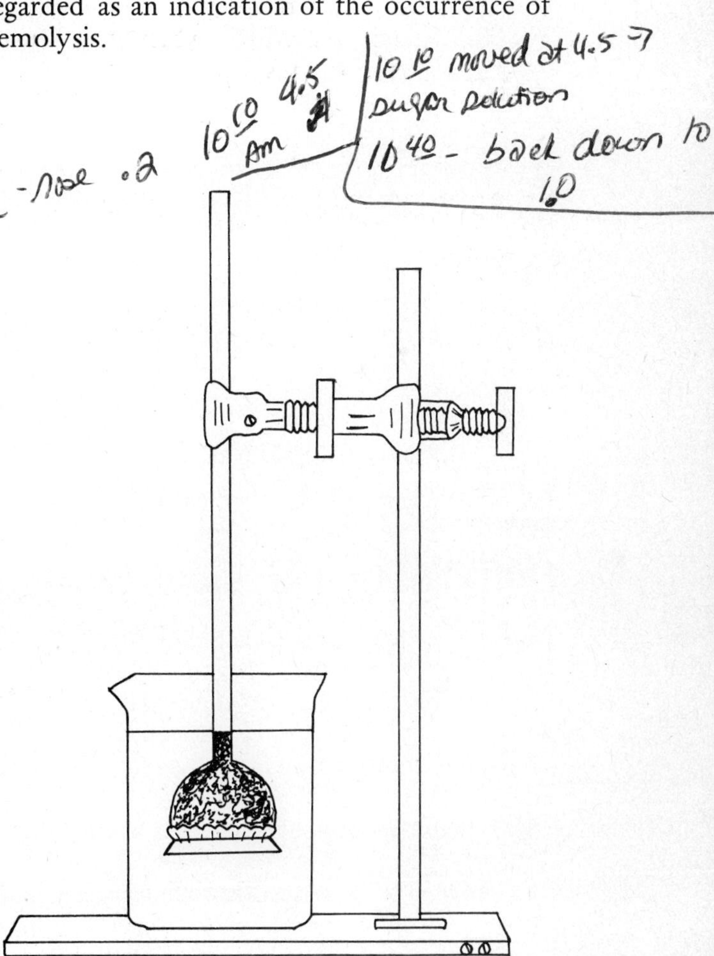

Figure 5-3. Apparatus for Osmosis Demonstration

When cells are placed in hypertonic solutions they may become wrinkled in appearance due to shrinking of the protoplasm. The shrinking is due to osmosis which results in a loss of water from the cell. This process is known as crenation.

Procedure

1. Add 3-5 drops of blood to a test tube containing 2 cc of 0.9% (or 0.15 M) NaCl solution.
2. Add the same amount of blood to a second test tube containing 2 cc of distilled water.
3. Add the same amount of blood to a third test tube containing 2 cc of 3% NaCl solution.
4. Compare the transparency of the three solutions by looking at a printed page through each tube. Record your observations under Results and Questions.
5. Mount one or two drops of the solution from each of the three test tubes on clean glass slides. Cover with a cover glass and observe with the microscope (on high power).
6. In the appropriate spaces below, draw three or four cells as observed from each test tube.

Exercise 5-G
Brownian Movement

Objective

To observe, indirectly, the movement of molecules as demonstrated by Brownian movement.

Materials

1% carmine dye; medicine dropper; slide; cover glass; and microscope.

Discussion

Each particle of carmine dye is an aggregate of many molecules. Motion is due to unequal bombardment by the molecules of the water environment. This action was first noticed by the botanist and army surgeon Robert Brown, for whom it is named.

Procedure

1. Take a small drop of 1% carmine suspension and place it on a clean slide by means of a medicine dropper.
2. Cover the suspension with a cover glass.
3. Observe under the high power of the microscope. (See Results and Questions.)

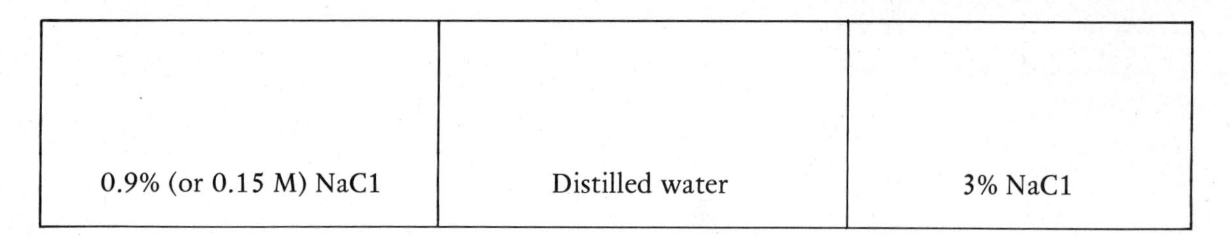

0.9% (or 0.15 M) NaC1	Distilled water	3% NaC1

Self-Test—Physical Transport of Materials

Directions: See Chapter 1, Self-Tests, p. 9.

1. In **active** transport, no energy is expended by the cell in the transfer of material.
2. Active transport involves movement of ions **against** a concentration gradient.
3. Hydrostatic pressure **increases** as the height of a column of fluid increases.
4. The rate of filtration through capillary walls is **the same** in all parts of the body.
5. Diffusion occurs as the result of **the constant motion of molecules.**
6. Osmotic pressure is **proportional to the number of particles** in the solution.

7. A **hypotonic** solution has a higher osmotic pressure than the blood.

8. The living cell is surrounded by a **permeable** membrane.

9. The principal **cation** in intracellular fluid is K^+.

10. In active transport, solutes pass from the **more** concentrated to the **less** concentrated solution.

11. Hydrostatic pressure is **higher** in the glomerulus than in other capillaries; this is largely responsible for filtration of blood in the kidneys.

12. Hemolysis of erythrocytes occurs in a **hypertonic** solution.

13. Plasmolysis is the rounding up of plant cell protoplasm due to a loss of **salt.**

14. The concentration of sodium chloride in a physiological saline solution is **0.9%.**

15. A **hypertonic** solution is one more dilute than the solution of a cell.

16. **Filtration** is the passage of water and dissolved substances through a membrane due to a difference in hydrostatic pressure on the two sides of the membrane.

17. Diffusion is the movement of molecules of a substance from a region of **lower** concentration (of the substance) to one of **higher** concentration.

18. A cell containing 0.5% NaCl is placed in a solution of 7% NaCl. Water leaves the cell.

19. An **isometric** solution is one that has the same concentration of dissolved substances as the cell.

20. Crenation occurs in a **hypertonic** solution.

21. Speed of diffusion is **directly** proportional to molecular weight.

22. Iodine is used to test for the presence of **albumin.**

23. **Protein** passes through the wall of a dialysis shell.

24. Particles forming **a suspension** will pass through filter paper.

25. Sugar is an example of a **nonelectrolyte.**

26. An **ion** is an electrically charged atom or molecule.

27. The central core of an atom is called the **nucleus.**

28. The negatively charged particles of an atom are called **protons.**

29. Atoms of the same element that exist in different forms (due to a variable number of neutrons) are called **isotopes.**

30. Colloidal suspensions of liquids within liquids are called **colloids.**

31. A colloidal suspension can be separated from a true solution by the process of **dialysis.**

32. Diffusion of water through a semipermeable membrane from a region of high water concentration to a region of low water concentration is called **osmosis.**

33. The pH of a material is a measure of its **hydroxyl** ion concentration.

34. The direction of diffusion is determined primarily by the **size** of the diffusing particles.

35. The direction in which a substance moves through a membrane by osmosis depends mainly upon differences **in concentration of water molecules.**

36. A colloid solute particle is **larger** than a crystalloid solute particle.

37. Only the solvent moves through a membrane by osmosis, but both solute and solvent may move through a membrane by **diffusion or filtration.**

38. **Diffusion, osmosis, and filtration** are related to the concentrations of the solution on both sides of a membrane.

39. The size of red blood cells immersed in 0.9% glucose would be **greater** than those immersed in 0.9% sodium chloride.

40. **A base** contains free hydrogen ions.

41. **A compound** contains only one kind of atom.

42. **A chemical compound** contains unlike atoms in definite proportion.

43. **A true solution** shows the Tyndall effect.

44. Protoplasm is a **colloidal suspension** that has all of the properties of life.

45. The shells of an atom contain **neutrons.**

46. The characteristics of a chemical compound **are the same as** those of the elements that compose it.

47. Diffusion of one substance through another is slowed down **by heating.**

48. A very **highly ionized acid** might have a pH of 1.

49. Glucose and fructose both have the empirical formula $C_6H_{12}O_6$; thus they are **isotopes.**

50. The **higher** the pH, the greater the hydrogen ion concentration.

51. **Table salt** is an example of an electrolyte.

KEY

1. passive
4. not the same
7. hypertonic
8. differentially (semi) permeable
10. less; more
12. hypotonic
13. water
15. hypotonic
17. higher; lower
19. isotonic
21. indirectly
22. starch
23. NaCl, glucose (crystalloids)
24. true solutions and colloids
28. electrons
30. emulsions
33. hydrogen
34. concentration
38. diffusion and osmosis
40. an acid
41. an element
43. a colloidal suspension
45. electrons
46. are different from
47. by freezing
49. isomers
50. lower

6

The Muscular System

Exercise 6-A The Gross Anatomy of the Major Superficial Human Muscles

Objectives

To become familiar with the location and action of selected major human muscles.

Materials

Text; reference books; anatomic charts; and dissectable torso.

Discussion

The basic function of muscle is to produce movement of the body or part of the body. In addition, muscles serve to protect the internal organs, maintain posture, and produce a large amount of body heat.

The muscles studied in this exercise are skeletal muscles. See Chapter 3 for the histology of skeletal, smooth, and cardiac muscle. The student should learn in this exercise the names of the major superficial muscles of the body and their actions.

Procedure

1. Use Figures 6-1, 6-2, 6-3, and 6-4 as self-tests. See Section J, p. xx (Illustrations as Self-Tests) for a suggested approach.
2. Compare the muscles indicated in Figures 6-1 and 6-2 with similar figures in your text, in references, on anatomic charts, and on the dissectable torso.

Exercise 6-B Dissection of the Muscular System of the Fetal Pig

Objectives

To identify selected muscles of the fetal pig and to compare them and their functions with corresponding human muscles.

Discussion

As you dissect the muscles of the pig, determine the major differences between them and those in man. As you locate the **origin** of the muscle (the site of attachment to a fixed bone) and the insertion (the attachment to a more freely movable bone), determine the function of the muscle. Remember that as a muscle contracts, it pulls the insertion toward the origin. Muscles are usually attached to bones by **tendons,** which are bands of dense white fibrous connective tissue. Some muscles insert by way of an aponeurosis.

KEY TO FIGURE 6-1

1. sternocleidomastoid
2. pectoralis major
3. biceps brachii
4. brachialis
5. iliopsoas
6. pectineus
7. adductor longus
8. gracilis
9. vastus medialis
10. gastrocnemius
11. deltoid
12. serratus anterior
13. external oblique
14. inguinal ligament
15. subcutaneous inguinal ring
16. sartorius
17. rectus femoris
18. vastus lateralis
19. patella

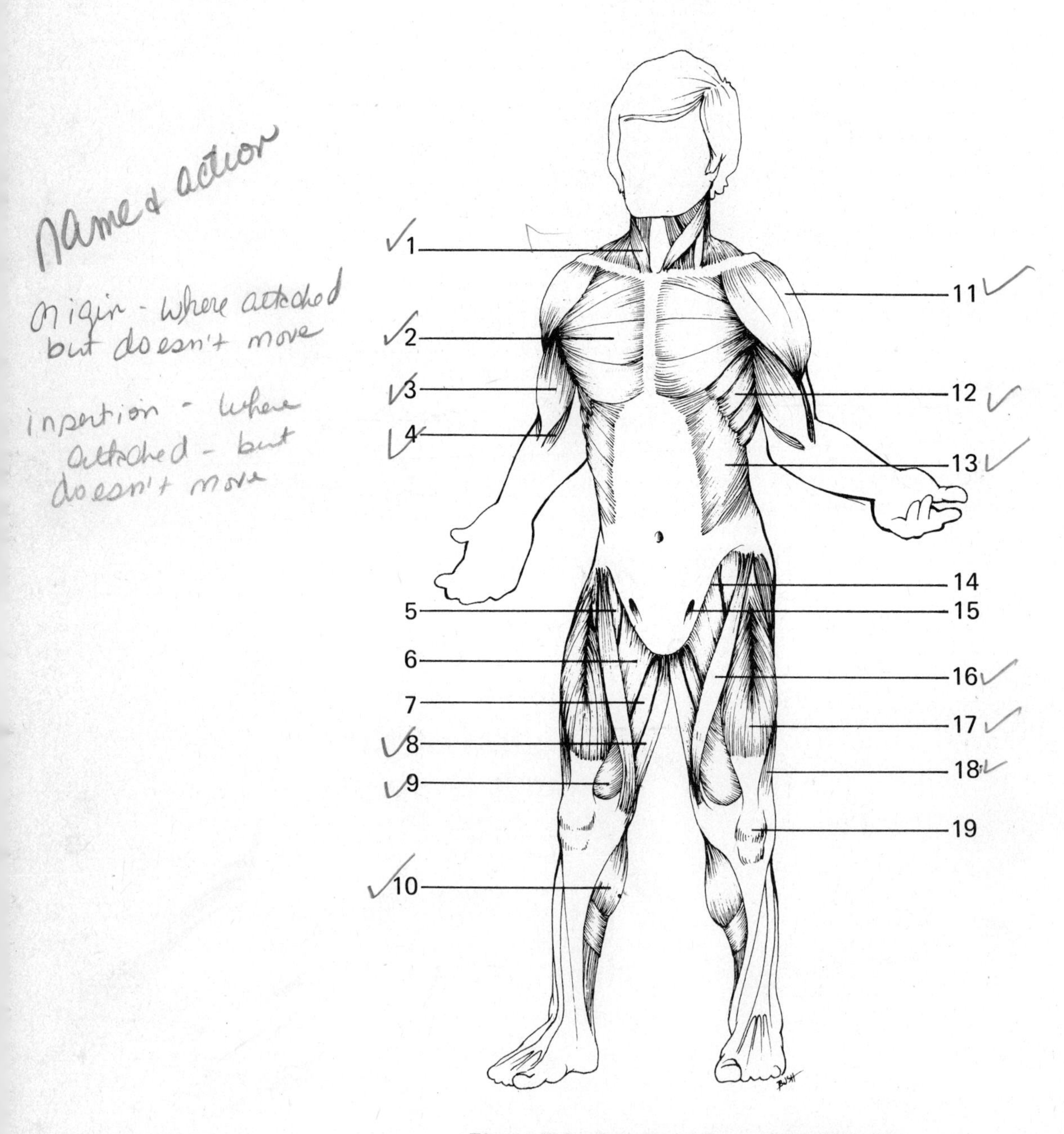

Figure 6-1. Anterior View of the Muscles

KEY TO FIGURE 6-2

1. deltoid
2. triceps brachii
3. gracilis
4. semimembranosus
5. semitendinosus
6. trapezius
7. latissimus dorsi
8. lumbodorsal fascia
9. gluteus medius
10. gluteus maximus
11. biceps femoris
12. popliteal fossa
13. gastrocnemius
14. soleus
15. Achilles tendon

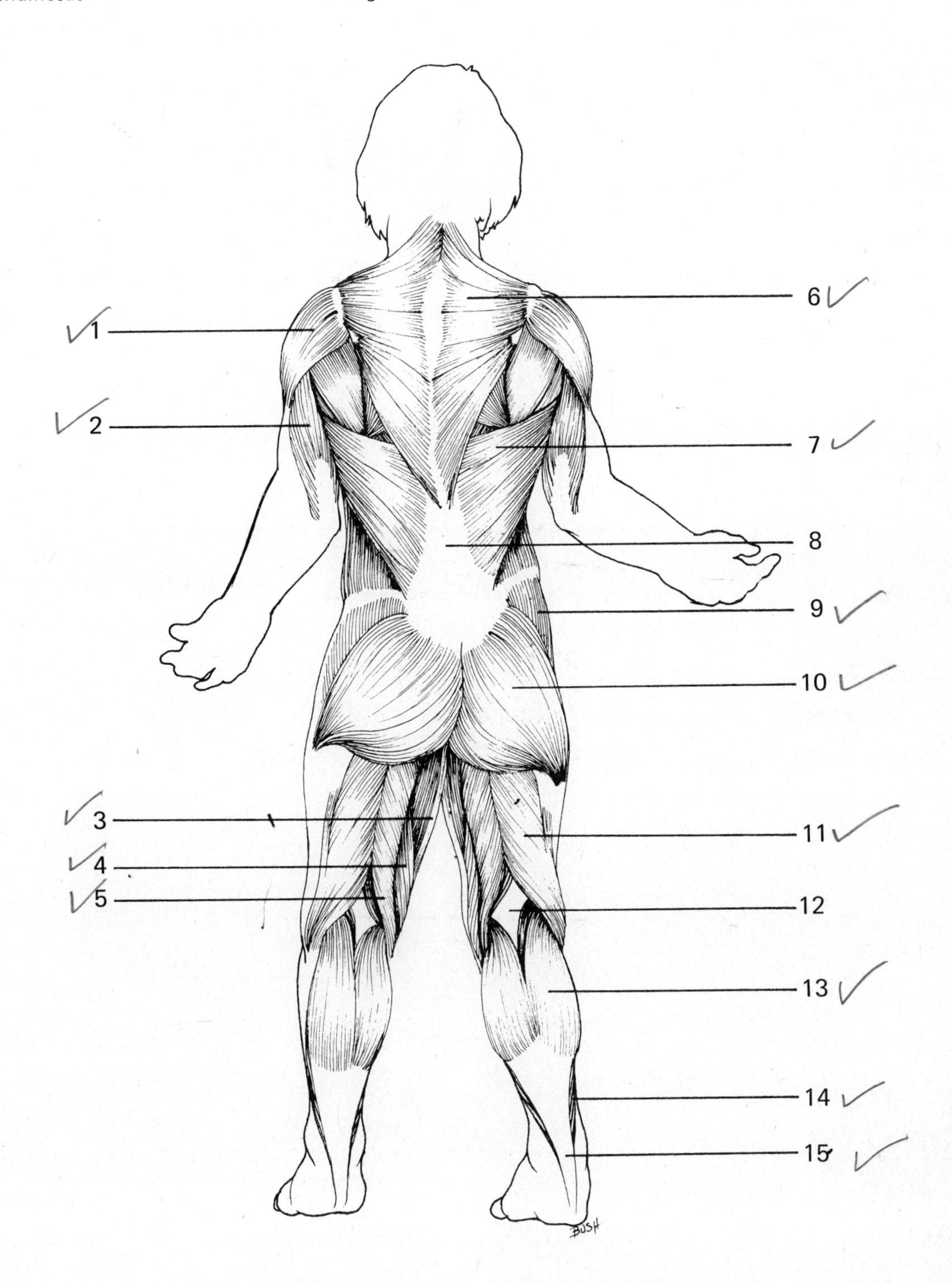

Figure 6-2. Posteri… v of the Muscles

KEY TO FIGURE 6-3

1. temporalis
2. occipitalis
3. zygomaticus
4. masseter
5. frontalis
6. corrugator
7. orbicularis oculi
8. quadratus labii superioris
9. buccinator
10. orbicularis oris
11. sternocleidomastoid

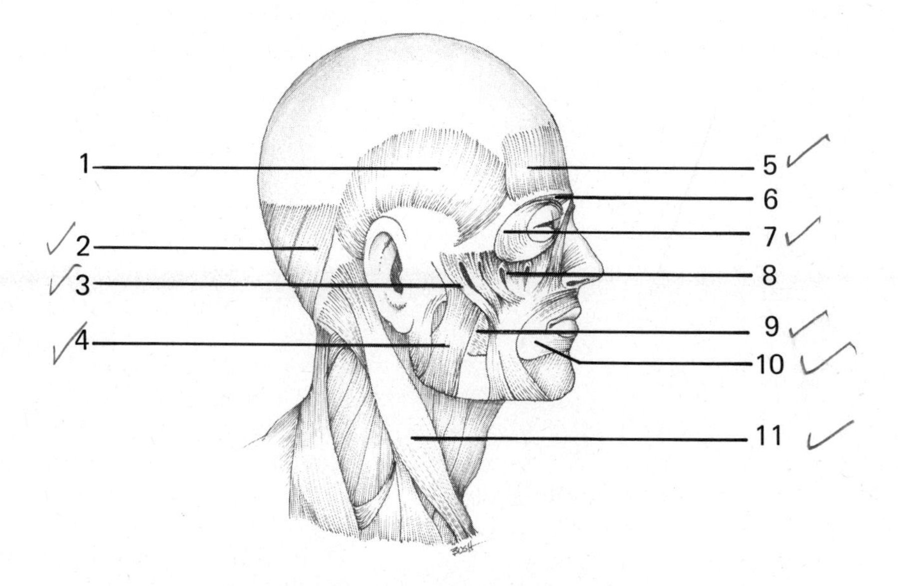

Figure 6-3. Lateral View of Major Muscles of Head and Neck

KEY TO FIGURE 6-4

1. frontalis
2. quadratus labii superioris
3. sternocleidomastoid
4. temporalis
5. orbicularis oculi
6. zygomaticus
7. orbicularis oris
8. platysma

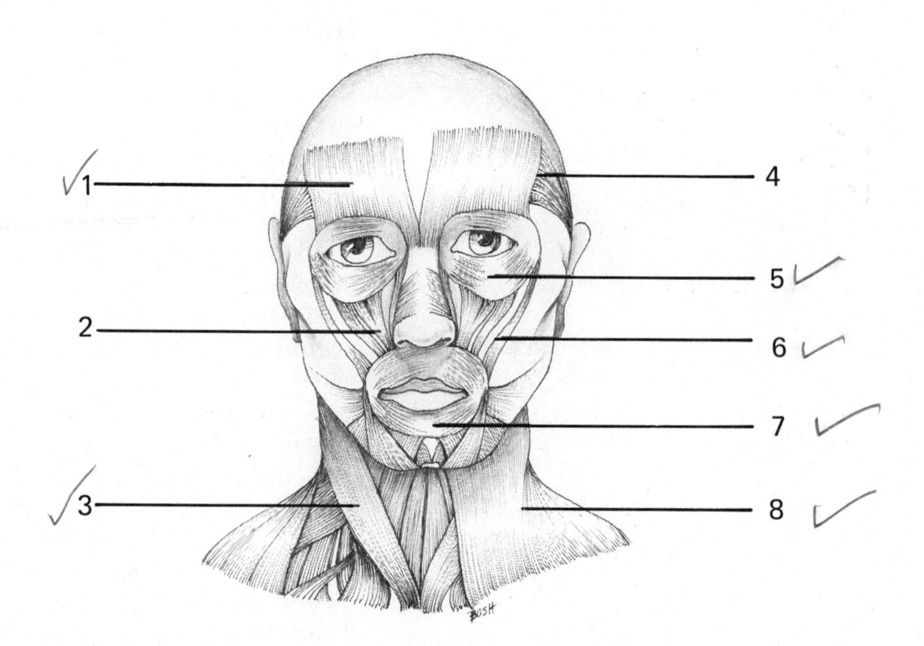

Figure 6-4. Anterior View of Major Muscles of Head and Neck

In each of the following steps, attempt to locate each muscle before making any incisions. It will be necessary to clean the surface of each of the muscles, by gently pulling off or cutting away superficial fascia and fat, in order to see the direction of the muscle fibers. Usually all the fibers of a muscle will run in one direction, while the fibers of an adjacent muscle will run in a different direction.

Do not cut into the fibers of the muscles you are separating. The groups of fibers comprising a muscle are surrounded by deep fascia. This fascia should not be removed from the muscle. If the muscle fibers are visible, it means that the muscle has been cut into, rather than being separated from adjacent muscles.

In order to examine the deep muscles, it is frequently necessary to sever a superficial muscle. Do not bisect a muscle unless directed to do so. To **bisect** means to cut into two parts. When bisecting a muscle, make an incision through it at right angles to the direction of the muscle fibers, and halfway between the origin and the insertion. In order to see the underlying muscles, it is then necessary to **reflect** the superficial muscle (pull the ends back—one toward the origin and the other toward the insertion). If these directions are followed, the origin and insertion of the muscle will be retained.

Procedure

1. To remove the skin from your pig, make a small longitudinal incision through the skin in the middorsal neck region. Continue the incision down the midline of the back to the tail, pulling the cut edges as you proceed. Be careful not to cut through the underlying muscles.
2. Make additional incisions through the skin around the neck, down the lateral surface of each leg, and around the wrist and ankles.
3. Beginning on the back, gradually separate the skin from the underlying muscles. This can be done most easily by pulling the skin away from the body with one hand, using the dull end of your scalpel or your fingers to separate the skin from fascia and muscles. The skin should be left on the head, tail, feet, and genital region (including the scrotum and penis in the male).
4. Note the fine brownish muscle fibers that attach to the undersurface of the skin in the region posterior to the armpit. These are part of the **cutaneous maximus,** a muscle that moves the skin of the pig but is not found in man.
5. Carefully remove the fat and connective tissue from the lateral surface of the left shoulder and neck.
6. Locate the **trapezius** muscle, a thin triangular muscle on the upper back (Fig. 6-5). Separate the anterior and posterior borders of this muscle from the underlying fascia. The trapezius moves the scapula dorsally and posteriorly.
7. The **brachiocephalic** is a thick, band-shaped muscle just anterior to the trapezius, extending from the back of the head behind the ear to the proximal humerus (Fig. 6-5). This muscle protracts the arm. Separate the muscle from the underlying fascia.
8. The **deltoid** extends from the spine of the scapula near the insertion of the trapezius to the proximal humerus next to the insertion of the brachiocephalic (Fig. 6-5). It is difficult to distinguish from adjacent muscles, but if the lateral shoulder region is carefully cleaned, the borders of the deltoid can be seen and freed from the underlying fascia. The deltoid flexes and abducts the arm.
9. The large triangular muscle on the back, posterior to the trapezius, is the **latissimus dorsi** (Fig. 6-5). Free both borders and separate the muscle from the underlying fascia. This muscle retracts the arm.
10. Place the pig on its back. Tie the arms out slightly, being careful not to tear the anterior chest muscles.
11. The **superficial pectoralis** extends from the sternum to the proximal humerus near the insertion of the brachiocephalic (Fig. 6-7). It adducts the arm. Bisect and reflect this muscle.
12. Deep to the superficial pectoralis is the **pectoralis profundus** (Fig. 6-7). It originates along the entire length of the sternum, the fibers extending anteriorly and laterally under the superficial pectoralis. This muscle may appear double since the anterior band of fibers extends in front of the scapula and along its anterior border, while the posterior fibers insert ventrally on the proximal humerus. This muscle adducts and retracts the forelimb.
13. Remove the skin and fascia from the left

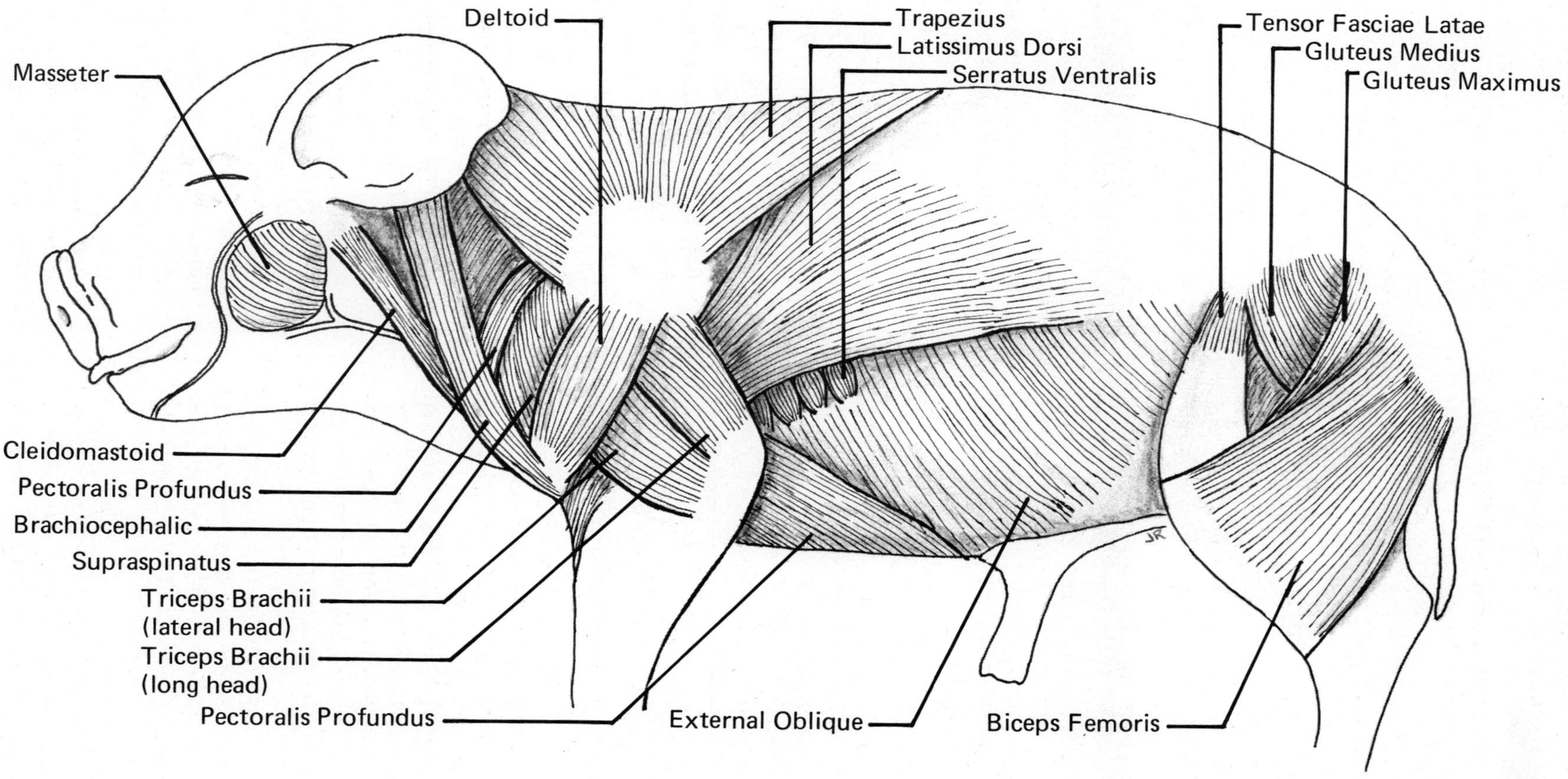

Figure 6-5. Lateral View of Superficial Muscles

side of the neck and face. Be careful not to remove the blood vessels, lymph nodes, and salivary glands in this region.

14. Examine the ventral surface of the neck. Locate the **sternomastoid**, the band-shaped muscle that extends from the sternum diagonally toward the mastoid region of the skull (Fig. 6-7). This muscle passes deep to the external jugular vein. The sternomastoid flexes the head. Free both borders of this muscle.
15. The **sternohyoids** are the narrow pair of muscles that lie along the midventral line of the neck (Fig. 6-7). Free both borders of the left sternohyoid. These muscles extend from the sternum to the hyoid bone, and depress the hyoid.
16. Locate the brachiocephalic muscle again. The thin band-shaped muscle just anterior to it is the **cleidomastoid** (Fig. 6-7). It is attached to the brachiocephalic near the point where the clavicle would be located if one were present, and extends diagonally towards the mastoid region of the skull. Free the cleidomastoid along both borders.
17. The **digastric** muscle is the superficial muscle extending along the inner surface of the mandible (Fig. 6-7). It runs from the base of the skull to the mandible, and is used to open the jaw.
18. The superficial muscle running transversely in the midline and passing deep to the digastric is the **mylohyoid** (Fig. 6-7). It is a thin sheetlike muscle that raises the floor of the mouth.
19. The large muscle mass located posterior to the angle of the jaw is the **masseter** muscle (Fig. 6-5). It extends from the zygomatic arch to the mandible and closes the jaw.
20. Lay the pig back on its side. Bisect and reflect the trapezius muscle (Fig. 6-6).
21. Deep to the trapezius you will see three muscles that extend from the dorsal border of the scapula to the spines of the cervical and thoracic vertebrae. From anterior to posterior these are the **rhomboideus capitis**, a straplike muscle extending forward along the side of the neck, the **rhomboideus cervicis**, and the **rhomboideus thoracis** (Fig. 6-6). These muscles pull the scapula dorsally.
22. Deep to the rhomboideus capitis is a large muscle, the **splenius**, that covers most of the dorsal and lateral surface of the neck. (Fig. 6-6). Its function is to extend the head.
23. Locate the spine of the scapula on its lateral surface. The thick muscle anterior to the spine and passing deep to the anterior end of the pectoralis profundus is the **supraspinatus** (Fig. 6-6). Its fibers extend from the supraspinous fossa to the proximal humerus. The supraspinatus protracts the arm.
24. The muscle posterior to the scapular spine is the **infraspinatus** (Fig. 6-6). Only part of this muscle can be seen since it passes deep to the deltoid. Lift the edge of the deltoid to more completely expose this muscle. The infraspinatus arises from the infraspinous fossa of the scapula and inserts on to the proximal humerus; it rotates the arm laterally.
25. Bisect and reflect the latissimus dorsi. The small muscle along the posterior border of the scapula is the **teres major** (Fig. 6-6). It inserts on the proximal humerus with the latissimus dorsi.
26. Pull the arm and scapula laterally. The large fan-shaped muscle originating by separate slips from the ribs, passing deep to the scapula, and inserting on its dorsal border is the **serratus ventralis** (Fig. 6-6). This muscle is also visible on the lateral surface of the body anterior and posterior to the scapula (Fig. 6-5).
27. The subscapular fossa of the scapula is filled with the **subscapularis** muscle (Fig. 6-7) which inserts on the humerus and rotates it medially.
28. Carefully clean the fascia from the lateral and medial surfaces of the arm (be careful not to cut the blood vessels under the scapula). The large muscle on the posterior arm is the **triceps brachii.** This muscle consists of three heads, all of which insert on the proximal end of the ulna and extend the forearm. The **long head**, located on the posterior surface of the humerus, is the largest (Fig. 6-5). Free both borders of this muscle.
29. Locate the **lateral head** of the triceps just anterior to the long head on the lateral surface of the arm (Fig. 6-5). Free both borders.
30. The **medial head** of the triceps can be seen deep between the long and lateral heads, or on the medial surface of the arm deep to

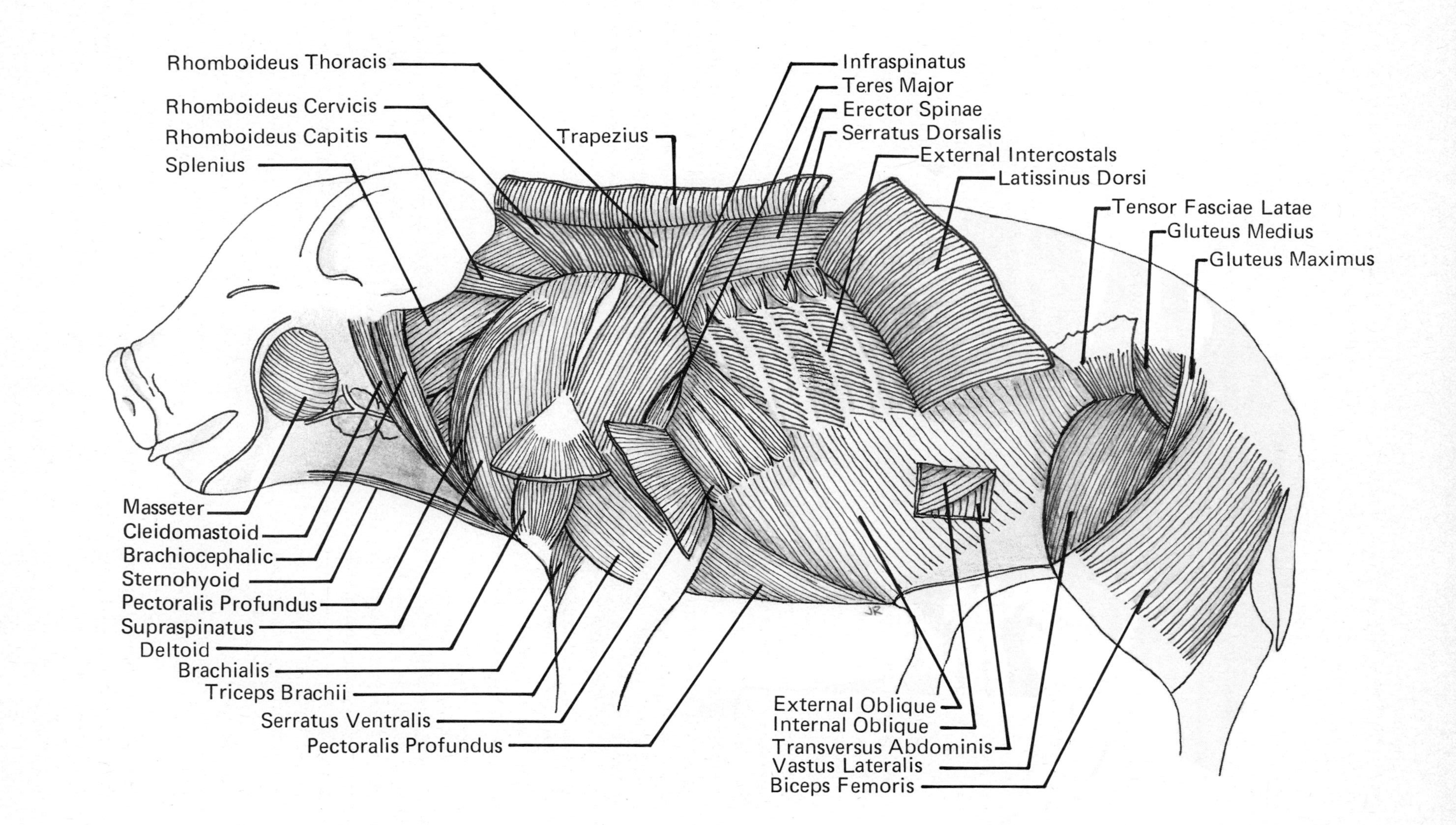

Figure 6-6. Ventral View of Deep Muscles

the large group of blood vessels and nerves (Fig. 6-7).

31. The **brachialis** is located on the ventrolateral surface of the humerus anterior to the lateral head of the triceps. It arises from the humerus and inserts on the proximal end of the ulna. The brachialis flexes the forearm (Fig. 6-7).
32. The **biceps brachii** lies on the ventromedial surface of the humerus. It can be seen by lifting the cut end of the superficial pectoralis (Fig. 6-7). The biceps brachii originates on the scapula and inserts on the radius; it flexes the forearm.
33. The small **coracobrachialis** can be seen just posterior to the proximal end of the biceps brachii (Fig. 6-7). It extends from the coracoid process of the scapula to the proximal humerus, and adducts the arm.
34. Clean off the ventral and lateral surfaces of the trunk. As in man, the lateral abdominal wall of the pig is composed of three layers of muscle. All three muscles help support and compress the abdomen.
35. The outermost of the lateral abdominal wall muscles is the **external oblique** (Fig. 6-6). Notice that the fibers of this muscle pass ventrally and posteriorly to insert along the **linea alba,** the midventral line of connective tissue formed by the union of the aponeuroses of the lateral abdominal wall muscles (Fig. 6-7).
36. Make a shallow incision about 1-1/2 inches long at right angles to the fibers of the external oblique on the lateral surface of the abdomen. Reflect back the cut edges of the external oblique to expose the **internal oblique** (Fig. 6-6). Notice that its fibers run ventral and anterior, or at right angles to those of the external oblique.
37. Make a shallow incision through the internal oblique at right angles to fibers to reveal the **transversus abdominis** (Fig. 6-6).
38. In order to locate the **rectus abdominis,** the longitudinal band of muscle lying lateral to the linea alba, reflect the external oblique. The rectus abdominis extends from the pubis to the sternum and costal cartilages and helps support the abdominal wall.
39. To study the intercostal muscles, raise the posterior border of the origin of the serratus ventralis in order to expose the ribs. The fibers of the **external intercostals** can be seen passing ventrally and posteriorly between the ribs (Fig. 6-6). The external intercostals lift the ribs in inspiration.
40. Bisect one external intercostal muscle to expose the **internal intercostals,** the fibers of which run at right angles to the external intercostals. The internal intercostals depress the ribs in forced expiration.
41. The thin muscle slips that arise from a middorsal aponeurosis and insert on the dorsal part of the ribs make up the **serratus dorsalis** (Fig. 6-6).
42. Reflect the serratus dorsalis by separating it from its origin to expose the **erector spinae** group of muscles (Fig. 6-6). The fibers of this muscle pass longitudinally on either side of the vertebral column and extend the spine.
43. Carefully remove the fat and fascia from the thigh and buttocks region.
44. A small triangular muscle, the **tensor fasciae latae,** can be seen on the anterolateral thigh (Fig. 6-5). It originates from the ilium and inserts on a layer of fascia (fascia lata) near the knee. The tensor fasciae latae tenses the fascia lata and extends the shank.
45. The broad muscle posterior to the tensor fasciae latae and covering most of the lateral surface of the thigh is the **biceps femoris** (Fig. 6-5). It originates on the sacrum and ischium and inserts on the tibia and fascia of the shank. It abducts and extends the thigh and flexes the shank. Free both borders of this muscle.
46. Locate the **gluteus maximus** between the biceps femoris and tensor fasciae latae (Fig. 6-5). Note that it may be partly fused to the biceps femoris.
47. Anterior to the gluteus maximus is the **gluteus medius,** whose fibers pass posteriorly and ventrally and go deep to those of the gluteus maximus (Fig. 6-5). It is a larger muscle than the gluteus maximus in the pig. The gluteus medius originates from the lateral ilium and inserts on the femur. It abducts and extends the thigh.
48. The straplike muscle arising from the ischial tuberosity and inserting on the tibia posterior and medial to the biceps femoris is the **semitendinosus** (Fig. 6-7). It extends the thigh and flexes the shank.
49. Carefully clean the fat and fascia from the medial surface of the thigh. The large, wide, flat muscle covering the posterior

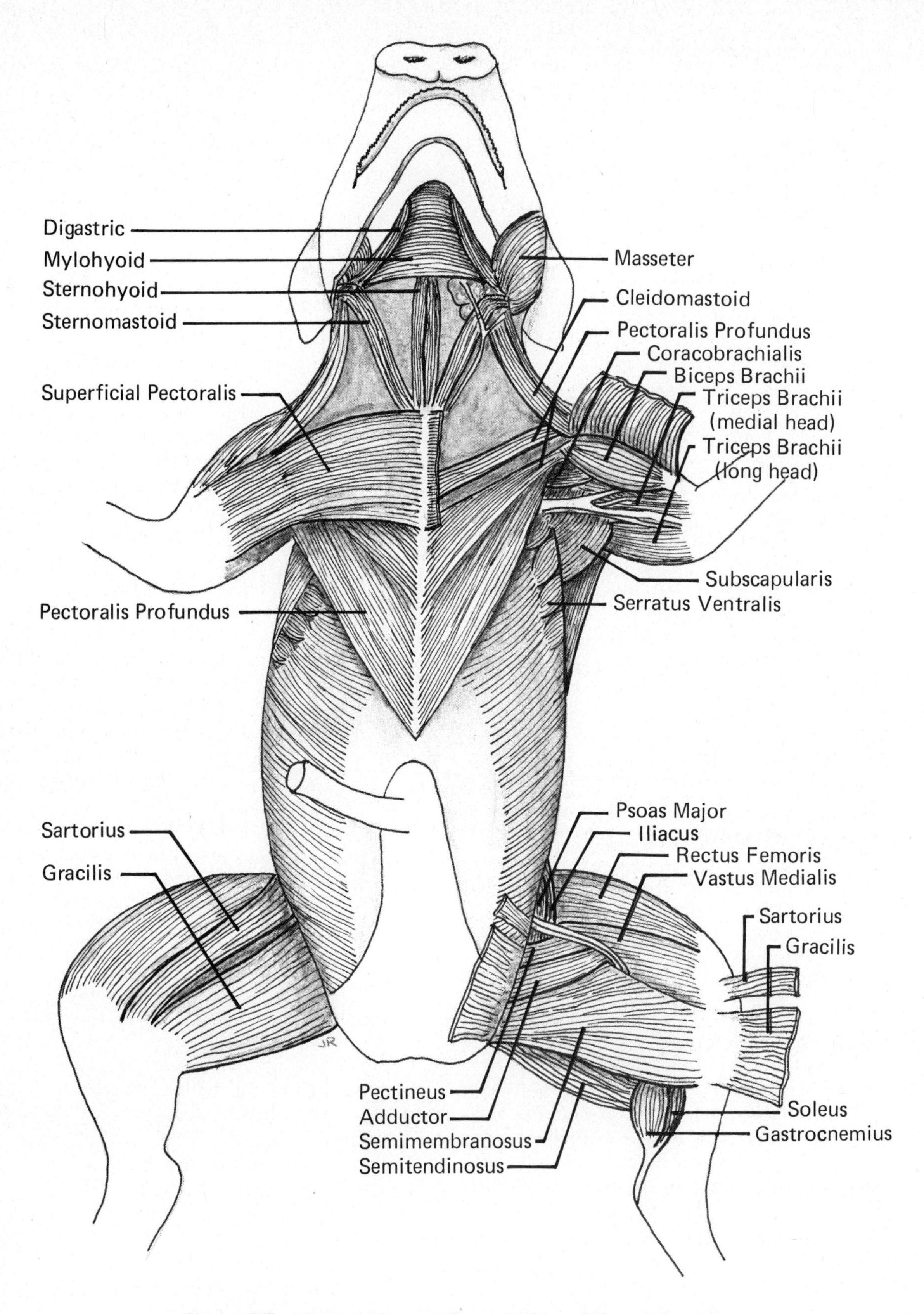

Figure 6-7. Ventral View of Superficial and Deep Muscles

half of the medial thigh is the **gracilis** (Fig. 6-7). Free both borders and then bisect and reflect this muscle. The gracilis originates on the pubis and inserts on the tibia and fascia of the shank. It adducts the thigh.

50. Anterior to the gracilis is the **sartorius**, a small band-shaped muscle (Fig. 6-7). Bisect and reflect the sartorius (this should expose the underlying femoral blood vessels). This muscle helps adduct the thigh.
51. The large muscle deep to the gracilis is the **semimembranosus** (Fig. 6-7). It originates on the ischium and inserts on the distal femur and proximal tibia. It extends and adducts the thigh. The semimembranosus, the semitendinosus, and the biceps femoris form the hamstring group of muscles.
52. The large triangular muscle just anterior to the semimembranosus is the **adductor** (Fig. 6-7). The origin of the adductor is on the ischium and pubis and it inserts on the femur; it adducts the thigh. The adductor muscle in the pig is not subdivided as it is in humans and some other mammals.
53. Locate the smaller **pectineus** muscle anterior to the adductor (Fig. 6-7). It also adducts the thigh.
54. Clean fascia from the region anterior and posterior to the proximal end of the femoral blood vessels. Two muscles can be seen passing deep to the blood vessels, the medial **psoas major** and the more lateral **iliacus** (Fig. 6-7). In man these two muscles fuse to form the iliopsoas. The psoas major arises from the lumbar vertebrae, and the iliacus from the ilium; together they insert on the proximal femur. These muscles rotate and flex the thigh.
55. In order to locate the **quadriceps femoris** group of muscles, bisect the fascia lata and reflect it and the tensor fasciae latae. The large muscle now exposed on the anterolateral surface of the thigh is the **vastus lateralis** (Fig. 6-6).
56. Medial to the vastus lateralis on the front of the thigh is the thick **rectus femoris** (Fig. 6-7).
57. The muscle on the ventral surface of the thigh medial to the rectus femoris is the **vastus medialis** (Fig. 6-7).
58. Carefully separate the vastus medialis from the rectus femoris. Bisect the rectus femoris. The fourth head of the quadriceps femoris, the **vastus intermedius**, can be seen deep to the rectus femoris. The three vastus muscles originate from the proximal femur, while the rectus femoris arises from the ilium; all four insert together by way of the patellar ligament on the anterior tibia. The entire complex extends the shank.
59. The large calf muscle, the **gastrocnemius**, is visible on the posterior leg (shank) (Fig. 6-7). It can be most easily seen by carefully separating the distal ends of the semitendinosus and semimembranosus. The gastrocnemius originates from the distal femur, and inserts by way of the thick **tendon of Achilles** on the calcaneus. The muscle plantar flexes the foot.
60. Examine the medial surface of the calf, ventral to the gastrocnemius, to locate the **soleus** (Fig. 6-7). It arises from the fibula and inserts by way of the tendon of Achilles (together with the gastrocnemius) on the calcaneus. It also plantar flexes the foot.

Self-Test—Muscular System

Directions: See Chapter 1, Self-Test, p. 9.

1. **Striated** muscle contains branched fibers.
2. A skeletal muscle has its insertion on a relatively **immovable** part of the skeleton.
3. The "hamstring" muscles are found **in the shoulder**.
4. The muscle of the calf of the leg that is most prominent when taking a step is the **deltoid**.
5. The insertion of the biceps brachii is on the **ulna**.

6. Muscles of expression include the **orbicularis oculi and orbicularis oris.**

7. When you reach upward for an object, the principal muscle involved in elevating the arm is the **trapezius.**

8. The pectoral muscles bring the arms **across the chest**.

9. Flexion of the trunk is brought about by the contraction of the **vastus lateralis** muscles.

10. In walking, the heel is elevated by the contraction of both the **gastrocnemius and soleus.**

11. The back muscles that extend the vertebral column are the **erector spinae** and the **semispinalis.**

12. The **triceps brachii** is a powerful extensor or antigravity muscle which aids in standing and in extending the leg in walking.

13. The pectoral muscles act in **adduction** of the arm.

14. Extension of the foot is bending the foot **upward**.

15. The membrane surrounding a single muscle cell is called the **sarcolemma.**

16. The cytoplasm of the skeletal muscle cell is called **myoplasm.**

17. An **aponeurosis** is a sheet of white fibrous tissue connecting a muscle to another muscle.

18. A **motor unit** consists of a ventral horn cell in the spinal cord, its axon, and all the muscle fibers it innervates.

19. Women cannot run as fast as men because the thigh bone is attached to the pelvis **at a more oblique angle.**

20. Muscular dystrophy is a **communicable** disease.

21. The broadest muscle of the back is the **trapezius**.

22. The **superior rectus** muscles turn the eyes toward the nose.

23. The deltoid muscle originates on the **occipital.**

24. The muscle filling the ventral surface of the scapula is the **serratus anterior**.

25. The quadriceps femoris **extends** the leg.

26. The **zygomaticus** produces the expression of smiling.

27. The biceps brachii **extends** the forearm.

28. The iliopsoas **extends** the thigh.

29. The biceps femoris, semimembranosus, and semitendinosus form the **quadriceps femoris** group of muscles.

30. The muscle covering the top of the skull is the **epicranius.**

31. The latissimus dorsi **flexes** the arm.

32. The external intercostals **elevate** the ribs.

33. The rhomboids **adduct** the scapulae.

34. The sternocleidomastoid **extends** the head.

KEY

1. cardiac
2. movable
3. on the posterior thigh
4. gastrocnemius
5. radius
6. deltoid
9. rectus abdominis
12. quadriceps femoris
14. downward
16. sarcoplasm
20. hereditary
21. latissimus dorsi
22. medial rectus
23. scapula and clavicle
24. subscapularis
27. flexes
28. flexes
29. hamstring
31. extends
34. flexes

Name ______________________

Results and Questions for Chapter 6

Exercise 6-A. The Gross Anatomy of the Major Superficial Human Muscles

1. Complete the following chart concerning muscle action. In the column headed Chief Antagonist, list a muscle that has an action antagonistic to the major action listed in the first column.

Muscle	*Major Action*	*Chief Antagonist*
Pectoralis major		
Biceps brachii		
Deltoid		
External intercostals		
Gluteus medius		
Gastrocnemius		
Triceps brachii		
Brachialis		
Quadriceps femoris		
Gluteus maximus		
Latissimus dorsi		
Rectus abdominis		

2. Contraction of which muscle will produce each of the following facial expressions?

   a. Smiling ______________________

   b. Frowning ______________________

   c. Pouting ______________________

   d. Contempt and disdain ______________________

3. Which muscle closes the eyes? ______________________

4. Which muscle opens the eyes? ______________________

5. Which muscle wrinkles the forehead to produce the expressions of fright, horror, and surprise?

   ____

6. Which muscle is called the trumpeter's muscle? ______________________

7. Which muscle is called the muscle of kissing? ______

8. Contraction of which muscle will pull down the corners of the lips? ______

9. Name the muscles of mastication. ______

______

10. Contraction of which muscle flexes the head laterally? ______

11. Contraction of which muscle would raise the head from a position of lateral flexion back to the vertical position? ______

12. Name the lateral abdominal wall muscles in order, from the outside in. ______

______

13. What is the difference between a tendon and a ligament? ______

______

14. Define the term "retroperitoneal." ______

______

15. What is the linea alba? ______

16. Define the term "aponeurosis." ______

______

17. Sometimes hard objects that cannot be chewed are found in bacon. What are these hard objects?

______

18. Name two muscles making up the calf of the leg. ______

______

19. What two muscles insert on the calcaneus by way of the tendon of Achilles? ______

______

7

Introduction to Stimulating and Recording Apparatus

Exercise 7-A Faradic Current

Objective

To demonstrate the characteristics of faradic current as a stimulating agent.

Materials

Induction stimulator and wires.

Discussion

Although a muscle can be stimulated in different ways, direct electrical stimulation of the muscle or of the nerve supplying the muscle is used most frequently in physiology laboratories. This is because the procedure is simple, the stimulus can be accurately measured, easily controlled as to strength and duration, closely resembles the physiological process of excitation, leaves the tissue relatively unharmed, and the results are reproducible.

Faradic current is an induced current. Closure of the key in the primary circuit induces a current in the secondary circuit which flows only when there is a change in the magnetic field; therefore, current flows only when the circuit is made or broken. When the key or switch is moved to close a circuit, it is referred to as the "make" ("making" the current); moving the key to open the circuit is known as the "break" ("breaking" the current). With faradic current, the break current is stronger than the make. Faradic current is used more frequently than galvanic current (direct current, a steadily flowing current) in physiology experiments.

The Model 330 Induction Stimulator contains a 1½ volt battery, a key, a vibrating stepup transformer, and a voltage control system. Three voltage ranges of 0-5, 0-50, and 0-500 volts can be selected by a range multiplier switch. Each range is variable from zero to its maximum voltage by means of a potentiometer. The approximate voltage of the **break** stimulus is indicated by the voltage dial setting (0 through 50 volts) multiplied by the range selector position (X.1, X1, X10). The make stimulus voltage is approximately 1/10 of that indicated by the voltage dial setting multiplied by the range selector position. The Model 330 Induction Stimulator is illustrated in Figure 7-1.

There are two pairs of terminals on the Induction Stimulator. A signal magnet may be

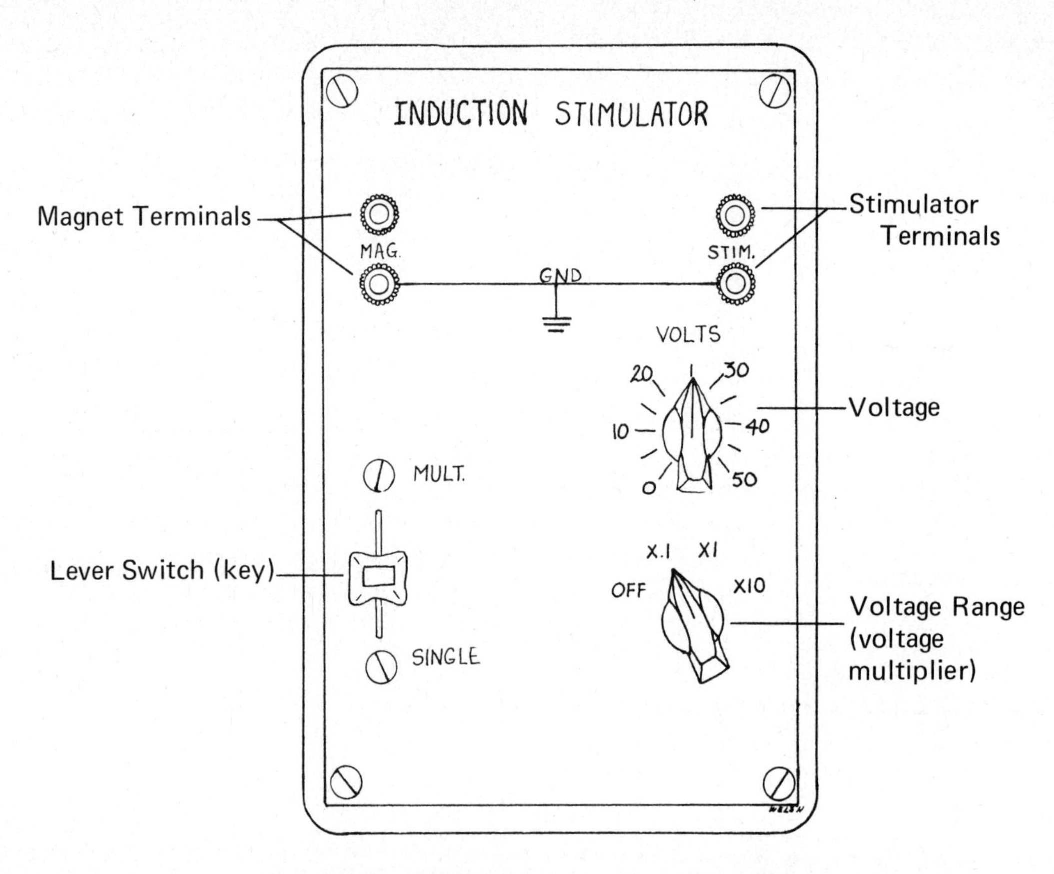

Figure 7-1. Model 330 Induction Stimulator

connected to the two terminals marked MAG. on the Model 330 Induction Stimulator. The signal magnet circuit delivers a fixed 1½ volts suitable for driving a signal magnet when the key is depressed. Insulated copper wires (with the ends scraped bare of insulation) or platinum electrodes may be connected to the other two terminals indicated by STIM. The variable voltage obtained is suitable for stimulating muscle or nerve.

Procedure

1. Connect two wires to the two terminals of the induction stimulator marked STIM. (Loosen the screw tops of the terminals and insert the wires in the holes in the terminals; retighten the screw tops.) Place the other ends of the wires on the tongue. Both wires must touch the tongue, but they should not be touching each other.
2. Turn the voltage range selector to X.1 and the voltage dial to 0.
3. At each position of the voltage dial indicator, pull the lever or key to SINGLE. Any current felt with the lever in this position is the **make** current. With the electrodes still in position, release the key. Any current now felt is the **break** shock.
4. Make and break the circuit with the voltage dial on 0. Did you feel a shock on either make or break? Should you feel a shock at these settings?
5. Increase the voltage until a shock is felt on the break. (It may be necessary to move the range selector to X1 or even X10.) The first break shock felt is called the **liminal** or **threshold** stimulus. Record the voltage, and in parentheses record the range used and the voltage dial reading under Results and Questions.
6. Continue increasing the voltage until a shock is also felt on the make. It may be necessary to remove the electrodes from the tongue before breaking the circuit in order to avoid sending a strong current through the tongue on the break. Record the voltage necessary for the liminal make shock under Results and Questions.
7. Note whether a current is felt during the time the key is depressed (between make and break).
8. Remove the electrodes from the tongue.

Have the voltage controls set where strong make and break shocks were felt. Depress the key.

9. Replace the electrodes on the tongue for 5 seconds and note whether any shock is felt.
10. Remove the electrodes from the tongue, then break the circuit.
11. Was a make or break shock felt in Steps 8-10?
12. In many experiments it is necessary to use constant strength current. Since the make and the break are not of the same intensity, it is necessary in such experiments to short circuit one or the other in order to elicit a number of shocks of the same intensity in fairly rapid succession.
13. In order to short circuit the make, set the voltage control to values that produced a good break and make shock. Place the electrodes on the tongue. Set the voltage control at 0. Make the circuit. Return the voltage control to the original setting. Break the circuit. If directions have been followed, a break shock should have been felt but not the make.

Exercise 7-B Tetanizing Stimulation

Objectives

To learn the characteristics of tetanizing current as a stimulating agent, and to compare the threshold stimulus of this with that obtained with single induction shocks.

Materials

Induction stimulator and wires.

Discussion

Tetanizing stimulation is used when it is desired to apply stimuli to a muscle in rapid succession. The shocks obtained with the induction stimulator using the Procedure of Exercise 7-A were single induction shocks. If the key is raised toward MULT. (see Fig. 7-1) the circuit is made and broken automatically 50 times per second, thus producing a frequency of stimulation of 50 per second.

Procedure

1. Using the same subject for this experiment as in the experiment with single shocks, determine the liminal stimulus for tetanizing current in the same manner as with the single induction shocks. A buzzing sound should be heard when the key is raised towards MULT.
2. Fill in the table under Results and Questions.

Exercise 7-C. Use of the Kymograph and Ink Recording Apparatus

Objectives

To learn the use of the kymograph and ink recording apparatus.

Materials

Kymograph; flat base stand; double clamp; a signal magnet with ink adapter; string; paper; scissors; tape.

Discussion

The function of the kymograph is to record motion and to make a permanent record of an experiment. It consists of a drum, a mechanism to move the drum, and a mechanism to change the speed of movement of the drum. The motion is transferred to a lever or stylus (pen), one end of which rests against a surface (kymograph paper) moving at a known rate of speed. The kymograph paper is wrapped around the drum which, when in operation, continuously rotates away from the point of contact with the stylus (see Fig. 7-2). The stylus is so constructed that motion transferred to it as a result of an experiment (e.g., the contraction of a muscle) causes it to move in directions

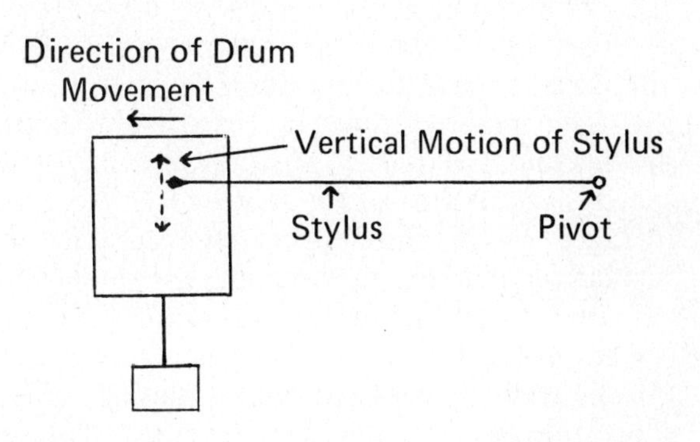

Figure 7-2. Movement of the Kymograph Drum

perpendicular to the motion of the paper (see Fig. 7-2). The stylus must be so aligned that it is in constant contact with the paper, otherwise a complete tracing of the process being recorded will not be obtained. Anything less is of limited value.

Figure 7-2 illustrates the direction in which the kymograph drum revolves, as well as the vertical motions made by the stylus (pen).

The kymograph described is the electric Model 600-000 (by Harvard Apparatus Co.), which consists of a gear train designed to transmit 12 separate rates of speed to the drum (multispeed transmission). When actuated by the **selector knob**, a selector gear comes into contact with one of the other gears (which are in constant motion while the kymograph is being operated). Depending on the position selected for the knob, the drum is either speeded or slowed in its revolving action.

Among the 12 possible selections (selector knob positions 1-12), drum speed varies from one revolution per second to one revolution per 5,000 seconds. Any rate of drum rotation selected may be repeatedly (and accurately) reproduced.

The reduction factors, speed of drum (seconds required to complete a revolution), as well as the number of centimeters the drum moves per second in a given amount of time are indicated by the following table:

events, such as many muscle twitches, is to be recorded.

Procedure

1. Screw the long steel drum-supporting rod into the hole located near the back on the upper surface of the kymograph.
2. Place the drum and its hollow cylinder, with the screw or lever on top, over this rod and rotate it until the pin at the lower end of the hollow cylinder fits into the depression provided for it on the kymograph.
3. Practice raising and lowering the drum. In order to do this, loosen the screw holding the drum to the rod, move the drum to the desired position, and retighten the screw.
4. To change speeds of the kymograph, the selector knob must be pulled upward (disengaging the selector gear) and turned to the desired position and released. **Do not attempt to change speeds without first pulling out the selector knob.** While all the gears can be shifted when the motor is running, it is recommended that the motor be shut off when shifting to positions 1, 2, or 3.
5. Practice starting and stopping the kymograph, and changing drum speeds.
6. In order to **rotate the drum by hand**, the motor may be left on, but the selector knob must be at position N (**neutral**).
7. For each selector knob position, the number of cm/sec the drum moves is known (see Table 7-1). For example, at knob position 7 the drum moves 0.5 cm/sec. Therefore, a recorded event that occupies 10 cm on the drum (easily measured with a ruler) must have taken 20 seconds.

Table 7-1
Kymograph Drum Speeds

Knob:	*1*	*2*	*3*	*4*	*5*	*6*	*7*	*8*	*9*	*10*	*11*	*12*
Reduction	1	.4	.2	.1	.05	.02	.01	.005	.002	.001	.0005	.0002
Sec/rev.	1	2.5	5	10	20	50	100	200	500	1000	2000	5000
Cm/sec	50	20	10	5	2.5	1	.5	.25	.1	.05	.025	.010

The drum of the kymograph is 50 cm in circumference. When the knob is at position 1, the drum revolves once (a distance of 50 cm) per second. At knob position 6, the drum speed is reduced to a rate of only 1 cm/sec, and it takes 50 seconds for it to complete one revolution. Therefore, the lower the **number** of the knob position, the more **rapid** the rate of revolution of the drum. The fast drum speeds (1 or 2) may be used when timing single, short-term events such as the muscle twitch. Slower speeds are used when a sequence of

The following equation may be used to determine the duration of an event:

$$E = \frac{L}{S}$$

E = Duration of the event in seconds.
L = Length of the recorded events in centimeters.
S = Drum (paper) speed in centimeters per second.

8. **Always** record the position of the selector knob used, in case the drum speed is to be used in the calculations.
9. The glazed side of the kymograph paper is used for ink writing. Lay a piece of kymograph paper on the table with the glazed surface facing downward.
10. Cut a piece of string approximately 1 foot long. Remove the drum and support cylinder and lay the drum in the center of the paper, **base towards the student**. Place the string on top of the drum so that the ends extend over both the upper and lower edge of the drum. (Do not let the drum roll onto the floor or allow anything to mar its surface. It is essential to have a true smooth surface.)
11. Wrap the paper tightly around the drum so that the junction of the ends of the paper is over the string. Tape the **left margin of the paper over the right**. This prevents the pen from dragging on the seam.
12. Shake the drum slightly to make sure the paper does not slide. If it does move, remove the paper and reapply it firmly enough to prevent the sliding.
13. To remove the paper from the drum at the conclusion of the experiment, move the string carefully to one side of the seam. To avoid dropping the paper when the junction is torn, press the string and at least one edge of the paper firmly against the lower part of the drum with the left hand. Now, grasp the upper part of the string with the right hand and carefully tear the paper by pulling the string.
14. The ink writing system consists of a stainless steel pen, adapter, ink tube, bottle, and bottle clamp. The bottle is of soft polyethylene, which allows the ink to be forced to the pen tip when the bottle is squeezed and the cap is tight. The fine pen tip provides a thin line and, because of its small bore, it ensures adequate capillary action.
15. In order to obtain good results with the kymograph, always observe the following precautions:
    a. Set up the equipment so the stand with its attached levers (recording apparatus) is to the right of the kymograph. This allows the drum to glide away from the pen point—not toward it.
    b. Make certain that the semicircular base (expanded portion) of the stand is towards the drum to prevent the stand from falling over when heavy equipment is used.
    c. The writing on the surface of the kymograph paper should always face the operator.
    d. The pen must always be horizontal and make a tangent with the recording surface (drum). This means it must touch the surface at right angles in the precise center of the front of the drum.
    e. The pen (or any other stylus used) must make a distinct mark over the entire surface covered by its vertical (up and down) movements. This means the pen must be in **constant** contact with the surface of the paper. This contact should not, however, be so firm that the vertical motion is restricted. To obtain these results, all levers must be exactly horizontal and at right angles (i.e., tangent) to the drum surface. **The ink adapter must also be in a vertical position.**
    f. Always begin each experiment (unless directed otherwise) with the **drum high** on its axis and the **recording equipment low**. Begin each tracing about one inch above the edge of the paper, then lower the drum when additional records are needed. The drum is lowered in preference to raising the position of the recording apparatus.
    g. Always begin the tracing about one inch to the right of the overlapping part of the paper. Any record on the overlapping portion will be lost when the paper is removed from the drum.
    h. Before removing the record from the drum, label it as fully as possible and show it to the instructor for approval. It is permissible to cut up the records so each student has a copy. The unused portions can then be discarded. There-

fore, label only good results, and place the student's name and experiment number on that part of the record.

i. Always paste the record in the appropriate position under Results and Questions. The record must always be discussed. If it is necessary for the entire group to use one tracing, indicate which student's results contain the tracing.

16. Mount the signal magnet on the flat base stand as directed in Exercise 7-D, making certain that the handle is on the **beveled** area in the double clamp and that it is oriented in accordance with the above instructions.
17. Connect the ink adapter with the shortest pen to the stylus of the signal magnet as directed in Exercise 7-D. If the ink bottle is less than half-full, use a funnel to refill it.
18. Clamp the bottle in a vertical position on the handle of the signal magnet. The ink level in the bottle should be at approximately the same level as the pen tip. If the ink does not flow out with sufficient pressure, the level of the bottle should be raised.
19. To protect the surfaces of laboratory tables, always place a paper towel below the kymograph drum before starting the ink flow. The towel may be removed during the experiment, but should be replaced when removing the ink from the pen.
20. To start the ink flow, screw down the bottle cap and gently squeeze the body of the bottle. The cap is then loosened while the bottle is under pressure. It may be necessary to occasionally squeeze the bottle to maintain the flow of ink.
21. At the end of an experiment, the ink should be withdrawn from the pen by reversing the procedure. With the bottle cap still slightly loosened, squeeze the bottle, tighten the cap, then release the pressure. The ink should flow back into the bottle. Next, detach the ink bottle and attach a second ink bottle filled with water. Force water through the tubes and pen until all the ink is removed. This procedure will prevent dried ink from clogging the pen.
22. Practice starting and reversing the flow of ink.
23. Position the pen so that it is at right angles to the recording surface. Use the least pressure on the tip consistent with a good recording.
24. Start the drum revolving and check the appearance of the line. Answer the questions under Results and Questions.

Exercise 7-D The Signal Magnet

Objectives

To demonstrate the advantages and disadvantages of the signal magnet as a timing device; to demonstrate the advantages and disadvantages of the signal magnet as a device for the indication of the precise time of application of a stimulus.

Materials

Kymograph; signal magnet with ink adapter; induction stimulator; stand; double clamp; wire; and a stop watch or watch with a second hand.

Discussion

The signal magnet is used to record the time of application of a stimulus, to record the opening and/or closing of a circuit, to record a

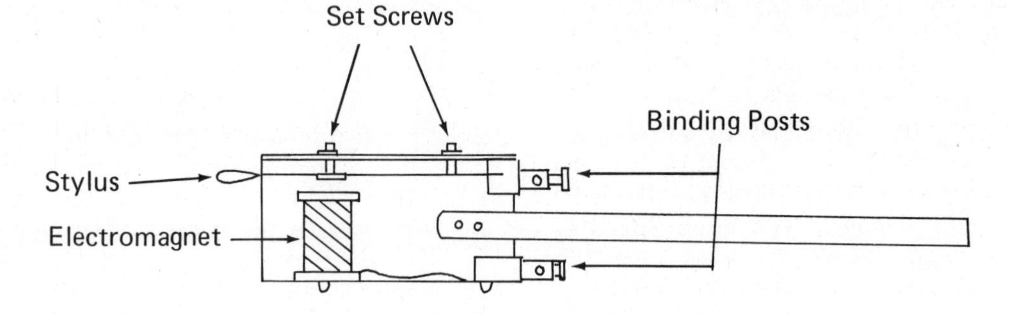

Figure 7-3. The Signal Magnet

change in experimental procedure, or it can be used as a timing device.

The signal magnet consists of an electromagnet (a coil of wire) and a vibrating, spring loaded armature extended out to form a stylus (see Fig. 7-3).

When current is applied ("make" = M), the stylus deflects, thus producing a permanent record of the event. When the circuit is broken ("break" = B), the stylus moves upward to its original position. Therefore, on a moving drum, the record appears as follows:

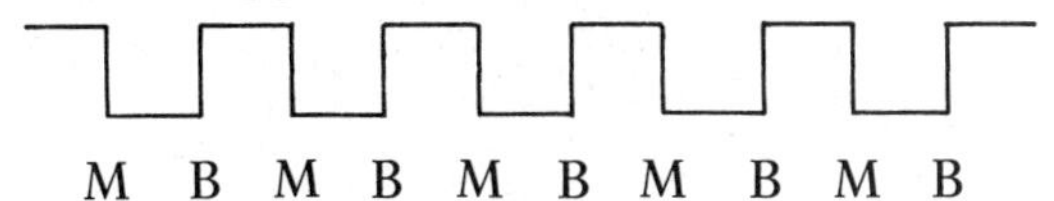

M B M B M B M B M B

If the signal magnet is used as an independent time base marker (independent of the induction stimulator), a 1½ volt battery is required to operate it. Generally, however, the signal magnet is used with the induction stimulator to record the time of application of a stimulus to a heart or a skeletal muscle, because it writes at the same instant the current passes to the muscle. This provides information as to the exact time the structure received the stimulus.

The signal magnet has two Allen head **set screws** on the upper surface for making adjustments. The set screw near the vibrating point is used to adjust the excursion of the point. The other set screw is used to adjust the frequency response (the latter should be touching the armature gently). Both adjustments have been set at the factory for 1½ volts and a frequency of 60 cycles per second. However, if the signal magnet stylus does not move, it may be necessary to adjust them. An Allen wrench is provided for making any necessary adjustments.

Procedure

1. Mount the signal magnet on the stand so the coil of wire faces the operator.
2. Connect two wires from the induction stimulator terminals marked MAG. to the binding posts on each side of the handle. Run the wire through the opening and then tighten the screw.
3. Make and break the current to determine whether the signal magnet is working properly. When the current is applied, the stylus should descend. When the current is withdrawn (broken), the stylus should return to its original position.
4. Now connect the ink adapter with the shortest pen to the stylus of the signal magnet by means of a small piece of **colophonium cement** softened by the heat of the fingers. Check again to make sure the pen falls and rises with the making and breaking of the current. The amount of wax may have to be adjusted so that the pen works properly.
5. Place the pen tangent to the drum and check it to be certain that it is writing correctly. It is important that the pen does not press too tightly against the drum, since this prevents rapid excursions of the pen.
6. Obtain an individual record for each student in the group for each part of the experiment.
7. At a drum speed of 6, hold the key down ("make") for 1 second, then release ("break") for 1 second. The intervals should be timed with a stop watch and the sequence should be continued for at least 10 seconds. Label the record with the time in seconds.
8. Repeat the procedure, first at a drum speed of 11 (for at least 20 seconds), then at a drum speed of 2 (which will result in just one make and one break mark on the kymograph paper). At drum speed 2, start the drum, make the circuit, break the circuit one second later, then immediately rotate the semicircular stand away from the drum so that the pen is removed from the paper. This prevents overlapping of the recording. Label both records.
9. Answer all questions and place all records under Results and Questions.

Exercise 7-E
The Tuning Fork

Objectives

To determine the advantages and disadvantages of the tuning fork as a timing device, and as a device to indicate the time of application of a stimulus.

Materials

Flat base stand; tuning fork with ink adapter; tuning fork starter; double clamp; and a kymograph.

Discussion

The tuning fork is used to measure the duration of events that occupy a short period of time. It is most useful with short-lived, single events, such as a muscle twitch, heart beat, or a single breath.

The tuning fork vibrates 100 times per second. An ink pen is attached to record the vibrations. The time elapsed from the crest of one vibration to the crest of the next (a-b, in the following diagram) is 1/100th second. The number of vibrations (a-b) should be multiplied by 0.01 to obtain the total time for any event measured.

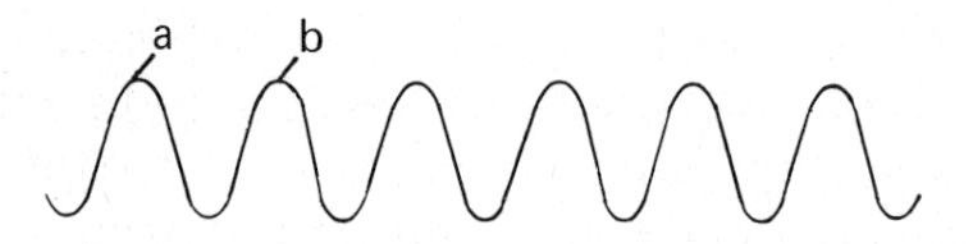

The setup for using a tuning fork is shown in Figure 7-4. Since the tuning fork is difficult to use, the known drum speed will ordinarily be used to calculate the duration of an event. The results obtained should be the same when either method is used.

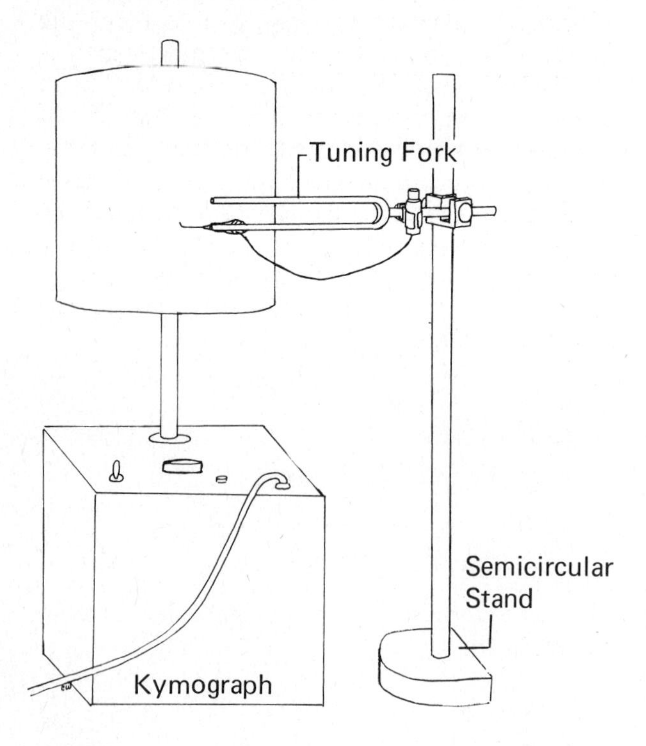

Figure 7-4. Diagram for Tuning Fork Setup

Procedure

1. Each student must obtain an individual record of the results in this experiment at drum speeds 6, 4, 2, and 1.
2. Attach the tuning fork to the semicircular stand by means of a double clamp (make sure the rod fits in the beveled portion of the clamp). Attach the pen to the back of the bottom prong of the tuning fork. The prongs of the tuning fork must be perfectly horizontal in relation to the drum, and the pen tangent to the kymograph.
3. Set the drum speed at 6 and start the flow of ink.
4. Near the end and on the very edge of the tuning fork place the tuning fork starter (the starter has a small handle attached to a U-shaped part which is used to compress the prongs of the tuning fork). A quick, sideways blow will knock the tuning fork starter off the tuning fork, causing it to vibrate. Practice this procedure several times, with the pen away from the drum. Do not allow the starter to hit the tuning fork after it is deflected.
5. Begin the writing operation 1 inch to the right of the taped portion of the paper. Start the drum and make certain that the pen is writing on the drum.
6. After the pen writes a base line approximately 1 inch long, deflect the tuning fork starter. Stop the drum after the tuning fork has ceased vibrating. Repeat at drum speed 4.
7. The drum moves so rapidly at the faster speeds (2 and 1) that it may be necessary to have the tuning fork vibrating before starting the drum. To prevent the record from overlapping at speeds of 2 and 1, be sure to stop the drum after one revolution. This can be done most easily, at fast speeds, by rotating the base of the semicircular stand away from the drum and, thus, disengaging the pen.
8. Label each of the four tracings with the drum speed. Label each record in which individual vibrations are visible with the time in 1/100th seconds.
9. Answer the questions and put your record in the appropriate place under Results and Questions.

Exercise 7-F
Use of the Physiograph

Objectives

To become familiar with the operation of the physiograph and the stimulator; to determine the characteristics of the current output from the stimulator; to calculate the duration of an event.

Materials

Physiograph with stimulator; photoelectric pulse pickup; pin electrodes; stimulator output extension cable (and transducer coupler with the DMP-4B).

Discussion

The physiograph is an instrument used extensively for the measurement and recording of physiological data. It is essentially a group of independently operated recording channels mounted in a main frame assembly which provides power distribution, control and monitoring circuitry, and facilities for mounting accessory plug-in modules. It can be operated as a single channel recorder for a single function, or it can be used to record simultaneously a variety of functions, such as heart beat, respiration, and blood pressure.

A typical recording channel consists of a transducer, an amplifier, a pen motor, and a recording stylus (see Fig. 7-5).

Transducers are devices which convert one form of energy into another form of energy. For example, a loudspeaker is a transducer which converts electrical energy into sound. In the operation of the physiograph, the transducer converts the physiological event into a proportional electrical signal. Transducers are available to convert blood pressure, muscle pull, and heart beat into proportional electrical signals. Some transducers connect directly to the input connector on the channel amplifier. Others operate in connection with preamplifiers. Typical transducers are the myograph, photelectric pulse pickup, and the heart sounds microphone.

Preamplifiers are used in place of transducers or with transducers to measure physiological potentials. Biopotentials are very small; consequently a preamplifier must be used to amplify the transducer output signal to match the input requirements of the amplifier. Sometimes the transducer and preamplifier are combined into the same unit. Typical preamplifiers are the impedance pneumograph, hi-gain preamplifier, and the electrosphygmograph.

A **coupler** is used in the DMP-4B model to match the transducer or the signal from the subject to the channel amplifier. It is also used to vary the sensitivity and to calibrate the system. Some of the transducers, preamplifiers, and couplers used in this manual are illustrated in Figures 7-6A to 7-6I.

Since transducers are normally designed to absorb as little energy as possible from the preparation measured, the output signal is small and must be amplified considerably in order to drive the pen motor. The function of the **amplifier** is to provide the required increase in signal strength, while retaining the proportional relationship of the electric signal and the physiological activity which it represents. The output of the amplifier drives the pen motor. The pen is numbered to correspond to the amplifier channel.

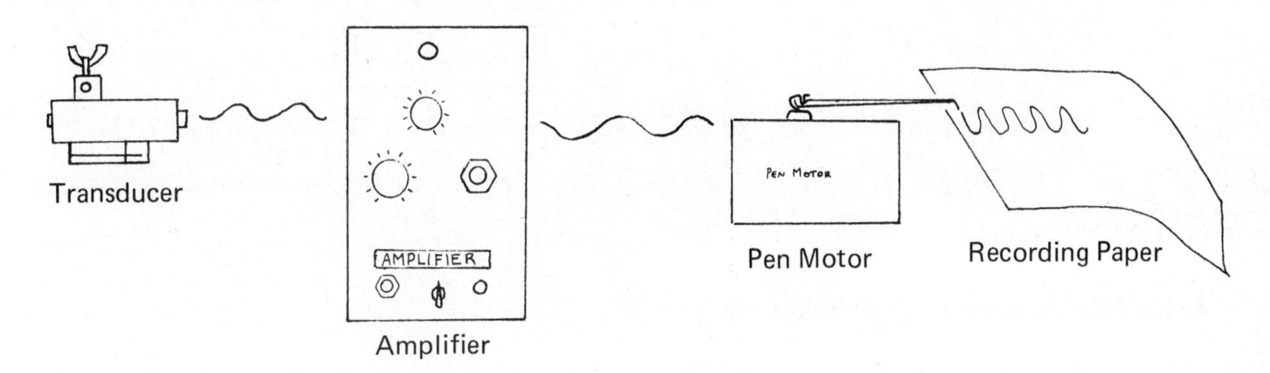

Figure 7-5. Physiograph Recording Channel

The **pen motor** converts electrical energy into rotary motion. It receives the electrical signals from the amplifier and drives the recording stylus. The excursions of the stylus across the moving chart are proportional to the physiological activity and they provide a permanent record of the experiments.

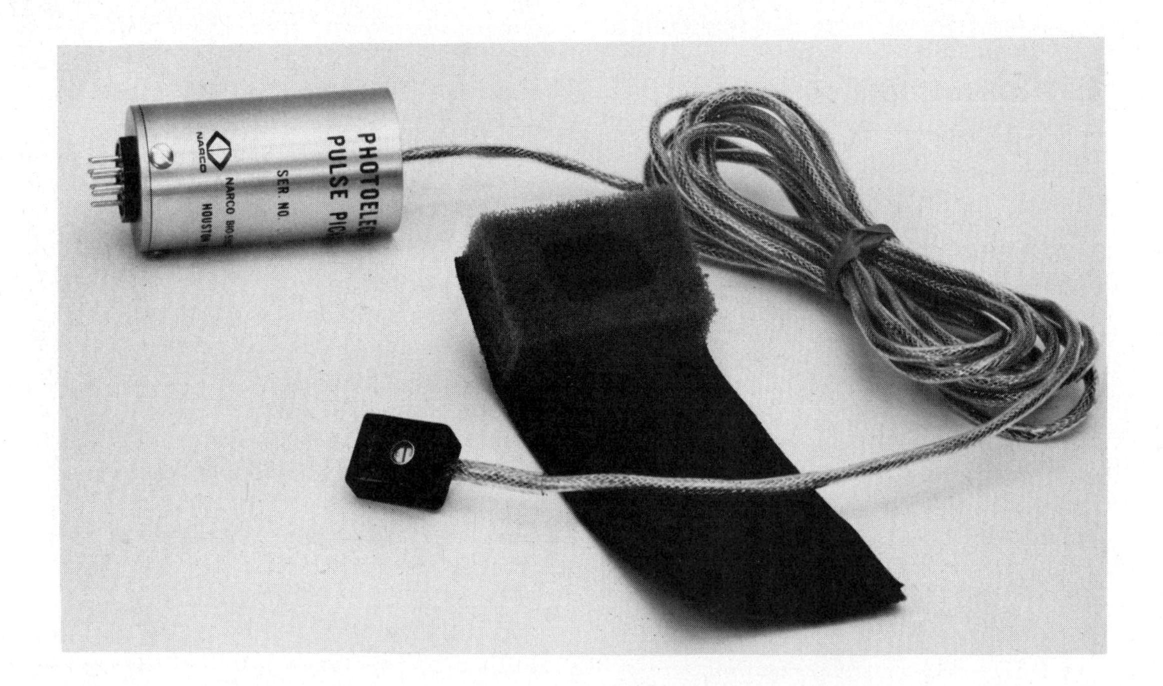

Figure 7-6A. TRANSDUCER—The Photoelectric Pulse Pickup to be used with DMP-4A and Physiograph Four-A

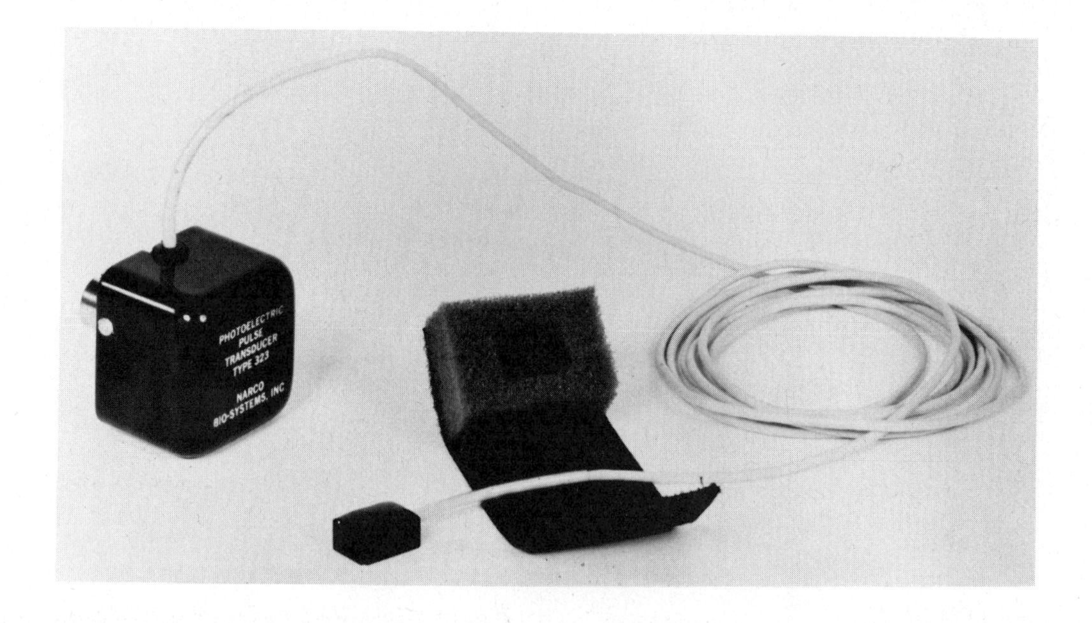

Figure 7-6B. TRANSDUCER—Photoelectric Pulse Pickup to be used with DMP-4B

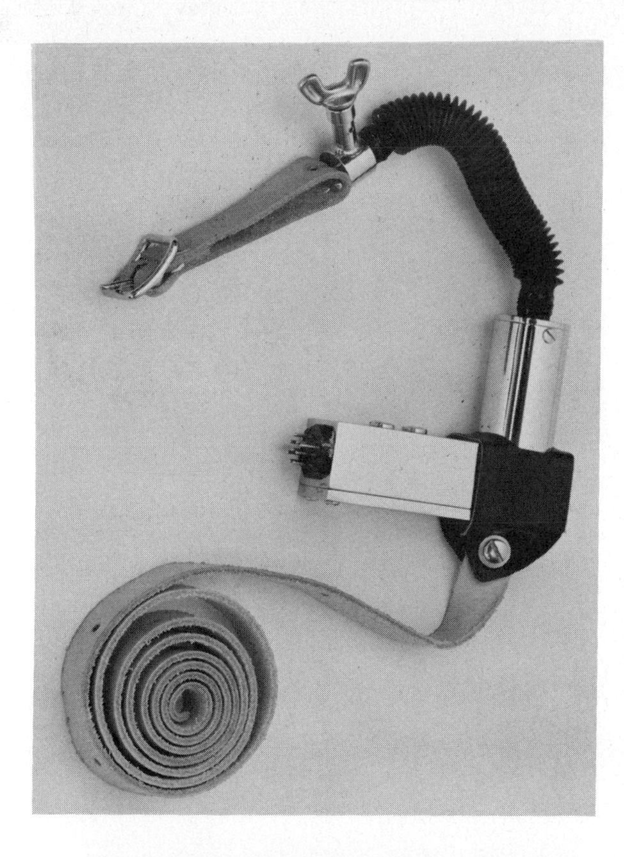

Figure 7-6C. TRANSDUCER—Bellows Pneumograph to be used with either the DMP-4A or DMP-4B Models

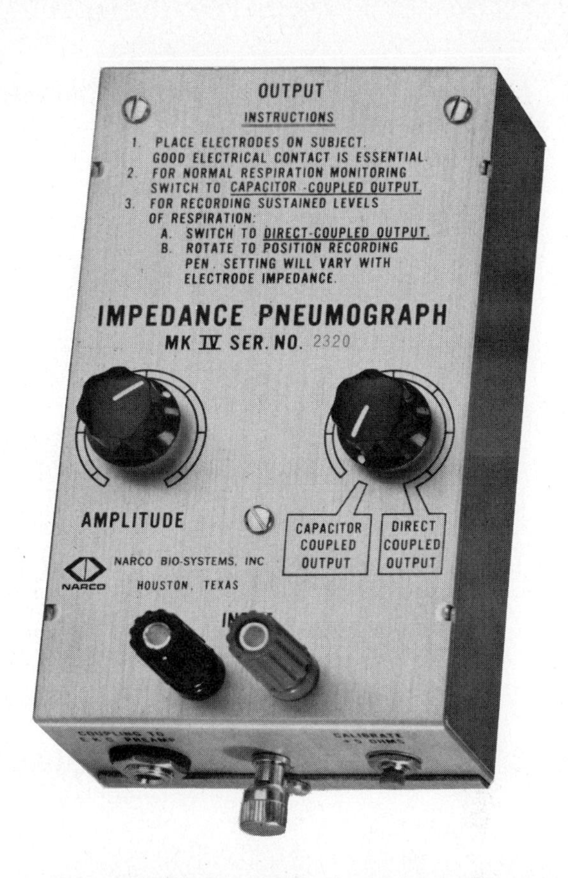

Figure 7-6D. PREAMPLIFIER—Impedance Pneumograph to be used with DMP-4A

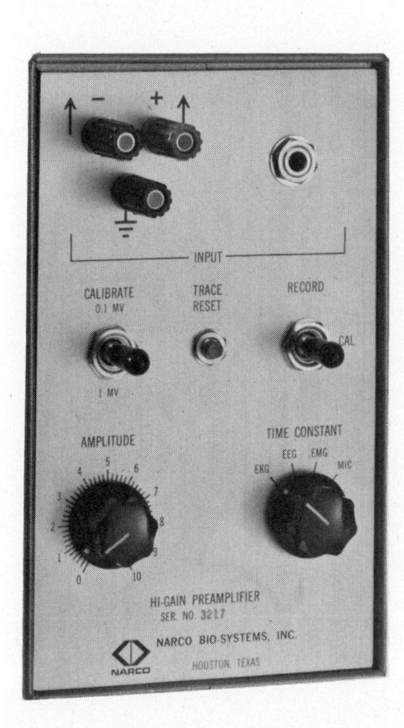

Figure 7-6E. PREAMPLIFIER—Hi-Gain Preamplifier to be used with DMP-4A

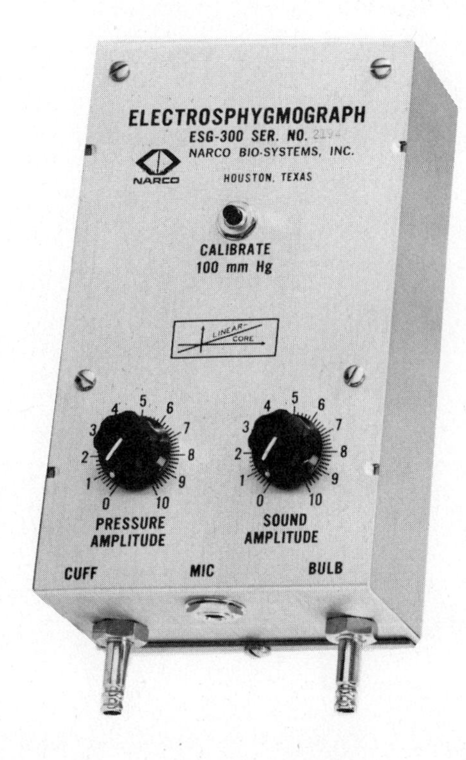

Figure 7-6F. PREAMPLIFIER—Electrosphygmograph to be used with DMP-4A

Figure 7-6G. COUPLER—Electrosphygmograph Coupler to be used with DMP-4B

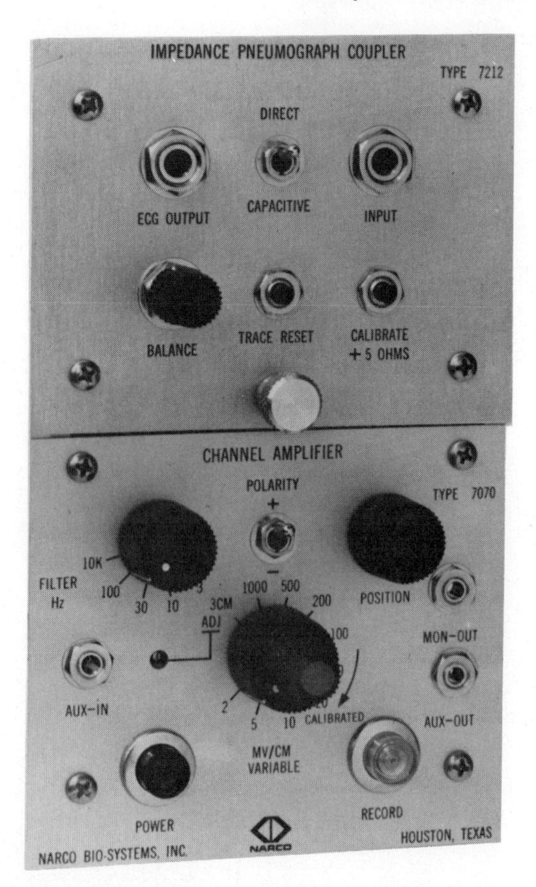

Figure 7-6H. COUPLER & CHANNEL AMPLIFIER—Impedance Pneumograph Coupler to be used with DMP-4B

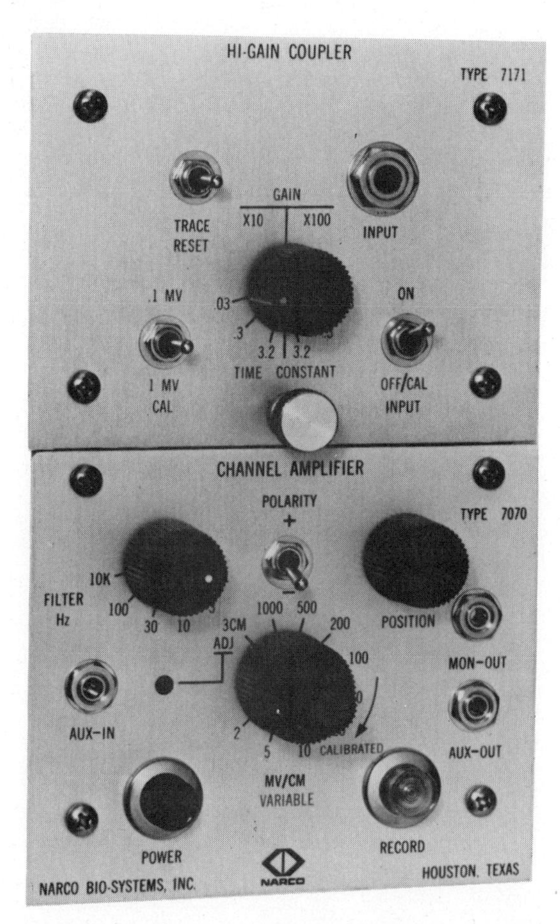

Figure 7-6I. COUPLER & CHANNEL AMPLIFIER—Hi-Gain Coupler to be used with DMP-4B

The basic layout for the physiograph can be diagrammed as follows and as illustrated in Figure 7-7.

The directions described in this section can be adapted to the Narco Bio-Systems Physiograph Four-A, the Desk Model Physiograph

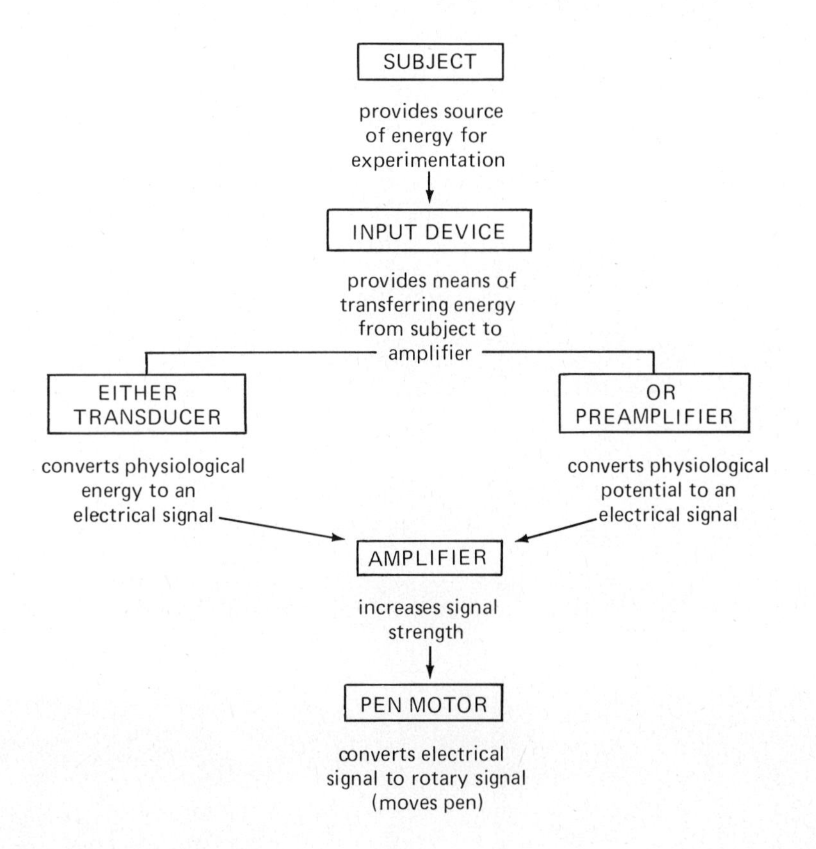

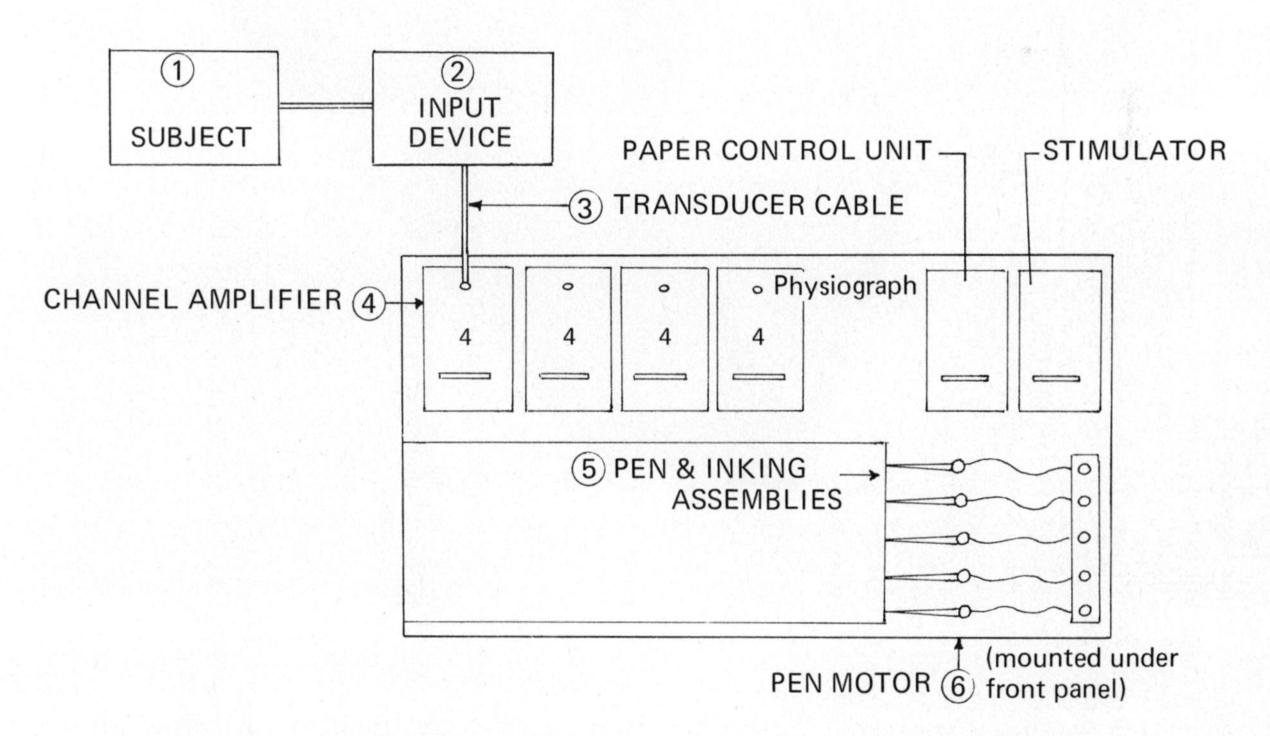

Figure 7-7. Physiograph Four-A with Recording Channels and Plug-in Models

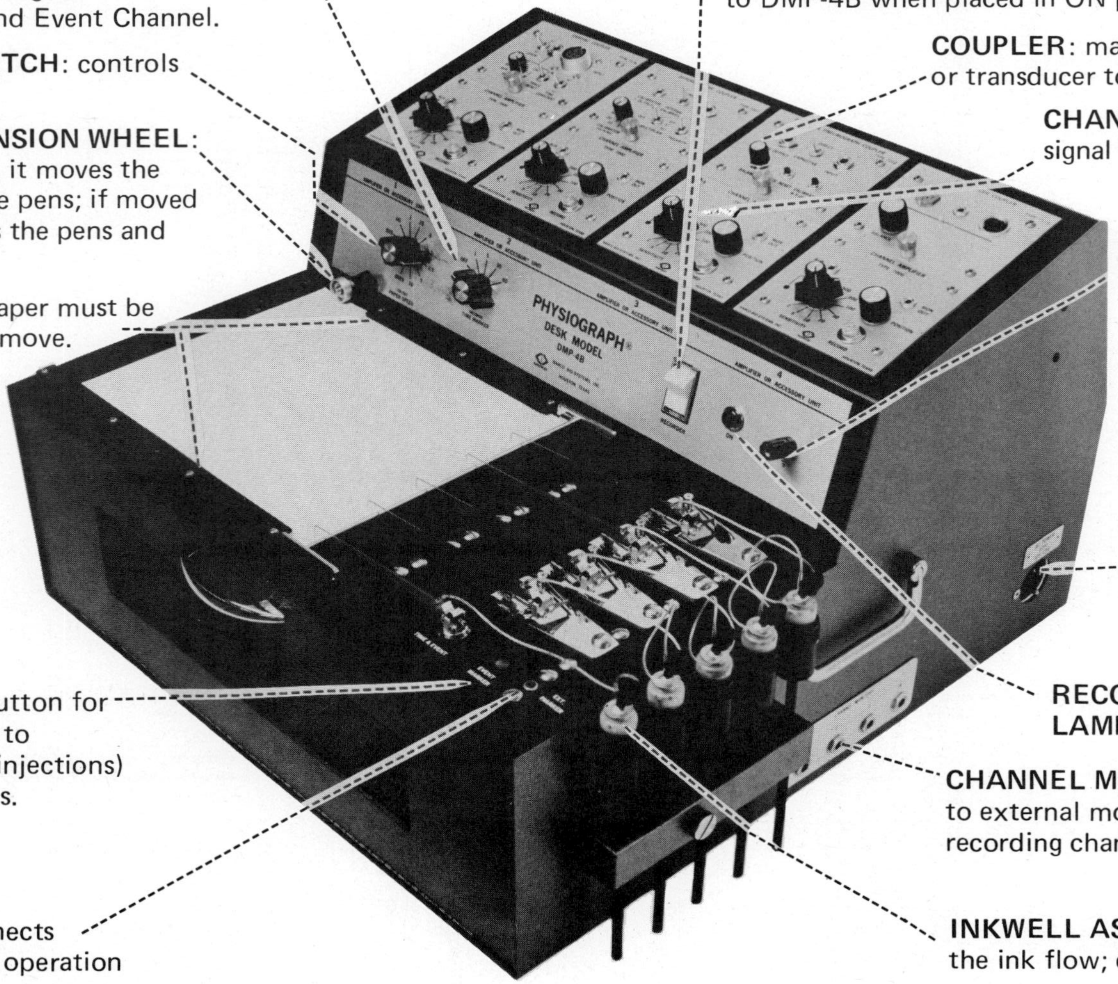

TIME MARKER SWITCH: selects interval between timing marks recorded on Time and Event Channel.

PAPER SPEED SWITCH: controls chart paper speed.

PAPER DRIVE TENSION WHEEL: if moved to the left, it moves the paper and lowers the pens; if moved to the right, it raises the pens and stops the paper.

PAPER GUIDES: paper must be under the guides to move.

RECORDER SWITCH: applies power to DMP-4B when placed in ON position.

COUPLER: matches signal from subject or transducer to the channel amplifier.

CHANNEL AMPLIFIER: amplifies signal from the subject.

GROUND TERMINAL: provides connection for grounding wire.

PLUG: connects to power cable.

EVENT MARKER: push-button for operating the event marker to correlate events (e.g., drug injections) with physiological responses.

RECORDER SWITCH INDICATOR LAMP: glows when power is on.

CHANNEL MONITOR: provides connections to external monitoring devices for each recording channel.

EXTERNAL EVENT: connects external switch for remote operation of event marker.

INKWELL ASSEMBLY: upward increases the ink flow; downward reduces ink flow.

Figure 7-8. The Desk Model Physiograph DMP-4B

DMP-4A, or the newer solid state model, the DMP-4B. The Physiograph Four-A has six compartments for plug-in modules: five for amplifiers or accessories and one for the paper control unit. The Desk Model Physiograph DMP-4A has four compartments for plug-in modules. Three recording channels can be used for recording events and one compartment for the stimulator. It will operate with all transducers, preamplifiers, and plug-in modules used for the Four-A with the exception of the paper control unit (which is built into the desk model physiograph). The DMP-4B is similar to the DMP-4A. (This model is illustrated in Figure 7-8.) Any directions specific for the DMP-4B will be given as an alternate procedure following the Four-A directions or illustrated in drawings. In order to perform all the experiments described in this manual it is assumed that there are three recording channels on the physiograph, a stimulator, and a time and event marker.

Procedure

Part 1: General Directions for the Use of the Physiograph

1. Place all ON-OFF switches (on all amplifiers and top panel) on OFF. Install the plug-in modules to be used in the experiment, if they are not installed. In Exercise 7-F the stimulator, one channel amplifier, the photoelectric pulse pickup, and (with the DMP-4B) the transducer coupler will be used. The transducer coupler should be installed in the physiograph. If not, remove one coupler by unscrewing the large screw on the lower center of the top panel of the coupler. Then lift the coupler straight up, insert the transducer coupler and tighten the large screw. Do not connect the photoelectric pulse pickup at this time.
2. Connect the power cable to the power connector found on the back of the physiograph and to a 115 volt power source. If a three-prong cable is not used, attach a grounding wire from the control panel ground terminal on the main frame to a water pipe, gas outlet, etc.
3. Check the paper supply to verify that there is sufficient paper in the physiograph for the experiment. The paper is packed (500 sheets to a box in a continuous folded chart) printed side up, but is fed into the physiograph printed side down. The paper is inserted somewhat differently with each model of the physiograph.
   a. When using the Physiograph Four-A, the paper box should be placed in the cabinet with the right edge of the paper box aligned with the main frame support. Raise the paper tension wheel (see Fig. 7-9). The right edge of the paper box should be aligned with the main frame support. Feed the paper through the paper feeder and underneath the paper guides and paper tension wheel. Lower the paper tension wheel.
   b. The paper loading procedure for the DMP-4A is similar except that the paper must be removed from the box and placed into the physiograph before starting the paper up through the top panel.
   c. In the DMP-4B, the paper box should be pushed all the way against the back of the paper compartment, with the bottom of the box against the left inner wall. Disengage the paper tension wheel

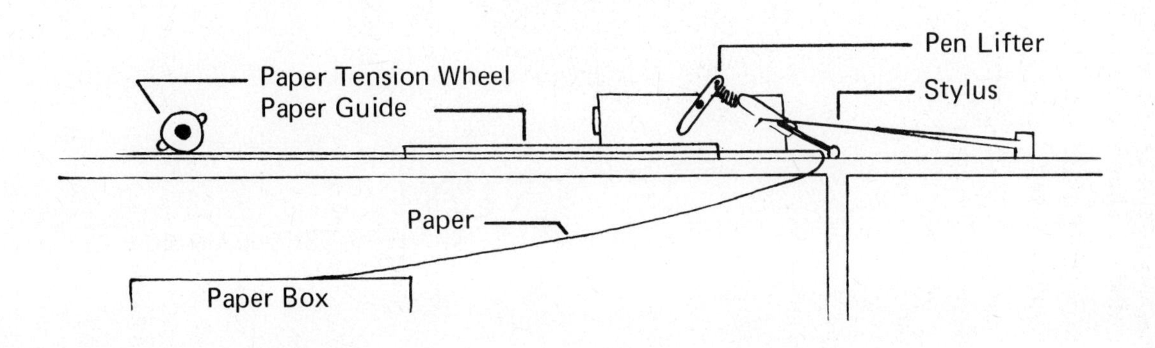

Figure 7-9. Paper Loading for the Physiograph Four-A

(by pressing it to the right) and insert the paper into the paper feed guide (see Fig. 7-10). When the paper emerges onto the upper surface of the physiograph, make certain that each edge of the paper is under the paper guides. Push the paper tension lever down to the left to engage the wheel and lower the pens.

4. If the pens are not installed, insert the stylus of the curvilinear pen horizontally **under** the retaining spring. The support pins should fit firmly into the stirrup cradle (see Fig. 7-11).
5. The inking assembly may be one of two types as illustrated in Figure 7-12 on the next page. If your assembly is of the type on the left it consists of a plastic inkwell (refillable), a mounting rod, and tubing. The flow of ink to the stylus is a function of the height of the inkwell, since this is a gravity feed system. To adjust the ink flow to the recording paper, move the inkwell up or down in the grommet hole. With this type of ink supply, start the flow of ink as follows:
   a. Place the finger over the hole in the inkwell and squeeze lightly on the inkwell.
   b. Wait for a drop of ink to appear on the tip of the stylus.
   c. When the drop appears, release the pressure on the inkwell and remove the finger from the hole.
6. If using the super ink cartridge (illustrated on the right), place the finger over the hole on the rubber bulb and squeeze lightly on the bulb. When a drop of ink appears on the tip of the stylus, release the pressure. Be careful not to lose the rubber bulbs as they fall off easily.
7. Set the main frame master switch (or the recorder switch on the DMP models) to ON. The indicator lamp should glow.
8. For the Physiograph Four-A and the DMP-4A, place the channel amplifier con-

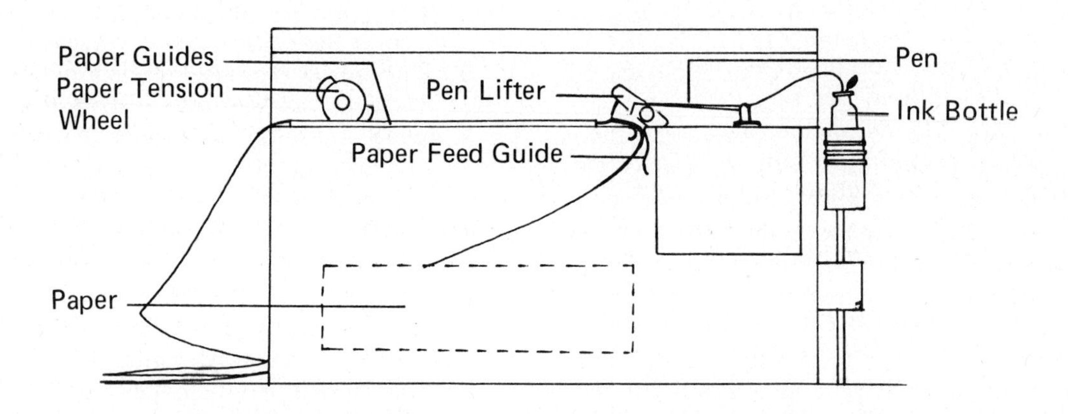

Figure 7-10. Paper Loading for the DMB-4B

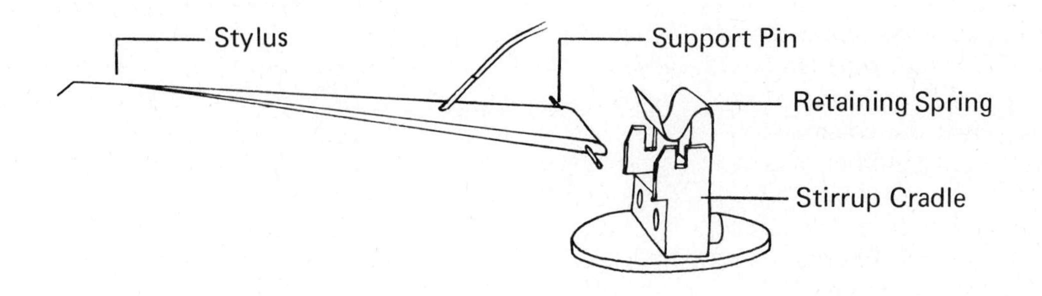

Figure 7-11. Pen Assembly

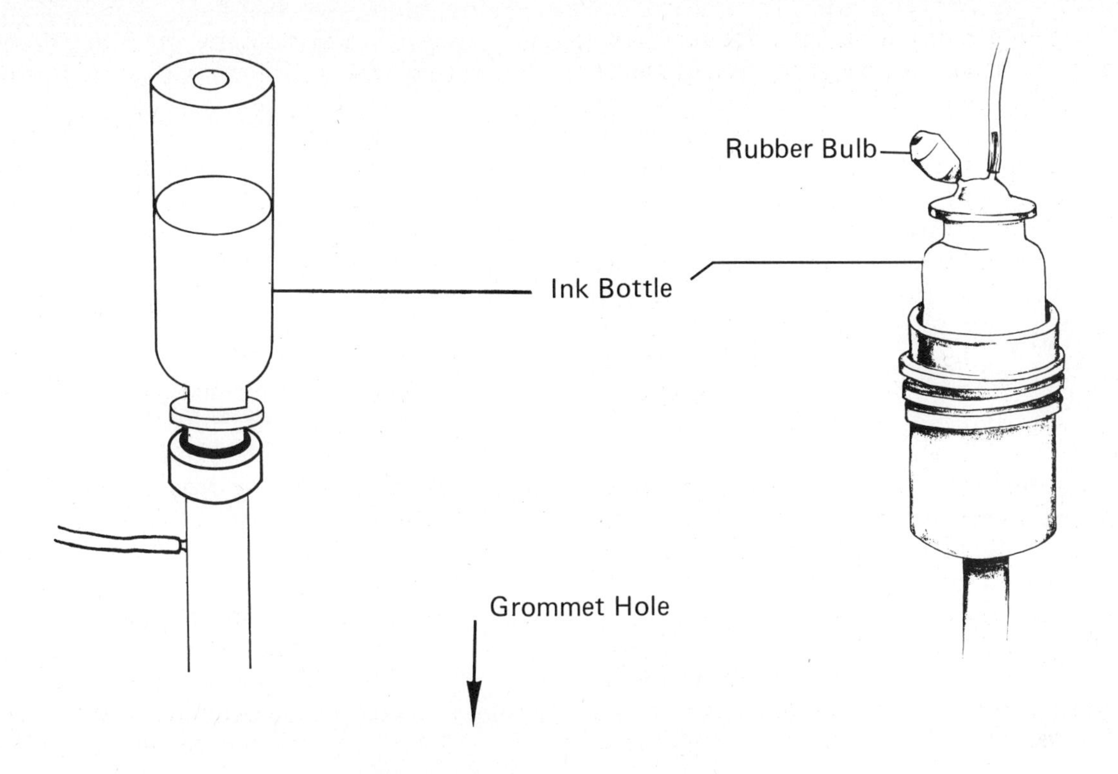

Figure 7-12. Inking Assemblies

trols in the following positions whenever a channel is to be used (see Fig. 7-13):

a. ON-OFF to ON. Allow a five minute warmup period. The pilot lamp on the amplifier should glow. Before making a connection from a transducer or preamplifier to the input of thc channel amplifier, **always turn the channel amplifier switch to OFF**.
b. RECORD-READY to READY. This is a standby position with the output to the recorder off. Always have the switch on READY before making a connection from the transducer or preamplifier to the input of the channel amplifier.
c. POSITION: to center (0) position.
d. AMPLITUDE to 0 (fully counterclockwise). The amplitude control is inoperative when the channel amplifier is used with a preamplifier or transducer which has an amplitude control. In that case, the preamplifier or transducer amplitude control determines the amplitude of the recorded signal.

9. For the Physiograph DMP-4B, place the channel amplifier controls in the following positions whenever a channel is to be used (see Fig. 7-14). Study the functions of each part of the channel amplifier as listed on Figure 7-14.

a. Press the POWER switch ON.
b. Press the RECORD switch OFF (if it is on). This switch must be off before any connections are made to the physiograph. It must be on only when actually recording.
c. Set the POLARITY to positive (+). All channels should routinely be set to positive at this time and left in that position.
d. Set the VARIABLE control (the upper knob of the double knob in the center) to its maximum clockwise position. All channels should be routinely set in this position at this time and left in this position.
e. Set the mV/CM switch to 1000 mV/cm position.
f. Set the recording pen 1.5 cm (3 green lines) below the center dark green line, using the POSITION control.
g. Set the mV/CM switch to 3 cm ADJ. The recording pen should write 1.5 cm

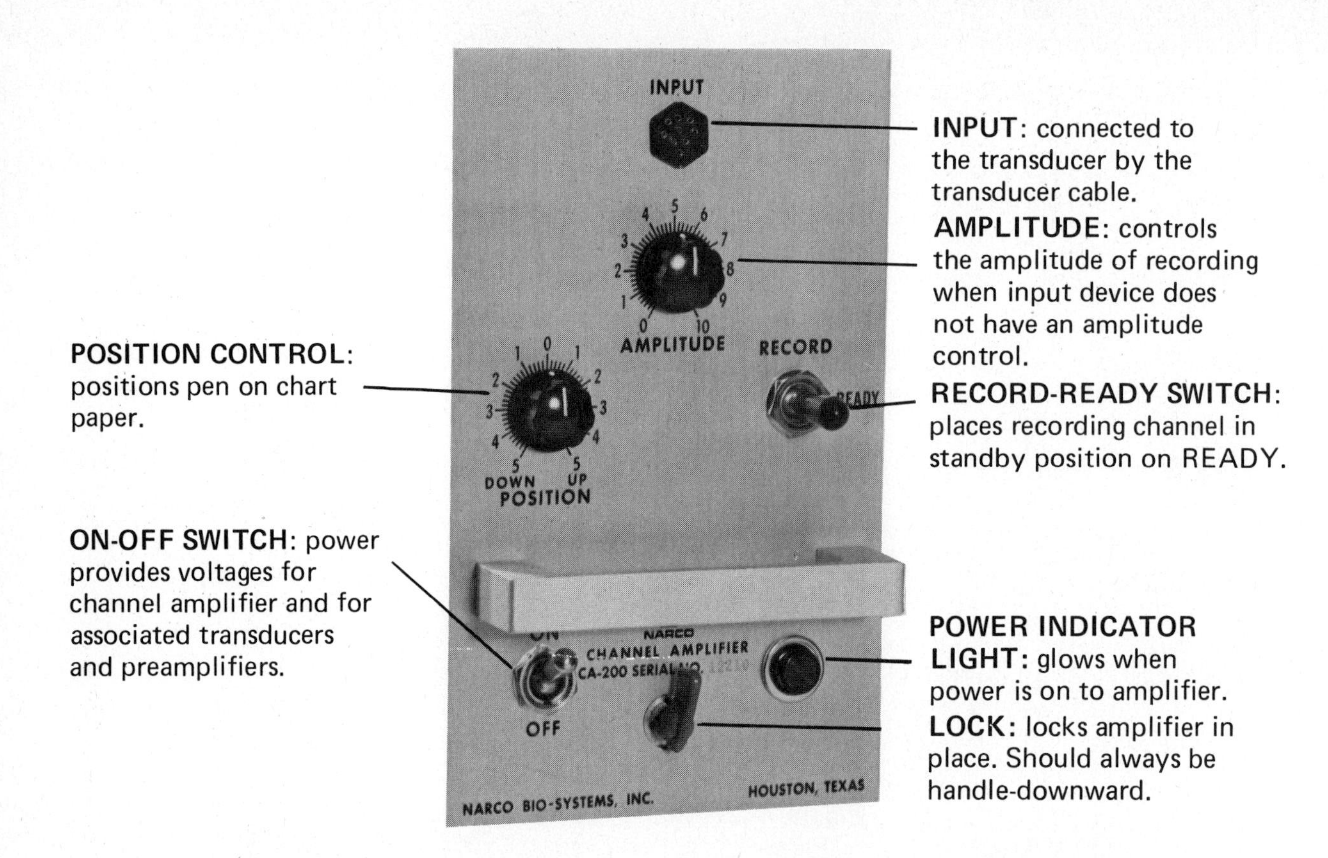

Figure 7-13. Channel Amplifier on the Physiograph Four-A and DMP-4A

(3 green lines) above the center dark green line.

h. If the pen does not deflect a total of 3 cm, adjust the inset multi-turn control (with a screwdriver) until the pen writes 1.5 cm below the center line when the mV/CM switch is on the 1000 mV/cm position and 1.5 cm above the center line when the mV/CM switch is on the 3 cm ADJ position. This adjustment should not have to be changed.

10. Set the desired paper speed and time marker. With the Physiograph Four-A, lower the paper tension wheel onto the paper, and turn the paper control unit ON-OFF switch to the ON position. With the DMP-4A and the DMP-4B, move the paper drive lever to the left. This also lowers the recording pens onto the chart paper.
11. Set the channel amplifier RECORD switch to RECORD. (Have this switch on record only when actually recording.)

Shutdown Procedure (go through these steps at the conclusion of today's experiments):

12. At the conclusion of any experiment, lift the paper tension wheel and turn the paper control unit to OFF or move the paper drive lever to the right to stop the paper movement on the desk model physiographs and raise the pens. With the Physiograph Four-A, set all amplifier switches to READY, set all module ON-OFF switches to the OFF position. Turn the master main frame switch to OFF. With the desk model physiographs, press all channel amplifier RECORD switches to OFF, set all channel amplifier POWER switches to OFF. Set the RECORDER SWITCH to the OFF position.
13. Push the inkwell assemblies down to the lowest level. Put the tips of the pens into the ink bottles if they are to be used within a few hours. If not, draw ink from the styluses and tubes as follows:
    a. Squeeze the inkwell.
    b. Place the finger over the hole in the inkwell and release the squeezing pressure.
    c. Repeat if necessary to draw all the ink from the pen.

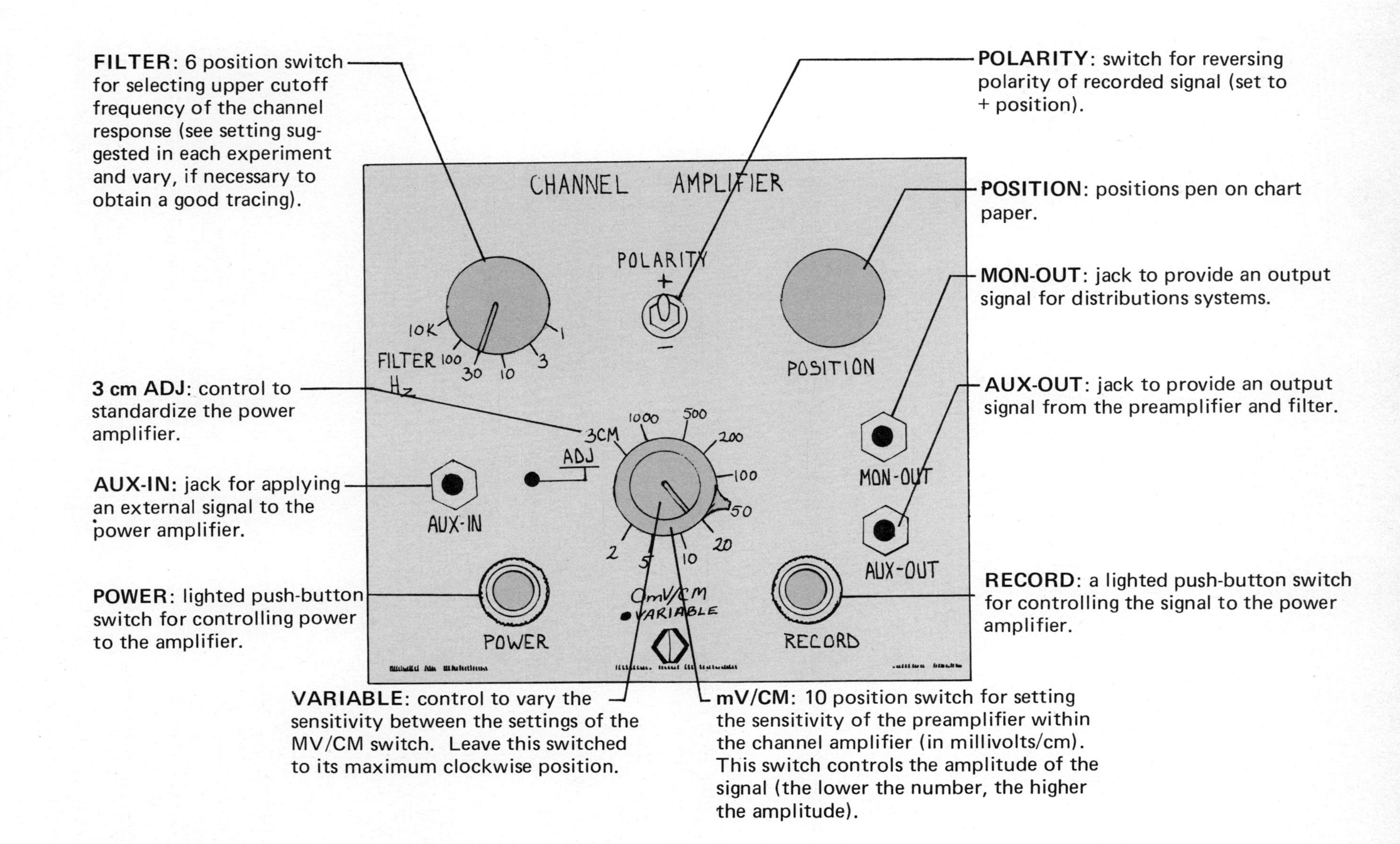

Figure 7-14. Channel Amplifier on theDMP-4B

With the super ink cartridges:

a. Squeeze the bulb.
b. Place the finger over the hole in the bulb and release the squeezing pressure.
c. Repeat if necessary to draw all the ink from the pen.

Thoroughly clean the stylus and the plastic tubes with alcohol before storing for extended periods of time. Clean the panels to remove ink stains.

14. Disconnect input devices. Connect the ends of the transducer cables for protection of the prongs. Do not touch the ends with the fingers.
15. Disconnect the power cable and ground; coil carefully.
16. Replace the protective dust cover.

Part 2: Procedure for the Use of the Physiograph Stimulator

1. Follow steps 1-7 of Part 1.
2. The pen for the time and event marker is the only pen needed in this part of the experiment.
3. Practice varying the paper control to become familiar with this operation. The paper moves at 12 fixed speeds from .01 to 5 cm/sec in the Four-A and from .0025 to 10 cm/sec in the DMP-4A and DMP-4B. The paper control unit provides time signals every 1, 5, 30, or 60 seconds, which are recorded as momentary downward deflections of the time and event channel pen. One time mark is omitted each minute in the one second position to indicate one minute intervals. The time and event channel pen deflects upward when actuated either by the panel-mounted event marker pushbutton, the stimulator, or an external marker switch.
4. The stimulator can deliver either a single pulse or continuous pulses. It is possible to adjust the voltage, frequency, and duration of the stimulus (see Figure 7-15). At this time the CONT-OFF-SINGLE switch should be on the OFF position. Turn the ON-OFF switch to ON. The pilot lamp will glow.
5. The frequency control regulates the number of impulses per second when using continuous stimulation. It has three ranges so that the frequency can be varied from .2 to 250 impulses per second. Set the frequency control to a frequency of 5 and the frequency multiplier to X10. The neon indicator on the front will provide a visual indication of the frequency and duration of the output pulses.
6. The duration control regulates the duration of each stimulus. It has a three range multiplier switch so that the duration can be varied from .1 to 120 milliseconds. Set the duration control and multiplier switches to a duration of 7 milliseconds, with the multiplier on X1.
7. The voltage can be varied from 0 to 120 volts with three ranges. Set the volts on 0 and the multiplier on X1.
8. Place the MONOPHASIC-BIPHASIC switch to MONOPHASIC.
9. Turn the paper speed to .5 cm/sec. Set the seconds on the paper control unit to 1.
10. The following three steps of the procedure should occupy approximately one piece of chart paper. Repeat the procedure for each member of the group.
    a. Start the paper and observe the time marks. Note that the marker omits one mark every 60 seconds. Depress the red event marker button next to the time and event pen. The pen should rise.
    b. Place the CONT-OFF-SINGLE switch in the SINGLE position. The neon indicator should flash once. The event marker pen will indicate the pulse with a single momentary upward deflection.
    c. Next, place the switch in the CONT position for three seconds.

    Repeated pulses are delivered in this position; the event marker pen will deflect upward and hold for the duration of the pulses.

    Now place the switch in the OFF position. Turn off the paper. Label the record and place it in the Results and Questions section.
11. Loosen the red and black binding posts on the stimulator, exposing an opening in each one. Connect one end of the stimulator output extension cable to this RF isolated output on the stimulator (see Figure 7-16) and tighten the red and black binding posts. Connect the other end of the stimulator output extension cable to a pair of pin electrodes (loosen the screws on the stimulator output extension cable, insert the pin electrodes through the openings, and tighten the screws).

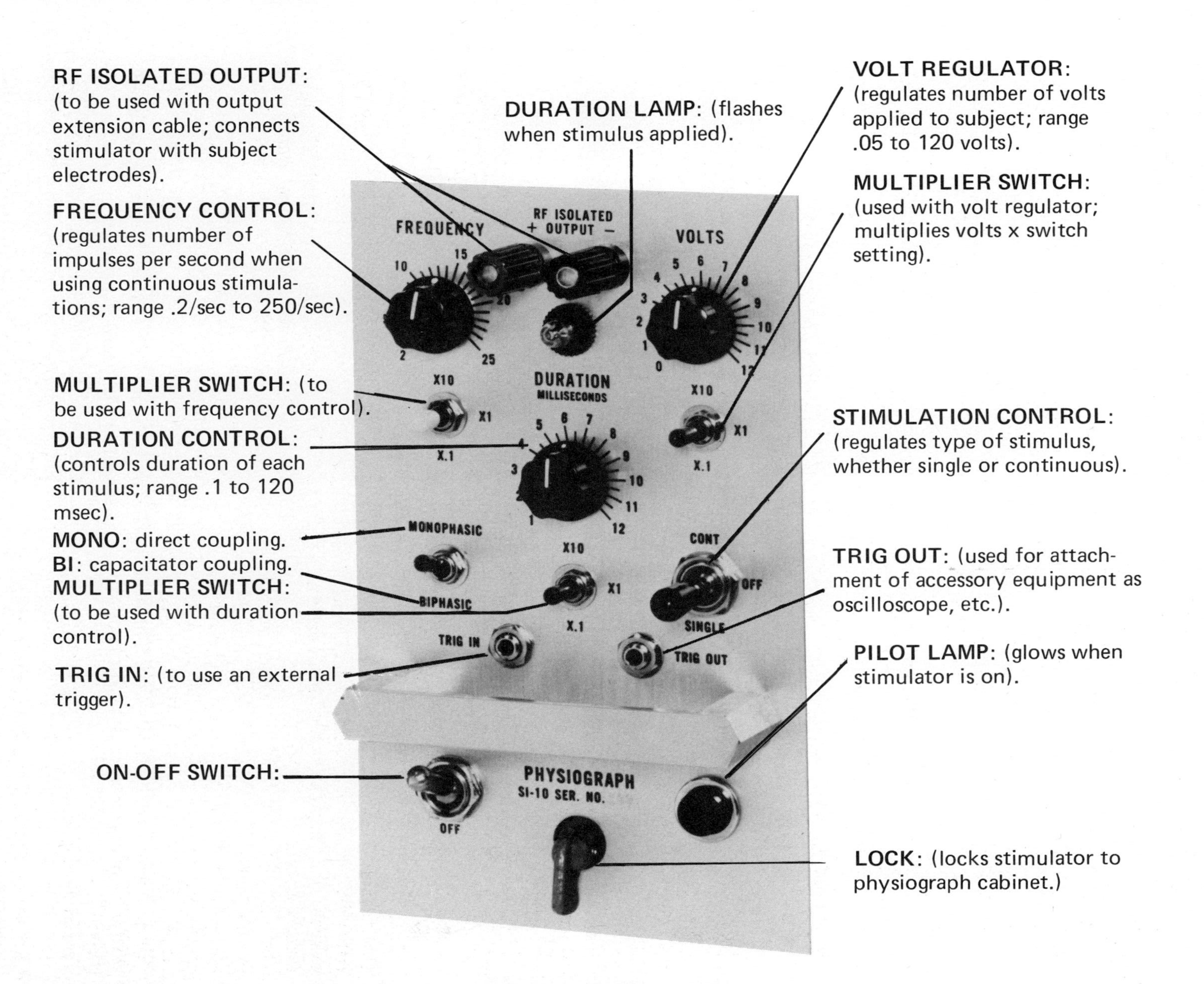

Figure 7-15. Stimulator

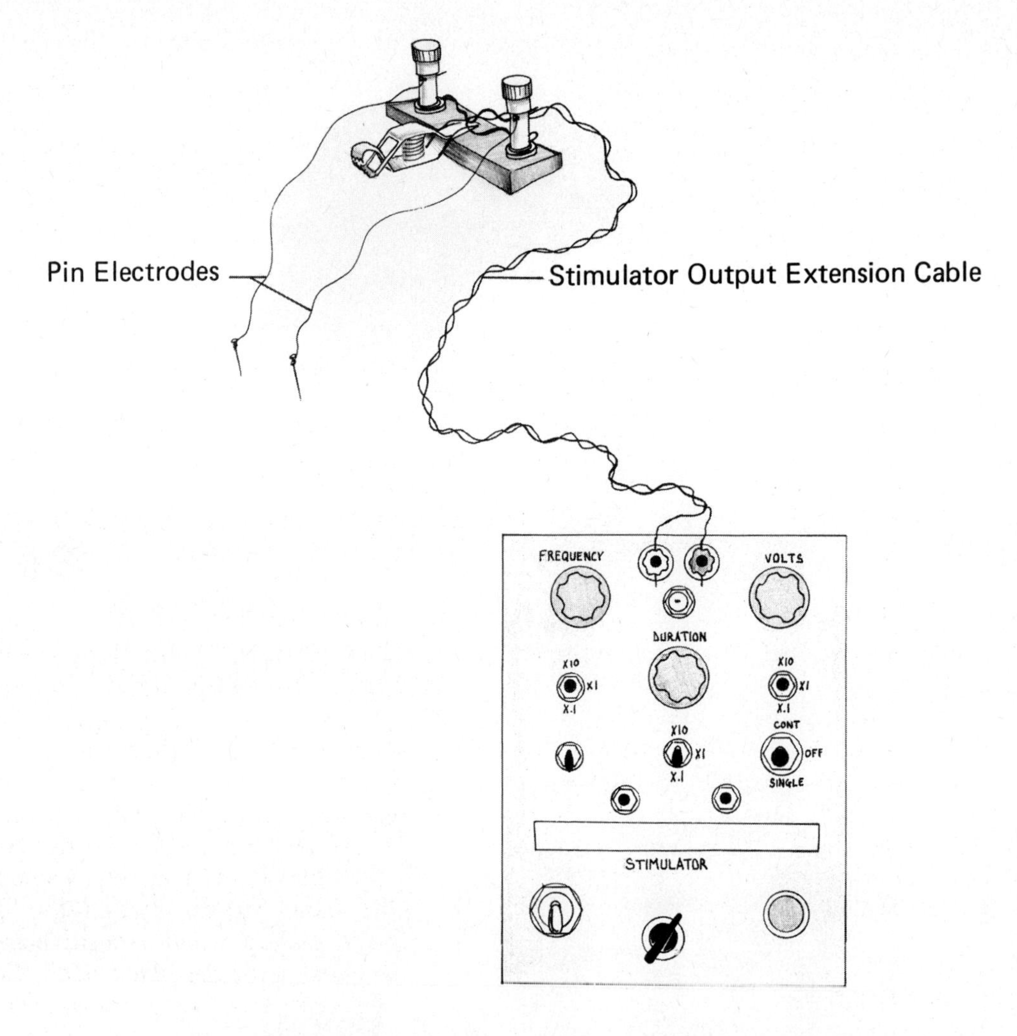

Figure 7-16. Setup for Stimulation of Tongue

12. To determine the characteristics of the output current from the stimulator, set the duration at 7 milliseconds (multiplier at X1), volts at 0 (multiplier on X1), frequency at 5 (multiplier at X10), and monophasic.
13. Place the pin electrodes on the tongue of the subject. Gradually increase the volts from 0, placing the switch in the SINGLE position for a single shock each time, until the initial shock is felt. Record the threshold voltage, the voltage at which the first shock is felt. ____________
14. Decrease the duration to 2 milliseconds (multiplier on X1) and determine again the volts needed for the threshold stimulus (reset the volts at 0 and **gradually** increase the voltage). ____________
15. Reset the volts at 0. Return the duration to 7 milliseconds. With the frequency at 50 per second (frequency at 5, multiplier at X10), place the switch in the CONT position for repeated pulses. Determine again the volts for the threshold stimulus. ______ Turn off the stimulator.

Part 3: Use of a Recording Channel and the Photoelectric Pulse Pickup

A recording of the pulse will be obtained in order to familiarize the student with the use of a recording channel and with one method of calculation of the duration of an event.

Procedure with the 4A models

1. Connect the photoelectric pulse pickup to the ventral surface of the finger (see Fig. 7-17) so that the small (4 mm in diameter) photoconductor is next to the finger. Fit the foam rubber black band over the pulse pickup and wrap the band snugly around

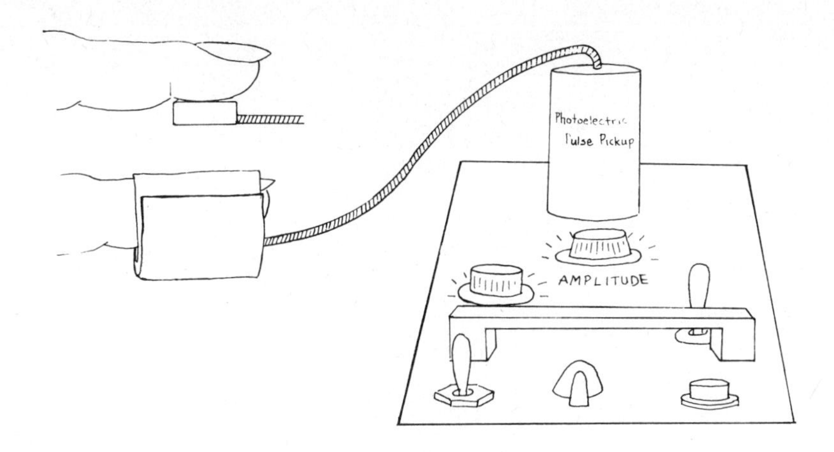

Figure 7-17. Photoelectric Pulse Pickup with Physiograph Four-A and DMP-4A

the finger. The pulse pickup is light sensitive and must be shielded with the black band.

2. Connect the pulse pickup cable plug to the input connector on the channel amplifier. A pen is needed in the appropriate channel and in the time and event channel.
3. Set the channel amplifier controls as follows:
   a. ON-OFF to ON. Allow 1 minute warm-up.
   b. AMPLITUDE to 0.
   c. POSITION TO 0 (center position).
   d. RECORD-READY to READY.
4. After the 1 minute warmup, place the RECORD-READY switch on RECORD. After the pen stabilizes its position, adjust the channel amplifier position control to position the recording tracing in the center of the pen's movement. Adjust the amplitude (the lower the number, the greater the amplitude). An amplitude of no more than 1 inch is desirable in most experiments. Set the paper speed at .5 cm/sec and the time marker at 1. If the recording does not resemble Figure 7-18, turn the channel amplifier control back to READY and reposition the pulse pickup on the finger.
5. Record the pulse for at least **70 seconds** at paper speed .5 cm/sec.
6. Increase the paper speed to 5 cm/sec and record the pulse for one entire piece of chart paper.
7. Withdraw the ink from the pens as directed in Part 1, Steps 12-16.

Procedure with the DMP-4B

1. Study Figure 7-19 to learn the function of each part of the transducer coupler.
2. Attach the detecting head of the photoelectric pulse transducer to the ventral surface of the finger so that the light source and the photocell are in contact with the finger (see Fig. 7-19). Fit the foam rubber black band over the pulse pickup and wrap the band tightly around the finger. The pulse pickup is light sensitive and must be shielded with the black band. Connect the photoelectric pulse transducer into the 7-pin INPUT connector of the transducer coupler.
3. Set the channel amplifier controls as follows:
   a. Press the POWER switch ON.
   b. Press the RECORD switch OFF (if it is glowing).
   c. Set the FILTER switch to 10K.
   d. Set the mV/CM switch to the 20 mV/cm position.
   e. Set the recording pen to the center dark green line, using the POSITION control.
4. Press the RECORD switch ON. Adjust the transducer coupler BALANCE control to restore the recording pen to the previously established baseline. After the pen stabilizes its position, turn the mV/CM switch until the recording is of the desired amplitude. An amplitude of no more than one inch is desirable in most experiments. Set the paper speed at .5 cm/sec and the time marker at 1. If the recording does not

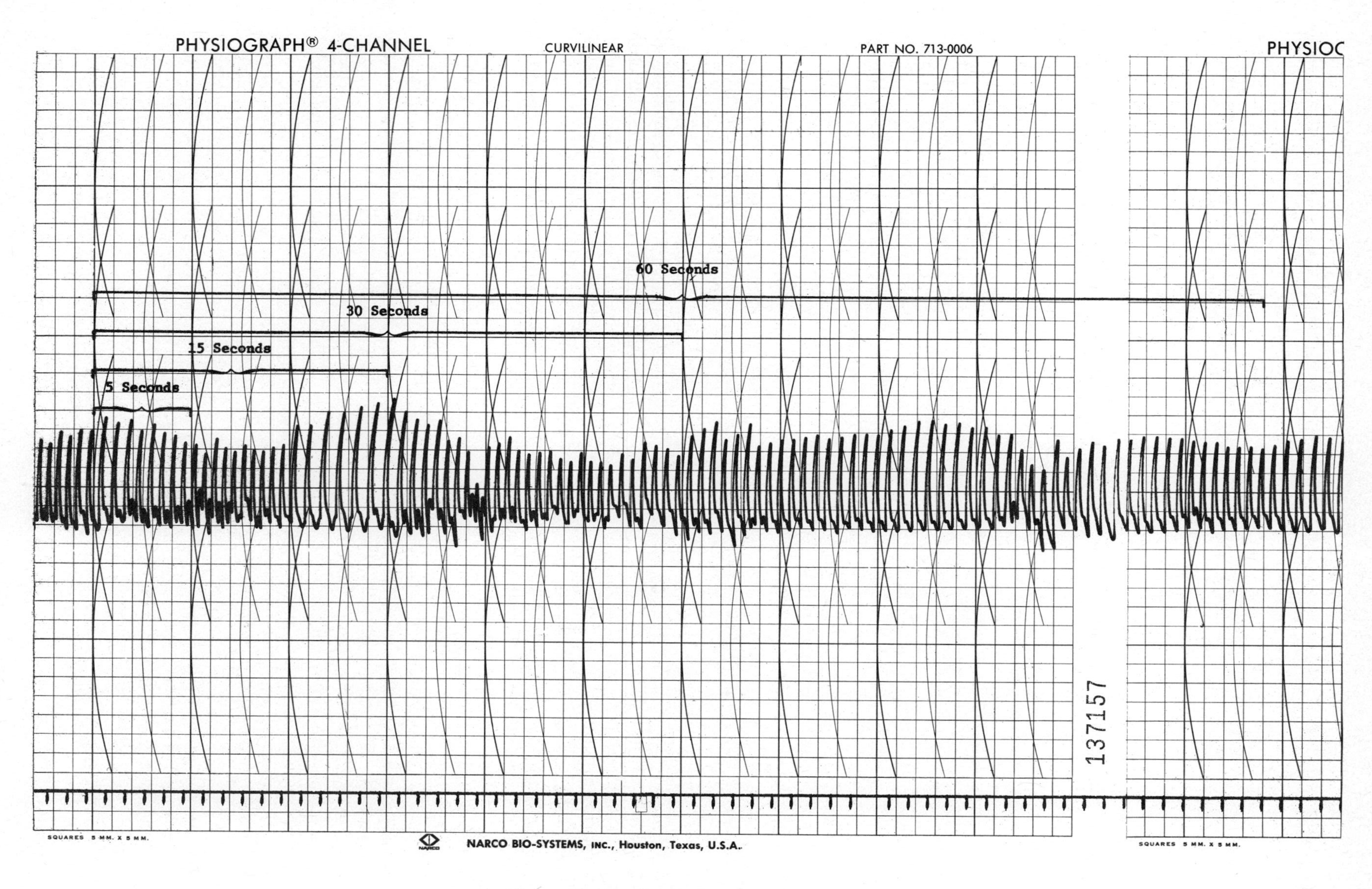

Figure 7-18. Recording of Pulse

resemble Figure 7-18 press the RECORD switch OFF and reposition the pulse pick-up on the finger.

5. Record the pulse for at least 70 seconds at paper speed .5 cm/sec. This will be approximately 1¼ pages.

6. Increase the paper speed to 5 cm/sec and record the pulse for one entire piece of chart paper.

7. Withdraw the ink from the pens as directed in Part 1, Step 13 and follow all the steps of the shutdown procedure (12-16).

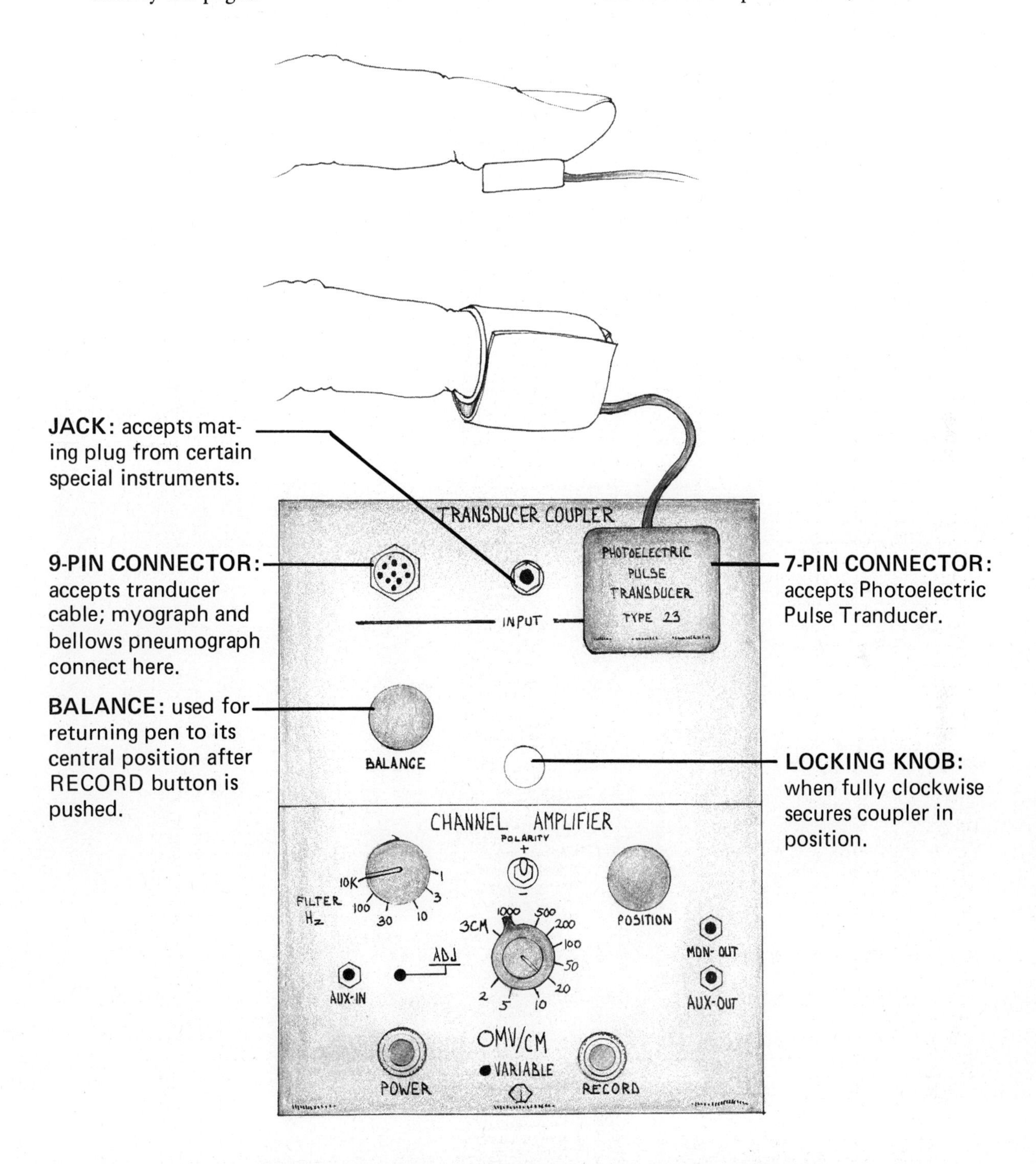

Figure 7-19. Photoelectric Pulse Pickup with the DMP-4B

Name ______________________

Results and Questions for Chapter 7

Exercise 7-A. Faradic Current

1. Did you feel a shock on either make or break with the voltage dial on 0? ______________

   Explain: ______________

2. Record the liminal stimulus for:

   make ______________

   break ______________

3. With faradic current, which is stronger: the make or break shocks? ______________

4. Was a shock felt during the time the key was depressed between the make and break? ______________

5. Was a shock felt in Steps 8-10? ______________ Explain: ______________

   ____

6. Does the faradic current flow while the key is held in a depressed position? ______________

7. Describe in detail how it would be possible to short circuit the break. ______________

   ____

   ____

Exercise 7-B. Tetanizing Stimulation

1. Fill in the following table using your results from Exercises 7-A and 7-B. Place the voltage range and the voltage dial setting in parentheses after the volts.

		Liminal Stimulus (volts)
Single Induction Faradic Shocks		
	Make	
	Break	
Tetanizing		

2. What conclusion can you draw concerning the relative intensity of the make single induction shock, the break single induction shock, and the tetanizing stimuli?

   ____

   ____

3. Why is tetanic stimulation stronger? ______

4. Are the shocks felt the entire time the key is on MULT.? ______

Exercise 7-C. Use of the Kymograph and Ink Recording Apparatus

1. At drum speed 6, an event was recorded on 10 cm of kymograph paper. What was the duration of this event?

Answer: ______

2. At drum speed 2, an event was recorded on 10 cm of kymograph paper. What was the duration of this event?

Answer: ______

3. At drum speed 3, the duration of an event was 15 seconds. How many centimeters did the drum move during the event?

Answer: ______

Exercise 7-D. The Signal Magnet

1. Fasten your kymograph records from this experiment in the spaces provided below.

   a. Drum speed 6, one second intervals.

   b. Drum speed 2, one second intervals.

c. Drum speed 11, one second intervals.

2. Using the record obtained with drum speed 6, measure and record the length of each one second interval. (Record the value on the record as well as here.)__________

   a. What is the average value?__________

   b. What is the range in values?__________

   c. Discuss the cause of any deviation from the expected value (1 cm).__________

3. Is the signal magnet more useful as a timing device at fast drum speeds (e.g., 2), at intermediate speeds (e.g., 6), or at slow drum speeds (e.g., 11)?
Explain:__________

4. Discuss the advantages and disadvantages of the signal magnet as a timing device.

5. Discuss the advantages and disadvantages of the signal magnet as a device for the indication of the precise time of application of a stimulus.

Exercise 7-E. The Tuning Fork

1. Fasten your kymograph records from this experiment in the spaces provided below.

   a. Drum speed 6.

   b. Drum speed 4.

c. Drum speed 2.

d. Drum speed 1.

2. At which drum speed(s) could the tuning fork be used as a timing device? Note: it is essential to have the crests separated in order to count them.

____

3. Can the tuning fork be used to time events recorded on a slow drum?______________

4. Can the tuning fork be used to indicate the time of application of a stimulus?__________
Explain:______________________________

5. How accurate is the tuning fork as a timing device?______________________

6. Which can be used to measure short intervals of time more accurately: the tuning fork or the signal magnet?______________________________

7. Which is easier to use in measuring longer intervals of time: the tuning fork or the signal magnet? ______________________________

8. Which can be used to indicate the time of application of a stimulus?______________

9. Which requires a rapidly moving drum?______________________________

Exercise 7-F. Use of the Physiograph

Part 2. Procedure for the Use of the Physiograph Stimulator

1. Insert the record here from Part 2, steps 9 and 10. Label the one second time marks, and the mark produced by the event marker, the single stimulus and the continuous stimulus.

Name ______________________

2. The paper speed used was .5 cm/sec. Measure the distance (in centimeters) between two dark green vertical lines. ______________

   How long a period of time elapsed as the pen traveled between two dark green vertical lines? (Divide the distance above by the paper speed.)

   ____

   Your answer should agree with the number of second marks minus one on that portion of the record. Study the record to verify this statement.

3. Mark and label on the record intervals of 5, 10, and 30 seconds.

4. Is it easier to use the known time between two dark vertical lines or the second markers to calculate the duration of an event? ______________

5. Fill in the following chart with your results from Exercise 7-F, Part 2.

	Threshold stimulus (volts)
Single shocks, duration 7 milliseconds	
Single shocks, duration 2 milliseconds	
Continuous shocks, frequency 50/second, duration 7 milliseconds	

6. Why was the threshold stimulus higher when the duration was 2 milliseconds as compared with 7? ______________

Part 3. Use of a Recording Channel and the Photoelectric Pulse Pickup

1. Fasten your record of your pulse here at paper speed .5 cm/sec. Mark the following time intervals on your record: 5, 15, 30, and 60 seconds. Note that one second mark is missing every 60 seconds.

2. Determine the number of heart beats during each of the time intervals.

	No. of heart beats in interval		No. of heart beats per minute
5 seconds	______	x 12 =	______
15 seconds	______	x 4 =	______
30 seconds	______	x 2 =	______
60 seconds			______

3. What might be a possible explanation of variation in the number of heart beats per minute in No. 2? ______

4. Which duration of determination of heart rate do you think is more accurate? ______

5. Fasten the record of your pulse recorded at paper speed 5 cm/sec in the space below.

6. Calculate the duration of one heart beat. Using a ruler, measure the length in centimeters of one heart beat, from the beginning of the beat to the beginning of the next beat. ______

   Divide this distance by the paper speed (5 cm/sec) to obtain the duration of one heart beat in seconds.

   ______

7. Determine the heart rate per minute, by setting up a proportion.

$$\frac{\text{1 heart beat}}{\text{duration from No. 6}} = \frac{\text{x heart beats}}{\text{60 seconds}}$$

heart rate per minute ______

8. Compare the answer obtained here for heart rate per minute with that obtained in No. 2 for 60 seconds. ______________________________

____

Which do you feel is more accurate? ______________ Why? ______________

8

Muscle Physiology

Exercise 8-A
Strength of Stimulus and Height of Contraction

Objectives

To demonstrate the characteristics of faradic current as a stimulating agent for muscle; to show the effect of the strength of the stimulus on the height of contraction; to determine whether skeletal muscle, as a whole, follows the all-or-none law.

Materials

Beaker; frog Ringer's solution; frog; dissecting set; watch glass; medicine dropper; and **either** wires, moist chamber, light muscle lever, induction stimulator, kymograph, stand, 10-gram weight, three double clamps and signal magnet (optional) **or** physiograph, frog board, clips, stand, myograph B, pin electrodes, stimulator output extension cable, and transducer cable.

Discussion

The gastrocnemius muscle, the large calf muscle of the frog, will be used in these muscle experiments. This muscle is easily prepared and has been extensively studied. It originates on the femur just above and behind the knee joint and terminates in a long tendon, the tendon of Achilles, which, in the frog, is inserted under the foot rather than on the calcaneus as in man.

The force of a muscle contraction depends, within limits, on the strength of the stimulus. If the stimulus is subliminal, no contraction results; however, as the strength of the stimulus is increased, the threshold of the most irritable muscle fibers is reached, and a feeble contraction occurs. As the strength of the stimulus is increased further, more muscle fibers contract, and the height of the contraction increases up to the point at which all the muscle fibers are contracting. A further increase in current will not produce an additional increase in the height of contraction, since all the fibers are contracting to their maximum.

Each of the following experiments is divided into two parts; one part for the use of the kymograph and one for the use of the physiograph. This procedure for the preparation of the frog and the setup of the equipment should be used in all experiments in this chapter unless directed otherwise.

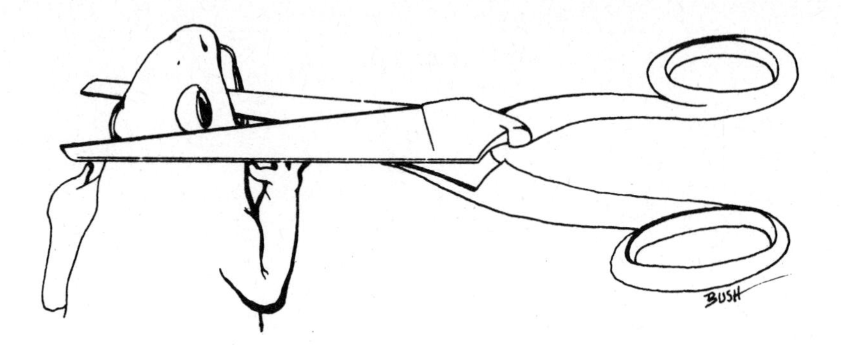

Figure 8-1. Decapitation of the Frog

Procedure using the Kymograph and Induction Stimulator

1. Decapitate a frog by placing the blade of the large shears in its mouth as far back as possible (see Fig. 8-1). With one cut (behind the eyes) remove the upper portion of the head with the brain. (If this is done rapidly the animal will not be caused unnecessary pain or discomfort.) Place the frog on a paper towel.
2. Cut completely across the trunk of the frog at the midabdominal region, discarding the upper half of the trunk.
3. Separate the two legs by cutting through the midline. Give one leg to another group of students.
4. The skin of the frog is loosely attached and can be removed by peeling. Cut around the skin at the ankle (cutting only through the skin). Now grasp the leg with one hand and peel back the skin with the other.
5. Lay this preparation on a watch glass. Do not touch the exposed tissues with the fingers unless absolutely necessary. Keep the preparation moist with Ringer's solution.
6. To facilitate placing the femur in the femur clamp, clean away the muscles from the bone. However, be careful at the knee joint not to remove the origin of the gastrocnemius muscle from the distal femur.
7. Slip the head of the femur out of the socket at the hip and discard the remainder (trunk portion) of the preparation.
8. Loosen the tendon of Achilles beneath the heel by inserting the blade of the scissors under it. Sever the tendon, keeping it as long as possible. Now lift up the tendon and free the gastrocemius muscle from underlying muscles, up to the knee joint.
9. Cut the tibio-fibula bone immediately below the knee joint, and discard this portion of the frog leg.
10. The preparation should now include only the entire femur and the gastrocnemius muscle with the tendon of Achilles. Moisten the preparation with Ringer's solution and place in the watch glass.
11. Set up the stimulating and recording apparatus as described in Exercise 7-C. Always have the muscle preparation to the right of the kymograph, with the writing facing the operator. Place the semicircular base of the stand toward the kymograph.
12. For recording a muscle contraction, use the muscle lever, which consists of a yoke supporting an axle to which a tapered metal tube with a double S hook is attached (see Fig. 8-2).
13. Attach three double clamps to the stand. Attach the moist chamber to the upper clamp, the light muscle lever to the middle clamp, and the signal magnet to the lowest clamp (see Fig. 8-3).
14. The muscle lever should be placed in the double clamp so that the after-loading screw (or "afterloader") is in a superior position (see Fig. 8-3). In this position the S hook should be vertical.
15. The after-loading screw should be raised so that it **does not touch the lever**. This ensures that the muscle is constantly held taut by the weight. A muscle thus prepared is called a free-loaded muscle.
16. Attach the intermediate (10 cm) length ink adapter and pen to the muscle lever. This length pen will magnify the height of muscle contraction. Attach the ink reservoir to the handle of the muscle lever.
17. Attach the shortest ink adapter to the

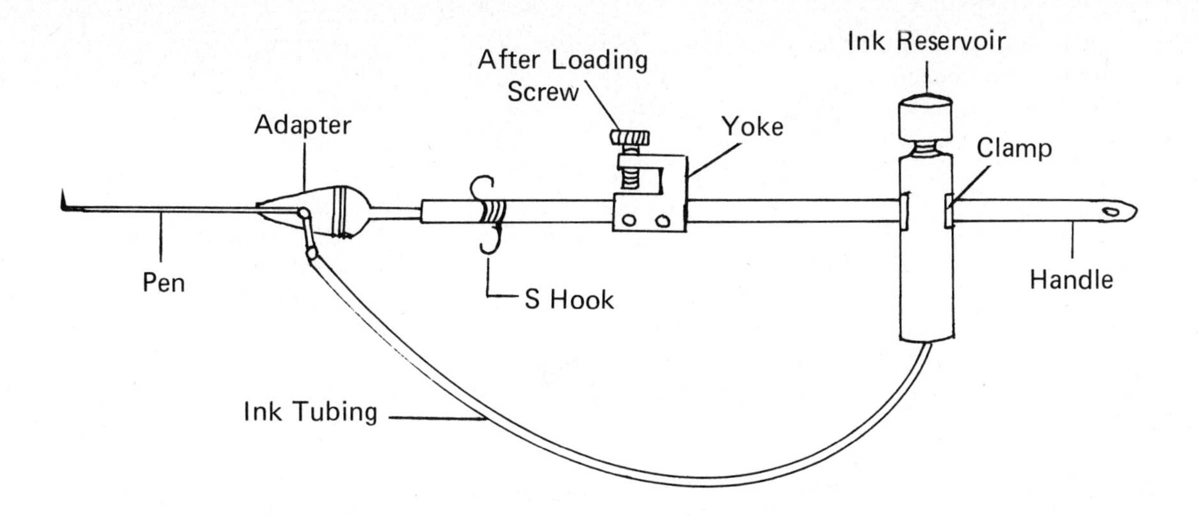

Figure 8-2. Muscle Lever with Ink Adapter

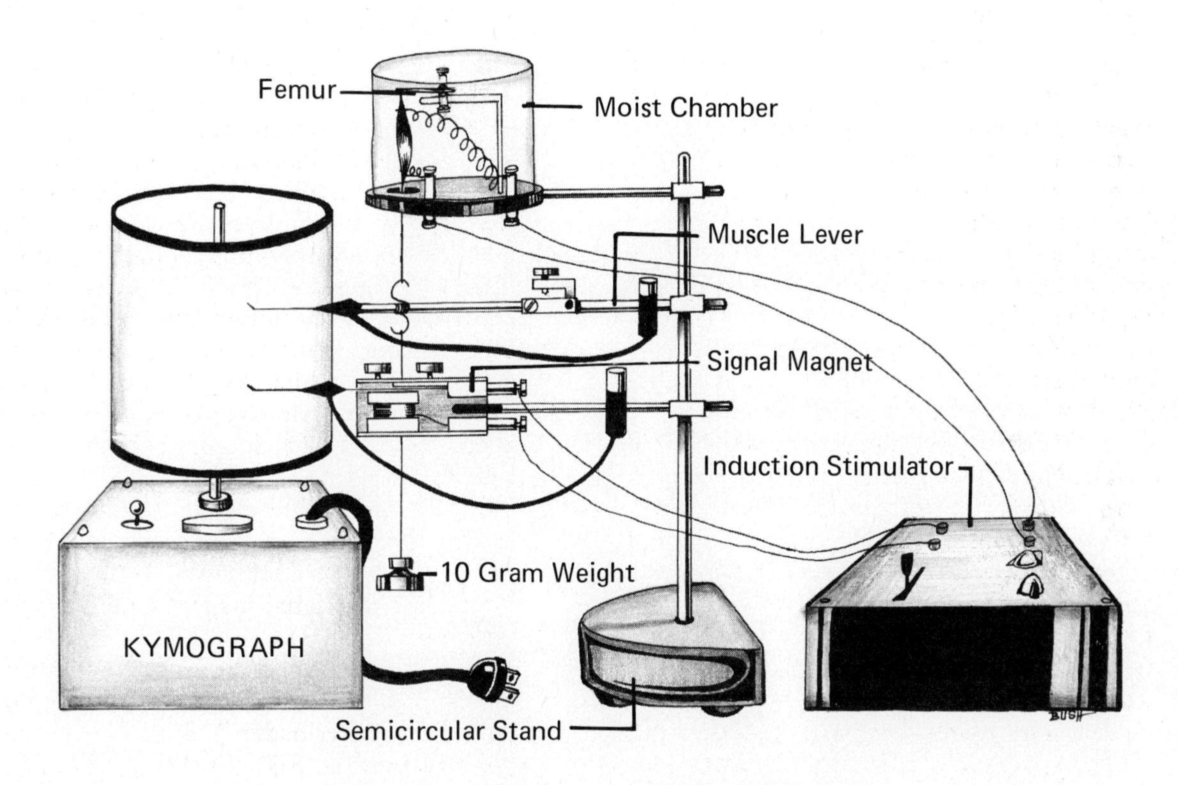

Figure 8-3. Faradic Stimulation of Muscle

stylus of the signal magnet with a small piece of colophonium cement.

18. Connect two insulated copper wires from the induction stimulator terminals, marked MAG., to the signal magnet. Connect two similar wires from the induction stimulator terminals marked STIM. to two terminals (any two of the four) on the inferior surface of the moist chamber.
19. Moisten the muscle with Ringer's solution. Fasten the femur horizontally into the femur clamp on the moist chamber. Position the femur so that the tendon of Achilles extends straight down through the hole on the bottom of the moist chamber.
20. Tie a thread from the tendon to the upper part of the double hook on the muscle lever. The resting position of the lever must

be horizontal and the string must extend vertically through the hole in the moist chamber to the hook. Reposition the muscle lever in the double clamp to obtain this condition. The string must be taut.

21. Tie a 10 gram (g) weight to the lower end of the S hook by means of a short piece of thread in order to slightly stretch the muscle.
22. Take two pieces of fine, uninsulated, copper wire of sufficient length to reach from the upper terminals of the moist chamber to the muscle.
23. Using the same two terminals on the moist chamber as in Step 18, run one piece of copper wire from a terminal on the moist chamber to the inferior end of the muscle belly. Wrap the wire around the muscle several times in order to make good contact. However, do not constrict the muscle with the wire. Extend the other wire from the second terminal to the upper end of the muscle and fasten similarly. The two wires must not touch each other.
24. Moisten a paper towel with water, fold it in thirds, and place it around the inner surface of the cover of the moist chamber, leaving a window in front.
25. Check the adjustment and the alignment of all the equipment as follows:
    a. Place the writing point of the signal magnet (which will indicate the point of introduction of the current into the muscle) immediately beneath the writing point of the muscle lever stylus.
    b. The muscle must be aligned vertically so that the tendon does not touch the sides of the hole in the base of the moist chamber.
    c. The S hook must be directly beneath the tendon so that the tendon is vertical.
    d. The position of the muscle lever must be adjusted so that there is no slack in the thread or the muscle. This can be done most readily by adjusting the position of the muscle lever.
    e. The after-loading screw must not touch the muscle lever.
26. Stimulate the muscle once, to make sure that the current passes to the muscle and the signal magnet, and that the pen writes during both the contraction and relaxation phases of the muscle twitch.
27. Before making any kymograph recordings, determine both the break threshold (the least intensity of stimulus which will cause a response on break) and the make threshold (the least intensity of stimulus which will cause a response on make), by gradually increasing the strength of the current. Record these values below and under Results and Questions.

    Make Threshold ______________________

    Break Threshold ______________________

28. Now increase the voltage and set the induction stimulator at a voltage which will produce **a submaximal make and break contraction**. (If the current is too strong, several twitches may result.)
29. Start the drum revolving slowly (speed 9). The speed must be such that the muscle contraction resembles a straight vertical line.
30. With the muscle at rest, let the pen write a short introductory **base line** on the drum.
31. Make the circuit (hold down the key). The muscle should contract and relax in the same vertical line. (**Do not** release the key.)
32. **Let the drum revolve approximately 1 cm.** Now break the circuit (release the key). The muscle should again contract and relax.
33. Repeat this procedure three times for each member of the group.
34. Label the record completely, including title, M (make) and B (break). Include the record in your results under Results and Questions.
35. It will be necessary to eliminate the make and to use a **constant strength break current** in this part of the experiment. (To do this, turn the voltage selector to zero, then make the circuit. While the key is depressed, reset the voltage to the desired value. Then break the circuit.)
36. Rotate the drum by hand in this experiment (**have drum speed selector on N**), so that the contracting muscle writes a vertical line. After each stimulus, rotate the drum 1 cm.
37. Before starting the experiment, determine which voltage will cause the muscle to barely contract on the break, then **reduce the voltage** below this value.

38. Stimulate the muscle with this reduced voltage. The muscle should **not** contract as this is a **subliminal** or **subthreshold stimulus.** Record this voltage (and subsequent voltages) on the record.
39. Increase the voltage by increments of 5 volts or less until the muscle barely responds (height of contraction about 1 mm) to the break shock. This is the break **threshold** or **liminal stimulus.**
40. Steadily increase the voltage (recording each voltage used). The muscle should contract to a progressively greater extent with an increase in voltage. These contractions are called **submaximal contractions.** When using the range selector to change voltage ranges, note that the 50 volt setting and the 5 volt setting on the next higher range produce identical voltages. For example, range X.1 and 50 volts = 5 volts; range X1 and 5 volts = 5 volts. Therefore, start above the 5 volt position on the new (higher) range.
41. Continue stimulation until there is no further increase in the height of the muscle contraction with increasing voltage. This is the **maximal stimulus.**
42. Continue increasing the strength of the stimulus (**supramaximal stimulus**) about five more times to see if there is a resultant decrease in height of contraction with a very strong current. Too strong a current may decrease the efficiency of the muscle as it may irritate it and increase fatigue.

Alternate Procedure using the Physiograph and Myograph

1. Detailed directions for the use of the physiograph can be found in Exercise 7-F, Part 1. The myograph is a photoelectric force transducer used primarily for quantitative measurements of smooth, cardiac, and skeletal muscle contractions. Three types of myographs with different sensitivities are available. Myograph B will be used in these experiments.
2. Connect the myograph to the laboratory stand with the mounting rod and clamp. Attach the frog board to the stand, as shown in Figure 8-4 for the Physiograph Four-A and Figure 8-5 for the DMP-4B. Attach the transducer cable to the myograph and to the input connector on the channel amplifier for the 4A models and to the 9-hole input connector on the transducer coupler on the 4B models.
3. Attach the stimulator output extension cable to the RF Isolated Output on the stimulator, red to red, black to black. Then, using the clip at the opposite end of the extension cable, attach the cable to the stand where the frog will be attached. Keep the clip near the bottom of the stand. This will minimize the effect of movement of the frog on the record.
4. Attach a set of pin electrodes (blue and yellow) to the clip electrodes attached to the stand. Color of attachment is not important. Make sure that the pin electrodes are fastened securely to the screw holders on the clip.
5. Double pith (see Exercise 12-G for procedure) or decapitate (see Exercise 8-A, Step 1 for procedure) the frog. Then place the frog on its ventral surface on the frog board.
6. Place the knee of the frog into the small clip on the frog board. It may be necessary to twist the body of the frog. As soon as it is inserted securely, insert a straight pin through the holes in the clip through the joint at the knee. (You may be able to pass it through the bone at this point.)
7. Remove the skin around the ankle so the Achilles tendon is visible. Pass the scissors underneath the Achilles tendon and draw a piece of thread under the tendon. Tie the thread to the tendon. After the thread is attached, cut through the tendon distal to the knot so that the foot will not contract when the muscle is stimulated.
8. Tie the other end of the thread in a fixed loop (not a slip knot) and slip the loop over the myograph leaf spring hook. Adjust the vertical position of the myograph to maintain a constant tension on the thread connected to the muscle, or use the tension adjustor to adjust the tension. The thread must be taut but not too tight. The muscles and tendon should be vertical and positioned directly below the hook on the myograph.
9. Insert the pin electrodes into the proximal and distal ends of the gastrocnemius muscle.
10. For the Physiograph Four-A and DMP-4A, place the channel amplifier controls in the following positions:

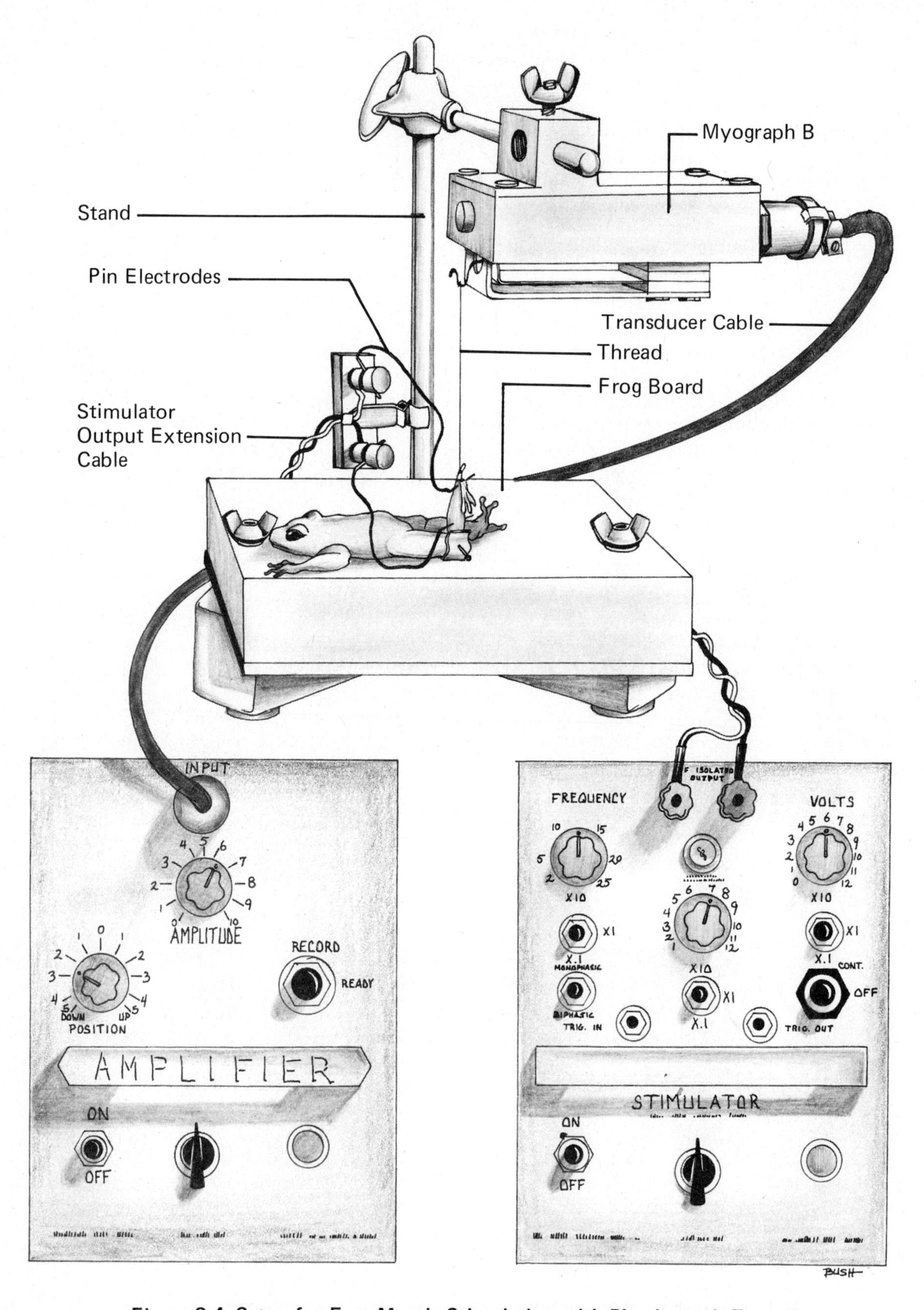

Figure 8-4. Setup for Frog Muscle Stimulation with Physiograph Four-A

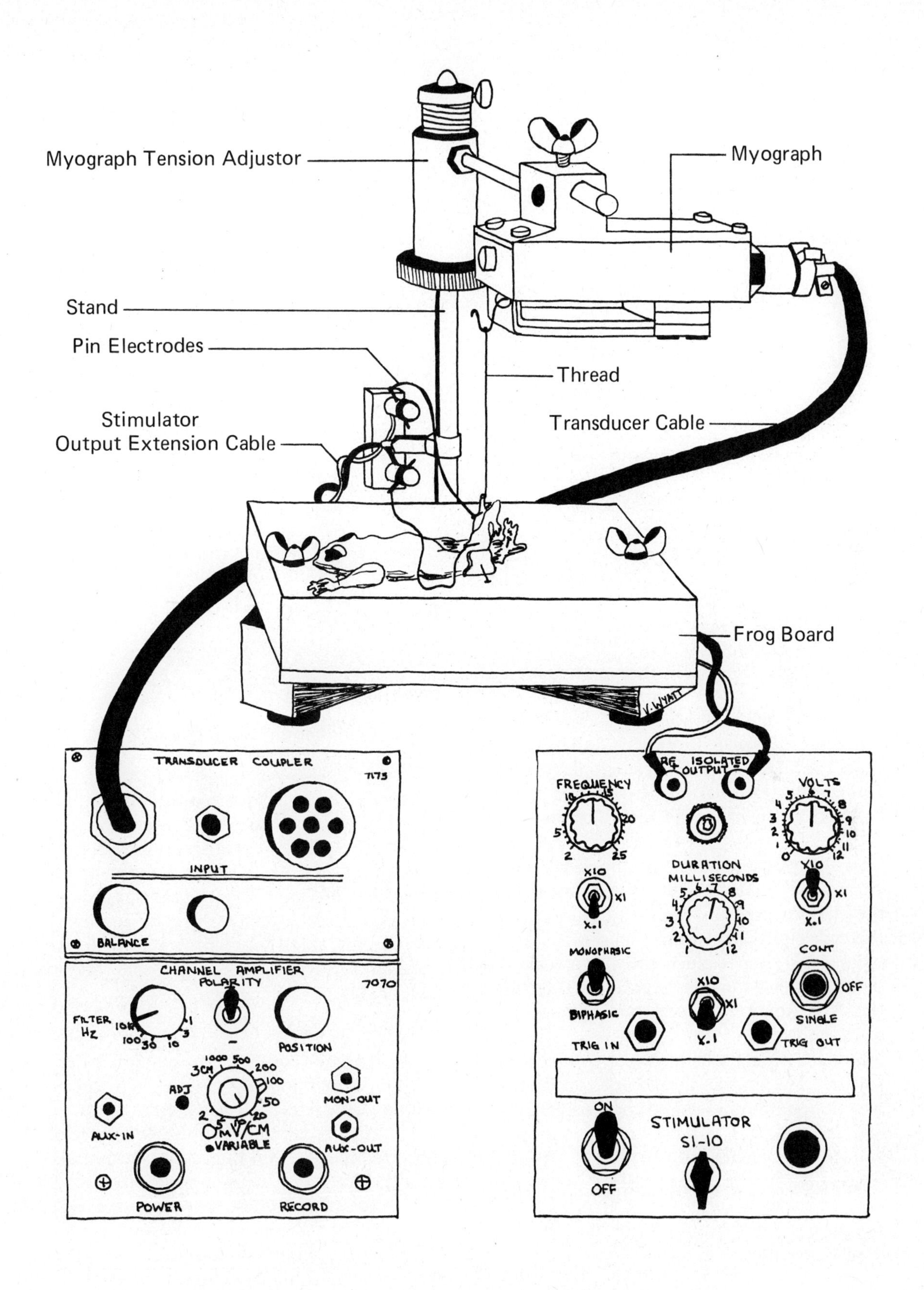

Figure 8-5. Setup for Frog Muscle Stimulation with DMP-4B

a. ON-OFF to ON. Allow a 5 minute warm-up time for maximum stability.
b. AMPLITUDE to 0.
c. POSITION to center 0 position.
d. RECORD-READY to READY.

11. For the Physiograph DMP-4B models set the controls as follows (recalibrate if necessary, following the procedure in Exercise 7-F, Part 1, Step 9):
    a. Press the channel amplifier POWER switch ON.
    b. Press the RECORD switch OFF (if it is glowing) and the filter to 10K.
    c. Set the mV/CM switch to 100. The amplitude may have to be readjusted later to produce a good height of contraction.
    d. Set the pen 1.5 cm below the center dark green line using the POSITION control.
12. Turn on the stimulator. Set the duration on 7, multiplier on X1. Set the volts on 0, multiplier on X.1. It is not necessary to set the frequency control because single stimuli will be used in this experiment. The CONT-OFF-SINGLE switch should be in the OFF position.
13. Start the ink flowing in the pen associated with the transducer coupler and the pen attached to the time-event marker.
14. Turn the channel amplifier to RECORD. With the DMP-4A and Physiograph Four-A, adjust the channel amplifier POSITION control to position the recording pen 1.5 cm below the pen centerline; with the DMP-4B physiograph, use the BALANCE control to reset the pen to 1.5 cm below the pen centerline.
15. Increase the voltage until the muscle barely contracts. Then reduce the voltage below this value. Set the paper speed at .1 cm/sec, the time marker to one per second, and start the paper.
16. Stimulate the muscle once with this reduced voltage. The muscle should not contract, as this is a **subliminal or subthreshold stimulus.** Record this voltage (and subsequent voltages) on the record.
17. Gradually increase the voltage until the muscle barely responds (height of contraction about 1 mm). This is the **threshold or liminal stimulus.**
18. Steadily increase the voltage (recording each voltage used). The muscle should contract to a progressively greater extent with an increase in voltage. These contractions are called **submaximal contractions.** When using the multiplier to change voltage ranges, note that the 10 volt setting and the 1 volt setting on the next higher range produce identical voltages. For example, range X.1 and 10 volts = 1 volt. Range X1 and 1 volt = 1 volt. Therefore, start above the 1 volt position on the new (higher) range.
19. Continue stimulation until there is no further increase in the height of the muscle contraction with increasing voltage. This is the **maximal stimulus.**
20. Continue increasing the strength of the stimulus about five more times to see if there is a resultant decrease (supramaximal stimulus) in the height of contraction with a very strong current. Too strong a current may decrease the efficiency of the muscle as it may irritate it and increase fatigue.

Exercise 8-B
Genesis of Tetanus

Objective

To demonstrate the genesis of tetanus.

Materials

Beaker; frog Ringer's solution; frog; dissecting set; watch glass; medicine dropper; and **either** wires, moist chamber, light muscle liver, kymograph, stand, 10-gram weight, induction stimulator Model 330 **or** stimulator Model 340, and two doulbe clamps, **or** physiograph, frog board, clips, stand, myograph B, pin electrodes, stimulator output extension cable, and transducer cable.

Discussion

A tetanic contraction occurs when the muscle is stimulated 30 to 100 times per second (the frequency needed to produce complete tetanus depending upon the muscle). At this frequency, the muscle is stimulated so rapidly that there is no opportunity for relaxation between twitches; therefore, the twitches fuse together, producing a smooth curve. The height of the tetanic contraction is greater than the height of single twitches.

Procedure Using the Kymograph

1. Use the same muscle preparation as in the previous experiment. It is not necessary to use the signal magnet in this experiment.
2. Use drum speed 9. At this speed the muscle contraction will produce a vertical line on the drum rather than a curve.
3. If the induction stimulator Model 330 is used, select a **maximal strength faradic stimulus**, so that the make and break stimulations produce the same height contraction.
4. Stimulate the muscle approximately five times at a rate of once per second. Each muscle twitch should be distinct, but close together. Let the muscle rest for 1 minute. Label these contractions single twitches.
5. Increase the rate of stimulation to two per second, for a period of 4 seconds. At this frequency the muscle may not have time to relax completely before the second stimulation is introduced. Allow the muscle to rest for one minute.
6. Continue to increase the rate of stimulation (maintaining the stimulation in each case for 3 to 4 seconds with a **1 minute rest period** in between) in order to produce incomplete tetanus. The individual twitches should be visible, but the muscle lever should not return to the base line between twitches.
7. Increase the rate of stimulation further in order to obtain complete tetanus. If possible, do this by making and breaking the circuit very rapidly.
8. If this is not possible, raise the key to the MULT. position. This produces a tetanic current, where the current is made and broken 50 times per second. This is of sufficient frequency to produce complete tetanus, but it results in rapid muscle fatigue. The record should show a complete fusion of twitches; individual contractions should not be visible.
9. Label single twitches, incomplete tetanus, and complete tetanus. Include the record under Results and Questions.
10. If the stimulator Model 340 (see Fig. 8-6) is used, attach the wires from the bottom of the moist chamber to the terminals on the stimulator marked STIM. Plug in the stimulator and set the voltage controls to produce a contraction between 1 and 2 inches in height. Note that this stimulator has a frequency control. The numbers indicate the frequency of stimulation per second. Set the lever marked SINGLE and MULTIPLE to MULTIPLE and follow the above directions (Steps 4-7), gradually increasing the frequency of stimulation from 1 per second until complete tetanus is obtained. The experiment can be performed on this instrument more easily and accurately than on the induction stimulator.

Procedure Using the Physiograph

1. Use the same muscle preparation and setup as in the previous experiment (see Exercise 8-A, Alternate Procedure, Step 10 for 4A model settings and Step 11 for the 4B model).
2. Turn on the stimulator. Set the frequency control on 5 and the multiplier on X.1. The CONT-OFF-SINGLE switch should be in the OFF position.
3. Press the RECORD button. Adjust the transducer coupler BALANCE control on the 4B model (the POSITION control on the 4A model) to reposition the pen 1.5 cm below the center dark green line if necessary.
4. Increase the voltage and adjust the amplitude (or mV/CM switch) to produce a single contraction 1-2 inches in height with

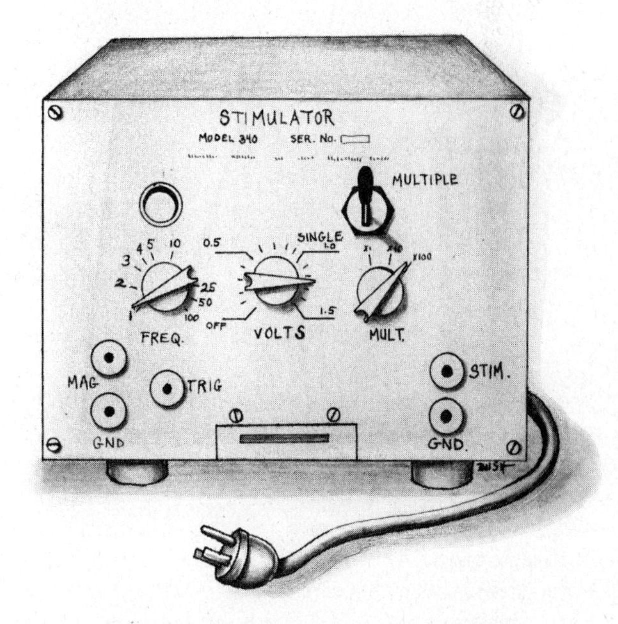

Figure 8-6. Stimulator Model 340

the CONT-OFF-SINGLE switch set on SINGLE.

5. Set the paper speed at .1 cm/sec and the time marker to one per second. Start the ink flowing. With a frequency of once every two seconds (frequency at 5, multiplier at X.1) move the CONT-OFF-SINGLE switch to CONT and stimulate the muscle 4-6 times. Increase the frequency to 1 per second (frequency on 10, multiplier at X.1) and stimulate the muscle approximately 6 times. Then turn the multiplier to X1 and the frequency to 2 (frequency 2 per second) and stimulate the muscle approximately 10 times. Gradually increase the frequency up to a frequency adequate to obtain complete tetanus. Each period of stimulation should occupy about 2 cm of chart paper.
6. Label the record completely with the frequency of each period of stimulation, single twitches, incomplete tetanus, and complete tetanus.

Exercise 8-C Duration of a Single Muscle Twitch

Objective

To measure the duration of a single muscle twitch.

Materials

Beaker, frog Ringer's solution; frog; dissecting set; watch glass; medicine dropper; and either wires, moist chamber, light muscle lever, kymograph, stand, 10-gram weight, induction stimulator, three double clamps, and signal magnet, or physiograph, frog board, clips, stand, myograph B, pin electrodes, stimulator output extension cable, and transducer cable.

Discussion

A single contraction of a muscle is called a single muscle twitch. There are three phases of the muscle twitch: the latent period, the period of contraction, and the period of relaxation. The latent period is the time between the introduction of the stimulus and beginning of the contraction. This is usually .01 sec or less. The contraction phase is the time during which the muscle shortens. This averages about .04 sec in the gastrocnemius muscle of the frog. The relaxation period is the time during which the muscle regains its original length. This may be .05 sec or longer. The entire muscle twitch lasts approximately .1 sec or longer. These figures vary greatly as the temperature of the muscle and the load to which it is subjected vary.

Procedure Using The Kymograph

1. Use the same equipment setup and frog as in Exercise 8-A.
2. Place the writing point of the signal magnet (which will indicate the point of introduction of the current into the muscle) immediately beneath the writing point of the muscle lever stylus.
3. To measure the duration of a single muscle twitch, use a rapidly rotating drum (speed 2). At this speed the drum moves 20 cm/sec.
4. Since only one contraction is desired, it is necessary to short circuit the break. In order to use only the make, depress the key at the proper time during the experiment. **Do not release the key** until after the conclusion of the experiment.
5. Start the drum rotating, then make the circuit. Rotate the stand away from the drum after one contraction of the muscle and before one complete revolution of the drum in order to prevent doubling up on the tracing. Then break the circuit.
6. Replace the muscle pen in its original position. Let it write a base line as illustrated in Figure 8-7.
7. To obtain the proper time relationship, turn the stand slightly away from the drum. Place the top of the pen at point c (see Fig. 8-7). Let the pen fall gently, describing an arc, to the base line.
8. Construct an appropriate vertical line from the signal magnet tracing to the base line, as indicated on Figure 8-7.
9. Label the curve as follows:
   a. point of introduction of the current
   b. beginning of contraction of the muscle
   c. end of contraction
   d. end of relaxation
   e. where arc crosses the base line.
10. Calculate the duration of each phase of the muscle twitch, as directed under Results and Questions. Include your kymograph record in Results.

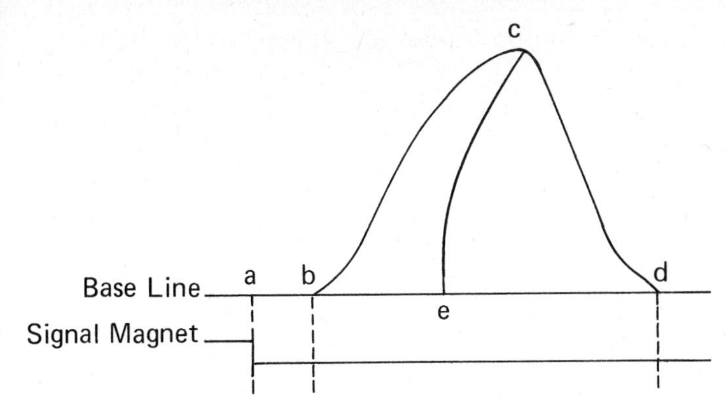

Figure 8-7. Duration of a Muscle Twitch

Procedure Using the Physiograph

1. Use the same equipment setup and frog as in Exercise 8-A. Use a paper speed of 10 cm/sec. (See Exercise 8-A, Alternate Procedure, Step 10 for the 4A model settings, Step 11 for the 4B model settings.)
2. Pens are needed for both the channel amplifier and the time-event channel.
3. Turn on the stimulator. Set the duration on 7 and the multiplier on X1. Set the volts on approximately 5 volts (voltage necessary to obtain an adequate height of contraction). The CONT-OFF-SINGLE switch should be in the OFF position.
4. Adjust the voltage and the amplitude (or mV/CM switch) to obtain an adequate height of contraction (approximately 1-2 inches in height).
5. After starting the paper, stimulate the muscle once. The time-event pen should indicate the precise point of stimulation.
6. Using a ruler, draw a baseline across the bottom of the recording. Draw vertical lines from the top of the curve to the baseline and from the point of introduction of the stimulus to the baseline.
7. Label the curve as in Step 9 with the kymograph and make the calculations as directed in Step 10.

Exercise 8-D Muscle Fatigue

Objective

To demonstrate the factors causing muscle fatigue and the restorative effect of rest.

Materials

Beaker; frog Ringer's solution; frog; dissecting set; watch glass; medicine dropper; and **either** wires, moist chamber, light muscle lever, kymograph, stand, 10-gram weight, induction or electronic stimulator, and two double clamps, **or** physiograph, frog board, clips, stand, myograph B, pin electrodes, stimulator output extension cable, and transducer cable.

Discussion

As a muscle continues to contract, it gradually becomes fatigued and loses its ability to contract. The time of appearance of fatigue depends upon the strength of the stimulus and the load. This experiment must be the last in any series with the frog muscle since it will not recover from the fatigue to any extent.

Procedure using the kymograph

1. Set up the recording and stimulating apparatus as in Exercise 8-A.
2. Load the muscle with 10 grams. Do not set the after-loading screw. Make certain that the resting position of the muscle lever is horizontal.
3. Use a slowly moving drum (speed 10) so that the muscle contraction produces a vertical line.
4. Either the Harvard Model 330 induction stimulator or the Harvard Model electronic stimulator can be used in this experiment. The electronic stimulator is easier to use since it can be set to stimulate the muscle automatically at a preset frequency. In order to use the electronic stimulator, connect two wires from the two terminals marked STIM to the two terminals on the inferior surface of the moist chamber. Set the frequency at once per second. Set the voltage at a level sufficient to produce a muscle contraction 1-2 inches in height. Start the drum and set the switch to MULTIPLE.
5. If the Model 330 Induction Stimulator is used, stimulate the muscle once per second with a constant strength, maximal break current. Short circuit the make. Practice this procedure before beginning the experiment.
6. Continue stimulating the muscle once per second until the muscle no longer contracts. Disregard contractions less than 1 mm in height on the drum.

7. **Continue the stimulation** (without a period of rest) for one more minute to see if there is any recovery. The muscle may not contract for 15-30 seconds, even though it is stimulated, and then it may contract slightly several times.
8. Allow the muscle to rest for 5 minutes; then, using the same strength and rate of stimulation, determine whether the muscle has recovered from fatigue.
9. Label the record completely and include it under Results and Questions.

Procedure Using the Physiograph

1. Set up the physiograph as described in Exercise 8-A. Use a paper speed of .05 cm/sec. Set the frequency of stimulation at 1 per second (frequency at 10, multiplier at X.1) and the duration at 7, multiplier at X1). Set the voltage of stimulation at a level sufficient to produce a muscle contraction 1-2 inches in height. (The muscle will fatigue faster at a higher voltage. Start the paper and set the switch to continuous stimulation.)
2. Follow steps 6-9 above.

Exercise 8-E Physiology of Human Muscle

Objectives

To demonstrate the change in size of skeletal muscle during contraction; to demonstrate human muscle fatigue.

Materials

Kymograph; ergograph; pen (intermediate length ink stylus, ball point pen or pencil); sphygmomanometer; tape measure.

Discussion

When a muscle contracts, it shortens, pulling the insertion toward the origin. The belly of the muscle becomes shorter and thicker. This change in size of the muscle will be measured in this exercise.

The ergograph is a device for measuring the amount of work performed in muscular action. It consists of an adjustable tension spring with a writing point and trigger, fastened in a support mounted for stability on a heavy board.

Repeated contraction of a muscle produces muscle fatigue. Fatigue is caused by the accumulation of waste products such as lactic acid, or by a deficiency in oxygen, glucose, and other raw materials. The compression of the brachial artery by the use of a sphygmomanometer in this experiment results in interference with the blood supply to the muscles of the fingers and the muscle should fatigue faster.

Procedure

1. Measure the circumference of your partner's arm with the forearm extended and with it flexed. Be certain to apply the tape measure to exactly the same place on the arm each time. Record your measurements under Results and Questions.
2. Set up the kymograph and ergograph as in Figure 8-8 with the ergograph to the right of the kymograph.
3. Use the bottom perforation on the spring housing, so that the tension is at its highest level.
4. Adjust the ink supply so that the ink does not flow out too freely. A ball point pen or lead pencil can be taped to the arm and used instead of the ink stylus.
5. Adjust the height of the kymograph so that the first tracings are approximately two inches from the bottom of the paper.
6. Use drum speed 9 throughout the experiment.
7. The subject should be seated comfortably and should grasp the trigger of the ergograph with the index finger of his dominant hand. It is important to keep the index finger in constant contact with the trigger.
8. Start the kymograph. Rhythmically squeeze the trigger at a rate of about two times per second until complete fatigue develops.
9. Allow the subject to rest exactly 30 seconds and repeat the above procedure.
10. Allow the subject to rest 5 minutes. Place a sphygmomanometer cuff on the arm immediately above the elbow. Inflate this to a pressure of 140 mm and repeat the above procedure.
11. Calculate the length of time necessary to fatigue the muscle initially, after 30 se-

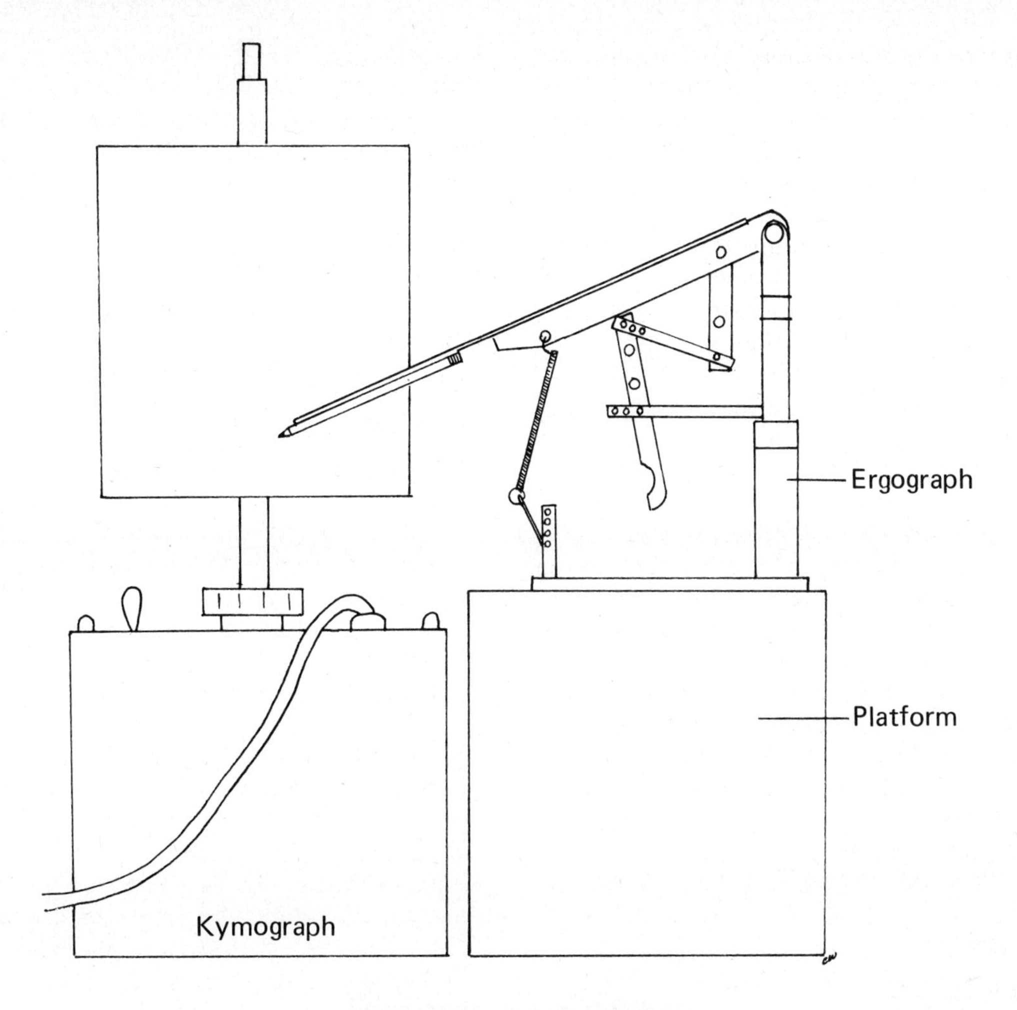

Figure 8-8. The Ergograph

conds rest, and with the sphygmomanometer. Compare the initial force of contraction with that after 30 seconds rest and with that produced with the use of the sphygmomanometer.

12. Compare the length of time necessary to produce fatigue on your record with that of your partner. Record your observations and answer the questions under Results and Questions.

Alternate Procedure

1. It is possible to perform approximately the same experiments without the ergograph. Lay your left forearm on the deck, palm up.
2. Flex and extend the fingers rapidly at the metacarpal-phalangeal joint, until they are fatigued. Record the time needed to cause fatigue.
3. Rest your fingers for 5 minutes.
4. Apply the blood pressure cuff immediately above the elbow. Inflate the sphygmomanometer to 140 mm. Repeat the procedure in Step 2, recording the time necessary to cause fatigue.

Self-Test—Muscle Physiology

Directions: See Chapter 1, Self-Test, p. 9. Also see the Self-Test at the end of Chapter 6.

1. During a single contraction of a muscle fiber, the fiber uses oxygen during the **contraction** period.

2. Myasthenia gravis involves an overproduction of **acetylcholine.**

3. Products formed during the contraction of a muscle include **ADP, creatine, and lactic acid.**

4. One result of muscular contractions is a change in the chemical composition of **the urine.**

5. Reciprocal innervation involves **both** stimulating and relaxing impulses.

6. A **tonic** spasm is a continued muscular contraction whereas a **clonic** spasm consists of a series of alternate contractions and relaxations.

7. **Fibrillation** is an ineffective quivering of muscle fibers.

8. **Flaccid paralysis** involves complete paralysis, abolishment of muscle tone, and "limp" extremities.

9. Muscle **spasm** is a steady, partial contraction present at all times in healthy muscles.

10. Visceral muscle reacts to stimuli **more** rapidly than does skeletal muscle.

11. Muscle tone is **lowest** during sleep, but it **is also reduced** in ill health and after prolonged bed rest.

12. Such drugs as Anectine, Intocostrin, and other curare-like compounds are used in the operating room to **prevent** muscular relaxation by blocking nerve impulses.

13. In terms of activity and the development and maintenance of the highest level of body efficiency and health, experts recommend **building up efficiency to age 30**, maintaining and preserving it thereafter.

14. **Choline acetylase** is a chemical needed to stimulate muscle fibers to action.

15. **Cholinesterase** inactivates acetylcholine, thus preventing continued action of the muscle fibers.

16. An individual poisoned by curare may die as a result of paralysis of the **diaphragm.**

17. Research with curare resulted in the use of **neostigmine** (which counteracts the affects of curare) for treatment of **myasthenics.**

18. Each skeletal muscle cell has **many** nuclei.

19. The advantage of recording muscular contractions on a moving surface such as the kymograph or oscillograph is that **the time element is amplified** so that the separate events are spread out and can be studied in detail.

20. The chemical and physical processes in a single muscle twitch of a muscle from a cold-blooded vertebrate will be carried out in **less time** than would be required for those in a mammalian muscle.

21. The endings of axons of an **afferent** nerve fiber on the muscle fibers they innervate are called motor end plates or myoneural junctions.

22. The number of muscle fibers in a motor unit is **unrelated** to the degree of precision with which a muscle is capable of acting.

23. Work **is** performed by isometric contraction.

24. Muscles **can** contract in the absence of oxygen.

25. In the chemical reactions associated with muscular contraction, the resynthesis of glycogen **can** take place anaerobically.

26. **Hemoglobin** is the name of the iron-containing protein in muscle.

27. Posture is maintained primarily by **isometric contraction** of muscles.

28. The slight muscular tonus or tension which continues, even under resting conditions, depends on **asynchronous contraction** of different muscle units.

29. Contraction of a muscle without shortening is called **isometric** contraction.

30. The warming up process in an isolated muscle that shows successive increases in the degree of shortening of the muscle in response to stimuli of constant strength is called **treppe**.

31. The time interval between application of a stimulus and the beginning of a response is called the **latent period**.

32. Work may be calculated by multiplying the **height** a load is lifted by the **weight** load.

33. When a muscle contracts, the blood within it becomes **slightly acid**.

34. The general term **muscle spasm** means a sudden, abnormal, involuntary muscular contraction.

KEY

1. recovery
2. cholinesterase
9. tonus
10. less
12. produce
14. acetylcholine
20. the same (approximately) time as
21. efferent
22. related
23. is not
25. cannot
26. myoglobin

Name ______________________

Results and Questions for Chapter 8

Exercise 8-A. Strength of Stimulus and Height of Contraction

Questions for Experiments with the Kymograph

1. Paste the kymograph record showing the make and break contractions in the space below. Label the make and break contractions.

2. Define the make threshold. ______________________

   What was your experimental value? ______________________

3. Define the break threshold. ______________________

   What was your experimental value? ______________________

4. Was the height of contraction with faradic stimulation greater on the make or on the break?

   ____

5. Was the current flowing between make and break when the key was depressed? ______________________

6. Did the muscle contract during the time the key was depressed between the make and the break contractions? ______________________

7. What is the general function of muscle? ______________________

Questions for both the Kymograph and Physiograph Experiments

8. Paste your fully labeled record in the space below. Label the voltage, subliminal, liminal, submaximal, maximal (and supramaximal, if evident) contractions.

9. Record the voltage at each contraction and the magnified height of contraction in the following table of results.

Voltage	*Magnified Height of Contraction in mm*

10. Plot the results of this experiment on the grid on the next page, using the magnified height in mm on the Y axis (vertical axis) and the voltage on the X axis (horizontal axis).

11. Define irritability. ______________________________

12. Define the all-or-none law. ______________________________

____

13. Did all muscle fibers in the muscle have the same degree of irritability? ______________________________

14. What caused the increase in height of contraction with increasing strength of stimulus?

____

Name ______________________

15. Does each individual muscle fiber follow the all-or-none law or does the muscle as a whole?

____

16. Was there any advantage to stimulating the muscle with a current stronger than maximal?_____ Any disadvantage?_______________

____

17. State Starling's Law of the Heart._______________

____

18. Does this law hold true for skeletal as well as cardiac muscle?_____Explain your answer:_____

____

Exercise 8-B. Genesis of Tetanus

1. Define muscle tetanus._______________

____

2. Paste your record, fully labeled, in the space below. Label single twitches, incomplete tetanus, complete tetanus, and the frequency on each record.

3. Was the height of the tetanic contraction greater or less than that obtained with the simple muscle twitches?______Explain: ____

____

4. What accounts for the smooth curve in complete tetanus?________________________________

____

5. Are most of the contractions in the human body simple twitches or tetanic contractions?

____

Exercise 8-C. A Simple Muscle Twitch

1. Paste the record of a simple muscle twitch in the space below. Label the record fully.

2. Fill in the following table of results by calculating the duration of each of the listed events. Divide the length of each period (in centimeters) by the paper speed. Then calculate the percent of time of the entire muscle twitch occupied by the latent period, the contraction period, and the relaxation period. In order to do this, substitute in the following equation:

$$\frac{\text{duration of the latent period}}{\text{duration of the entire muscle twitch}} \times 100 = ________$$

Repeat the calculation for each phase of the muscle twitch.

Name ____________________

Stages in a Muscle Twitch	*Length of Each Phase in Cm.*	*Duration (sec)*	*% of Time of Entire Twitch*
Latent period (a - b)			
Contraction (b - e)			
Relaxation (e - d)			
Entire muscle twitch (a - d)			100%

3. How do the values obtained in this experiment agree with the expected values for the duration of each event in a muscle twitch? ____________________

4. Define the latent period. ____________________

5. How would a decrease in the temperature of the muscle affect each of the following?

   a. Duration of the latent period. ____________________

   b. Height of contraction. ____________________

   c. Duration of the relaxation period. ____________________

Exercise 8-D. Muscle Fatigue

1. Fasten your record from muscle fatigue in the space below. Label treppe, contracture, fatigue, rest period, and any recovery from fatigue.

2. List the possible causes of muscle fatigue. ______________________________

____

3. Fatigue inhibits the complete relaxation of a muscle. Did you see any evidence of such inhibition in this experiment? ______________________________

4. Define contracture. ______________________________

5. Is the pH of a fatigued muscle likely to be more acid or more alkaline than usual? ______________ Explain: ______________________________

____

Exercise 8-E. Physiology of Human Muscle

1. Record the size of the arm below:

	Size (in inches)
With forearm extended	
With forearm flexed	

2. What caused the above changes in size? ______________________________

____

3. Fasten your ergograph record below. Label each portion of the record.

Name ______________________

4. Calculate the duration of time required to cause fatigue by dividing the length of the record by the paper speed. Record all calculations in the following table. Fill in the results in the table.

	Time in Seconds	*Initial Force of Contraction in cm*
Time required for initial fatigue		
Time required for fatigue after a 30 second rest		
Time required for initial fatigue of partner		
Time required for fatigue of partner after a 30 second rest		
Time required for fatigue with sphygmomanometer		
Time required for fatigue of partner with sphygmomanometer		

5. What caused the difference in time needed for fatigue in each part of the experiment? ______________________

____

____

6. Describe the chemical steps that occur during muscle contraction. ______________________

____

____

____

7. Describe the chemical steps that occur during recovery from fatigue. ______________________

____

____

8. Assume the subject's hands were placed in ice water immediately preceding the rapid flexion and extension of the fingers. What effect would you expect this to have on the length of time necessary for inducing muscle fatigue? ______________________

____

____

9. Define an isometric contraction. ______________________

List one example of where this type of contraction occurs in the body. ______________________

____

____

10. Define an isotonic contraction. ______________________________

List one example of where this type of contraction occurs in the body. ______________________________

____

11. In which of the above is work performed? ______________________________

9

The Digestive System

Exercise 9-A Anatomy of the Human Digestive System

Objectives

To become familiar with the location and functions of the true and accessory human digestive organs.

Materials

Dissectable torso; anatomic charts; and reference books.

Discussion

The digestive system is concerned with receiving, transporting, digesting, and absorbing food. The alimentary canal or digestive tract, a tube approximately 30 feet long, is subdivided into the following regions: mouth, pharynx, esophagus, stomach, small intestine, and large intestine, all of which are considered to be true digestive organs. In addition, other organs are connected to the alimentary canal and are considered to be part of the digestive system. These accessory digestive organs include the tongue, teeth, salivary glands, liver, gall bladder, and pancreas.

For food to be absorbed into the blood stream, it must first be broken down into smaller, diffusable substances. This chemical breakdown of food is accomplished by digestive enzymes which break down complex foods into simple compounds which are readily absorbed from the digestive tract. This type of chemical reaction, which involves the addition of water, is called hydrolysis. The end products of digestion are absorbed into the blood stream, primarily through the walls of the small intestine, by the processes of dialysis and active absorption.

Procedure

1. Use Figure 9-1 as a self-test. See Section J, p. xx (Illustrations as Self-Tests) for a suggested approach.
2. Locate the structures listed in the key to Figure 9-1 on the dissectable torso and determine at least one important function of each.
3. Learn the parts of a typical tooth and the various types of teeth by studying Figures 9-2, 9-3, and 9-4. These figures may be used as self-tests.

KEY TO FIGURE 9-1

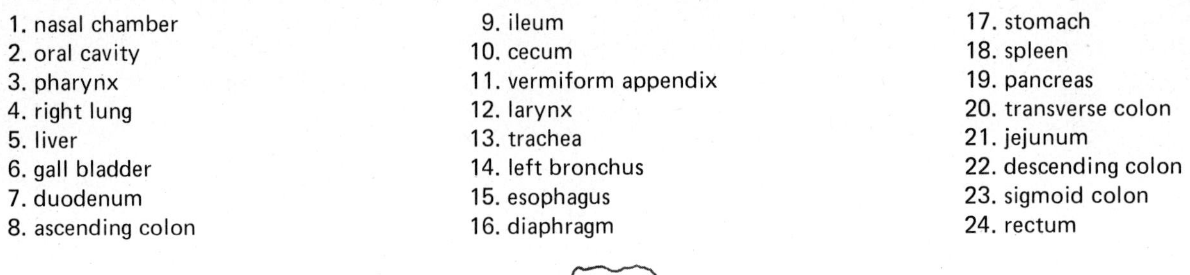

1. nasal chamber
2. oral cavity
3. pharynx
4. right lung
5. liver
6. gall bladder
7. duodenum
8. ascending colon
9. ileum
10. cecum
11. vermiform appendix
12. larynx
13. trachea
14. left bronchus
15. esophagus
16. diaphragm
17. stomach
18. spleen
19. pancreas
20. transverse colon
21. jejunum
22. descending colon
23. sigmoid colon
24. rectum

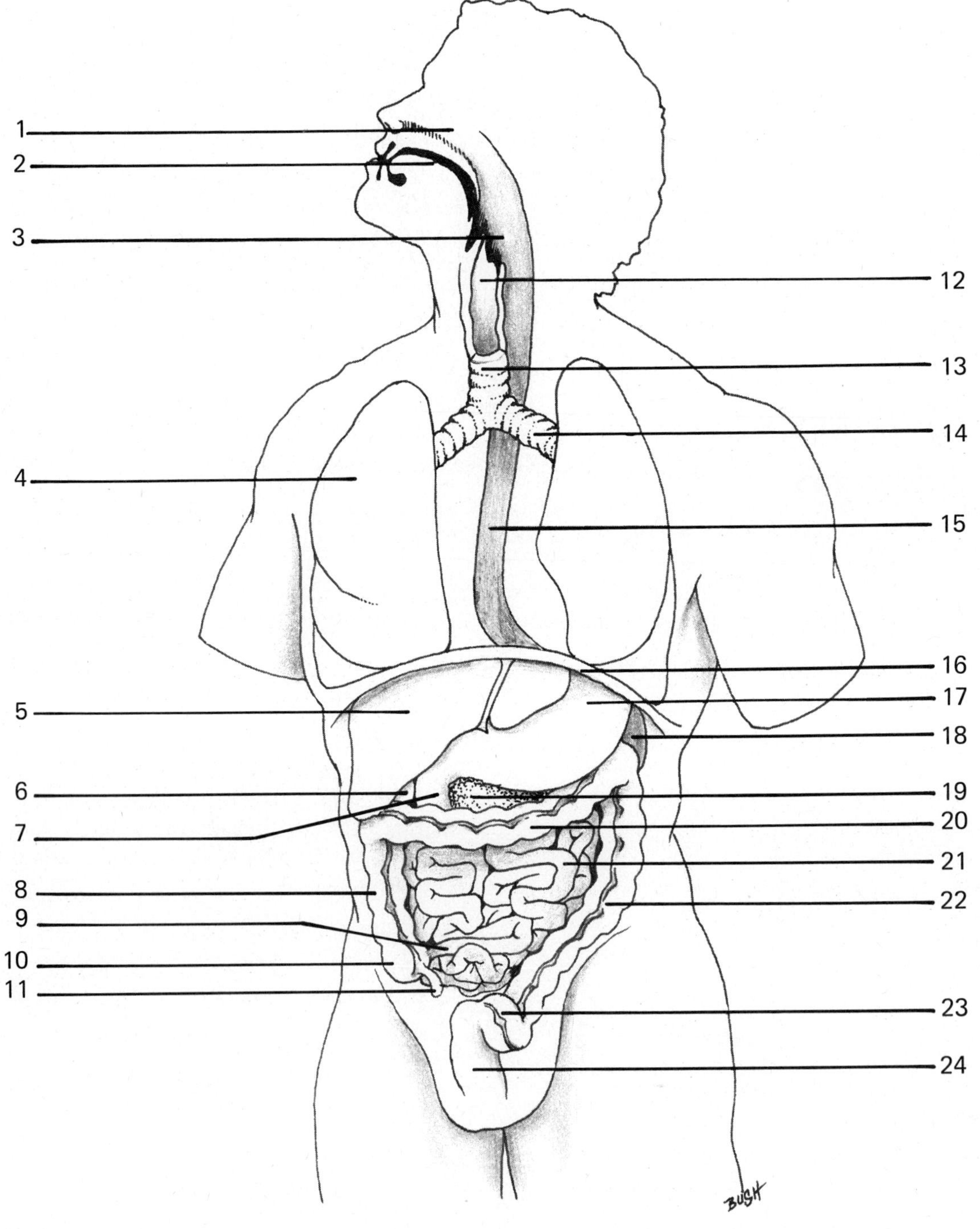

Figure 9-1. Digestive Organs

KEY TO FIGURE 9-2

1. crown
2. neck
3. root
4. enamel
5. dentin
6. pulp cavity
7. gingiva
8. cementum
9. root canal
10. periodontal membrane
11. bone

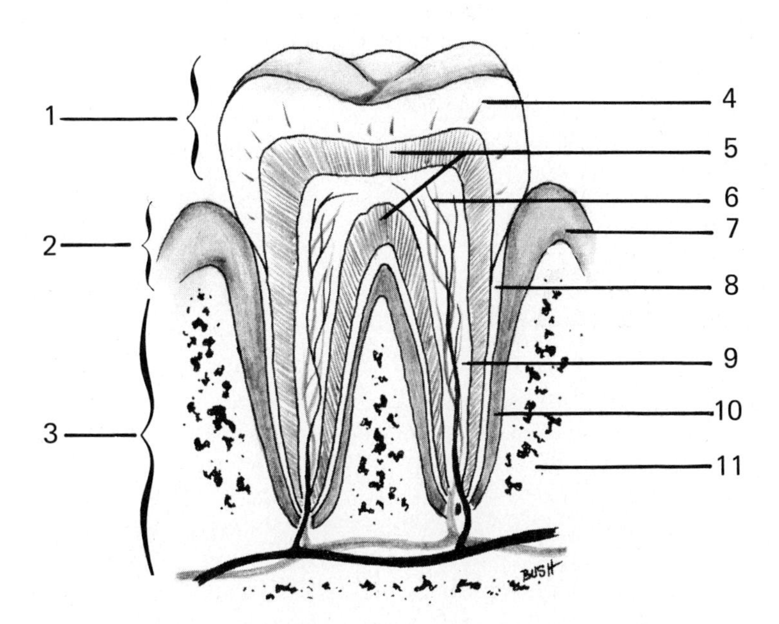

Figure 9-2. Longitudinal Section through a Molar

KEY TO FIGURE 9-3

1. second molar
2. first molar
3. cuspid
4. lateral incisor
5. central incisor

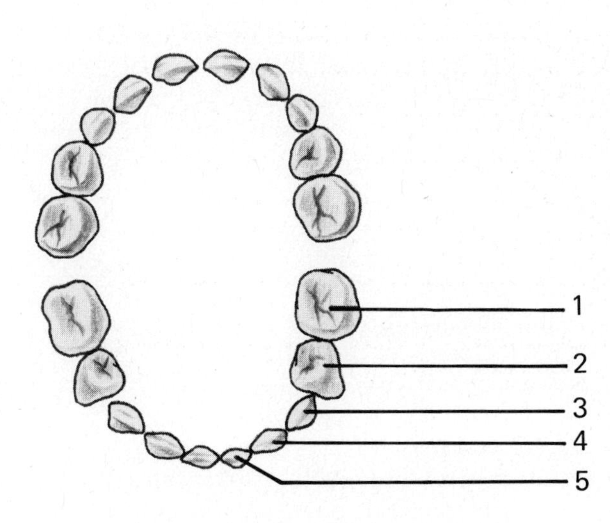

Figure 9-3. The Deciduous Teeth

KEY TO FIGURE 9-4

1. central incisor
2. lateral incisor
3. cuspid
4. first premolar
5. second premolar
6. first molar
7. second molar
8. third molar

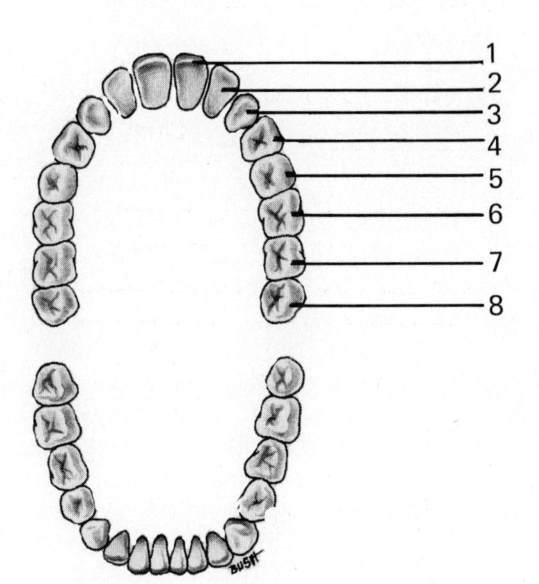

Figure 9-4. The Permanent Teeth

Exercise 9-B Dissection of the Digestive System of the Fetal Pig

Objectives

To identify the digestive organs of the fetal pig and to compare them with the corresponding human structures.

Materials

Fetal pig; dissecting board or tray; dissecting set.

Discussion

The digestive organs of the fetal pig are very similar to those of the human. Observe carefully the differences between the two organisms, such as the arrangement of the colon, and the absence of both the vermiform appendix and the uvula in the pig. Be prepared to trace the path of food through the digestive tract.

Procedure

1. Lay the pig, ventral surface up, on the dissecting board. To spread the legs of the specimen, tie a length of cord from each ankle to the nails on the sides of the board, or loop the cord underneath the board and tie it.
2. Use Figure 9-5 as a guide for the following dissection:
   a. Make an incision with the scalpel through the skin from the small, hairy **papilla** on the upper part of the throat down the midline to a point just anterior (cranial) to the umbilical cord.
   b. Continue to deepen the incision very carefully, cutting through the cartilaginous sternum, until the body cavity is reached. Be careful not to cut the large veins and arteries.
   c. Cut around the cord (as illustrated) and continue the incision laterally on each side to the head of the femur in the ventral thigh region. If the directions are followed, a flap will be produced which includes the umbilical cord and some urogenital organs.
   d. If the body cavity is filled with a dark fluid, flush it out with water; be careful not to damage any organs or tear the umbilical vein.
3. Pull the flap that contains the umbilical vessels back carefully and observe the **umbilical vein**, which runs from the umbilical cord to the liver. The umbilical vein continues on as the **ductus venosus**, which passes through the liver and then enters the posterior vena cava.
4. Tie a thread around the umbilical vein and cut it between the thread and the flap. The thread will enable quick location of the severed vessel.
   a. Pull the ventral flap further back and observe the two **umbilical arteries**, injected with red latex, on the inner surface.
   b. Locate the **urinary bladder**, the large sac situated between the two umbilical arteries, at the posterior (caudal) portion of the abdominal cavity.
5. To enlarge the exposed cavities, make the following incisions, which will facilitate subsequent dissections and at the same time leave the diaphragm intact:
   a. Cut laterally, on each side, immediately above the **diaphragm**.
   b. Make a second incision on each side directly below the diaphragm.
6. As in the human, there are three pairs of salivary glands in the pig. However, their size and shape is variable in the fetal pig, making dissection and accurate identification difficult. Using Figure 9-6 as a guide, limit your dissection to the location of the parotid and submandibular glands. Use the following instructions for this dissection:
   a. Starting at the base of the ear, inferior to the external auditory meatus, make a shallow incision through the skin.
   b. Continue the incision down the side of the face to the shoulder.
   c. Separate the skin carefully from the musculature and glands on either side of the incision. Locate the masseter muscle.
   d. The **parotid gland** is a large, thin, light colored, triangular gland, which usually extends from the base of the ear to the shoulder. It is underdeveloped in the fetal pig.
   e. **Stensen's duct**, the thin, white duct of the parotid gland, is located on the anteroventral border of the masseter

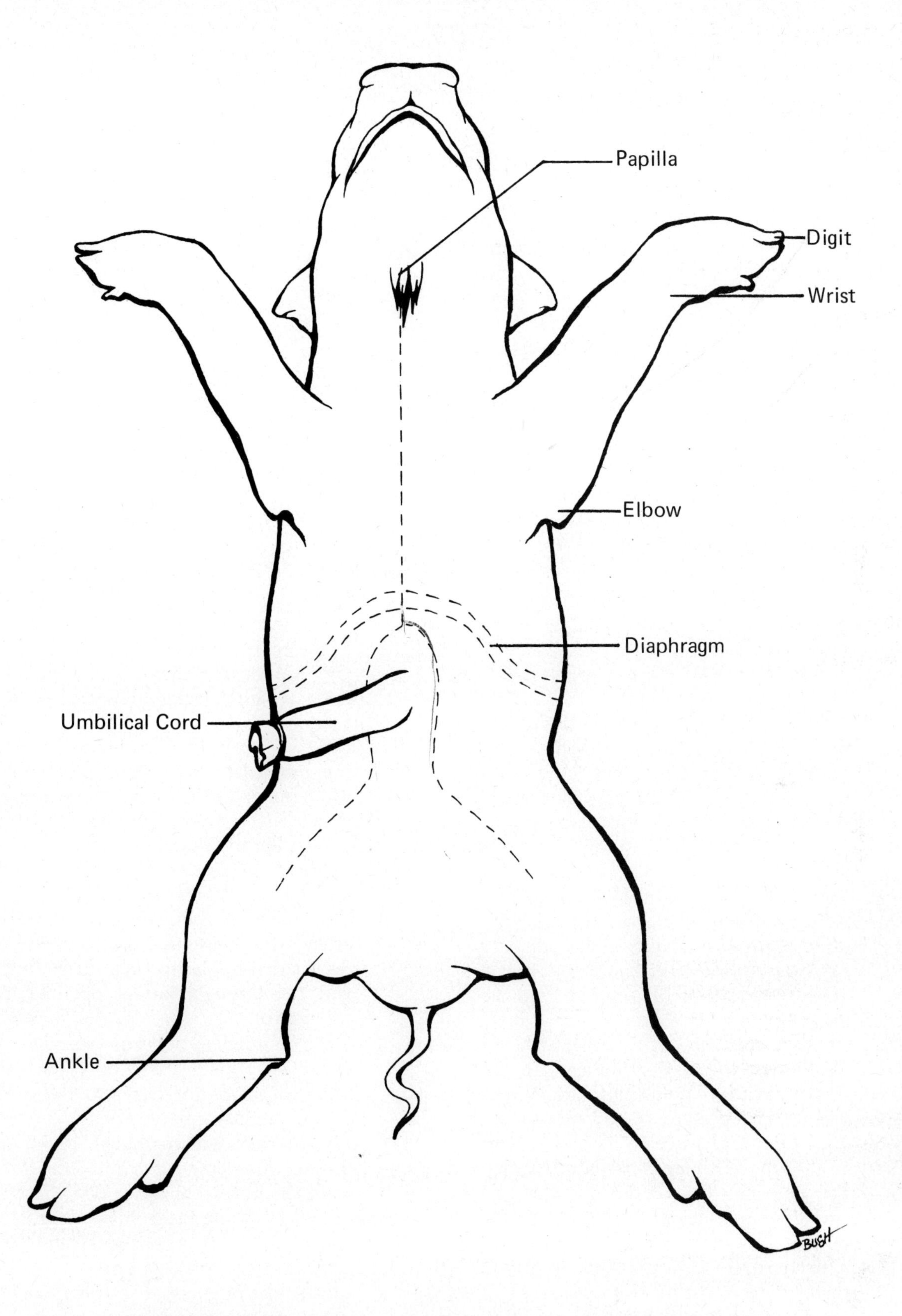

Figure 9-5. Diagram Showing Initial Incisions for Fetal Pig Dissection

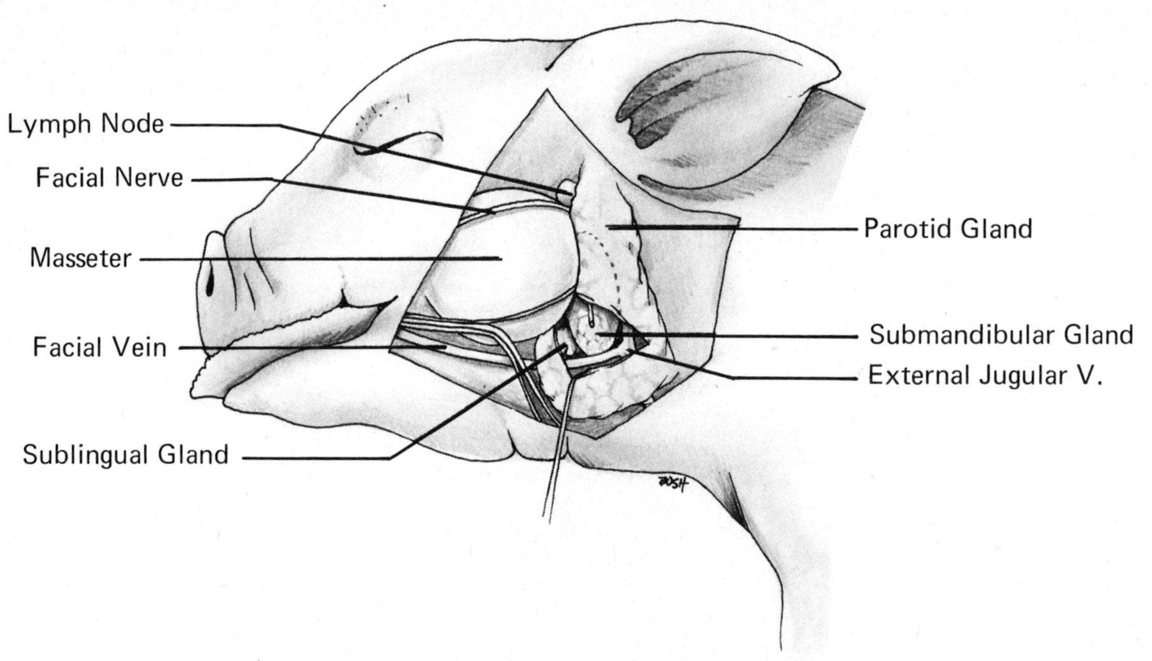

Figure 9-6. Dissection of the Salivary Glands

muscle, the large muscle anterior to the parotid gland covering the angle of the lower jaw. The duct opens into the oral cavity by the fourth upper premolar.

f. The **submandibular** gland lies deep to the parotid gland. To locate this gland, make a shallow incision through the parotid gland at the level of the inferior border of the masseter muscle. Reflect the lower portion of the parotid. The submandibular gland should now be visible.

g. The **sublingual gland** is located anterior to the inferior border of the submandibular gland. A small **lymph node** may be located by separating the dorsal anterior border of the parotid from the masseter.

7. Cut through the angle of each jaw, toward the ears, until it is possible to see the **epiglottis** and the opening into the throat (pharynx). The interior of the mouth can now be examined (see Fig. 9-7).

   a. Locate the **vestibule**, the space between the lips and the teeth, and the **oral cavity**, the principal cavity inside the mouth.

   b. The **tongue** is attached ventrally throughout most of its length by a membrane called the **lingual fenulum**.

   c. The surface of the tongue contains numerous **papillae**. Taste buds are associated with the papillae.

   d. If the **teeth** are not visible, cut into the jaw and expose them. The dental formula for the deciduous set of teeth in the pig is 3 1 4 0.

   e. Observe thc bony ridged **hard plate** and the muscular **soft palate**, posterior to the hard palate.The pig lacks the uvula, which is the posterior extension of the soft palate in humans.

   f. Identify the **fauces**, the opening of the oral cavity into the **oropharynx**.

   g. The pharynx is divided into three regions: the **nasopharynx** behind the nose, the **oropharynx** behind the mouth, and the **laryngopharynx** opening into the larynx.

   h. Make a median incision through the entire soft palate to expose the nasopharynx. (Do not cut into the dorsal wall of the nasopharynx.)

   i. Locate the two small slits, about one mm in diameter, in the lateral dorsal walls of the nasopharynx. These are the openings of the **Eustachian tubes** (auditory tubes). The **internal nares** also open into the nasopharynx.

   j. Locate the opening of the pharynx into the **esophagus**, dorsal to the larynx.

   k. Locate the **glottis**, the opening into the

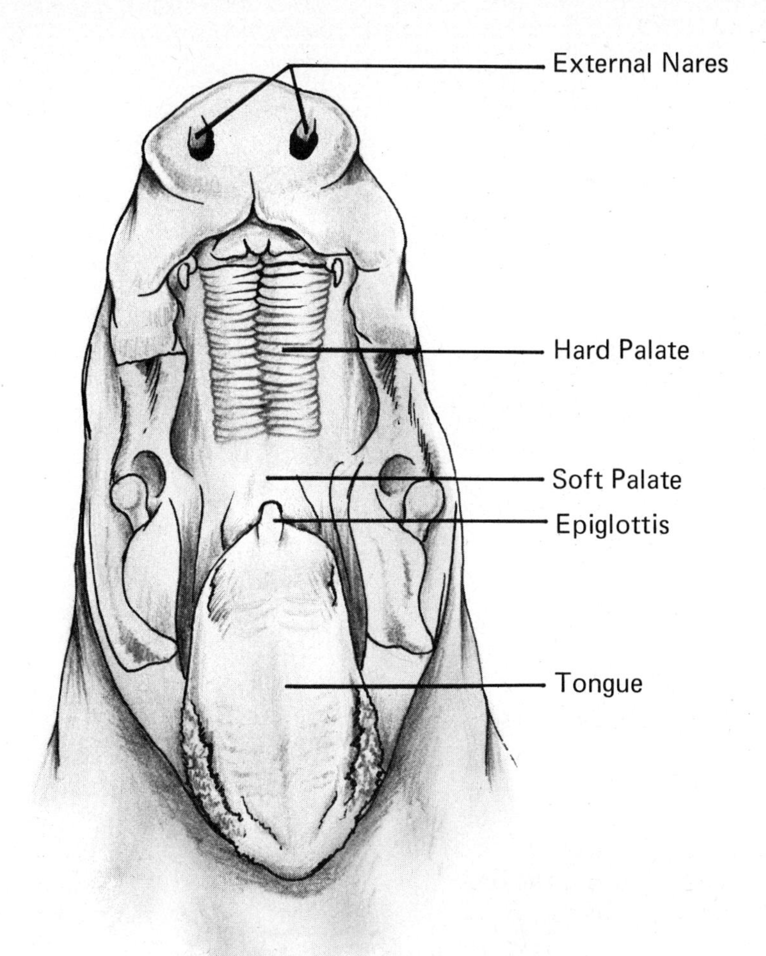

Figure 9-7. The Oral Cavity

larynx. The **epiglottis** can be seen as a small tonguelike flap at the entrance to the larynx.

8. Locate the esophagus again in the thoracic cavity in order to determine its position relative to the heart.
9. Locate the **diaphragm**, which separates the thoracic cavity from the abdominal cavity.
10. Use Figures 9-8 and 9-9 to assist you in locating the following organs. Compare these structures with those of the human indicated in Figure 9-10 (which may be used as a self-test).
    a. Locate the large, reddish-brown colored liver posterior to the diaphragm. Note that the superior surface of the liver is convex to match the concavity of the diaphragm.
    b. Count the number of lobes in the liver. The pig liver is divided into five lobes: the **right lateral, right central, left central, left lateral,** and **caudate**. Compare this with the situation in humans.
    c. Lift up the right lobe of the liver and locate the **gall bladder**, the small, pear-shaped sac embedded in the right central lobe.
    d. The umbilical vein can be found entering the liver to the left of the gall bladder.
    e. The **cystic duct** from the gall bladder and the **hepatic duct** from the liver unite to form the **common bile duct**, which empties into the duodenum (see Fig. 9-9). In order to locate these structures, gently tease away the **lesser omentum** between the stomach and liver. First locate the green cystic duct from the gall bladder and the common bile duct. To locate the hepatic duct, trace the common bile duct upward to the point where the cystic duct enters.

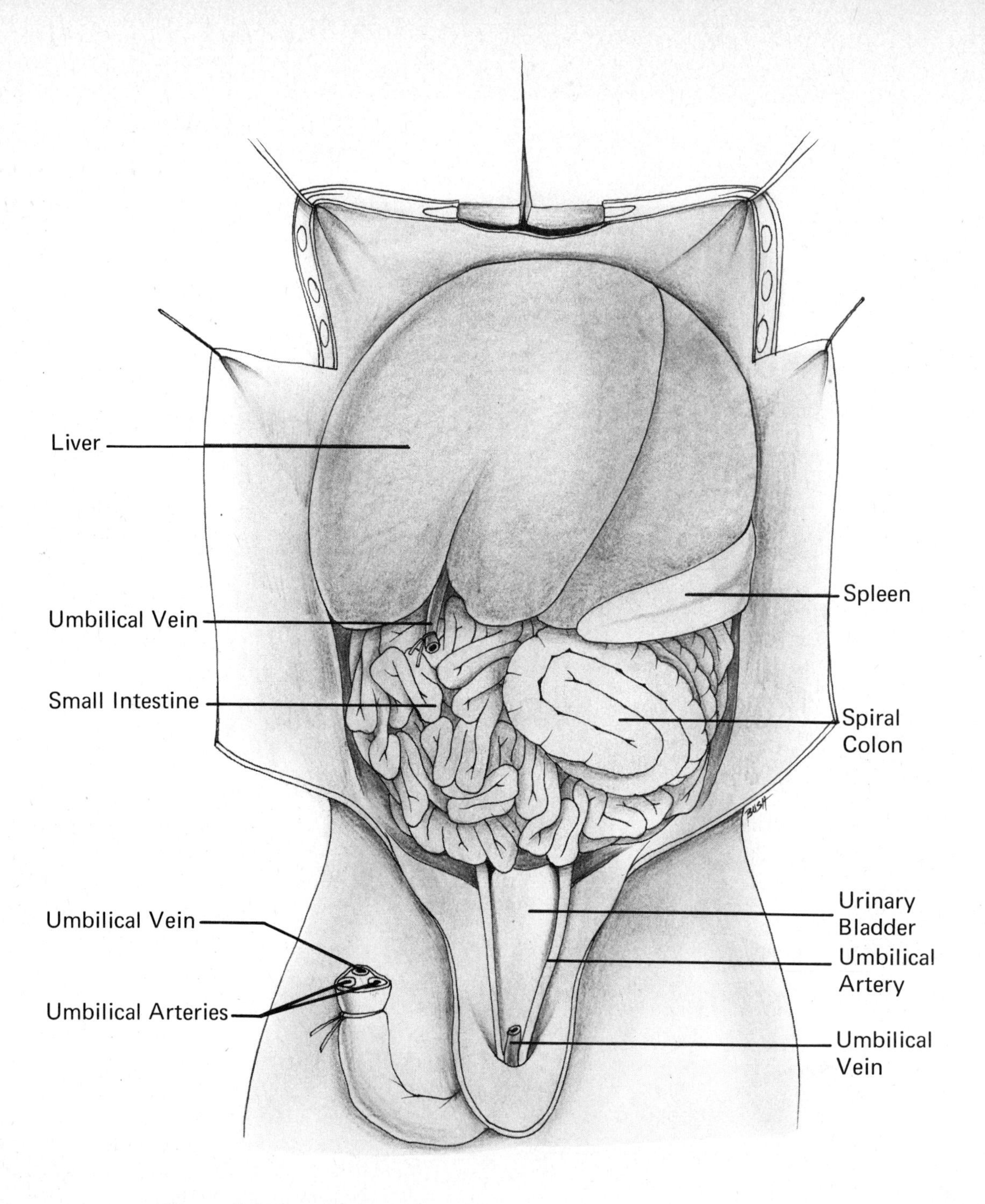

Figure 9-8. Superficial View of Digestive Organs of the Fetal Pig

The duct branching to the left is the hepatic duct. It is necessary to dissect carefully to avoid destroying these structures.

11. Lift up the liver to expose the **stomach**, which is located on the left side of the abdominal cavity.
    a. Locate the entrance of the esophagus into the stomach.
    b. Identify the following gross regions of the stomach: the **fundus, body, greater curvature, lesser curvature, cardiac region,** and **pyloric region.**
    c. Make an incision along the greater curvature of the stomach from the pyloric region to the cardiac region. Wash out the stomach. The green debris found here and elsewhere in the digestive tract

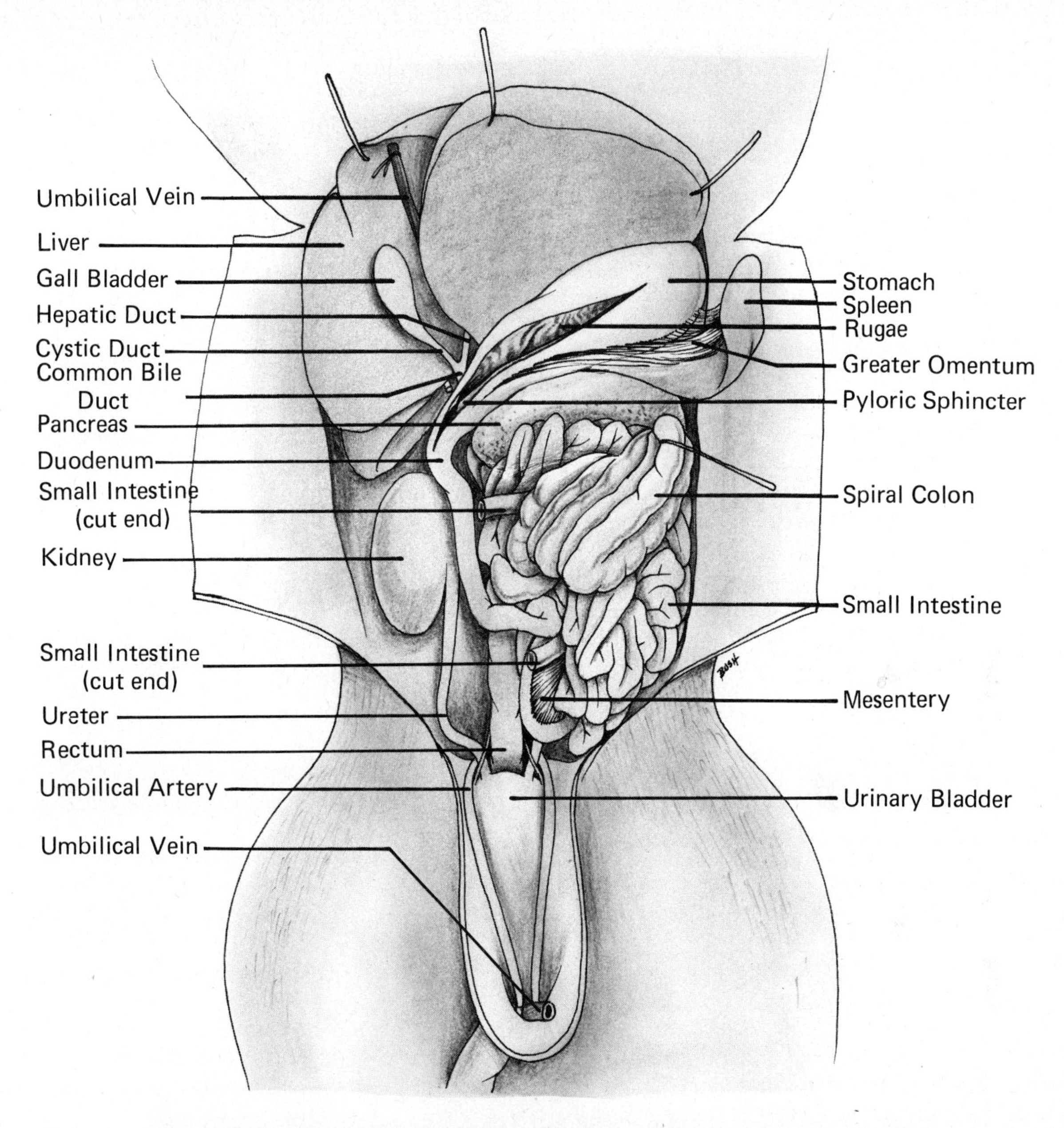

Figure 9-9. Digestive Organs of the Fetal Pig (with liver and spleen pulled back, and small intestine removed on the right side)

is called **meconium**. It is bile-stained mucus, epithelial cells sloughed off from the skin and lining of the digestive tract, and amnionic fluid swallowed by the fetus. It is discharged in the first bowel movements of the newborn.

d. Observe the **gastric mucosa** lining the stomach and the **rugae**, the longitudinal folds visible in the interior of the stomach.

e. Continue the longitudinal incision through the **cardiac sphincter**, a circular ring of smooth muscle surrounding the opening of the esophagus into the stomach. Note that the sphincter is tightly closed. In this area there is a small **diverticulum** of the stomach near the esophagus. This is believed to be the rudiment of the complex chambering of the stomach of the cow.

f. Continue the same longitudinal incision through the **pyloric sphincter**, the sphincter regulating the passage of food into the duodenum.

12. Locate the **small intestine**, beginning at the posterior end of the stomach. The small intestine is a long coiled tube, divided into three regions: the **duodenum, jejunum**, and **ileum.**
    a. The anterior curved portion of the small intestine leaving the stomach is the duodenum. This portion is approximately one cm long. The common bile duct can be seen entering the duodenum from the liver.
    b. Open the duodenum by continuing the longitudinal incision through its wall from the pyloric sphincter, on the side away from the opening of the common bile duct. Note that the common bile duct empties into the duodenum through a small papilla, the **duodenal papilla.**
    c. Observe the velvety texture of the interior of the small intestine. This texture is due to small fingerlike projections called **villi** that greatly increase the absorptive surface of the small intestine.
    d. The two remaining, approximately equal, portions of the small intestine, the **jejunum** and the **ileum,** have no readily distinguishable boundary. The jejunum is the middle region of the small intestine, and the ileum is the latter half that enters the large intestine.
13. Locate the **spleen,** the long dark organ to the left of the stomach. It is attached to the greater curvature of the stomach by means of the **greater omentum,** a specialized fold of the peritoneum. Compare the structure of the greater omentum in the pig with that in humans. The spleen is considered to be part of the circulatory system.
14. The **pancreas** lies in the angle between the curve of the stomach and the duodenum. The greater part of the gland is located behind the stomach. The pancreas is connected to the duodenum by the pancreatic duct. This duct is small and need not be dissected out.
15. Trace the ileum to its point of attachment with the **large intestine.**
    a. The ileum opens into the side of the colon, forming a blind pouch, the **cecum,** at the beginning of the **colon.** In man, the vermiform appendix is located below the cecum. This is not present in the pig.
    b. Cut into the cecum, wash out the contents, and observe the **ileocecal sphincter.**
    c. The first part of the large intestine in the fetal pig is called the **spiral colon.** It is visible as a compact coiled mass on the left side of the abdominal cavity. This structure is characteristic of the pig.
    d. The posterior portion of the large intestine is the **rectum.** Locate this structure passing from the spiral colon as a straight tube into the pelvic cavity. The external opening of the rectum is the **anus.**
16. Locate the **peritoneum,** the double membrane lining the abdominal cavity.
    a. The parietal layer of the peritoneum lines the body wall; the visceral layer covers the abdominal organs.
    b. Locate the **mesentery,** the double layer of the peritoneum extending from the dorsal wall of the abdominal cavity to the small intestine. The mesentery contains blood vessels, lymphatics, and nerves (see Fig. 9-9).
    c. Observe the mesentery near its point of contact with the dorsal body wall and locate a row of small white bulges. These are **lymph nodes.**
    d. If not done previously, locate the **greater omentum** extending between the stomach and the spleen, and the **lesser omentum,** extending between the stomach and duodenum to the liver.

Exercise 9-C The Action of Salivary Amylase

Objectives

To determine the action of salivary amylase on starch.

Materials

Test tubes; saliva; 1% boiled starch; paper cups; Benedict's solution; Lugol's solution; graduate; beakers; thermometer; tripod; wire gauze; paraffin wax; and distilled water.

KEY TO FIGURE 9-10

1. liver
2. gall bladder
3. hepatic duct
4. cystic duct
5. hepatic artery
6. portal vein
7. duodenum
8. common bile duct
9. accessory pancreatic duct
10. pancreatic duct
11. duodenal papilla
12. sup. mesenteric V.
13. R. hepatic duct
14. L. hepatic duct
15. gastric artery
16. celiac artery
17. adrenal gland
18. left kidney
19. splenic artery
20. spleen
21. aorta
22. pancreas
23. sup. mesenteric A.
24. ampulla of Vater

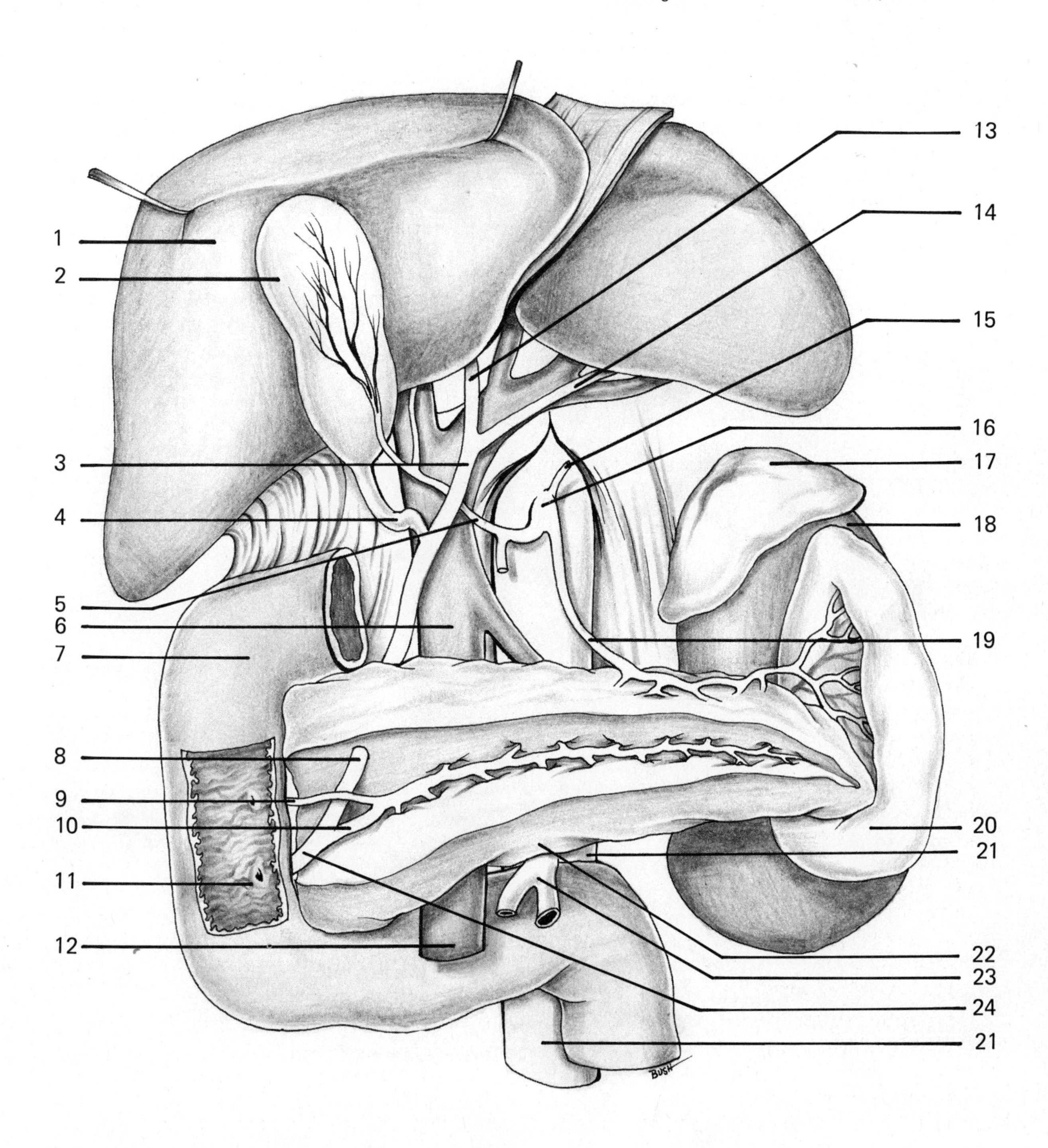

Figure 9-10. Human Abdominal Blood Vessels and Ducts

Discussion

Salivary amylase (or ptyalin) is produced by the parotid, submandibular, and sublingual glands, which empty their secretions by way of ducts into the mouth. Amylase acts on cooked starch and glycogen, hydrolyzing them to erythrodextrin, achromodextrin, and finally to maltose. Although the optimum pH of salivary amylase is 6.8 to 7, salivary digestion proceeds in the stomach until the acid gastric juice is mixed thoroughly with the bolus of food.

In this experiment, Benedict's test and the iodine test (Lugol's solution) will be used to follow the course of salivary digestion.

Benedict's test is one of the most commonly used tests for the detection of reducing sugars (sugars which contain a free aldehyde or ketone group). Glucose, or any other reducing sugar, reacts with Benedict's solution, which contains alkaline copper sulfate solution with sodium citrate. The sugar reduces the blue cupric ion, forming insoluble red cuprous oxide. The precipitate of cuprous oxide can usually be seen in the bottom of the tube upon standing. Depending upon the amount of reducing sugar present, Benedict's solution turns green, yellow, orange, or red.

Lugol's solution is a 1% solution of iodine and potassium iodide (1 part iodine, 2 parts potassium iodide, and 300 parts distilled water). Certain polysaccharides give characteristic colors when treated with iodine. Starch, for example, gives a deep blue to "black" color with Lugol's solution (the black is really a concentrated blue color). Glycogen gives a red-brown color with Lugol's solution, while erythrodextrin gives a red color. Cellulose, monosaccharides, and disaccharides do not react. A negative test is indicated by the yellow to brown color of the solution itself, or some other color (other than blue to black) resulting from pigments present in the substance being tested. The iodine test will not work in hot solutions.

Procedure

1. Collect some saliva by rinsing the mouth thoroughly with distilled water two or three times. Then chew a piece of paraffin wax and expectorate about 5 cc of saliva into a paper cup.
2. Add 5 cc of 1% boiled starch solution to each of the two test tubes. Label these tubes 1 and 2.
3. Add 4 drops of Lugol's solution to each tube (or add Lugol's solution until the starch solution turns blue).
4. Add 8 drops of saliva to tube 1. Note the time at which the saliva is added. Add 8 drops of distilled water to tube 2.
5. Immediately place both tubes in a 37° C. water bath.
6. Observe the changes in color in the tubes. Note the time it takes tube 1 to become colorless (achromic).
7. After tube 1 becomes colorless, test both tubes with Benedict's solution (add 5 cc of Benedict's solution to each tube and place the tubes in a boiling water bath for two minutes).
8. Observe for any of the following color changes:

-	solution remains blue
+	green solution
++	yellow solution
+++	orange solution
++++	red solution or precipitate

9. Tabulate your results under Results and Questions. Use the symbols indicated in No. 8 where appropriate.

Exercise 9-D Protein Digestion

Objectives

To demonstrate the effect of pepsin on protein and to study the effect of pH and boiling on the rate of action of pepsin.

Materials

0.5% pepsin; fibrin; 0.8% HCl; 0.5% NaOH; Hydrion paper (range 1-11); rack; beakers; graduate; thermometer; water bath; and test tubes.

Discussion

Pepsin is the proteolytic enzyme secreted by the chief cells in the lining of the stomach. It digests proteins (fibrin in this experiment) to proteoses and peptones. The activity of pepsin depends upon the pH of the solution, the optimum pH being 1.5-2.5. Pepsin is almost completely inactive in neutral or alkaline solutions. Pepsin is secreted in the inactive form, pepsinogen it is activated by HCl in the gastric juice.

Procedure

1. Prepare five test tubes numbering them from 1-5, and add the following solutions to them (**measure the quantity of each solution carefully in this experiment**):
   Tube 1: 5 cc of 0.5% pepsin; 5 cc of 0.8% HCl.
   Tube 2: 5 cc of 0.5% pepsin; 5 cc of water.
   Tube 3: 5 cc of 0.5% pepsin; boiled for 10 minutes in a water bath; 5 cc of 0.8% HCl
   Tube 4: 5 cc of 0.5% pepsin; 5 cc of 0.5% NaOH.
   Tube 5: 5 cc of water; 5 cc of 0.8% HCl.
2. Determine the approximate pH of each test tube with Hydrion paper (range 1-11) and include the data in the table under Results and Questions.
3. Place the same amount of fibrin (about the size of a pea) in each test tube. Shake well.
4. Place the tubes in a 40° C. water bath and shake them from time to time (do not erase the numbers on the tubes while shaking them). Maintain the temperature close to 40° C.
5. Keep the tubes in the water bath for at least 1½ hours. Note the changes that the fibrin undergoes. The swelling (imbibition) that occurs in some tubes should not be confused with digestion. If the fibrin is digested, it becomes transparent and gradually disappears (dissolves), as the protein is digested to soluble proteoses and peptones.
6. Conclude the experiment when the fibrin is digested in one of the five tubes.
7. Record your observations under Results and Questions.

Exercise 9-E The Effect of Rennin on Milk

Objective

To determine the effect of rennin on milk.

Materials

Rennin; milk; test tubes; beaker; thermometer; and Benedict's solution.

Discussion

The gastric juice of young animals contains rennin, an enzyme which causes milk to curdle. This enzyme, while not present in the gastric juice of human adults, is present in the stomach of young infants. Rennin is manufactured by the chief cells in the inactive form, prorennin, and is converted by HCl in the lumen of the stomach to the active form, rennin. Rennin converts casein (milk protein) to paracasein; the optimum pH for this reaction is 5.4. Calcium ions then react with the paracasein to form insoluble calcium paracaseinate (a white curd). Pepsin assumes this function in adults.

Procedure

1. Break a rennin tablet in half and use one piece.
2. Dissolve this piece in 3 cc of water.
3. Add the rennin to 10 cc of skimmed milk. Mix thoroughly and place the tube in a 37° C. water bath. Let the tube stand until it can be inverted without spilling the contents. Note the time needed for the curd to form.
4. Let the tube stand for an additional half hour or more, until liquid appears on the top of the tube. This supernatant fluid is called the whey, and the remainder of the tube's contents is the curd.
5. Perform Benedict's test on the whey.
6. Record your observations under Results and Questions.

Exercise 9-F Pancreatic Digestion of Fat

Objectives

To determine the effect of pancreatic juice on fat and to determine the effect of cold on enzyme activity.

Materials

Pancreatin (1% solution of pancreatin in 0.2% Na_2CO_3); cream; test tubes; powdered litmus; beakers; graduate; thermometer; pipette; and ice.

Discussion

Pancreatin contains all of the enzymes present in pancreatic juice. Since the optimum pH of the pancreatic enzymes ranges between 7.0 and 8.8, the pancreatin is prepared in sodium carbonate. The enzyme active in this experiment is lipase, which digests fat to fatty acids and glycerol. The fatty acids produced change the color of blue litmus to red.

Procedure

1. Place 5 cc of litmus cream (cream to which powdered litmus has been added to give it a blue color) in a test tube, and place it in a 40° C. water bath.
2. Place another 5 cc portion of litmus cream in a second test tube and put it in an ice bath.
3. After these tubes have attained the temperature of the water baths, add 5 cc of pancreatin to each tube and replace the tubes in the water baths until a color change occurs in one tube.
4. Complete the table under Results and Questions.

Exercise 9-G Bile

Objectives

To determine the effect of bile on surface tension; and to demonstrate Gmelin's test for bile.

Materials

Bile; test tubes; concentrated nitric acid (HNO_3); sulfur; beakers; and distilled water.

Discussion

Bile is the digestive juice secreted by the liver. Although it contains no digestive enzymes, it is important in digestion since it contains bile salts, which emulsify fats. This emulsification of fats facilitates their digestion, since the very small fat droplets provide a greater surface area for the action of lipase on fat. Bile salts are also important in the absorption of digested fats. Bile also contains bile pigments, which are waste products from the breakdown of erythrocytes. Gmelin's test for bile involves the oxidation of the bile pigments by nitric acid, with a resultant formation of colored rings. This test can be used to test urine for the presence of bile pigments.

Procedure

1. Observe the color and taste of bile.
2. To perform Gmelin's test for bile, place about 3 cc of concentrated nitric acid in a test tube.
3. Dilute 5 cc of bile with an equal volume of water. Carefully overlay the nitric acid with this dilute solution of bile by slowly pouring it down the side of the tube containing the acid.
4. If this test is properly performed, colored rings should appear at the interface between the two liquids; yellow nearest the acid, then red, violet, blue, and green nearest the top. These colored compounds constitute Gmelin's test for bile.
5. To demonstrate the effect of bile upon surface tension, place about an inch of distilled water in a clean beaker.
6. Sprinkle a small amount of sulfur upon the surface of the water.
7. Observe the behavior of the sulfur. Do the particles fall to the bottom of the beaker through the water or do they remain floating on the surface?
8. Repeat the experiment by replacing the water with dilute bile (10 cc bile, 10 cc water). Compare the results of the addition of sulfur with those obtained when distilled water was used.

Self-Test—Digestive System

Directions: See Chapter 1, Self-Test, p. 9.

1. The hardest substance in the body is **enamel**.

2. The **root** is the part of the tooth which is embedded in bone.

3. The **periodontal** membrane attaches the tooth to the bone.

4. The major nerves and blood vessels supplying a tooth are found in the **periodontal membrane.**

5. The enamel is formed by **odontoblasts.**

6. The **crown** is the exposed surface of the tooth.

7. When teeth fail to meet properly for eating, the condition is known as **gingivitis.**

8. The human adult set of teeth normally contains a total number of **32** teeth.

9. Dental **caries** are decayed areas in teeth.

10. **Deciduous** teeth come in after the primary teeth are shed.

11. The eight teeth in the front of the mouth are called **incisors.**

12. The dental formula for the deciduous teeth is **2 1 0 2.**

13. Saliva is secreted by **three pairs** of salivary glands.

14. The enzyme present in saliva is called either **salivary amylase or ptyalin.**

15. Salivary amylase starts the digestion of **monosaccharides.**

16. An example of a polysaccharide is **galactose.**

17. Proteins are made up of large numbers of **amino acid units,** which are joined together by peptide linkages.

18. Benedict's test for reducing sugars provides a positive test with **starch.**

19. **Both starch and cellulose** are made up of glucose units.

20. Table sugar is an example of a **monosaccharide.**

21. Fat molecules consist of **three molecules** of glycerol and **one molecule** of fatty acid.

22. A positive test for starch is provided by application of **Lugol's iodine solution.**

23. Proteoses and peptones are reduced to polypeptides by **trypsin.**

24. The main enzymes found in protein digestion are pepsin in gastric juice, trypsin in pancreatic juice, and peptidases in **pancreatic juice.**

25. A chemical compound which increases the rate of reaction without actually entering into the reaction is called a **catalyst.**

26. Organic catalysts made by living cells are called **enzymes.**

27. Enzymes which catalyze the hydrolysis of proteins are called **proteases.**

28. Pancreatic juice is the **only** digestive juice with strong amylases, proteases, and lipases.

29. The enzyme **rennin** assists in the curdling of milk.

30. In the **small intestine** the enzyme pepsin acts to change proteins to peptones.

31. Pepsinogen is converted by **HCl** to pepsin.

32. Each disaccharide has its **own specific enzyme** for conversion to monosaccharides.

33. Swallowing is completed when the **pyloric** sphincter opens.

34. A portion of the roof of the mouth cavity is formed by the **hard palate.**

35. A baby with a **cleft palate** cannot take milk from a bottle because the mouth opens into the nasal cavities.

36. The esophagus is lined with **stratified squamous epithelium.**

37. The pyloric sphincter separates the **duodenum** from the stomach.

38. The vomiting center is located in the **medulla oblongata.**

39. A lesion of the inner wall of the stomach or first part of the small intestine is called a **peptic ulcer.**

40. Lining the small intestine are many fine, velvetlike or fingerlike projections called **rugae.**

41. Enlarged veins in the **stomach** are referred to as hemorrhoids.

42. The appendix is a cul-de-sac attached to the **descending colon.**

43. A **hernia** is a protrusion of part of an organ through an abdominal opening.

44. A hernia occurring through the navel is called **an umbilical hernia.**

45. The membrane lining the abdominal cavity is called the **pleura.**

46. The scientific name for the throat is the **larynx.**

47. **Peritonitis** is an inflammation of the lining of the abdominal cavity.

48. Peristalsis occurs in the stomach and intestines, **but not** in the esophagus.

49. The two regions in which the liver is normally located are the **epigastrium and right hypochondrium.**

50. The common bile duct enters the small intestine at the **ampulla of Vater.**

51. The temporary folds in the mucosa of the stomach are called **papillae.**

52. That part of metabolism involving the building up of tissues is called **catabolism.**

53. The **enzymes** in bile aid in the digestion of fats.

54. The principal elements of protein molecules are **carbon, hydrogen, oxygen, and nitrogen.**

55. The production of bile by the liver represents both **secretory and excretory activities.**

56. The contents of the intestines are largely **fluid** upon entering the cecum.

57. The most important single factor in maintaining normal defecation is the development of **regular habits that include conditioned reflexes.**

58. **Hydrolysis** is the principal chemical reaction involved in the digestion of food.

59. Saliva has a **lower** pH than gastric juice.

60. **Bile salts** emulsify fat particles and thus increase their total surface area.

61. In the process of digestion, **bile** splits fats into fatty acids and glycerol.

62. The tongue has nerve endings capable of detecting **hundreds of flavors.**

63. The mouth cavity opens posteriorly into the **nasopharynx.**

64. **Pyorrhea** is a condition of the gums and periodontal membrane which is characterized by inflammation and bleeding.

65. The liver receives blood (which it processes and refines) from the **hepatic vein.**

66. The lining of the small intestine is the **only area** from which food is absorbed.

67. Movements occurring in the **stomach** cause hunger pangs.

68. In the liver, urea is formed from **fat** metabolism.

69. In the absence of bile, from one-third to two-thirds of the ingested fat is lost in the **urine.**

70. The cystic duct leads from the gall bladder and opens into the **common bile duct.**

71. The colon is divided into segments called **haustra.**

72. **Kupffer cells** located in the stomach are responsible for the secretion of hydrochloric acid.

73. Anger during the course of a meal increases the rate of gastric motility and thus **hastens** emptying of the stomach.

74. Constipation is a condition usually arising from a **lack of tonus** in the muscular layers of the colon.

75. Detoxification is a process occurring in the **liver** by which toxic materials absorbed through the walls of the gastrointestinal tract are rendered harmless.

76. Deficiency of **vitamin A** results in night blindness.

77. **Vitamin K** is essential for normal clotting of the blood.

78. Scurvy is caused by a deficiency of **riboflavin.**

79. Proteins are called complete if they contain all of the nutritionally essential **amino acids.**

80. In fats the proportion of **oxygen** is less than in sugars and starches.

81. Vitamins A and D are **water** soluble.

82. **Saturated** fatty acid molecules contain all the hydrogen they are able to hold.

83. The most important processes in digestion and absorption take place in the **stomach.**

84. The first teeth of the permanent set to erupt are the **six-year molars.**

85. Gingivitis is commonly associated with a dietary deficiency of **ascorbic acid.**

86. Weight-for-weight proteins provide **twice** as many calories as fats.

87. Gallstones are formed from substances in **bile.**

88. The first set of teeth includes **no** bicuspids.

89. When giving an enema the rectal tube should be inserted **2 or 3 inches** because the length of the anal canal is about **1½ inches.**

90. During an enema the patient should lie on his left side because the **descending colon** is on this side and the fluid will flow in easily.

91. Parotitis or "mumps" is an inflammation of the **parotid** gland.

92. A **gastroenterostomy** is the creation of an artificial passage between the stomach and the intestine.

93. **Sclerosis** is an excessive formation of the interstitial connective tissue that causes a hardening of an organ such as the liver.

94. Cholecystitis is an inflammation of the **urinary bladder.**

95. **Cholelithiasis** is the formation of stones in the gall bladder or ducts.

96. Clay-colored stools are produced if the common bile duct is blocked, because the **bile pigments** are missing from the feces.

97. In the preceding case the doctor might order a low-fat diet, because bile is important for **emulsification** of fats.

98. Lacteals are lymph vessels that absorb food from the **large intestine.**

99. Digestion of food may be **speeded** by as much as one hour if tobacco is used before eating.

100. Tooth decay in children may be reduced as much as **60%** with the use of sodium fluoride in water.

101. The **longitudinal** layer of muscle in the large intestine forms the taenia coli.

102. The peritoneal cavity is the term used to describe the spaces between the **parietal** and **visceral** layers of peritoneum.

103. A surgeon can readily distinguish between the large and small intestine, by sense of touch, due to the presence or absence of **haustra** and the **epiploic appendages.**

104. The **plasma proteins** formed by the liver play an important role in maintaining the osmotic pressure of the blood.

105. Bile has a **higher** pH than gastric juice.

106. Residues of a particular food may require **more than 24 hours** to be excreted from the digestive tract.

107. Glycerol is one of the end products of **carbohydrate** metabolism.

108. Gastric lipase **is** essential for digestion of fats.

109. Succus entericus is secreted by the glands of the **small intestine.**

110. The **greater and lesser** omenta are folds of the visceral layer of the peritoneum.

111. The greater omentum hangs from the **lesser** curvature of the stomach.

112. The mesentery of the **small intestine** serves as a support and holds blood vessels, lymphatics, and nerves.

113. The gastric contents are reduced to a more or less homogenous creamy mass called **the bolus.**

114. Secretin is a hormone which stimulates the secretion of **pancreatic juice.**

115. The hormone which is specific for stimulating contraction of the gall bladder is **chole-cystokinin.**

116. The most important cause of **chronic constipation** is the taking of laxatives.

117. The stomach is concerned with digestion of **carbohydrates.**

118. The presence of sodium bicarbonate in **pancreatic juice** neutralizes the acid chyme.

119. **Osmosis** is involved in the absorption of water from the colon.

120. Proteins **are stored** in muscles and other parts of the body.

121. Most people **need to** supplement their diet with vitamin pills.

122. Animal protein **is more** likely to be a complete protein than vegetable protein.

123. Rickets is a deficiency disease of childhood caused by a diet poor in **vitamin D.**

124. Endemic goiter is related to a **lack of iodine** in the soil.

125. **Vitamin B_1** is used to treat pernicious anemia.

KEY

4. root canal (pulp cavity)
5. ameloblasts
7. malocclusion
10. permanent
15. polysaccharides
16. starch, cellulose, or glycogen
18. monosaccharides and most disaccharides
20. disaccharide
21. 1 molecule; 3 molecules
24. intestinal juice
30. stomach
33. cardiac (esophageal)
40. villi
41. rectum
42. cecum
45. peritoneum
46. pharynx
48. as well as
51. rugae
52. anabolism
53. bile salts
59. higher
61. lipase
62. 4 basic taste sensations
63. oropharynx
65. portal vein
66. major area
68. protein
69. feces
72. parietal cells
78. ascorbic acid (vitamin C)
81. fat
83. small intestine
86. half
93. cirrhosis
94. gall bladder
98. small intestine
99. slowed
107. fat
108. is not
111. greater
113. chyme
117. proteins
120. are not stored
121. do not need to
125. Vitamin B_{12}

Name ____________________

Results and Questions for Chapter 9

Exercise 9-A. Anatomy of the Human Digestive System

1. What is the dental formula for the deciduous set of teeth? ____________________

2. What is the dental formula for the permanent set of teeth? ____________________

3. What is another name for the cuspids? ____________________

4. What is another name for the premolars? ____________________

5. List the major functions of saliva. ____________________

____

6. What is the function of the cecum? ____________________

7. Trace the path of food through the digestive tract to the point of absorption of the food into the blood stream. ____________________

____

8. In which layer of the tooth are the nerves and blood vessels located? ____________________

Exercise 9-B. Dissection of the Digestive System of the Fetal Pig

1. Name the most abundant tissue in the tongue. ____________________

2. Compare the dental formula for the deciduous teeth of the pig and human.

____________________ ____________________

Pig Human

3. Name the bones that comprise the hard palate. ____________________

____

4. What is the posterior extension of the soft palate called in humans? ____________________

5. Does the pig have this structure? ____________________

6. What is the function of the Eustachian tubes? ____________________

____

7. What is the function of the common bile duct? ____________________

____

8. How many lobes are in the human liver? ____________________

9. How many lobes are in the pig liver? ____________________

10. What is the function of the gall bladder?____

11. Name the sphincter surrounding the opening of the common bile duct.____________________

12. Name the sphincter between the small and large intestine.___________________________

13. List the functions of the large intestine.____

____

14. Name the microscopic fingerlike projections present in the lining of the small intestine.________

____

15. What is the function of the structures in No. 14?___________________________________

____

16. Name the blood vessel that absorbs digested food from the digestive tract._________________

____

17. Name the three regions of the small intestine.______________________________________

18. To which curvature of the stomach is the greater omentum attached?_____________________

Exercise 9-C. The Action of Salivary Amylase

1. Complete the following table. In the column headed "Explanation," explain the changes in color of Lugol's solution and the appearance of Benedict's test.

Tube No.	*Contents*	*Initial Lugol's Color*	*Final Lugol's Color*	*Achromic Time*	*Results of Benedict's Test*	*Explanation*
1						
2						

2. Did you see any evidence of the formation of dextrins as an intermediate product?______Explain:____

____

3. What is the purpose of performing Benedict's test after tube 1 becomes achromic?____________

____

4. Write the equation for the action of salivary amylase on starch.________________________

____

Name ____________________

5. Based on your observations in this experiment, list two ways in which it was proven that digestion of starch occurred. ____________________

____

____

6. What other digestive organ synthesizes amylase? ____________________

Exercise 9-D. Protein Digestion

1. Complete the following table.

Tube No.	*Contents of Tube*	*pH*	*Digestion (+ or −)*	*Explanation of Results*
1				
2				
3				
4				
5				

2. Which organ synthesizes pepsin? ____________________

3. Write the equation for the action of pepsin on protein. ____________________

____

4. What is the approximate percentage of HCl in gastric juice? ____________________

5. What is the average pH of the stomach? (Assume this is the optimum pH for pepsin in discussing this experiment.) ____________________

6. List the functions of HCl in gastric juice. ____________________

____

____

7. What was the function of HCl in this experiment? ____________________

8. To which class of chemical compound does fibrin belong? ____________________

9. Why was 40° C. used for the temperature of the water bath in this experiment rather than 100° C.? ____________________

Exercise 9-E. The Effect of Rennin on Milk

1. How long did it take for the milk to curdle in Step 3? ____________________

2. Write the equation for the action of rennin. ______________________

____

3. What is the main constituent of the solid curd? ______________________

4. Name the substances present in the whey. ______________________

5. What results were obtained with Benedict's test on whey? ______________________
   Explain: ______________________

6. Name the digestive organ that synthesizes rennin. ______________________

7. Would rennin be effective in the adult stomach, if present? ______________________
   Explain: ______________________

8. How is rennin used commercially? ______________________

Exercise 9-F. Pancreatic Digestion of Fat

1. Complete the following table:

Tube No.	*Temp. of Water Bath*	*Change in pH (color)*	*Explanation of Results*
1			
2			

2. Name the enzymes present in pancreatin. ______________________

____

3. Which enzyme was active in this experiment? ______________________

4. Write the equation for the action of the enzyme in No. 3. ______________________

____

5. In what other digestive juice may the enzyme in No. 3 be located? ______________________

6. What was the substrate of the same enzyme in this experiment? ______________________

7. What color is litmus in alkaline solution? ______________________

8. What color is litmus in acid solution? ______________________

9. What was the color of the freshly prepared litmus cream? ______________________

10. What caused the change in color of the litmus cream in this experiment? ______________________

____

Name ______________________

Exercise 9-G. Bile

1. What was the taste of the bile sample? ______________________

2. What color was the bile sample? ______________________

3. To what is this color due? ______________________

4. What results were obtained in Gmelin's test for bile? ______________________

____

5. Where are bile pigments produced in the body? ______________________

6. Name the two common bile pigments. ______________________

7. Of what value are the bile pigments? ______________________

8. Did the particles of sulfur fall through the distilled water? ______________________

   Explain your answer: ______________________

9. Did the particles of sulfur fall through the diluted bile? ______________________

   Bile salts decrease the surface tension of the water in which they are dissolved. Use this to explain the behavior of the sulfur in the beaker containing dilute bile. ______________________

   ____

   ____

10. How does emulsification of fat by bile salts aid in the digestion of fat by lipase? ______________________

    ____

11. Name the two common bile salts. ______________________

12. List the major functions of bile salts. ______________________

    ____

    ____

10

Metabolism

Exercise 10-A
Basal Metabolic Rate

Objective

To determine theoretical basal metabolic rate.

Materials

None.

Discussion

The surface of the body is the principal region from which heat is lost. This area varies with the height and weight of the individual. The relation between height and weight and body surface area is shown by means of a chart such as the Dubois Body Surface Chart in Appendix 3-A.

In health, the rate of heat loss over the surface of the body equals the amount of heat produced in excess of that required to maintain normal body temperature. Both age and sex affect this rate. No satisfactory reason has as yet been established for the difference between the sexes. The table below gives the average rate at which heat is lost in terms of calories per minute per square meter of body surface area. The figures provide a means of estimating basal metabolic rate.

TABLE 10-1
Heat Loss per Square Meter of Body Surface

	Males		*Females*	
	Calories per		*Calories per*	
Age	*Min.*	*Hour*	*Min.*	*Hour*
5	0.891	53.0	0.860	51.6
6	0.878	52.7	0.845	50.7
7	0.866	52.0	0.825	49.5
8	0.853	51.2	0.801	48.1
9	0.840	50.4	0.781	46.9
10	0.825	49.5	0.763	45.8
11	0.810	48.6	0.743	44.6
12	0.796	47.8	0.723	43.4
13	0.785	47.1	0.700	42.0
14	0.770	46.2	0.683	41.0
15	0.755	45.3	0.660	39.6
16	0.745	44.7	0.641	38.5
17	0.728	43.7	0.623	37.4
18	0.715	42.9	0.621	37.3
19	0.701	42.1	0.620	37.2

(table continued on next page)

Age	Males Calories per Min.	Hour	Females Calories per Min.	Hour
20-24	0.683	41.0	0.615	36.9
25-29	0.671	40.3	0.610	36.6
30-35	0.663	39.8	0.603	36.2
35-39	0.653	39.2	0.596	35.8
40-44	0.638	38.3	0.588	35.3
45-49	0.630	37.8	0.583	35.0
50-54	0.620	37.2	0.575	34.5
55-59	0.610	36.6	0.568	34.1
60-64	0.600	36.0	0.563	33.8
65-69	0.588	35.3	0.556	33.4
70-74	0.580	34.8	0.546	32.8
75-79	0.570	34.2	0.538	32.3

Procedure

1. Record your height, weight, age, and sex under Results and Questions.
2. Using the Dubois Body Surface Chart (located in Appendix 3-A) determine the approximate surface area of your body (BSA) and record under Results and Questions. Follow the directions on the chart.
3. Using Table 10-1 (Heat Loss per Square Meter of Body Surface), determine your theoretical basal metabolic rate for one minute, one hour, and one day and record under Results and Questions. To do this, multiply the heat loss for your age and sex, in Calories per minute, times your body surface area.

Exercise 10-B Diurnal Variations in Body Temperature

Objective

To observe the variation in body temperature during a period of 24 hours.

Materials

Clinical thermometer (Fahrenheit); 70% alcohol; and colored pencils.

Discussion

Body temperature is the result of metabolism. In fact, the oxidation of nutrients within body cells accounts for almost all of the heat produced. However, metabolic rate is greater in some tissues than others, and it is in these that the major production of heat occurs. Muscles make up over half of the soft tissues of the body; oxidation in the muscles supplies the largest quantity of heat.

Body temperature varies with the external environment in **poikilothermic** animals, the cold-blooded animals, such as fish and reptiles. However, in **homoiothermic** (warm-blooded) animals, such as mammals and birds, temperature is maintained at a relatively constant level regardless of fluctuations in external temperatures. Animals of the latter type (including man) have a heat-regulating center located in the hypothalamus. This center consists of two distinct areas, one of which regulates loss of heat while the other regulates heat production.

A fever is thought to be due to the failure of the hypothalamus to function properly. This failure may result from the action of bacteria or viruses, or from serious emotional disturbances.

Part 1. Instructions for use of the Thermometer

1. Clean the thermometer in 70% alcohol.
2. Shake the thermometer in order to move the mercury below the 96° F. mark.
3. Place the thermometer under the tongue, with the lips closed, for at least 2 minutes. Be sure the lips are closed during the entire 2 minute period. If the reading on the thermometer is to be accurate, the subject should not attempt to talk until the thermometer has been removed. Furthermore, **the temperature should not be taken immediately after drinking any kind of beverage.**
4. Read the thermometer by holding it at or near eye level in a horizontal position, with the numbers down and the marks in a superior position. Slowly rotate the thermometer back and forth until the line of mercury is visible.
5. The distance between each short mark on the thermometer represents 0.2° F., while the distance between long marks represents 1° F. Most thermometers also use a long mark to indicate 98.6° F., "normal" temperature.

Part 2. Body Temperature Variations during a 24 Hour Period

1. Determine your **basal temperature** by taking your temperature upon waking but **before arising**. Record under Results and Questions.
2. Take your oral temperature at 4 hour intervals during a continuous 24 hour period (e.g., at 7 a.m., 11 a.m., 3 p.m., 7 p.m., 11 p.m., 3 a.m., and again at 7 a.m.). Record the data in the table provided under Results and Questions.
3. Collect the temperature measurements obtained by each member of the class during a 24 hour period at four hour intervals, and record in the table under Results and Questions.
4. Plot your personal temperature data on the grid provided under Results and Questions, with time on the horizontal axis and temperature (starting at 96° F.) on the vertical axis. Use either blue or black ink (or blue or black pencil) in plotting this data.
5. From the data in the table, determine the average temperature of the class as a whole for each time interval indicated, and plot the data on the grid using either red or green ink (or red or green pencil).

Exercise 10-C Effect of Exercise on Temperature

Objective

To determine the effect of exercise on body temperature.

Materials

Clinical thermometers; 70% alcohol; and (optional) bicycle ergometer.

Procedure

1. Enter all data from this experiment in the table under Results and Questions.
2. Take your temperature before exercising ("normal" temperature).
3. Take your temperature immediately following one minute of mild exercise (running in place for one minute).
4. Take your temperature immediately following severe exercise (running around the building rapidly) or bicycling 2 minutes on the bicycle ergometer (belt setting 2 for women, 3 for men, 35 kilometers per hour).
5. Wait 10 minutes (in resting position while seated working on other laboratory experiments) and again take your temperature.

Exercise 10-D Mapping of Sweat Glands

Objectives

To demonstrate the presence, activity, and distribution of sweat glands in the skin.

Materials

Iodine; 1 inch squares of bond paper; and masking tape.

Procedure

1. Clean a 2 inch square on the forehead, neck, and arm with alcohol; let it dry, then paint the area with iodine. Select areas of the skin that are not unduly hirsute (hairy), but where sweating readily occurs. Allow the iodine to dry completely.
2. Hold a square piece of bond paper over the painted square for precisely 2 minutes. The paper should be held or taped firmly in place and not allowed to slip.
3. Remove the paper and count the number of blue (or purple) dots. Each dot represents the location of an active sweat gland pore. The size of each dot varies with the amount of perspiration secreted by that sweat gland; the number of dots on the square gives the number of **active** sweat glands per square inch, at rest.
4. Repeat this experiment under conditions of mild exercise (running in place for 2 minutes) and severe exercise (running around the building or bicycling).
5. Enter all data in the table provided under Results and Questions.

Self-Test—Metabolism

Directions: See Chapter 1, Self-Test, p. 9.

1. **Muscle tone, along with metabolic rate,** is lowest while an individual is asleep.

2. A person who has difficulty getting started in the morning usually has a **subnormal body temperature.**

3. **Catabolism** is the process of growth and repair of body tissues.

4. **Reduction** is the process by which body tissues release energy from food which has been digested.

5. The basic fuel of the body is **glucose or some other hexose.**

6. The source of energy for the body is the **chemical energy** obtained from the consumption of foods.

7. Changes in water temperature have a **greater** effect on body temperature than changes in air temperature.

8. The ratio of carbon dioxide given off to the volume of oxygen consumed is of value because it indicates whether a carbohydrate, fat, or mixed diet is being **oxidized** by the body.

9. Normal basal metabolic rate is **higher** for women than for men.

10. Carbon dioxide and water are **end products** in the oxidation of proteins, carbohydrates, and fats.

11. The formation of glycogen from glucose is called **glyconeogenesis.**

12. **Exercise, food, and cold** are all powerful stimulators of metabolism.

13. A gram of carbohydrate, when oxidized in the body, yields about **4 Calories.**

14. A gram of fat oxidized in the body yields about **9 Calories.**

15. In protein metabolism, when an increase in the release of energy is greater than can be derived from protein alone, it has been shown that the "extra" energy is derived from simultaneous oxidation of other foodstuffs. This response is known as the **specific dynamic action of protein.**

16. A basal metabolism test indicates the amount of energy required to maintain your body functions **when you are at complete rest, but not sound asleep.**

17. The presence of food in the digestive tract **has no effect** on body temperature.

18. The energy produced by one gram of protein is **considerably less than that** produced by one gram of carbohydrate.

19. Basal metabolic rate **increases** with age.

20. The energy requirements of the body are usually measured in units of **small calories.**

21. Energy requirements of the body are **influenced by hormones.**

22. Measurements of metabolic rates are **almost always indirect** measurements.

23. The metabolic rate of **homoiothermic animals** increases as body temperature rises, due to the increase in the rate of chemical reactions that results with the increase in temperature.

24. **Variations between 98° and 99° F. may be normal,** when body temperature is taken orally, even if the variations occur in the same person.

25. Fever is a pathologic condition in which the hypothalamus (or "thermostat") is "set" at a **higher level.**

26. The **thalamus** is known to play a major role in the regulation of body temperature.

27. Daily variations in body temperature are referred to as **diurnal variation.**

28. Heat stroke is characterized by a **fall** in body temperature.

29. A normal rectal temperature is approximately **98.6° F.**

30. Loss of heat in excretory products plays **an important role** in the control of body temperature.

31. Two external factors important in the regulation of body temperature are the **humidity and the movement of surrounding air.**

32. Successful efforts to reduce a fever result in **constriction** of skin capillaries.

33. Body temperature is **already well regulated** at birth.

34. In cool environmental temperatures, **conduction** accounts for most of the heat loss from skin.

35. Control of body temperature can be upset by **bacteria or viruses.**

36. Most of the heat of the body is produced by **smooth muscle tissue of the viscera.**

37. Heat production in the body is a **catabolic** process involving oxidation of nutrients during cellular respiration.

38. **Sunstroke** may occur as a result of high temperature and humidity. The body fails to lose excess heat and body temperature rises, along with the pulse and blood pressure.

39. The basal metabolic rate is related to **body surface area.**

40. Treat sunstroke by using such things as **wet sheets and ice packs to keep the patient cool.**

KEY

3. anabolism
4. oxidation
9. lower
11. glycogenesis
17. tends to elevate
18. approximately the same as that
19. decreases
20. large Calories
26. hypothalamus
28. rise
29. 99° F.
30. a limited role
32. dilation
33. poorly regulated
34. radiation
36. skeletal muscles

Name ____________________

Results and Questions for Chapter 10

Exercise 10-A. Basal Metabolic Rate

1. Record the following personal data:

   Height ____________________

   Weight ____________________

   Age ____________________

   Sex ____________________

2. As determined by the Dubois Body Surface Chart, record your BSA. ____________________

3. Calculate your basal metabolic rate.

   a. Calories/min/meter^2 body surface area ____ x BSA ____ = Calories/min ____

   b. Calories/min ____ x 60 = Calories/hour ____

   c. Calories/hour ____ x 24 = Calories/day ____

   The last figure, Calories per day, is the calculated standard basal metabolic rate.

4. The calculated standard metabolic rate may not correspond with your actual basal metabolic rate as determined with the respirometer (or some other method). Explain why this is true.

   ____

5. What is a Calorie? ____________________

   ____

6. Which would you expect to be greater in a young child: caloric intake or expenditure? ________ Explain: ____________________

   ____

7. Which of the two in Question 6 would you expect to be greater in an adult of constant weight? ______ Explain: ____________________

   ____

8. Using the BMR calculated in Question 3, solve the following problems:

   a. Assume you are typing an assignment (e.g., a term paper) and that this activity increases your BMR by 51%. Determine the number of calories you would consume in this activity for 20 minutes. Show all calculations.

   Answer ____________

b. Assume you are running up the stairs of a 20 story building, and that this activity increases your BMR by 780%. If it took you three minutes to reach the top of the building, how many calories would you consume for the activity? Show all calculations.

Answer________

Exercise 10-B. Diurnal Variations in Body Temperature

Part 2. Body Temperature Variations during a 24 Hour Period

1. What was your basal temperature?________

2. Record the seven temperature readings you took during a period of 24 hours on the first line of the table on the following page. If you took your temperature at 7 a.m., this would be recorded under the 4-8 a.m. range, an 11 a.m. reading would be recorded under the 8-12 a.m. range, etc.

3. Record the temperature readings obtained by each member of the class in the remaining spaces in the table. Make certain that the temperatures are recorded in the appropriate time range. Add each column and determine the average. (Check in the spaces provided whether the data was obtained from a male or a female.)

4. Plot your personal temperature data and that obtained from determining class averages, on the grid on p. 200 in accordance with the instructions given under Procedure, Steps 4 and 5.

5. If variations in body temperature were observed, were they cyclic?________

6. If the variations were cyclic, how would you explain this fact?________

____

____

7. If some observations did not fit the cyclic pattern, how can you account for them?________

____

8. Describe and account for any variations in body temperature of males when compared with those of females.________

____

9. In general, at what time interval are temperatures lowest for the class?________When highest?________Explain these results:________

____

____

Exercise 10-C. Effect of Exercise on Temperature

1. Record the data obtained in accordance with Steps 2-5 under Procedure, in the table on p. 200. To obtain the percentages required for the table, use the equations on p. 200, showing all

Temperature Readings Taken at 4 Hour Intervals

	A. M. 4 - 8	*A. M. 8 - 12*	*P. M. 12 - 4*	*P. M. 4 - 8*	*P. M. 8 - 12*	*A. M. 12 - 4*	*A. M. 4 - 8*	♀	♂
Personal Data									
1									
2									
3									
4									
5									
6									
7									
8									
9									
10									
11									
12									
13									
14									
15									
16									
17									
18									
19									
20									
21									
22									
23									
24									
25									
26									
27									
28									
29									
30									
31									
32									
TOTALS									
AVERAGES									

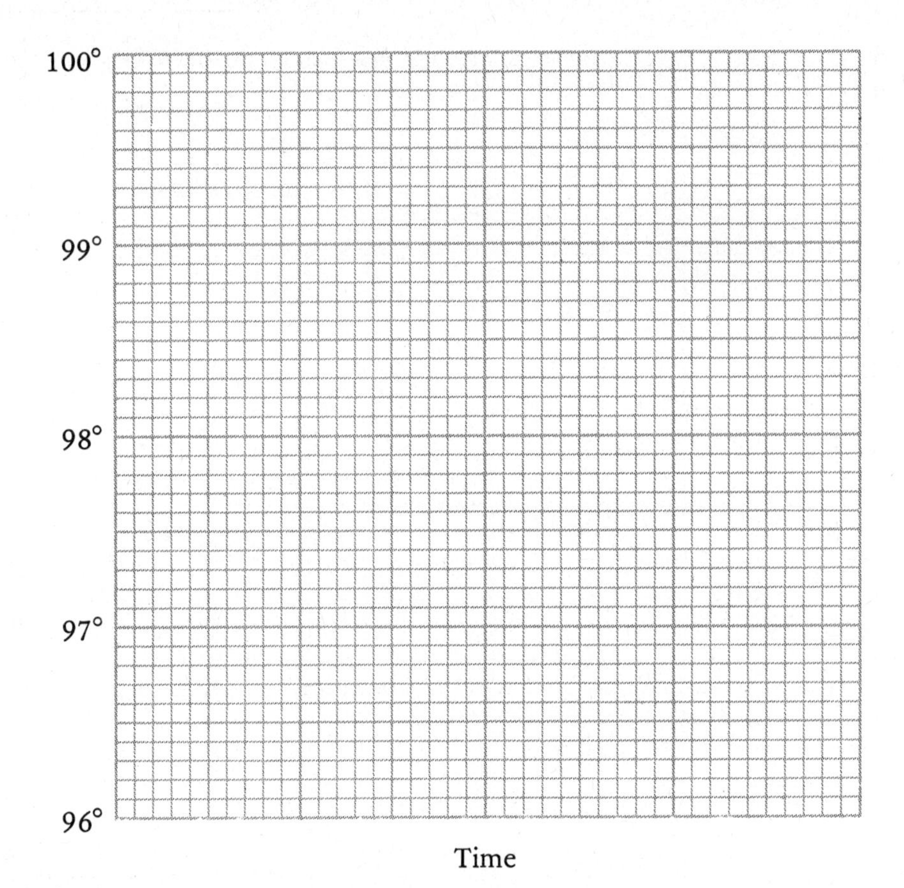

calculations (use temperature obtained before exercise as the "normal" temperature):

$$\% \text{ Increase} = \frac{\text{T after exercise} - \text{normal T}}{\text{normal T}} \times 100 = ________$$

$$\% \text{ Decrease} = \frac{\text{normal T} - \text{T after exercise}}{\text{normal T}} \times 100 = ________$$

Body Condition	*Subject's Temperature*	*% Increase or % Decrease*
Before exercise "normal temperature"		X
After mild exercise		
After severe exercise		
Ten minutes after severe exercise		X

2. What caused the changes that occurred in body temperature upon exercise?________

____

Compare your results with those of four other class members.________

____

Name ______________________

3. Upon waiting 10 minutes after severe exercise, did your temperature return to normal?______If it did, explain why. If it did not, explain why it didn't.______________________

____

4. On the basis of your results in this series of experiments on body temperature, comment on the figure 98.6° F. as the "normal" oral body temperature.______________________

____

Exercise 10-D. Mapping of Sweat Glands

1. Record the data obtained in the experiment in the following table:

Body State	*No. of Active Sweat Glands per Square Inch*		
	Forehead	Neck	Arm
At rest			
Mild exercise			
Severe exercise			

2. Explain how the colored dots were transferred to the bond paper.______________________

____

3. Explain any differences observed in the number of active sweat glands as a result of the three experimental conditions (rest, mild, and severe exercise).______________________

____

____

4. Why was it necessary to use bond paper rather than filter paper in this experiment?______________________

____

11

The Respiratory System

Exercise 11-A Anatomy of the Human Respiratory System

Objectives

To become familiar with the location and function of the human respiratory organs.

Materials

Dissectable torso; anatomic charts; model of larynx; and reference books.

Discussion

The respiratory system is concerned with the transportation of gases to and from the lungs and with the exchange of gases between the lungs and the pulmonary capillaries. Oxygen moves by diffusion from the air in the alveoli in the lungs into the pulmonary capillaries, and carbon dioxide moves in the reverse direction. The circulatory system transports these gases between the pulmonary capillaries and the systemic capillaries. Oxygen diffuses through the walls of the systemic capillaries, through the tissue fluid, into the cells. There it is used up in cellular respiration, the process by which energy is released from food. Carbon dioxide is a waste product of this reaction.

Breathing movements must occur in order for the gases to move along the respiratory tract into the lungs. Breathing is usually controlled automatically by the respiratory center in the medulla oblongata. The nerve impulses to the muscles in the thorax pass down the spinal cord to the cervical and thoracic regions. The phrenic nerve then carries these impulses to the diaphragm, and the intercostal nerves transmit the impulses to the external intercostal muscles. During inspiration the size of the thorax increases, resulting in an increase in the size of the lungs, causing air to rush in. During expiration the reverse occurs.

Procedure

1. Use Figures 11-1 and 11-2, two views of the larynx, as self-tests. See Section J, p. xx (Illustrations as Self-Tests) for suggested approaches.
2. Locate the thyroid cartilage on your own larynx.
3. Hold the larynx down and try to swallow. Now release the larynx and swallow.

KEY TO FIGURE 11-1

1. epiglottis
2. hyoid bone
3. thyrohyoid membrane
4. thyroid notch
5. thyroid cartilage
6. laryngeal prominence
7. cricoid cartilage
8. trachea

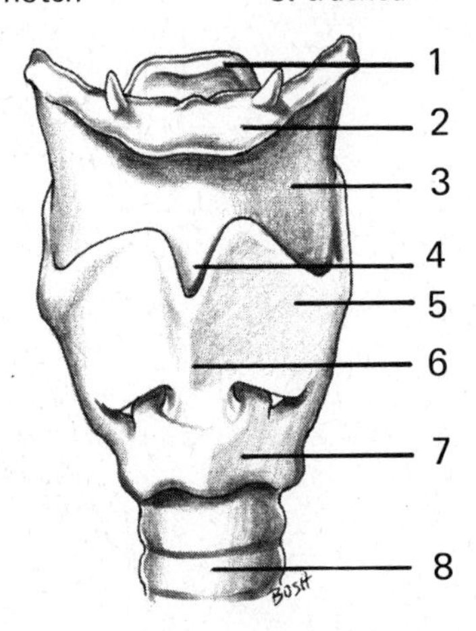

Figure 11-1. Larynx (anterior view)

KEY TO FIGURE 11-2

1. epiglottis
2. hyoid bone
3. thyrohyoid membrane
4. cricoid cartilage
5. trachea

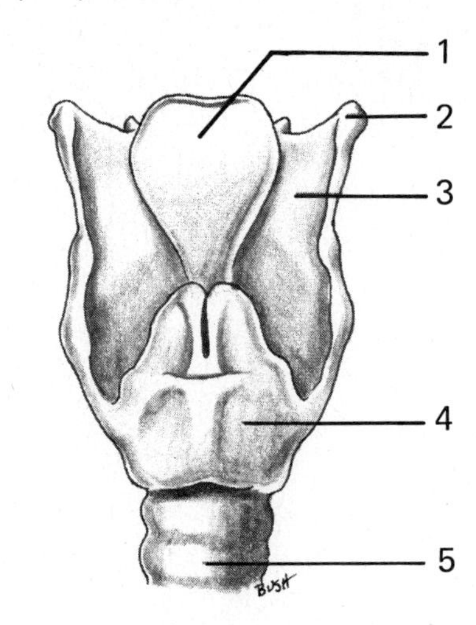

Figure 11-2. Larynx (posterior view)

KEY TO FIGURE 11-3

1. frontal bone
2. frontal sinus
3. superior, middle, & inf. nasal conchae
4. external nares
5. hard palate
6. tongue
7. mandible
8. hyoid bone
9. thyroid cartilage
10. cricoid cartilage
11. sella turcica
12. sphenoidal sinus
13. eustachian tube opening
14. solf palate
15. nasopharynx
16. uvula
17. oropharynx
18. epiglottis
19. cricoid cartilage
20. larynx
21. trachea

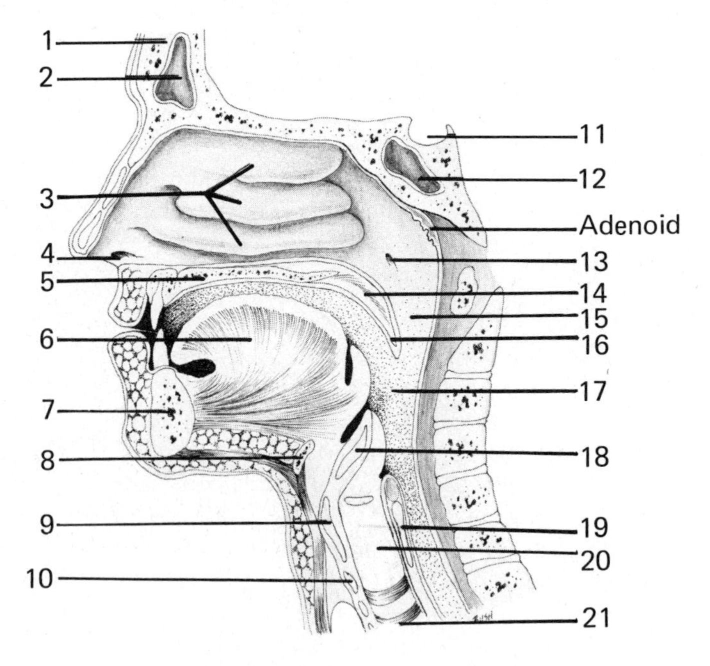

Figure 11-3. Sagittal Section through the Head

4. Use Figure 11-3 as a self-test. The structures indicated in the key should also be located on the dissectable torso and the model of a sagittal section through the head.
5. Figure 11-4 includes three drawings illustrating the relationship of alveoli to the lungs. Note the anatomical and histological features which provide the mechanical basis for gas exchange. This figure may be used as a self-test.

KEY TO FIGURE 11-4

1. thyroid cartilage
2. cricoid cartilage
3. trachea
4. right bronchus
5. secondary bronchi
6. apex (left lung)
7. rib
8. left bronchus
9. pulmonary artery
10. pulmonary vein
11. respiratory bronchiole
12. alveolar duct
13. alveoli
14. alveolar sac

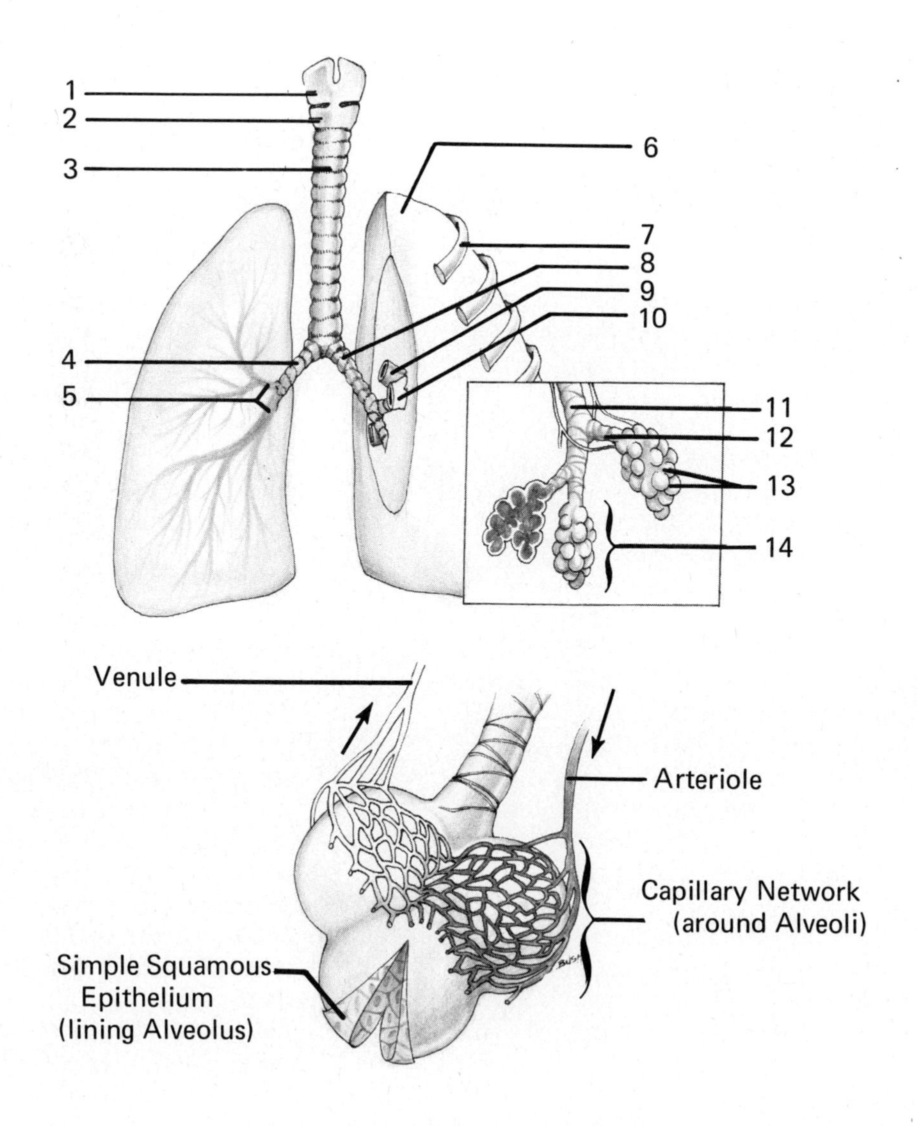

Figure 11-4. Respiratory Tract

Exercise 11-B Dissection of the Respiratory System of the Fetal Pig

Objectives

To identify the respiratory organs of the fetal pig and to compare them with the corresponding human structures.

Materials

Fetal pig; dissecting board or tray; and dissecting set.

Procedure

1. Make a median longitudinal incision through the muscles in the neck in order to expose the **larynx** and **trachea.** Do not sever the blood vessels or nerves located on either side of the trachea. Use Figure 11-5 as a guide for identification of these structures.
   a. The trachea contains rings of cartilage in its walls. Determine whether these rings are complete on the dorsal surface of the trachea.
   b. Remove muscular tissue from the larynx. Make a longitudinal incision through the ventral wall of the larynx and locate the **vocal folds,** which are two small, shelflike membranes. These are poorly developed in the fetal pig.
   c. Locate the **hyoid bone** anterior to the larynx.
   d. Remove the parotid gland. The sublingual and submandibular glands are now visible adjacent to the larynx.
2. Identify the **thymus gland,** the large gland ventral to the heart. This gland consists of two major lobes which extend anteriorly into the neck region on either side of the trachea. The thymus is relatively large in the fetus.
3. The ventral neck muscles and the cervical part of the thymus gland cover the **thyroid gland,** the small, dark gland which lies on the upper trachea. Part the muscles and remove part of the thymus to expose this gland (see Fig. 11-6).
4. Observe the large right and left **common carotid arteries** and the **internal jugular veins** on each side of the trachea.
5. The **vagus nerve** is the conspicuous white band that is bound to the dorsal surface of the common carotid artery. This nerve supplies many of the thoracic and abdominal organs.
6. Free the trachea, laterally, from the preceding blood vessels and nerves. Lying along the trachea, and attached to it, are the two slender **inferior laryngeal nerves.** These nerves which are essential for speech in humans originate from the vagus nerve and, although they are small and delicate, are easily seen against the trachea on either side.
7. Locate the **esophagus,** the muscular tube dorsal to the trachea.
8. Examine the interior of the thoracic cavity.
   a. Note that the thoracic cavity is divided into two lateral **pleural cavities,** which contain the lungs. The **pericardial sac,** which contains the **heart,** is located in the space (mediastinum) between the lungs.
   b. The **pleura** is a double-layered serous membrane which lines the thorax. That portion of the pleura lining the thoracic wall is called the **parietal pleura;** that which covers and adheres to the lungs is called the **visceral pleura.**
   c. The **pericardium,** the membrane surrounding the heart, is also composed of two layers: the outer parietal layer and the inner visceral layer. Much of the parietal pleura forming the medial walls of the pleural cavities is tightly bound to the parietal paricardium.
9. Remove thymus tissue in the thoracic cavity in order to study the **lungs.**
   a. Note that the lung is attached to other structures in the thorax only by the root. The **root of the lung** is formed by the bronchus, pulmonary artery and vein, bronchial arteries and veins, nerves, lymphatic vessels, and bronchial lymph nodes, all encircled by pleura.
   b. Determine the number of lobes in each lung. Each lung is divided into three major lobes: **apical, cardiac,** and **diaphragmatic.** The right lung has an **intermediate lobe** beneath the apex of the heart (see Fig. 11-6).
   c. Cut off a small section of the left lung, and note the density of the lung. The lungs have not yet filled with air, since

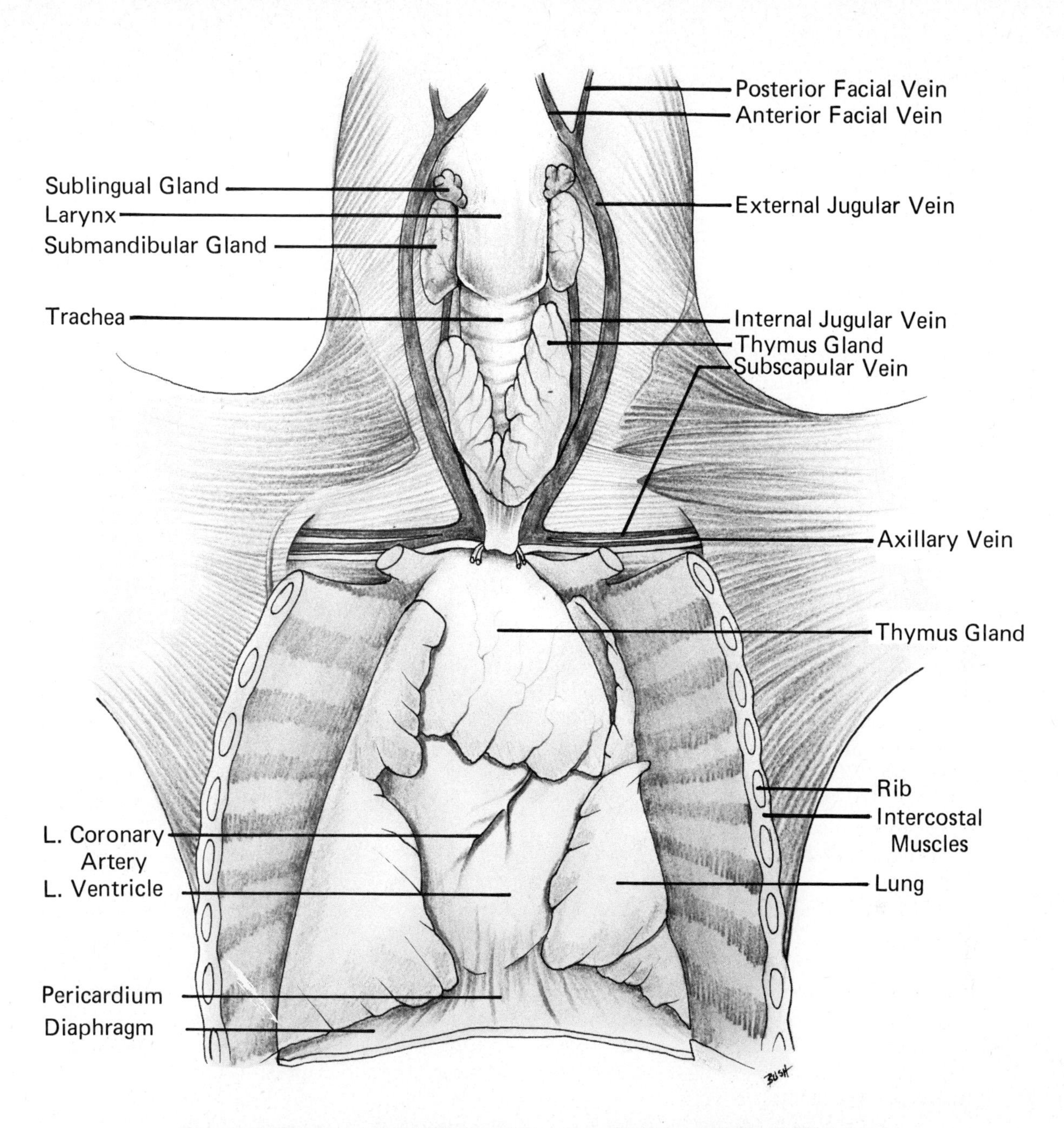

Figure 11-5. Superficial View of Thoracic Cavity with Neck Dissected

they are nonfunctional before birth.

10. The trachea branches into a right and left **bronchus** dorsal to the heart. In order to locate the right bronchus, push the heart to the left side of the thoracic cavity (see Fig. 11-6); then locate the inferior end of the trachea dorsal to the heart and right pulmonary blood vessels. Locate the **apical bronchus** which leaves the trachea anterior to its termination and supplies the right apical lobe. Note the right **main** (primary) **bronchus,** which supplies the right cardiac and diaphragmatic lobes and the small branch of the bronchus which supplies the intermediate lobe. Then scrape away the right cardiac lobe of the lung, bit by bit, noting the organization of the bronchial tree and blood vessels. Leave the vessels intact. The branches of the bronchi can be distinguished from blood vessels because of the presence of cartilage in their walls and absence of colored latex.
11. Locate the **phrenic nerve,** the conspicuous white line that passes along the pericardium

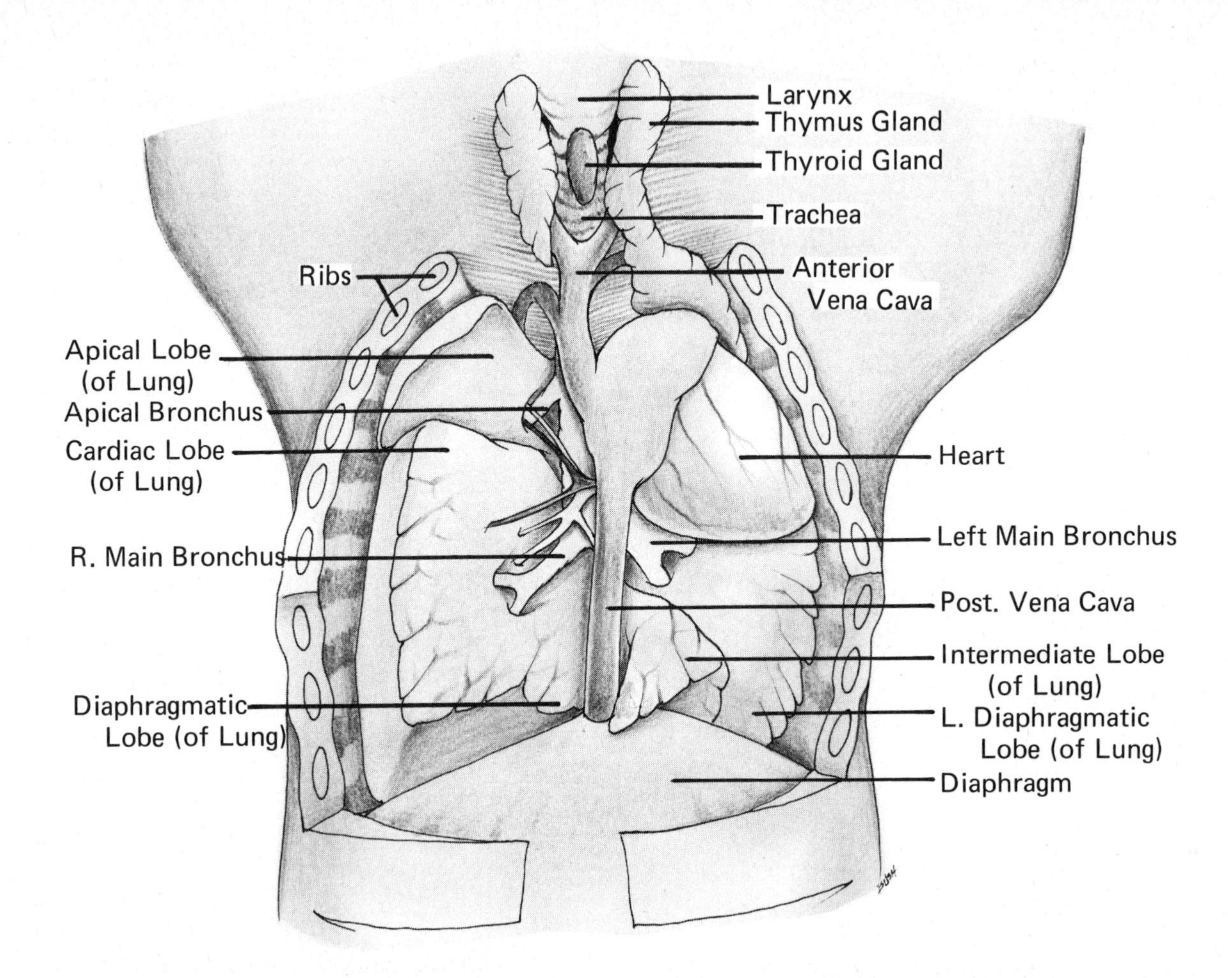

Figure 11-6. Dissection of the Organs of the Thoracic Cavity of the Fetal Pig

to the diaphragm on either the right or left side of the heart.

12. Lift up the left lung and remove some of the parietal pleura dorsal to the lung to locate the **esophagus**. Follow the esophagus to the diaphragm.

Exercise 11-C Chest Measurements during Respiration

Objective

To determine the variation in chest measurements during respiration.

Materials

Tape measure and lung demonstration model.

Discussion

In inspiration, due to the contraction of muscles in the thorax, the size of the thoracic cavity increases. This causes an increase in the size of the lungs, lowering the air pressure inside the lungs. The decrease in air pressure inside the lungs causes air to enter the lungs. This constitutes inspiration. During expiration, the muscles of the thorax relax, the chest size decreases, and air leaves the lungs.

Procedure

1. Using the tape measure, determine the circumference of the chest in inches at the axillae. To do this, place the tape around the chest as high up under the armpits as possible during each of the following conditions: at the end of a quiet inspiration, at the end of a quiet expiration, at the end of the deepest possible inspiration, and at the end of the deepest possible expiration.
2. Record the data in the table under Results and Questions.
3. Examine the lung demonstration model, which consists of a bell jar, rubber sheeting, two balloons, a rubber stopper, and glass

tubing. Compare each of the parts of the model with the corresponding part of the human thorax.
4. Pull down on the rubber sheeting. Observe the change in the balloons.
5. Allow the rubber sheeting to return to its original position. Observe the results.

Exercise 11-D Respiratory Volumes

Objective

To determine respiratory volumes with the spirometer.

Materials

Spirometer, beaker, and 70% alcohol.

Discussion

The amount of air inhaled or exhaled from the lungs can be measured with a spirometer. This instrument consists of a double set of tanks into which air can be inhaled or exhaled. The outer tank contains water and has a tube running through it to carry air above the water line. The inner tank is inverted over the air tube and rises a measured distance when air is exhaled into the tank. The amount of air inhaled or exhaled is indicated by the rise or fall of the pointer along the scale.

Procedure

1. Before using the mouthpiece of the spirometer, clean it in 70% alcohol.
2. The subject should stand in front of the spirometer.
3. Since a single measurement is usually not reliable, each determination should be made a total of three times (unless other instructions are given) and an average calculated. Data should be recorded in the chart under Results and Questions.
4. Determine, before beginning the experiment, how to read the scale on the spirometer. Use the scale graduated in liters. Each number from 1 through 6 indicates the number of liters. Each small mark between the liter marks indicates 0.1 liter. All answers should be recorded in milliliters (ml); to do this, muliply the answer in liters by 1000.
5. Do not let any air escape through the nose in any of these experiments. If necessary, hold the nostrils closed.
6. Tidal Volume (Tidal Air)
   a. Tidal Volume (TV) is the volume of air moved in and out of the lungs with each breath in normal quiet breathing. It averages approximately 500 ml for men and 400 ml for women.
   b. Make certain that the indicator (pointer) is at the zero mark on the scale.
   c. Take a normal quiet inspiration of room air.
   d. Now exhale to the usual extent into the spirometer (do not look at the scale). The tank should rise slightly. Record the tidal volume.
   e. Return the pointer to the zero position by pressing down on the bell (remove the rubber stopper from the tank, if one is present, before applying pressure to the bell). Water will flow from the top of the instrument if this maneuver is not performed **carefully**.
   f. Make two additional determinations of the tidal air.
7. Expiratory Reserve Volume (Supplemental Air)
   a. Expiratory Reserve Volume (ERV) is the amount of air that can be forcibly exhaled from the lungs after a normal quiet expiration (tidal air), following a normal inspiration. This averages about 1000 ml in males.
   b. Inhale normally. Then immediately exhale a normal quiet breath into the room air. Now, exhale as deeply as possible into the spirometer as much of the remaining air in the lungs as you can. This value is the ERV.
   c. Return the pointer to the zero position.
   d. Repeat the above procedure two additional times.
8. Inspiratory Reserve Volume (Complemental Air)
   a. IRV is the amount of air that can be forcibly inhaled after a normal quiet inspiration. It averages 3000 ml in males. This measurement can be determined directly with the spirometer, or it can be calculated as follows:

      $$IRV = VC - (TV + ERV)$$

   b. The bell of the spirometer should be raised to the 5 liter mark.
   c. Inhale normally. Then insert the mouthpiece and inhale as much addi-

tional air as possible from the spirometer. Record this figure and subtract from the 5 liters. This is the IRV.

   d. Repeat the above procedure two additional times.

9. Vital Capacity
   a. To determine vital capacity (VC), inhale as large a quantity of air as possible.
   b. Exhale into the spirometer to the fullest extent (try to completely empty the lungs). The exhalation should be done in a smooth, steady manner. The figure obtained is the vital capacity.
   c. Repeat this determination two additional times.
10. Determine the respiratory rate of the subject per minute by counting the number of breaths per minute.

Exercise 11-E Respiratory Movements

Objective

To demonstrate respiratory movements with the pneumograph.

Materials

Pneumograph; Marey tambour; kymograph; ink writing apparatus; semicircular stand; clamp (or bellows or impedance pneumograph, transducer cable, physiograph); paper cup; straw.

Discussion

The pneumograph and Marey tambour are used to record the rhythmical movements of the chest wall. The pneumograph consists of a coiled spring covered with a rubber tube, the ends of which are connected by an adjustable chain. The Marey tambour consists of a metal pan covered with a diaphragm of rubber, with a metal tube attached. The diaphragm rises whenever the air pressure increases within the tambour. As a subject breathes, the size of the chest varies and changes the air pressure within the pneumograph, which transmits these changes to the tambour. The same experiments can be performed with the bellows pneumograph and the physiograph. The bellows pneumograph is a photoelectric transducer for recording rate and pattern of respiration.

Procedure Using Marey Tambour and Kymograph

1. Select for the subject a student whose chest shows a change in circumference as he breathes. The same subject can be used throughout the experiment. Three copies of each record should be obtained so that each student has a copy.
2. With the rubber tube in an anterior position, fasten the pneumograph around the chest of the subject at approximately the sixth or seventh rib (where the greatest amount of chest movement occurs).
3. Clamp the metal tube of the Marey tambour to the semicircular stand so that the diaphragm side of the tambour side is up. Attach the yoke to the tambour (see Fig. 11-7).
4. Attach the Marey tambour pen to the yoke (see Fig. 11-7). The long arm of the stylus with the pen should extend out toward the kymograph. Insert the short end of the stylus through the hole in the axle. Tighten the check nut to hold it in position. The short arm of the stylus must rest in the notch on the metal glued to the diaphragm. The ink adapter must not touch the tambour.
5. Connect the rubber tubing extending from the pneumograph to one end of a Y tube. Connect a second arm of the Y tube to the tambour. At the end of a quiet inspiration close the third arm with a clamp. It may be necessary to regulate the pressure in the pneumograph and tambour by letting air in or out by means of this clamp.
6. Throughout the experiment the subject should attempt to breathe naturally. He should not look at the drum as the record is being obtained.
7. Using drum speed 5, obtain a tracing of two normal breaths. Label inspiration and expiration on the record. Note that the pen **falls** during inspiration.
8. Calculate the duration of inspiration and that of expiration by dividing the length of the appropriate portion of the record, in centimeters, by the drum speed.
9. Using drum speed 9, make a tracing of the subject's respiratory movements under the following experimental conditions. (Precede and follow each event with normal breathing.)

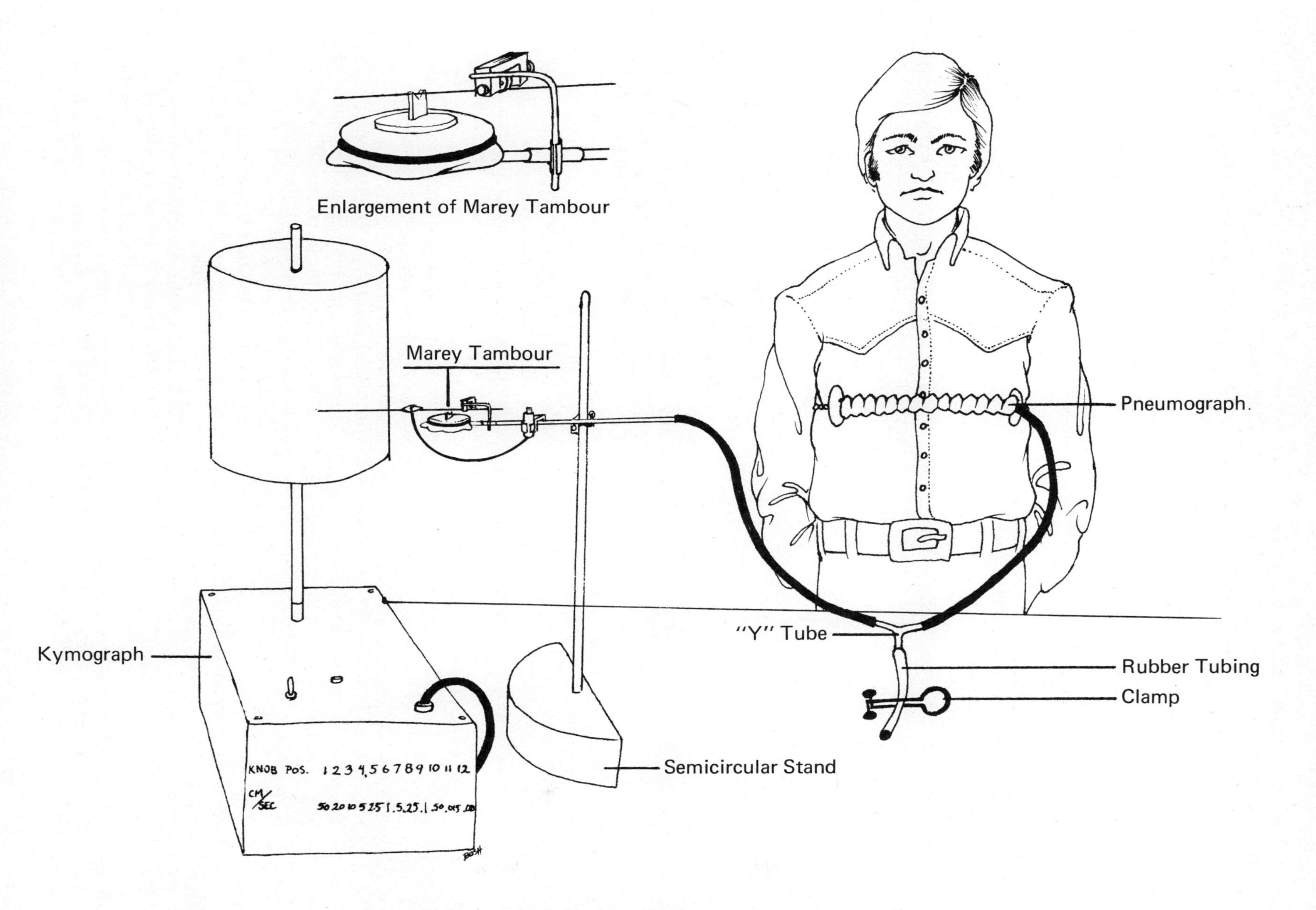

Figure 11-7. Diagram of Pneumograph Setup

a. The subject breathes quietly for ½ minute, then hyperventilates 20 times, following the hyperventilation with a return to normal breathing for approximately another ½ minute.
b. The subject breathes normally several times, then reads aloud from the textbook. The reading should continue for several minutes, with the subject trying to read for a lengthy period of time before taking each breath. At the conclusion of the reading the subject should breathe normally several times.
c. Record the respiratory pattern while the subject is **mentally** calculating the fifth power of some small number suggested by a nonsubject. The subject should not use paper or talk aloud while calculating.
d. The subject breathes normally, then drinks an entire paper cup of water without stopping, then breathes normally again for several breaths.

10. Label each record with the correct letter, a short title, and a brief explanation of the results.

Procedure Using the Bellows Pneumograph and Physiograph

1. The subject must wear a shirt to prevent multiple grounding through the transducer.
2. Open the bellows pneumograph valve by turning the screw one-half turn counterclockwise. Fasten the strap around the chest (high up under the arms) with the transducer plug facing upward, positioned as shown in Figure 11-8. Tighten the strap sufficiently so that the bellows pneumograph remains in position under the arms, but it should not fit too tightly.

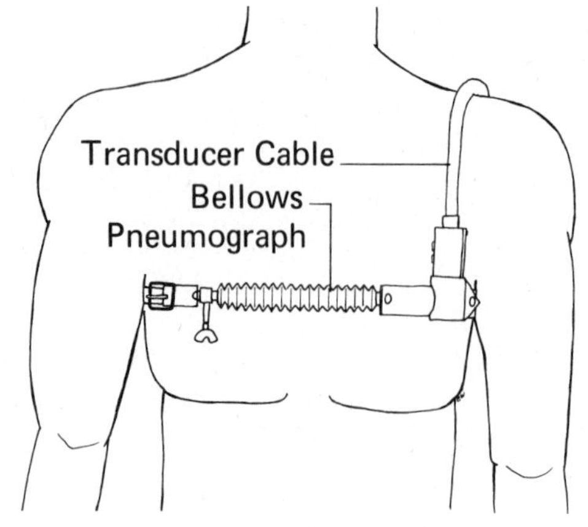

Figure 11-8. Diagram Showing Position of Bellows Pneumograph

3. Compress the bellows **slightly** and close the transducer valve by rotating the wing screw clockwise. Release the bellows **after** the valve is closed to produce a slight internal negative pressure.
4. Connect the bellows pneumograph to the input connector of the channel amplifier of the Physiograph Four-A or the DMP-4A by means of the transducer cable. With the Physiograph DMP-4B, connect the output of the bellows pneumograph to the 9-pin connector of the transducer coupler by means of the transducer cable. Refer to Exercise 7-F, Part 1, for operating instructions for the physiograph.
5. With the Physiograph Four-A and the DMP-4A, position the channel amplifier controls as follows:
   a. ON-OFF to ON.
   b. RECORD-READY to READY.
   c. AMPLITUDE to 0.
   d. POSITION to center 0 position.

   After 5 minutes of warmup, place the RECORD-READY switch on RECORD. Adjust the amplitude control for the desired height of the recording and adjust the position control to place the recorded signal symmetrically about the pen center line. (If the amplitude switch does not work, the internal negative pressure in the bellows may be too high. Open the valve and release the pressure, then follow Step 3 again, compressing the bellows less than before. If this does not work, loosen the strap of the bellows pneumograph slightly and repeat the procedure.)
6. With the Physiograph DMP-4B, place the channel amplifier controls in the following positions (refer to Exercise 7-F, Part 1, Step 9, for recalibration, if necessary):
   a. Press the POWER switch ON.
   b. Press the RECORD switch OFF if it is glowing. This must be on only when actually recording.
   c. Set the mV/CM switch to the 50 mV/CM position and the filter to 10K.
   d. Set the recording pen to the center dark green line, using the POSITION control.
   e. Press the channel amplifier RECORD switch ON.

f. Adjust the transducer coupler BALANCE control to restore the recording pen to the previously established baseline, if necessary.

g. If the signal cannot be returned to the baseline with the BALANCE control, open the valve on the bellows pneumograph and release the pressure. Then follow Step 3 as before, compressing the bellows **less** than before or loosening the chest strap slightly.

h. Adjust for the desired signal amplitude (height of the recording) by using the channel amplifier mV/CM control.

7. Using a paper speed of 2.5 cm/sec and with the time-event marker set to mark one second intervals, obtain a tracing of three normal breaths. Label inspiration and expiration on the record. Note that the pen rises on inspiration and falls during expiration (see Fig. 11-9).
8. Calculate the duration of inspiration and that of expiration by dividing the length of the appropriate portion of the record, in centimeters, by the paper speed (2.5 cm/sec).
9. Using a paper speed of .1 cm per second, make a tracing of the subject's respiratory movements as directed in Step 9 of the Procedure in the first part of the experiment (Step 9a-d).
10. Label each record with the correct letter, a title, and a brief explanation of the results.
11. Repeat the above procedure on each member of the experimental team.

Alternate Procedure Using the Impedance Pneumograph

1. It is possible to perform the same experiments with the impedance pneumograph instead of the bellows pneumograph. A small alternating current of approximately 2 microamps is passed through the electrodes attached to the subject, with the voltage across the electrodes directly proportional to the subject impedance. (The impedance or conductivity to the passage of a current is reduced when air is brought into the lungs and increased as air is exhaled. The magnitude of the changes is proportional to the amount of air entering and leaving the lungs.) These small voltage variations resulting from the impedance changes are amplified and detected as a signal for recording on the physiograph.
2. Apply a salt solution to two plate electrodes and attach them to the interspace between the fifth and sixth ribs near the sides of the chest, fastening them by means of the long rubber strap. Attach an additional plate electrode on the center of the subject's chest for grounding.
3. With the Physiograph Four-A and DMP-4A, mount the impedance pneumograph on the stand (see Fig. 11-10). Connect the output on the top of the impedance pneumograph

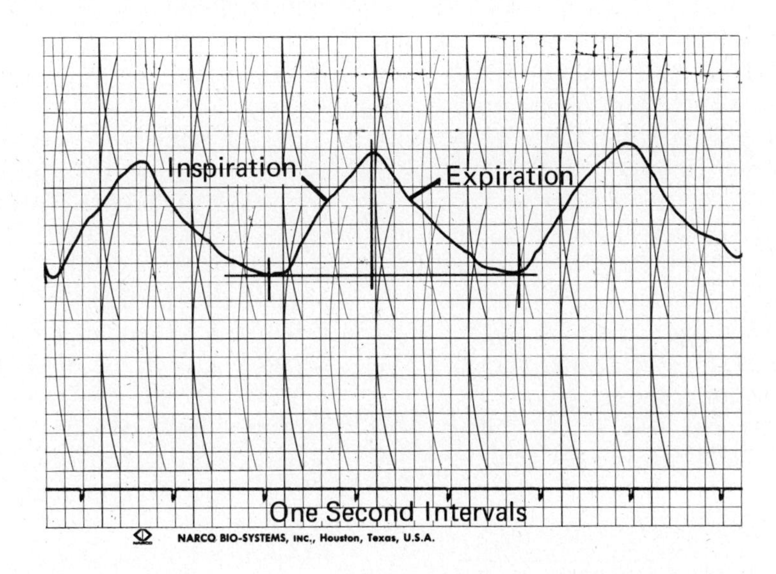

Figure 11-9. Tracing of Two Normal Breaths at Paper Speed 2.5 cm/sec

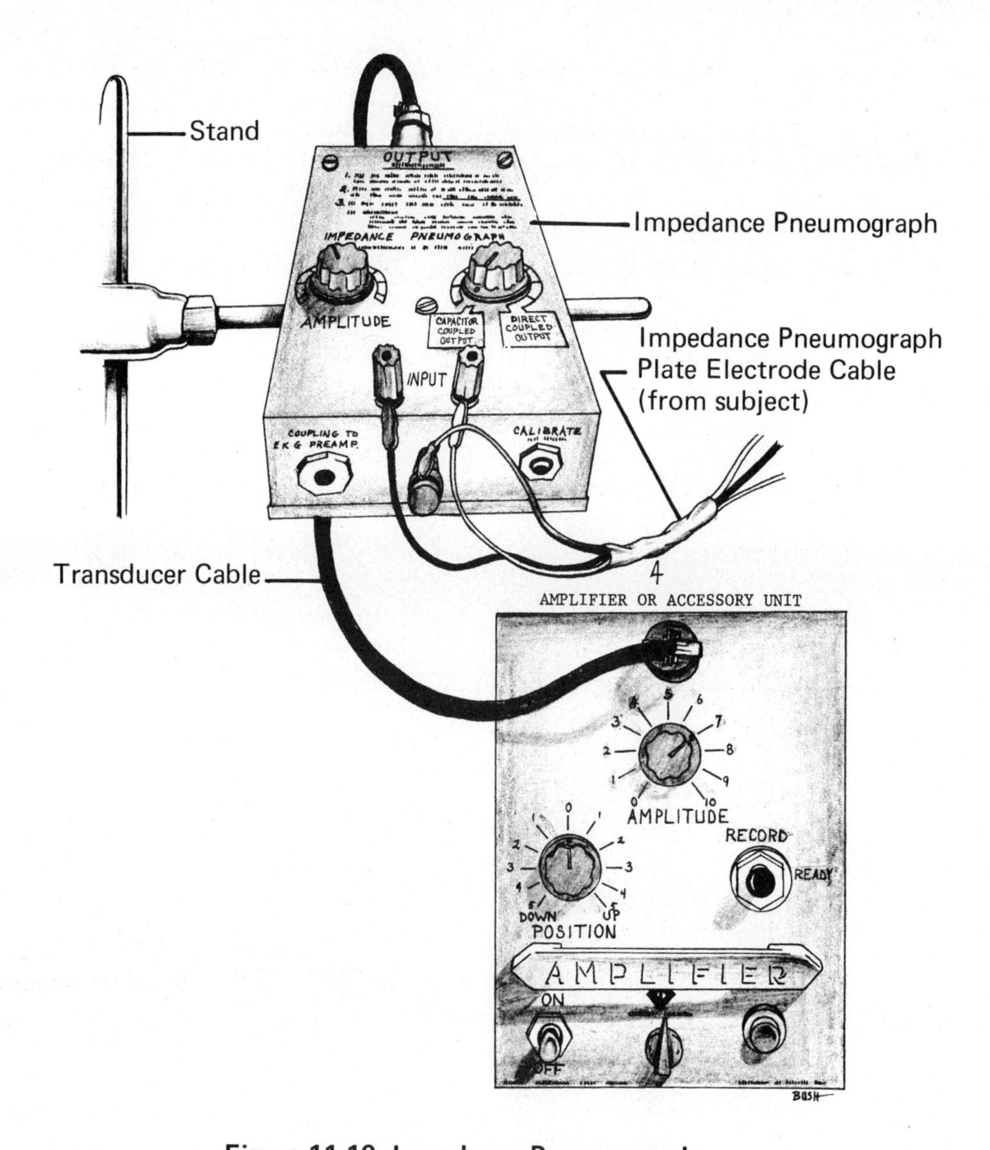

Figure 11-10. Impedance Pneumograph

to the input connection of the physiograph amplifier channel with a transducer cable. Set the physiograph amplifier controls as follows: ON-OFF switch to ON, POSITION to center (0) position, RECORD-READY to READY (AMPLITUDE is inoperative on the physiograph). Allow a five minute warm-up period.

4. Use the impedance pneumograph plate electrode cable (red, black, and green wires). Attach the long, loose ends to the subject, red on the left side of the chest, black on the right, and green to the ground in the middle. Attach the short twisted ends to the front of the impedance pneumograph, the black electrode to the black input terminal, the red electrode to the red input terminal, the green to the ground terminal on the lower surface (see Fig. 11-10).
5. Turn the impedance pneumograph AMPLITUDE control fully counterclockwise and the CAPACITOR COUPLED OUTPUT-DIRECT COUPLED OUTPUT to CAPACITOR COUPLED OUTPUT. The knob should click into position.
6. Place the channel amplifier RECORD-READY switch on RECORD. Adjust the position of the pen to the centerline by turning the POSITION knob. Adjust the height of the record by turning the AMPLITUDE knob on the impedance pneumograph slowly clockwise until the recorded patterns are of the desired magnitude.
7. Follow steps 7-11 of the procedure with the bellows pneumograph.

8. With the Physiograph DMP-4B, set the channel amplifier controls in the following positions (see Fig. 11-12 and Exercise 7-F, Part 1, Step 9, for general directions for recalibration if necessary):

a. Press the POWER switch ON.
b. Press the RECORD switch OFF, if it is glowing.
c. Set the FILTER switch to 30.
d. Set the mV/CM switch to 10 mV/cm.

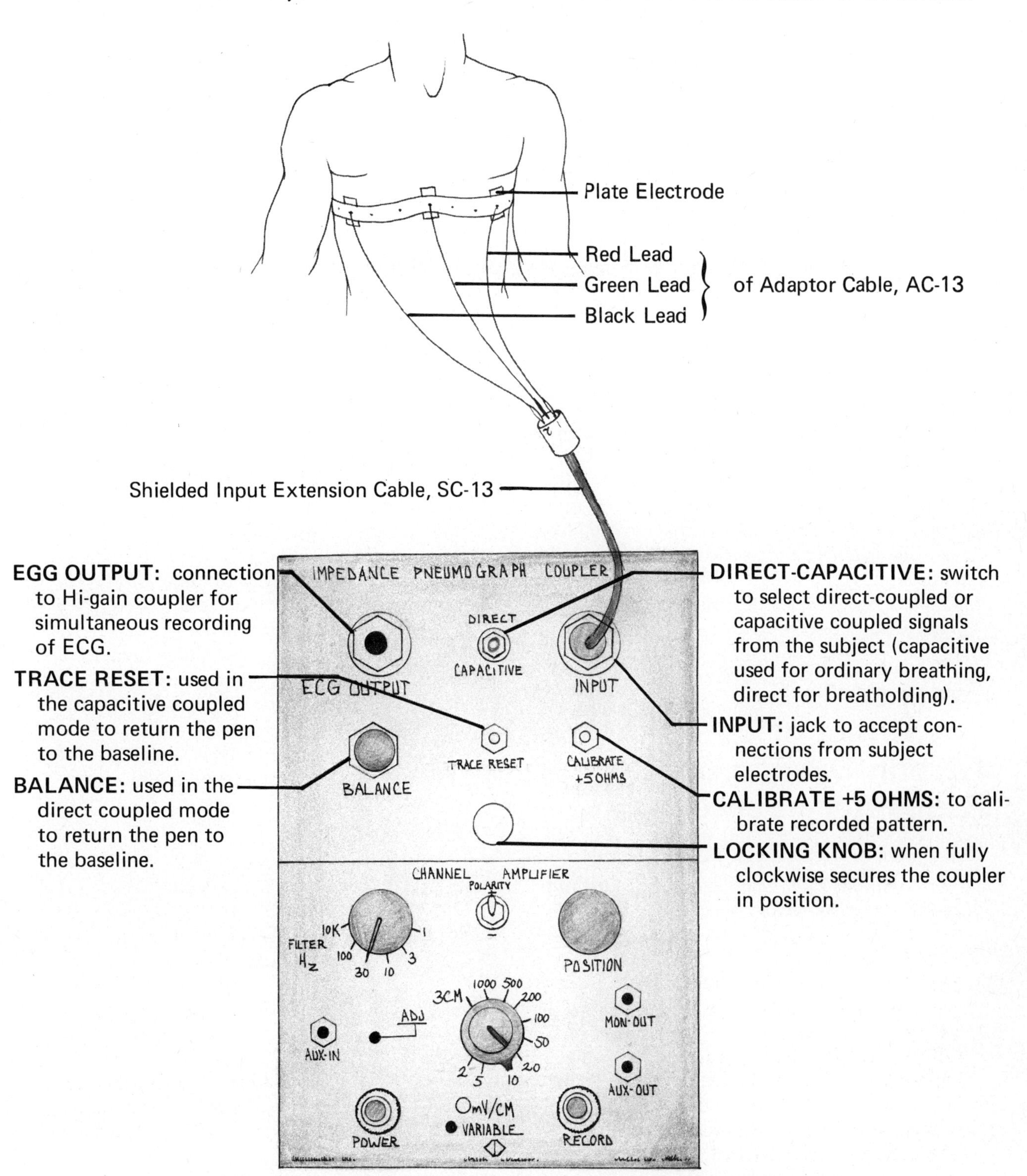

Figure 11-11. Impedance Pneumograph, with Physiograph DMP-4B

e. Set the recording pen to the center dark green line using the POSITION control.

9. Insert the Shielded Input Extension Cable SC-13 to the input jack of the impedance pneumograph coupler.
10. Attach the long loose ends of the Adaptor Cable AC-13 to the subject, red on the left side, black on the right side, and green to the center plate electrode on the chest. Connect the short twisted ends of the Adaptor Cable AC-13 to the Shielded Input Extension Cable SC-13. Connect the red wire to the 1 position, the black to the 2 position and the green to the G position.
11. Set the impedance pneumograph coupler DIRECT-CAPACITIVE switch to CAPACITIVE.
12. Press the channel amplifier RECORD switch to ON.
13. Return the recording pen to the previously established baseline using the TRACE RESET button (the BALANCE control is not operative in the capacitive model).
14. A tracing of the breathing pattern should now be obtained. To increase the amplitude of the recording, turn the mV/CM switch clockwise; to decrease the amplitude, turn the switch counterclockwise.
15. Follow steps 7-11 of the procedure with the bellows pneumograph.

Exercise 11-F The Effect of Carbon Dioxide on the Respiratory Center

Objective

To demonstrate the effect of carbon dioxide on the respiratory center.

Materials

Paper (or plastic) bag; straws; and a beaker or paper cup.

Discussion

Hyperventilation washes much of the carbon dioxide out of the blood. The increase in the oxygen content of the arterial blood after hyperventilation is negligible, but the carbon dioxide level may fall from a normal value of 44 mm Hg to as low as 15 mm.

Hyperventilation may result in a feeling of dizziness because of cerebral anoxia (due to either decreased blood pressure, an increase in pH, or the constriction of cerebral vessels caused by the diminished carbon dioxide content of the blood).

Procedure

1. Breathe quietly for 3 minutes and then note how long the breath can be held after a quiet inspiration.
2. Place a straw in your mouth, with the other end in a beaker of water. After a quiet inspiration, hold your breath as long as possible. Just before it is necessary to take a breath, begin to sip water. Continue timing, and determine the total time elapsed after the original quiet inspiration until it is necessary to take a breath.
3. Hyperventilate 20 times (breathe deeply and rapidly 20 times) and then determine how long the breath can be held. Do not continue the experiment if a pronounced feeling of dizziness results.
4. Place a large paper (or plastic) bag over the mouth and nose. Hyperventilate into the bag 20 times. Note how long the breath can be held after hyperventilation into the bag.
5. Run around the building rapidly or bicycle for 2 minutes on the bicycle ergometer (see instructor for settings). Determine how long the breath can be held immediately after the cessation of this exercise.
6. Enter all times in the table under Results and Questions. In explaining the results obtained in this experiment, make use of the fact that the respiratory center is very sensitive to slight changes in the carbon dioxide content of the blood.

Exercise 11-G Temperature and Composition of Exhaled Air

Objective

To determine the temperature and composition of exhaled air.

Materials

Thermometer; glass plate; test tube; lime water; and a straw.

Discussion

When converting from Fahrenheit to Centigrade, substitute the thermometer reading in the following formula:

$$\text{Centigrade} = \frac{5}{9}(\text{F}^\circ - 32)$$

When converting from Centigrade to Fahrenheit, substitute the thermometer reading in the following formula:

$$\text{Fahrenheit} = \frac{9}{5}\text{C}^\circ + 32$$

Procedure

1. Using a chemical thermometer, determine room temperature.
   Results in degrees Centigrade?__________
   Results in degrees Fahrenheit?__________
2. Now, hold the bulb of the thermometer between the lips and approximately an inch into the mouth. The bulb should not be touching any part of your body.
3. Exhale air rapidly over the bulb of the thermometer and **immediately** take a reading of the temperature.
   Results in Centigrade?__________
   Results in Fahrenheit?__________
4. Exhale air on a glass plate. Observe what collects on the surface of the glass plate.
5. Place a prepared solution of lime water in a test tube. Gently blow air into the lime water through a straw. A white precipitate of calcium carbonate should form in the test tube.
6. Record your observations under Results and Questions.

Self-Test—Respiratory System

Directions: See Chapter 1, Self-Test p. 9.

1. Carbon dioxide **is essential** to human life.
2. Zero temperature of sleeping quarters **causes** a person to develop respiratory infection.
3. Mountain sickness at 15,000 feet above sea level occurs because the **oxygen content** of the blood is below normal.
4. Bends or caisson disease, associated with high pressure, is due to an increased amount of **oxygen** in the blood.
5. Caisson disease is prevented by **gradual decompression.**
6. Artificial respiration **need not be** administered to someone unconscious from carbon monoxide poisoning if the individual is placed in fresh air.
7. **Cilia** aid in the movement of dust particles along the breathing passages.
8. **Glucose** is the fuel used by cells of the body during cellular respiration.
9. The structure which is common to both the digestive and respiratory systems is the **trachea.**
10. The structures in cytoplasm containing the enzymes of cellular respiration are called **mitochondria.**
11. The **glottis** prevents food from entering the larynx.
12. The opening between the two vocal folds is called the **laryngeal aperture or glottis.**
13. Inspiration is an **active process** involving primarily the contraction of the diaphragm and external intercostal muscles.

14. The gaseous exchange across the respiratory membrane from the alveoli of the lungs to the capillary circulation is called **internal** respiration.

15. The hemoglobin picks up oxygen which has been **dissolved in the fluids of the lung tissue**.

16. **Sinus infection** is an extension of an infection from the nasal passages.

17. An illustration of a homeostatic reaction is the **increase in the rate of breathing** during exercise.

18. Bones making up the **lateral** wall of the nasal cavity include the inferior nasal concha, the maxilla, and the ethmoid.

19. The **pharyngeal** tonsils are also called the adenoids.

20. The tonsils are located within the **laryngeal aperture.**

21. The volume of air in excess of tidal air that can be inhaled by the deepest possible inspiration is called the **inspiratory reserve volume.**

22. Passive expiration is primarily due to the **contraction** of the muscles of inspiration.

23. The lungs expand during inspiration because the **two layers of the pleura adhere** to each other.

24. The site of greatest CO_2 concentration would be found in the **lungs.**

25. The **pulmonary veins** contain a greater concentration of oxygen than the **alveoli.**

26. Most of the oxygen carried in the body has combined with **hemoglobin** to form **oxyhemoglobin.**

27. Most of the carbon dioxide in the body is present in the form of sodium bicarbonate in the blood **plasma.**

28. The cartilaginous rings of the **trachea** are incomplete behind.

29. The **left** bronchus is shorter, larger in diameter, and more nearly vertical than the **right.**

30. The pleural cavities in which the lungs lie are formed by invagination of **two serous sacs.**

31. The maximum volume of air which can be exchanged in a single, voluntary respiratory cycle, (i.e., inspiration and expiration) is called the **vital capacity**.

32. Average vital capacities for women are **approximately the same as** those for men.

33. If an opening exists connecting two volumes of gas, the gas will always flow from the area of **higher pressure** to the area of **lower pressure.**

34. **Intrapleural** pressure refers to either the absolute or relative value of pressure between the visceral and parietal layers of pleura.

35. **Intrapulmonary** pressure refers to the pressure within the respiratory passages of the lungs.

36. The **intrapleural** cavity is a true cavity only when accident or disease results in an opening between the potential **intrapleural** space and the atmosphere.

37. During inspiration intrapulmonary pressure is **greater** than atmospheric pressure.

38. The partial pressure of oxygen in the lungs is approximately **100 mm Hg.**

39. In combining with oxygen, hemoglobin is approximately 95% saturated at an oxygen partial pressure of 100 mm Hg.

40. The speed of reaction of carbon dioxide with water is increased approximately 600 times by the enzyme, **carbonic anhydrase.**

41. Carbon dioxide is much **less** soluble than oxygen in blood plasma.

42. Reduced oxygen tension in the blood stimulates **chemoreceptors** which influence respiratory rates.

43. **Hypoventilation** may induce apnea.

44. Involuntary mechanisms of the control of respiration **may be taken over or modified** by the higher brain centers.

45. Normally the oxygen content of blood is approximately **20 cc per 100 cc** of arterial blood.

46. The principal **afferent** nerve in respiration is the phrenic nerve.

47. Inspiratory neurons of the medulla are subject to **constant chemical stimulation.**

48. **Eupnea** is normal breathing.

49. The amount of oxygen in the atmosphere is approximately **21%.**

50. The atmosphere contains from 0.03-0.049% of carbon dioxide, but air exhaled from the lungs contains about **5%** carbon dioxide.

51. The respiratory center, which controls the breathing rate, acts primarily in response to the **amount of carbon dioxide** in the blood.

52. A substance which combines with the hemoglobin of the blood to form a stable compound and prevents oxygen from reaching the tissues is **carbon monoxide.**

53. The respiratory center is located in the **hypothalamus.**

54. Pleurisy is an inflammation of the **peritoneum.**

55. Bronchitis is an inflammation of the **lungs.**

56. A pneumonectomy is the removal of **one lung**, while a lobectomy is the removal of **a lobe of the lung.**

57. **Bronchoscopy** is visual examination of the bronchi with a special instrument.

58. A laryngectomy is the removal of the **vocal cords.**

59. Pulmonary emphysema is **overdistention** of the air spaces in the lungs.

60. A doctor may prescribe high percentage **oxygen** inhalations after surgery to increase the rate and depth of breathing.

61. A patient has tuberculosis of the left lung. In order to rest this lung and promote recovery, the doctor injects a measured amount of air into the **interpleural** space, inducing artificial pneumothorax.

62. Artificial pneumothorax **is not permanent** because the air is finally absorbed.

63. Crushing the **phrenic** nerve prevents movement of the diaphragm **on the side of the crushed nerve**. This procedure allows the lung to "rest" since only limited inflation is possible.

64. Drugs, such as Isuprel or epinephrine (adrenalin), provide asthmatics relief because they **stimulate** smooth muscle contractions in bronchioles, allowing them to dilate.

65. Anoxia (hypoxia) is a **reduced** level of oxygen in the blood.

66. A person with **orthopnea** must sit erect or stand in order to breathe without discomfort.

67. **Dyspnea** is the temporary cessation of breathing due to lack of stimulation of the respiratory center.

68. Asphyxia is the cessation of breathing due to a lack of **carbon dioxide.**

69. **Cheyne-Stokes** respiration consists of alternating periods of dyspnea and apnea.

70. **Atelectasis** refers to collapse of the lungs.

KEY

2. predisposes
4. nitrogen
6. should be
9. pharynx
11. epiglottis
14. external
20. pharynx
22. relaxation
24. cells
25. alveoli—pulmonary veins
29. right—left
32. less than
37. less
41. more
43. hyperventilation
46. efferent
53. medulla oblongata
54. pleura
55. bronchi
58. larynx
60. carbon dioxide
61. intrapleural
64. inhibit
67. apnea
68. oxygen

Name ______________________

Results and Questions for Chapter 11

Exercise 11-A. Anatomy of the Human Respiratory System

1. Name the bones in the skull that possess paranasal sinuses. ______________________

____

2. Into which part of the respiratory tract do these sinuses open? ______________________

3. What is the function of the paranasal sinuses? ______________________

____

4. Name the specific cartilage(s) in the larynx:

   a. that serves as a lid for the larynx. ______________________

   b. to which the vocal folds are attached. ______________________

   c. shaped like a signet ring. ______________________

   d. shaped like a shield. ______________________

   e. that forms the Adam's apple. ______________________

5. What are the functions of the larynx? ______________________

____

6. What is the name for the opening between the vocal folds? ______________________

7. Describe the location of the trachea with respect to the esophagus. ______________________

____

8. Trace the path of air from a point outside the nose to the point of exchange of gases with the pulmonary capillaries. ______________________

____

____

9. Why was it impossible to swallow while depressing the larynx? ______________________

____

10. Describe the events that occur during the swallowing process. ______________________

____

____

____

Exercise 11-B. Dissection of the Respiratory System of the Fetal Pig

1. What is the common name for the trachea? ______________________

2. What is the function of the cartilage in the wall of the trachea? ______________________

____

3. Are the rings of cartilage in the trachea incomplete dorsally or ventrally? ______________________

4. What is the common name for the larynx? ______________________

5. What is the function of the thymus gland? ______________________

____

6. What is the function of the thyroid gland? ______________________

____

7. Name three serous membranes. ______________________

8. What is the name given to the space in between the lungs? ______________________

9. Name five structures located in this space. ______________________

____

10. What is the function of the phrenic nerve? ______________________

11. Compare the number of lobes in the lungs of the pig with the number present in the human lungs. ______________________

12. What is the function of the esophagus? ______________________

13. Does the trachea divide into two bronchi dorsal or ventral to the heart? ______________________

Exercise 11-C. Chest Measurements during Respiration

1. Record the chest measurements in the following table.

Chest Measurement	*Quiet Breathing*		*Forced Breathing*	
	Inspiration	*Expiration*	*Inspiration*	*Expiration*
Circumference (in inches)				

2. Do the above results agree with the theoretical results? __________ Explain: ______________________

____

3. Name the muscles that contract during quiet inspiration. ______________________

____

Name ____________________

4. Name the additional muscles that contract during forced inspiration. ____________________

____

5. Name the muscles used in forced expiration. ____________________

6. Which dimension of the thorax is increased by contraction of the diaphragm? ____________________

____

7. To which structure in the human respiratory system does each of the following structures in the lung demonstration model (bottle thorax) correspond?

   a. Balloons ____________________ c. Glass Y tube ____________________

   b. Rubber sheeting ____________________ d. Bell jar ____________________

8. Explain why the balloons expand when the rubber sheeting is lowered. ____________________

____

Exercise 11-D. Respiratory Volumes

1. Record the respiratory volumes (in ml) in the following data chart. Compare your results with the average (theoretical) value for each respiratory volume for your sex.

Respiratory Volumes (ml.)	*I*	*II*	*III*	*Average*	*Theoretical Average*
Tidal Volume					
Expiratory Reserve Vol.					
Inspiratory Reserve Vol.					
Vital Capacity					

2. List any possible causes of differences between your values (as recorded in the data chart) and the theoretical. ____________________

____

3. Calculate your combined vital capacity by adding the average tidal air, expiratory reserve volume, and inspiratory reserve volume. Record the calculated and measured vital capacity in the spaces provided.

   Combined VC: ____________________

   Measured VC: ____________________

   The combined vital capacity is often larger than the actual vital capacity. This may be due to the fact that patients with some kind of respiratory obstruction (particularly an obstruction resulting from bronchial asthma and pulmonary emphysema) are unable to move their thoracic cage from fullest inspiration to maximal expiration in a single maneuver.

4. What conclusion can you draw from a comparison of your combined and measured vital capacity? ____

____

5. Compare your measured IRV with that calculated from the formula:

   IRV = VC – (TV + ERV) ____

6. Calculate the total amount of air inhaled (in liters) in one day.

   a. Respiratory rate per minute ____

   b. Average tidal air (in ml) ____

   c. Average tidal air (in liters) ____

   d. Minute volume ____

   e. Amount inhaled per hour ____

   f. Amount inhaled per day ____

7. What is the predicted vital capacity for your sex, age, and height? (See Appendix 3, Sections B and C, for predicted vital capacity.)

   a. Predicted vital capacity ____

   b. Actual vital capacity ____

8. Discuss possible causes of any difference between the predicted and actual vital capacity noted in No. 7. ____

____

____

9. Why would you expect males to have a larger vital capacity than females for a given age and height? ____

____

Exercise 11-E. Respiratory Movements

1. Paste the record of three normal breaths below. Label inspiration and expiration on this record. Draw a horizontal baseline beneath the record. Draw a vertical line from the end of inspiration to the baseline. See Figure 11-9.

Name ______________________

2. Calculate the duration of both inspiration and expiration. Show all calculations in the space below. Measure the distance along the baseline in centimeters for both inspiration and expiration. Divide the length (in cm) of inspiration and of expiration by the paper speed.

Paper Speed __________ Length of Inspiration __________ Length of Expiration __________

Duration of inspiration __________________

Duration of expiration __________________

Total duration of one breath ______________

3. Calculate the breathing rate per minute by substituting in the following equation:

$$\frac{\text{1 breath}}{\text{duration 1 breath from Q. 2}} = \frac{\text{X breaths}}{\text{60 sec.}}$$

rate of breathing per minute _____

4. Separate the parts of the record obtained in Step 9 of the procedure. Paste each part in the appropriate space. Next to each record discuss in detail what that record illustrates.

a. Hyperventilation

b. Reading aloud

c. Mental calculations

d. Sipping water

5. Does the hyperventilation record show a period of apnea before the rate and depth of breathing returned to normal? ________________

Exercise 11-F. The Effect of Carbon Dioxide on the Respiratory Center

1. What is the importance of the glossopharyngeal reflex (deglutition apnea)? ________________

____

____

2. Explain how hyperventilation causes an increase in the pH of the blood. ________________

____

____

Name ______________________

3. Record all data and a short explanation of your results in the following table.

Respiratory Activity	*Time of Breath Holding*	*Explanation of Results*
After quiet inspiration		
After water sipping		
After hyperventilation		
After hyperventilation into bag		
After exercise		

4. Why is it easier to hyperventilate using a paper or plastic bag? ______________________

____

5. Why might a person have no desire to breathe after a period of hyperventilation? ______________________

____

Exercise 11-G. Temperature and Composition of Exhaled Air

1. What was temperature of the inhaled air (room temperature)? ______________________

2. What was the temperature of the exhaled air? ______________________

3. Explain what caused the difference between the temperature of inhaled and exhaled air. ______________________

____

4. What collected on the surface of the glass plate? ______________________

5. What is the origin of this material? ______________________

6. Explain the formation of the precipitate in Step 5 of the Procedure. ______________________

____

7. What functions of the respiratory tract are demonstrated in this experiment? ______________________

____

12

The Circulatory System

Exercise 12-A Blood Tests

Part 1. Determination of the ABO and Rh Blood Types

Objectives

To determine the ABO and Rh blood types.

Materials

Sterile cotton; lancets; 70% alcohol; glass slides; wax pencil; toothpicks or applicators; Anti-A, Anti-B, Group O, (Anti-A, B), and Anti-Rh_0 (Anti-D) sera; and a light-heat box.

Discussion

The ABO and Rh blood type of each student will be determined in this exercise. The antigens (agglutinogens) that determine blood type are located on the plasma membrane of the erythrocytes, the antibodies (agglutinins) in the plasma. The antibodies cause the agglutination (clumping) of the erythrocytes carrying the corresponding antigen.

To determine the ABO blood type of the student, Anti-A and Anti-B antisera (which contain the a and b antibodies, respectively) will be used. If the blood is type A, there should be agglutination of the RBC's by the Anti-A serum; if type B, agglutination of the cells by the Anti-B serum; if type AB, there should be agglutination by both antisera; if type O, there should be no agglutination by either antisera.

The Rh blood type will be determined in a similar manner. Although, there are several Rh agglutinogens, the blood will be tested only for the presence of the D (Rh_0) antigen, the most common antigen. If the blood reacts with the Anti-D (Rh_0) serum, the blood type is called Rh positive (Rh+); if the antiserum does not cause agglutination of the cells, the blood is called Rh–. Approximately 85% of all Caucasians are Rh+ and approximately 15% Rh–.

Procedure

1. Using a wax pencil (or felt-tip pen), draw two complete circles on a glass slide (as illustrated on the next page), one at each end of the slide (the circles prevent blood and antisera specimens from running together).

2. Place the letter A under one circle; B under the other.
3. Place one drop of Anti-A serum in the A circle; one drop of Anti-B serum in the B circle.
4. Clean the finger with 70% alcohol and allow it to dry.
5. Puncture the finger tip with a sterile lancet.

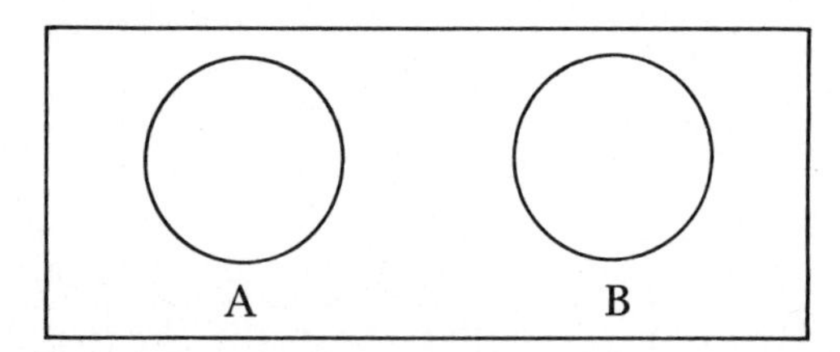

6. Place one drop of blood next to each drop of antiserum.
7. Mix the blood with the antiserum by means of a toothpick or applicator. To avoid contamination use a separate toothpick for each stirring operation.
8. Tip the slide carefully back and forth for one minute (do not let the two samples run together).
9. Agglutination visible with the naked eye should begin almost immediately and should be complete by the end of one minute. Examine the slide with a microscope for verification of agglutination.
10. If the blood type is O, confirm this by using Group O (Anti-A, B) serum, which contains both the Anti-A and Anti-B agglutinins. Add one drop of the Group O serum to the slide, followed by one drop of blood, and mix as before. No agglutination should result if the blood is type O; the serum will agglutinate the cells of Groups A, B, and AB.
11. In order to determine the Rh blood type, place one drop of Anti-Rh_0 (Anti-D) serum on a clean and dry slide that has been warmed on the light-heat box.
12. Add one drop of blood.
13. Mix the blood and antiserum thoroughly and spread over an area about an inch in diameter.
14. Place the slide on a warm (40° C.) light-heat box (or a microscope substage lamp) and tilt back and forth for no more than two minutes. The reaction usually appears within 30 seconds and is complete within two minutes. Readings should be taken only with the naked eye. The readings must be made before the preparation dries.
15. If the blood is Rh negative, the mixture will remain homogeneous. If the blood is Rh positive, coarse clumps will form promptly and increase in size with time.
16. Record the results of the entire class in the table under Results and Questions.

Part 2. Determination of the Hemoglobin Content of Blood

Objectives

To determine the amount of hemoglobin in a personal sample of blood; to compare the results obtained with the Tallquist scale with those obtained by means of a hemoglobinometer.

Materials

Sterile cotton; lancet; 70% alcohol; Tallquist scale; Spencer hemoglobinometer; and hemolysis applicators.

Discussion

The Tallquist scale is used to estimate the grams of hemoglobin per 100 cc blood. This involves the comparison of the color of a drop of blood with a standard color chart. The Spencer Hb-meter measures the oxyhemoglobin directly as a result of its light absorption in the green portion of the spectrum. A suitable filter is used and the resulting color is matched with a colored glass standard. In both methods of determining the amount of hemoglobin, 15.6g/100 cc of blood is taken as 100%.

Procedure A: The Tallquist Method of Determining Hemoglobin Content of Blood

1. Remove one square of absorbent paper from the Tallquist test booklet.
2. Clean the finger with 70% alcohol. **Let all the alcohol evaporate.**
3. Puncture the finger with a sterile lancet.
4. Wipe away the first drop of blood with the edge of the absorbent paper (it contains too much tissue fluid).
5. Place the second drop of blood in the center of the paper. The drop must be free flowing. Do not squeeze the finger as this forces out tissue fluid.
6. In a few **seconds** the blood stain will lose its glossiness and comparison should then be made immediately under direct light (not artificial). **Do not let the blood dry to a brown color**, as inaccurate values will be obtained.

7. Place the specimen under the color comparison chart and move it along so that the blood stain appears at the apertures.
8. Compare the color of the drop with the colors on the Tallquist scale. Since the color differences on the chart represent 10%. variations in the hemoglobin content, it will be necessary to estimate the intermediate percentages.
9. Record the data (in grams hemoglobin, percent hemoglobin, and stage of anemia) in the table under Results and Questions.

Procedure B: Determining Hemoglobin Content of Blood Using a Spencer Hb-Meter

1. In order to use the Spencer Hb-meter, assemble the following three parts: a double-sided chamber with an H-shaped moat, a cover plate, and a clip. If necessary, clean the parts with distilled water and dry with lens paper.
2. Offset the chamber (which should be beneath the cover glass) from the cover glass so that half of the H is visible.
3. Obtain a drop of blood using the method described in Procedure A.
4. Place a sizable drop of blood on the chamber surface.
5. Hemolyze the blood by agitating it gently with the tip of a hemolysis applicator, which has been treated with a hemolytic reagent. The blood will change from cloudy to transparent when hemolysis is complete. Thirty to 45 seconds are required for this.
6. Once hemolysis is complete, push the chamber into the clip.
7. Insert this chamber into the slot on the left side of the instrument.
8. Hold the instrument to the eye with the left hand in such a manner that the left thumb rests on the light switch on the bottom. When this button is pressed, a green split field appears in the instrument. With the other hand, move the slide button on the right side of the instrument until the two halves of the field appear as a single field.
9. The position of the index mark on the slider knob indicates the hemoglobin content. Record the grams of hemoglobin per 100 cc of blood and the percent hemoglobin, using the scale marked 15.6.
10. Clean and dry the chamber.
11. Record your data in the appropriate place under Results and Questions.
12. Obtain the results in grams of hemoglobin, from each member of the class, using the results obtained with the Spencer Hb-meter. Average the sexes separately and record the averages under Results and Questions.

Part 3. Coagulation Time and Bleeding Time

Objectives

To determine the coagulation time of blood by the slide method and by the capillary tube method; to determine the bleeding time of blood.

Materials

Lancets; sterile cotton; 70% alcohol; glass slide; pins; filter paper; and a capillary tube.

Discussion

The bleeding time (normal range from 1 to 3 minutes) is taken as the time for a small sharp incision to stop bleeding. However, the time depends upon the depth of the wound and the degree of hyperemia in the finger as well as upon the presence of adequate amounts of clotting factors. The coagulation (or clotting) time is the time required for a sample of blood removed from the body to clot. The usual way to determine coagulation time is to allow the blood to flow from an incision into a glass capillary tube. The tube is broken at ½ minute intervals to determine when the clot formed (normal clotting time ranging from 2 to 6 minutes). The diameter of the capillary tube influences the coagulation time.

Procedure A: Coagulation Time—Slide Method

1. Clean the finger with 70% alcohol and allow it to dry.
2. Puncture the finger tip with a sterile lancet. Do not use the first drop of blood since it clots abnormally fast.
3. **Note the time** that the drop of blood to be used first appears at the finger puncture. This time is used as the beginning of the experiment.
4. Place the drop of blood on a glass slide.
5. At 1/2 minute intervals **slowly** draw a straight pin through the blood and observe the point of the pin.

6. Repeat Step 5 until fine red threads can be detected on the end of the pin. This is the end point. Sometimes the entire mass forms a gel, which is also considered the end point.
7. The time between the appearance of the second drop of blood and the appearance of the first thread is the coagulation time. Using this method of determination, the normal range of coagulation is from 2 to 8 minutes.
8. Record coagulation time in the table under Results and Questions.

Procedure B: Coagulation Time—Capillary Tube Method

1. Use the method in Procedure A for obtaining the drop of blood. The drop of blood must be large.
2. Note the time the drop of blood to be used appears. This time is used as the beginning of the experiment.
3. Place one end of the capillary tube into the drop of blood (keep the other end of the tube open). Hold the tube horizontally (do not let the tip touch the skin) and allow the tube to fill by capillarity.
4. After exactly one minute (use a longer period of time if the blood took a long period of time to coagulate in the slide method), carefully break off a small piece (about one cm) of the capillary tube and determine whether a thread of coagulated blood is visible between the two pieces of tubing.
5. Repeat Step 4 every 30 seconds until such a thread is obtained.
6. Record the coagulation time in the table under Results and Questions.

Procedure C: Bleeding Time

1. Puncture the finger as directed earlier. Secure a free flow of blood so that the first drop on the filter paper will be one centimeter or more in diameter.
2. Note the time of appearance of the first drop of blood. Blot the blood with filter paper every 30 seconds until the bleeding stops.
3. Divide the total number of drops on the filter paper by two. This gives the bleeding time in minutes.
4. Record bleeding time in the table under Results and Questions.

Exercise 12-B Structure of an Artery, Vein and a Lymph Vessel Valve

Objectives

To observe differences in the structure of an artery and a vein; to observe the structure of a valve in the wall of a lymphatic vessel.

Materials

Microscope; slides of a cross section of an artery and a vein; and a longitudinal section of a lymphatic vessel containing valves.

Discussion

An artery may be defined as any vessel carrying blood away from the heart; a vein as any vessel returning blood from the capillary network to the heart. The lymphatic vessels form a different route for the return of fluid to the veins.

The wall of an artery is composed of three layers: the innermost **tunica intima**, a thin coat composed of one layer of endothelium, a small amount of connective tissue, and an internal elastic membrane; the thick **tunica media**, composed of smooth muscle arranged perpendicular to the length of the vessel and elastic connective tissue; and the outermost coat, the **tunica adventitia**, composed of loose connective tissue and adipose tissue. This external layer contains the vasa vasorum, the nutrient blood vessels nourishing the wall of all arteries and veins larger than 1 mm in diameter.

The veins contain the same three coats; however, the layers are thinner. The muscular and elastic tissues are much more poorly developed in the veins than in the arteries, while the connective tissue predominates in the veins. Many veins (and lymphatic vessels) contain valves in their walls in order to assist the return of fluid against gravity. Each valve consists of two semilunar pouches which open to let blood return to the heart and close to prevent backward flow of the blood.

Procedure

1. Examine the prepared slide of a cross section through an artery using the low power objective. Locate the three coats in the wall. Diagram the artery wall, showing the three layers in the correct proportions.
2. Draw one small section through the entire

wall in detail. In order to see tissue in the tunica intima it will be necessary to use the high power objective of the microscope. The innermost layer of the tunica intima is the endothelium. Observe the internal elastic membrane, the wavy purple line which forms the boundary between the tunica intima and media.
3. Returning to low power, examine the tunica media. This layer contains both smooth muscle, which stains pink, and purple-staining elastic connective tissue. It is not necessary to draw cellular detail in the smooth muscle. Locate the external elastic membrane which separates the tunica media from the tunica adventitia.
4. Locate the vasa vasorum and adipose tissue in the tunica adventitia. Include these in your drawing.
5. Examine a prepared slide of a cross section through a vein. Diagram the same three layers through the wall of the vein, showing differences in thickness of the wall. It is not necessary to include cells on this diagram.
6. Examine a slide of a longitudinal section of a lymphatic vessel with a valve. Diagram a section of the vessel with a valve; include the shape of the valve.

Exercise 12-C Gross Anatomy of the Human Circulatory System

Objective

To become familiar with the location and function of the major blood vessels in the body.

Materials

Anatomic charts and reference books.

Discussion

The circulatory system consists of the heart, blood vessels, and blood. Its primary function is to transport materials around the body. It transports food and oxygen to the cells, wastes away from the cells, hormones, antibodies, enzymes, etc.

The heart is a double pump located in the thoracic cavity. The right side of the heart pumps blood to the lungs, the left to the rest of the body. Be prepared in your study of the heart and blood vessels to trace the path of blood around the body.

Procedure

1. Use your text, charts, and references for orientation while studying the following figures.
2. Use Figures 12-1, 12-2, 12-3, and 12-4 as self-tests. See Section J, p. xx (Illustrations as Self-Tests) for a suggested approach. The arteries in the 2 drawings in this chapter are colored red, the veins blue.

Exercise 12-D The Dissection of the Sheep Heart

Objectives

To dissect the sheep heart, and to compare this with the human heart.

Materials

Preserved sheep, human, and horse hearts; dissecting set; dissecting pan; and models of the human heart.

Discussion

The anatomy of the sheep heart is very similar to that of the human. In Exercise 12-E the anatomy of the fetal pig heart will be compared with that of the adult sheep heart. Observe carefully in this exercise the remnants of structures important in fetal circulation. Directions for this exercise are given with the heart in anatomic position.

Procedure

1. Locate the following structures on the models of the human heart:

   a. aorta
   b. aortic semilunar valve
   c. bicuspid (mitral) valve
   d. chordae tendineae
   e. inferior vena cava
   f. left atrium
   g. left auricle
   h. left ventricle
   i. papillary muscle
   j. pulmonary artery
   k. pulmonary semilunar valve
   l. pulmonary vein
   m. right atrium
   n. right auricle
   o. right ventricle
   p. superior vena cava
   q. tricuspid valve

2. The same structures listed under Step 1 should also be located on the sheep heart. Rinse the sheep heart in water to remove as much preservative as possible.
3. The **pericardium** (the fibro-serous membrane surrounding the heart) has been largely removed in preparing the sheep heart. Observe the remnants of this membrane attached to the large blood vessels above the heart.
4. Separate a small portion of the **epicardium** (the visceral layer of the pericardium) from the **myocardium** by careful dissection with a scalpel. The third layer of the heart, the

KEY TO FIGURE 12-1

1. gall bladder
2. cystic duct
3. hepatic duct
4. liver
5. common bile duct
6. duodenum
7. superior mesenteric vein
8. ascending colon
9. vermiform appendix
10. ileum
11. left hepatic duct
12. portal vein
13. coronary vein
14. stomach
15. splenic vein
16. spleen
17. pancreas
18. inferior mesenteric vein
19. descending colon
20. rectum

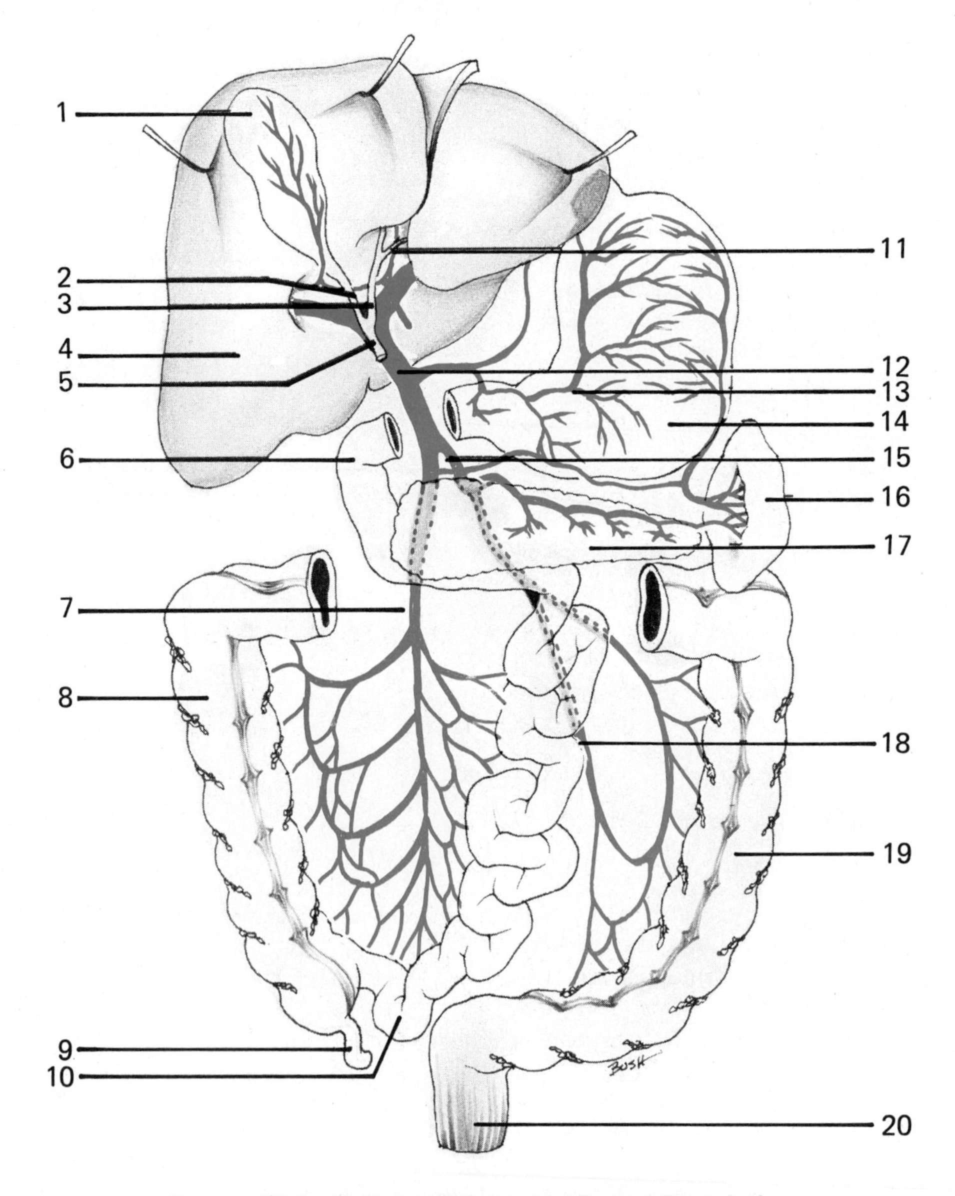

Figure 12-1. Digestive Organs and Portal Circulation

KEY TO FIGURE 12-2
(arteries)

1. R. common carotid
2. R. subclavian
3. brachiocephalic
4. axillary
5. aortic arch
6. brachial
7. superior mesenteric
8. radial
9. ulnar
10. palmar arch
11. middle sacral
12. digital
13. anterior tibial
14. dorsalis pedis
15. arcuate
16. L. common carotid
17. L. subclavian
18. celiac
19. renal
20. spermatic
21. inferior mesenteric
22. L. common iliac
23. internal iliac
24. external iliac
25. femoral
26. popliteal
27. posterior tibial

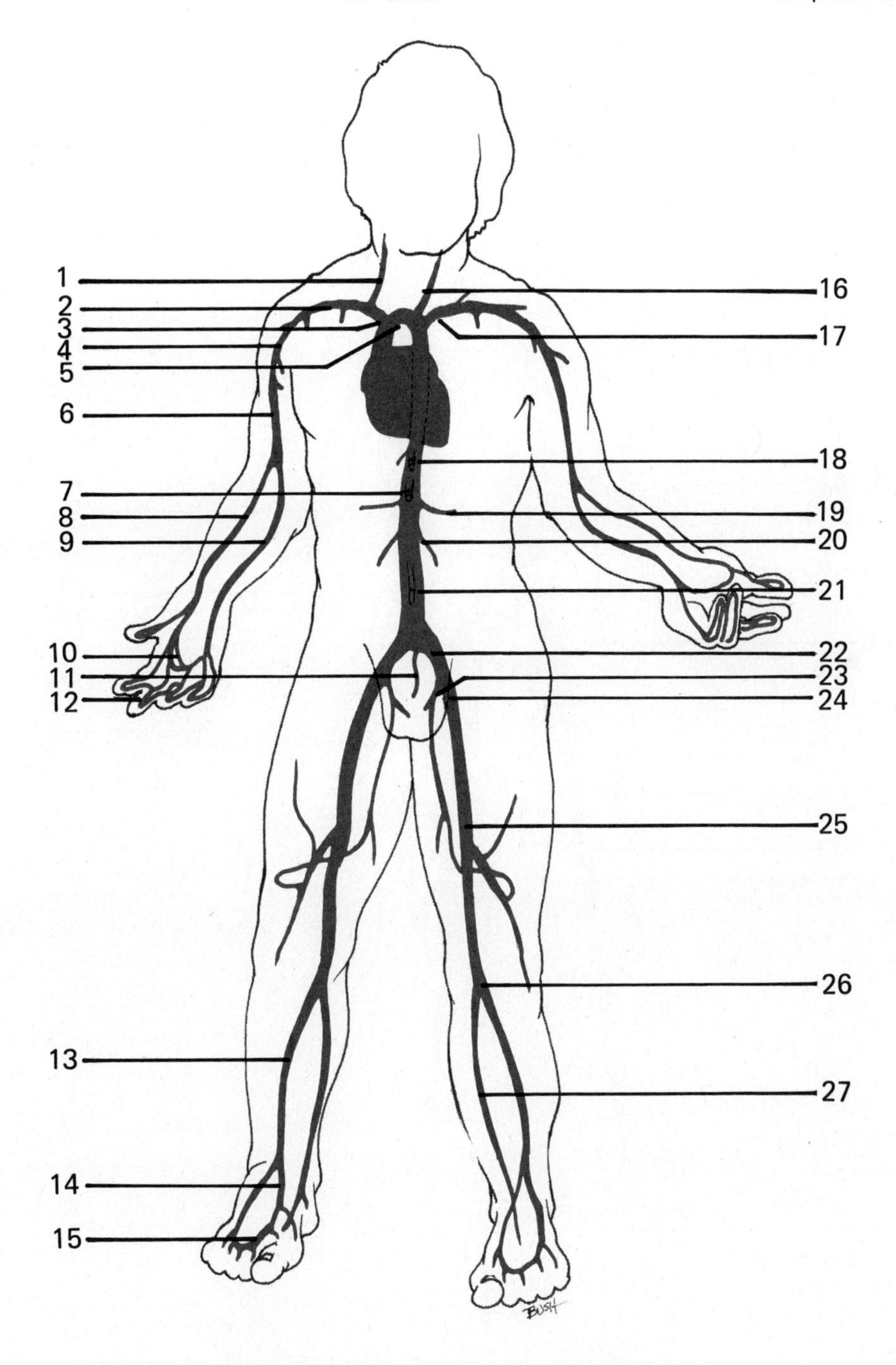

Figure 12-2. Major Arteries of the Body

KEY TO FIGURE 12-3
(veins)

1. external jugular
2. internal jugular
3. subclavian
4. axillary
5. brachial
6. cephalic
7. basilic
8. internal iliac
9. external iliac
10. great saphenous
11. brachiocephalic
12. superior vena cava
13. inferior vena cava
14. hepatic
15. renal
16. L. common iliac
17. middle sacral

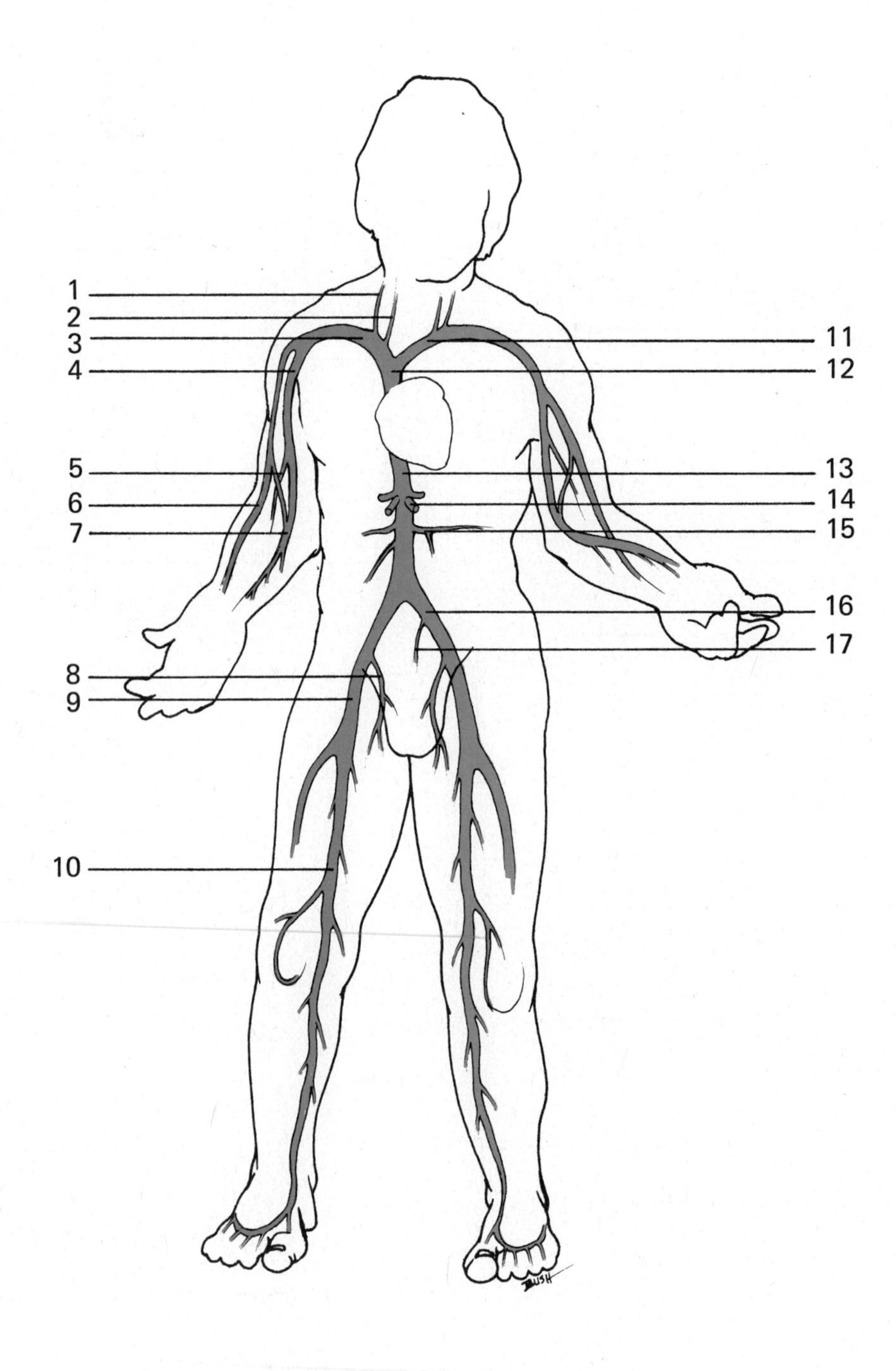

Figure 12-3. Major Veins of the Body

KEY TO FIGURE 12-4

1. external jugular vein
2. internal jugular vein
3. common carotid artery
4. brachiocephalic artery
5. right pulmonary artery
6. right pulmonary veins
7. superior vena cava
8. aortic semilunar valve
9. right atrium
10. tricuspid valve
11. right ventricle
12. inferior vena cava
13. axillary vein
14. subclavian vein
15. brachiocephalic vein
16. aortic arch
17. L. pulmonary artery
18. pulmonary artery
19. L. pulmonary veins
20. pulmonary semilunar valve
21. left atrium
22. mitral valve
23. chordae tendineae
24. left ventricle
25. papillary muscle
26. interventricular septum

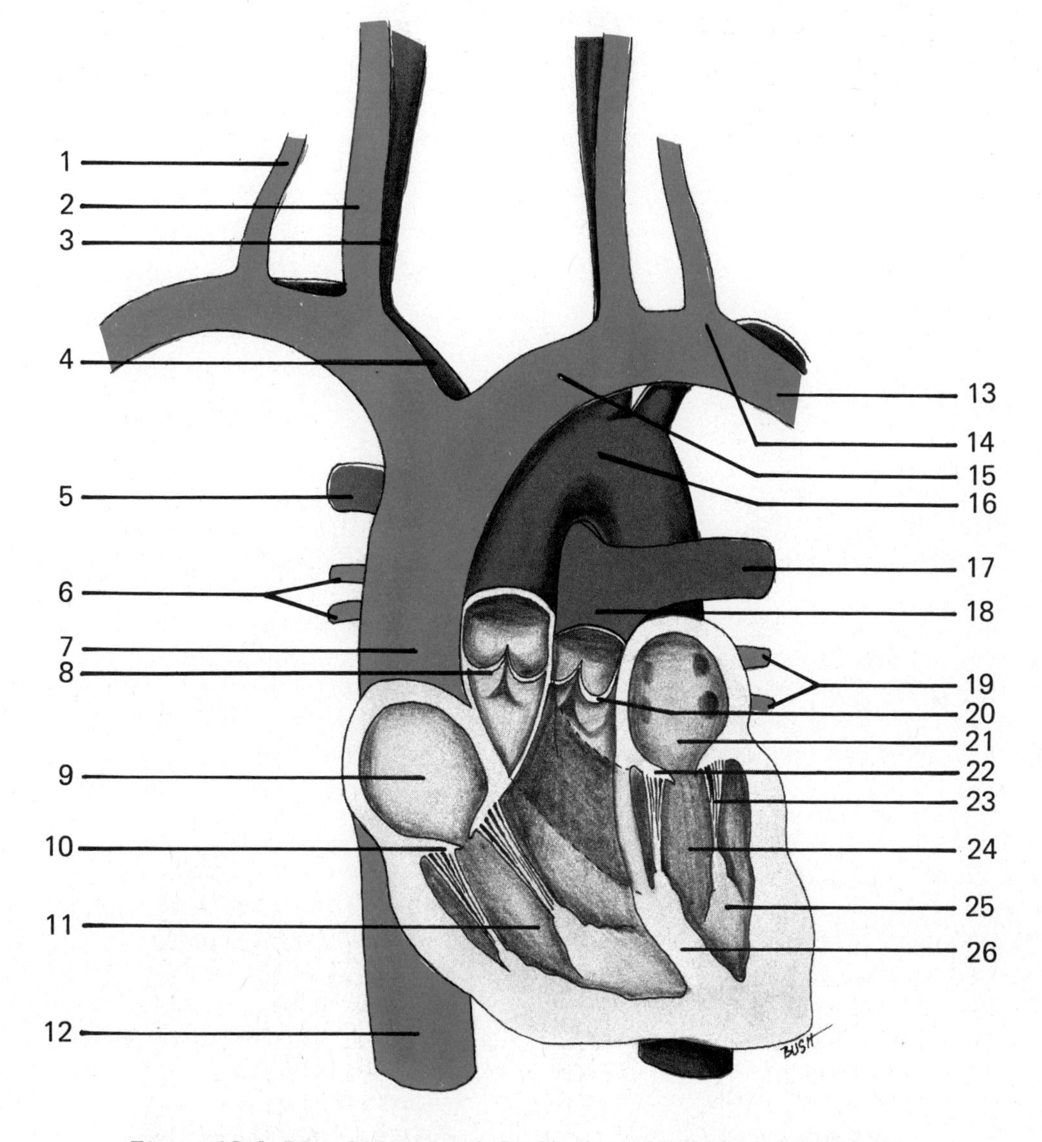

Figure 12-4. Blood Vessels of the Neck and Interior of the Heart

endocardium, will be visible when the heart is opened.

5. Locate the **pulmonary artery** on the ventral surface of the heart. This artery emerges from the anterior ventral surface of the heart, medial to the **left auricle**. (To determine which is the ventral surface of the heart, compare the sheep heart with Figure 12-5.
6. Note the **anterior longitudinal sulcus** separating the right ventricle from the left ventricle. The **coronary blood vessels** can be seen in this sulcus.
7. Compare the thickness of the wall of the

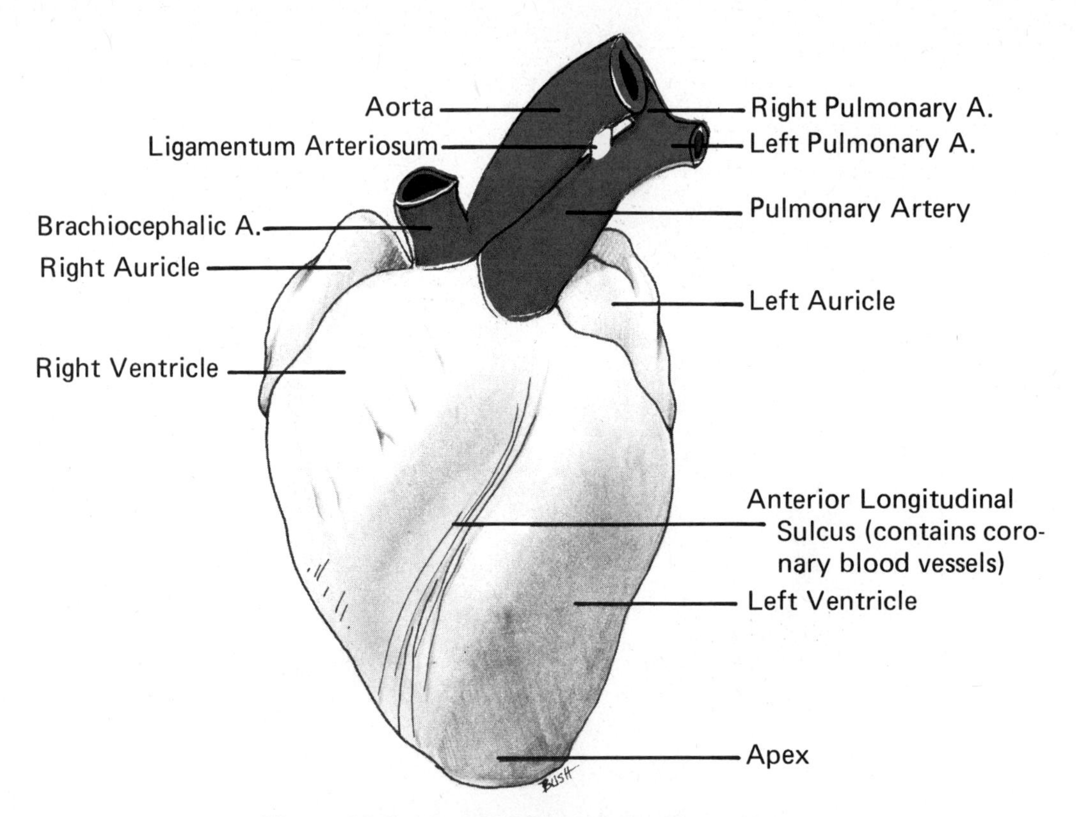

Figure 12-5. Ventral View of the Sheep Heart

right ventricle with the **left ventricle** by pressing the wall of each ventricle between your thumb and forefinger.

8. If there is a demonstration dissected sheep heart on display, observe the incisions before proceeding. Using great care, make an incision through the ventral wall of the pulmonary artery and the right ventricle **parallel to, and 3/4 of an inch to the right of**, the anterior longitudinal sulcus. Do not dissect so deeply that you cut into the dorsal surface of either the pulmonary artery or right ventricle. (Continue the incision forward anteriorly to the point where the pulmonary artery branches into the **right and left pulmonary arteries.** Remove as much adipose tissue as necessary to expose the branches of the pulmonary artery.
9. Open the pulmonary artery and note the **pulmonary semilunar valve.** Observe the number of pouches in this valve. Remove any coagulated blood present in the blood vessels or heart.
10. Continue the original incision through the right ventricle wall (**keeping parallel to the anterior longitudinal sulcus**) around and through the dorsal wall of the right ventricle until reaching the **interventricular septum.**
11. Examine the dorsal surface of the heart. The thin-walled **superior vena cava** can be seen above the right auricle, extending straight down into the right atrium. Make a longitudinal incision through the dorsal wall of the superior vena cava, continuing down through the right atrium (immediately to the left of the right auricle). Continue this incision down the dorsal right ventricle wall to the point of juncture with the first incision.
12. If the incisions were properly made, it should now be possible to spread open the superior vena cava, right atrium, and right ventricle. Compare your dissection with Figure 12-6.
13. Determine whether there is a valve present at the entrance of the superior vena cava into the right atrium.
14. Observe the internal structure of the right auricle. The muscle visible in the interior of the auricle is called the **pectinate muscle**, since it resembles a comb (pecten).
15. Locate the large opening of the **inferior vena cava** on the left side of the interior of the right atrium. Insert a finger or probe

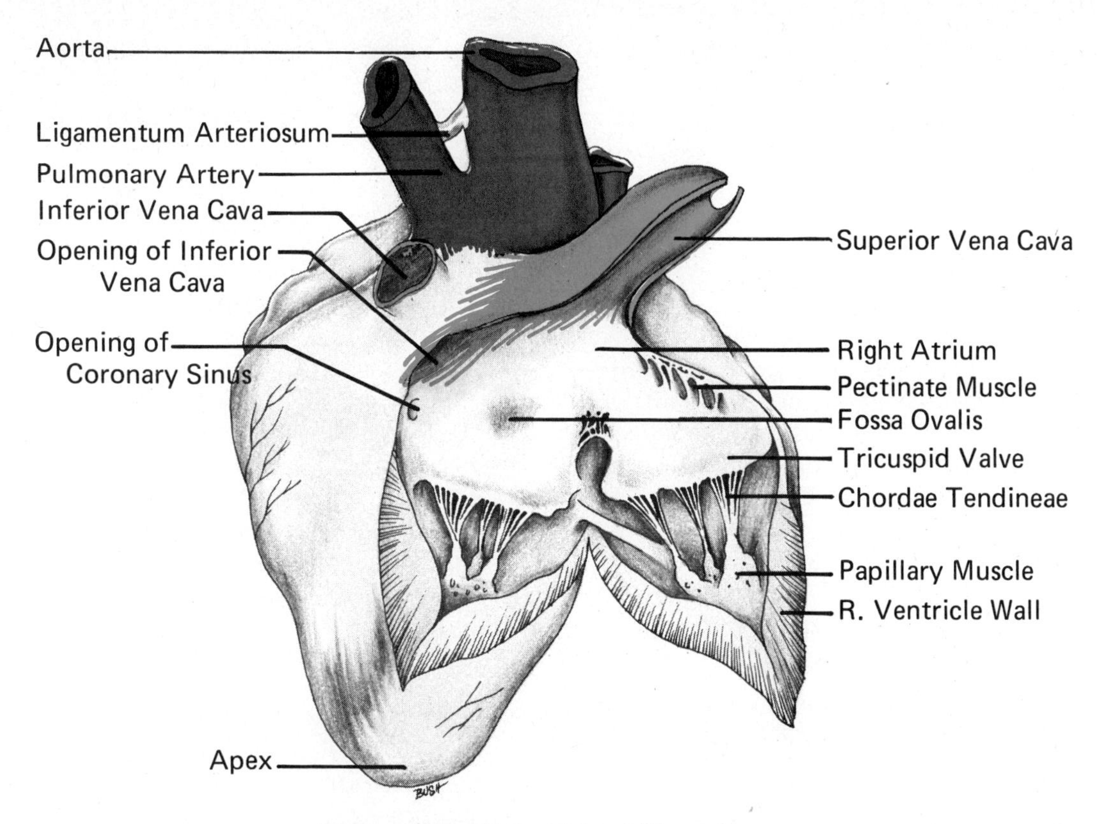

Figure 12-6. Right Side of Sheep Heart

through the mouth of this vein in order to locate the external opening on the dorsal surface of the heart. Compare the location of this vein with the corresponding vein in humans.

16. Locate the orifice of the **coronary sinus** posterior to the opening of the inferior vena cava. Locate this vessel on the dorsal surface of the heart.
17. Locate the **interatrial septum**, the wall that separates the two atria. Examine this septum from the interior of the right atrium. Locate the **fossa ovalis**, the oval-shaped depression, ventral to the entrance of the inferior vena cava.
18. Examine the **tricuspid valve** between the right atrium and the right ventricle. Determine the number of cusps that form this valve.
19. Locate the **papillary muscles** and **chordae tendineae** in the wall of the right ventricle. Determine the number of papillary muscles.
20. Locate the **moderator band** crossing the lumen of the right ventricle. This is believed to prevent overdistension of the ventricle.
21. Try to locate the **four pulmonary veins** entering the **left atrium**. These vessels are visible on the dorsal surface of the left atrium. Whether all four vessels can be located depends on how they were cut.
22. Make a longitudinal incision through the most lateral of the pulmonary veins. Continue this down through the wall of the left atrium and the left ventricle to the **apex** of the heart.
23. Spread open the left side of the heart. Compare the wall thickness of the left ventricle with that of the right.
24. Observe the **bicuspid valve**. Determine how many major cusps comprise this valve. Determine which of the following are present in the left ventricle: moderator band, chordae tendineae, or papillary muscles.
25. Insert a finger up along the midline of the left ventricle into the **aorta**. Cut along this line until the **aortic semilunar valve** is visible. Count the number of pouches in this valve.
26. Returning to the external surface of the heart, locate the thick-walled aorta above the heart, arching to the left. Locate the **brachiocephalic artery** which branches from the aortic arch. This vessel later branches into the subclavian and common

carotid arteries, which supply the arms and head. In humans, three large blood vessels branch from the aortic arch.

27. Open the aorta as it emerges from the network of blood vessels anterior to the heart. Observe the smooth lining of this vessel.
28. If the pulmonary artery has not been severed too close to the heart when the heart was removed from the sheep you will be able to locate the **ligamentum arteriosum**, the remnant of the **ductus arteriosus**. Carefully dissect away the adipose tissue between the pulmonary artery and the aorta anterior to the heart. Be careful not to sever the ligamentum arteriosum when removing the adipose tissue.
29. Locate all the preceding structures on the preserved horse and human hearts.

Exercise 12-E The Dissection of the Circulatory System of the Fetal Pig

Objectives

To dissect the circulatory system of the fetal pig and to compare it with the corresponding structures in the human circulatory system; to become familiar with the pathway of blood in fetal circulation.

Materials

Fetal pig; dissecting set; dissecting tray; and model of human fetal circulation.

Discussion

The blood vessels of a fetal mammal, such as the pig, closely resemble those of the human adult. Modifications for fetal life include a placental circulation by way of the umbilical cord and two devices to bypass the lungs, since the lungs are not functional before birth. The arteries of the fetal pig have been injected with red latex; the veins with blue.

Procedure

1. Observe the **pericardium** surrounding the heart. After determining the structure to which it is attached, remove the parietal layer of the pericardium. The visceral layer of the pericardium forms the **epicardium** of the heart, the outermost layer of the heart.
2. Note that the **apex** of the heart is directed toward the left. The heart is tilted so that the greater part of the **right ventricle** lies directly in front, along the ventral surface of the heart. The **left ventricle** forms the apex of the heart.
3. The atria lie anterior to the ventricles. Each atrium has a conspicuous earlike appendage called the **auricle** on the ventral surface.
4. A groove, the **coronary sulcus**, separates the right atrium from the right ventricle. The **anterior longitudinal sulcus** is the groove that separates the right ventricle from the left ventricle. Dorsal to this sulcus is the **interventricular septum**. The **coronary blood vessels** are located in these grooves (the right coronary artery in the coronary sulcus, the left in the anterior longitudinal sulcus).
5. Observe the **anterior vena cava** (precaval vein), the large vessel entering the anterior part of the right atrium. (This vein is called the superior vena cava in humans.) The anterior vena cava drains the head, neck, and arms.
6. Trace this vessel forward and note that it is formed by the union of the two **brachiocephalic veins** (see Fig. 12-7).
7. Trace the left brachiocephalic vein forward. This vein is formed by the union of three veins: the small **left internal jugular vein,** which lies next to the **left common carotid artery**, the larger **left external jugular vein,** which lies lateral to this, and the subclavian **vein**, which drains the arm. In humans the internal jugular is larger than the external jugular.
8. Trace the subclavian vein through the chest wall. In the axillary region this vein is known as the **axillary vein.** On the arm it becomes the **brachial vein.** In order to follow the vein on the arm, slit the skin and muscles on the ventral surface of the arm.
9. Locate the **posterior vena cava** (postcaval vein) posterior to the heart and trace it forward to the point where it drains into the right atrium. This large vein, called the inferior vena cava in humans, drains the lower portion of the body.
10. Trace the postcaval vein back through the **diaphragm** into the abdominal cavity, where it lies to the right of the aorta. In order to see the vein and its tributaries, it

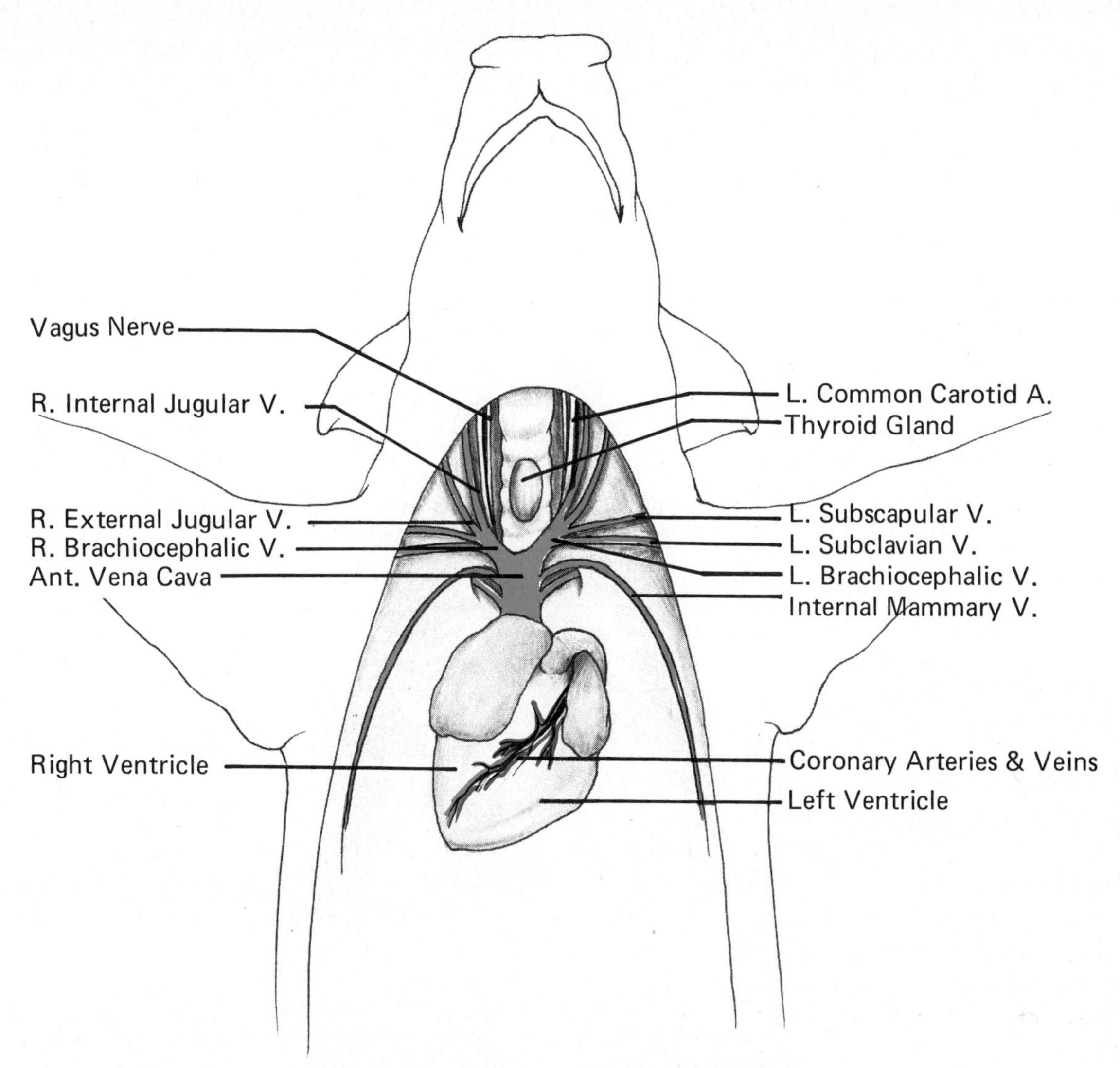

Figure 12-7. Veins of the Thorax and Neck

will be necessary to dissect away the peritoneum, since the posterior vena cava and the aorta are retroperitoneal.

11. The **hepatic veins** drain blood from the liver into the posterior vena cava. To locate these veins, gently scrape away tissue on the right anterior surface of the liver. Several hepatic veins may be located in this manner. The **umbilical vein** (carrying fresh oxygenated blood from the placenta) passes through the liver and connects with one of the larger hepatic veins.
12. Locate the **renal veins**, which carry blood from the kidneys into the posterior vena cava.
13. Returning to the thoracic cavity, push the left lung toward the right side of the body. Locate the **hemiazygos vein** which receives blood from the **intercostal veins**. The hemiazygos vein enters the dorsal surface of the right atrium.
14. The other major veins will be dissected with the arteries.
15. Locate the **pulmonary artery** on the ventral surface of the heart. Trace it down to its origin in the right ventricle; then follow it toward the lungs, noting that it branches into a **right** and **left pulmonary artery**.
16. At the point of branching of the pulmonary artery, the large **ductus arteriosus** passes anteriorly to connect to the aorta. The ductus arteriosus is larger than either the right or left pulmonary artery.
17. The **dorsal aorta** arises from the left ventricle. Locate this vessel dorsal and anterior to the pulmonary artery. The first branches of

the dorsal aorta are the small **right** and **left coronary arteries**, which arise from the base of the aorta. The left coronary artery is visible on the ventral surface of the heart in the anterior longitudinal sulcus; the right coronary artery in the coronary sulcus.

18. The dorsal aorta passes anteriorly for a short distance and then turns to the left. This region of the aorta is called the **aortic arch**.
19. To enable you to see the arteries branching off the aortic arch, free the anterior vena cava from the arteries beneath.
20. The first branch off the aortic arch is the **brachiocephalic artery**. This artery first gives off the **right subclavian artery** and then the **right** and **left common carotid arteries**.
21. Trace the common carotids toward the head along each side of the trachea. These arteries branch to form the **external** and **internal carotid arteries** at the anterior border of the larynx.
22. Returning to the aortic arch, locate the left **subclavian artery**, which supplies the left side of the chest and the left arm. Locate the **right** and **left internal mammary (sternal) arteries**, which supply the pectoralis muscles and mammary glands. These vessels arise from the subclavian arteries on either side of the sternum.
23. The subclavian artery becomes the **axillary artery** as it crosses the axillary space, and then the **brachial artery** on the upper arm. Follow the brachial artery until it branches to form the **radial** and **ulnar arteries**, which supply the forearm.
24. Pull the viscera in the thorax gently to the right to expose the dorsal aorta. As this vessel passes through the thorax, it is called the **thoracic aorta**. Remove the pleura to expose the aorta in the thorax.
25. Note the **intercostal arteries** emerging from the thoracic aorta. These supply the intercostal muscles.
26. Trace the descending aorta through the diaphragm. The first major branch from the abdominal aorta is the **celiac artery** (see Fig. 12-8). In order to locate this artery it will be necessary to remove the peritoneum covering the anterior end of the abdominal aorta immediately beneath the diaphragm. This large artery supplies the liver, pancreas, spleen, and duodenum.
27. Locate the **anterior mesenteric artery**, the unpaired vessel located a short distance posterior to the origin of the celiac. This vessel is also retroperitoneal. It supplies the small intestine and a portion of the large intestine.
28. Locate the **renal arteries** (which supply the kidney) posterior to the anterior mesenteric artery.
29. The right and left genital arteries (**testicular** or **ovarian**) are small vessels that emerge from the ventral surface of the aorta, posterior to the renals near the base of the aorta. If your specimen is a male, follow the testicular artery to the **inguinal canal**.
30. Turn the descending portion of the colon to one side to see the **posterior mesenteric artery**. The posterior mesenteric is a small, unpaired vessel, arising from the aorta immediately posterior to the ovarian or testicular arteries. It divides almost immediately, with one branch proceeding anteriorly and one posteriorly along the dorsal surface of the colon.
31. The paired **external iliac arteries** arise from the aorta lateral to the posterior mesenteric artery. They continue downward on each side to become the **femoral artery**. Locate this vessel and the femoral vein by teasing away the ventral thigh muscle after removing the skin.
32. Locate the **internal iliac arteries** below the point at which the external iliacs arise from the aorta. These give rise to the large **umbilical arteries**, which pass lateral to the bladder, to the umbilical cord, and then continue on downward as the smaller **internal iliac arteries**.
33. The small, **median sacral artery** can be seen emerging from the base of the aorta between the two internal iliac arteries.
34. Locate the two **common iliac veins**, which unite to form the posterior vena cava. Each common iliac vein is formed by the union of the **internal** and **external iliac veins**. These veins can be located next to the corresponding artery.
35. To dissect the **fetal heart**, make an incision through the lateral wall of each atrium and auricle. Carefully remove the latex and any blood that is present.
36. Observe the point of entrance of the anterior and posterior venae cavae into the right atrium.

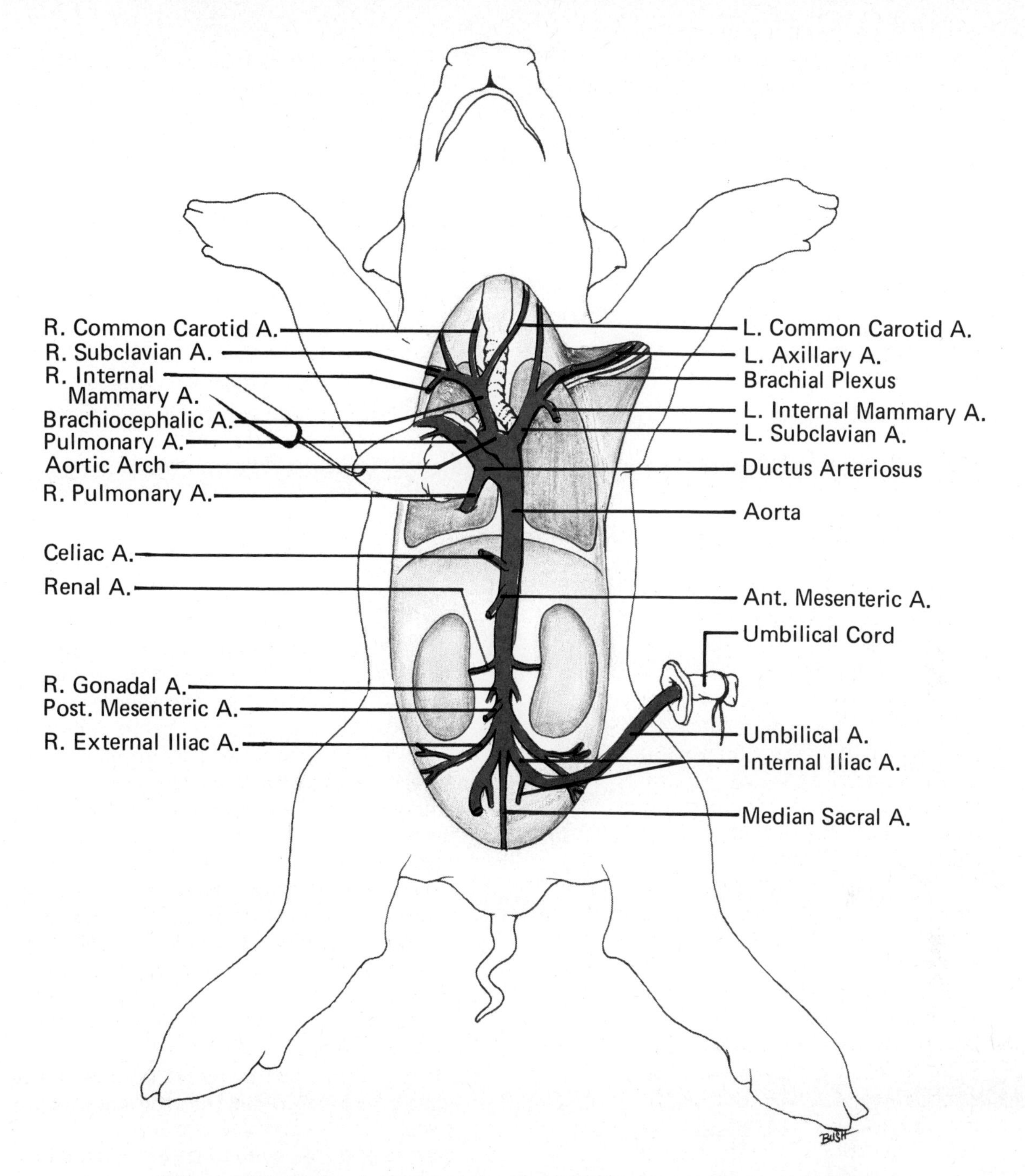

Figure 12-8. Major Arteries of the Fetal Pig (heart pulled to the right)

37. Locate the **foramen ovale** (the hole in the **interatial septum**) near the dorsal wall of the heart, just anterior to the entrance of the posterior vena cava. At this late time in fetal life, the opening is quite small. Pass a probe through the foramen ovale (see Fig. 12-9). Blood returning to the heart by way of the posterior vena cava passes from the right atrium directly to the left atrium, bypassing the lungs. This structure closes after birth, leaving the depression, the fossa ovalis, seen in the sheep heart.
38. Continue the lateral incision down on each side of the heart in order to examine the interior of the ventricles. Since the remainder of the anatomy of the fetal pig heart is the same as that of the sheep and human heart, it will not be described fur-

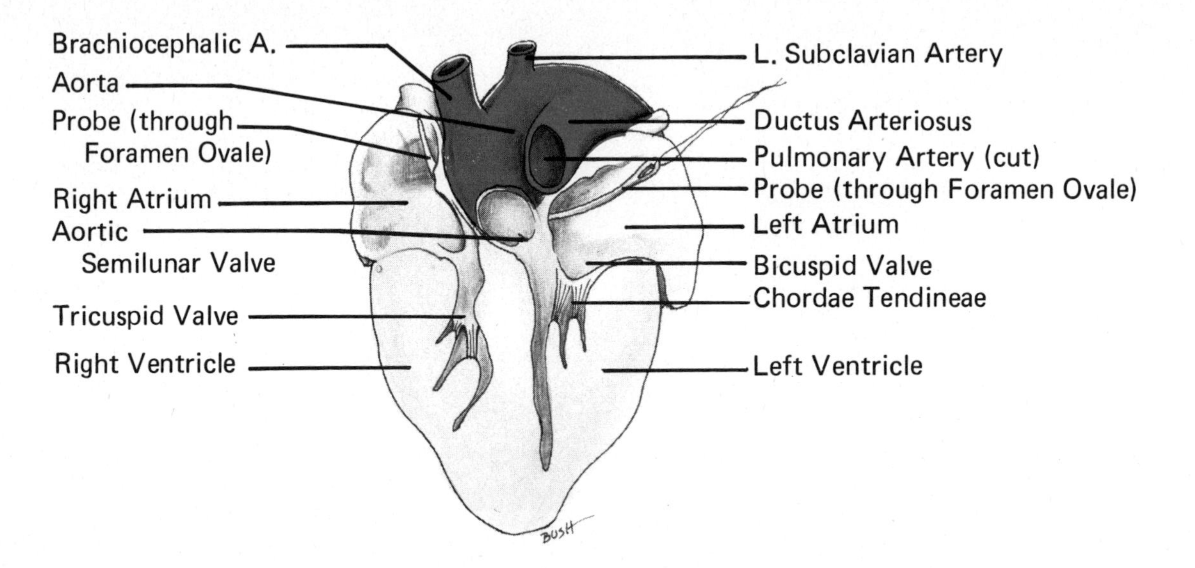

Figure 12-9. Coronal Section through Fetal Pig Heart

ther. Locate all structures listed in Exercise 12-D.

39. Locate the following on the model of human fetal circulation:
    a. aorta
    b. ductus arteriosus
    c. ductus venosus
    d. foramen ovale
    e. inferior vena cava
    f. internal iliac artery
    h. pulmonary artery
    i. superior vena cava
    j. umbilical artery
    k. umbilical cord
    l. umbilical vein

Exercise 12-F Physiology of the Frog Heart

Part 1. Frog Anatomy

Objective

To become familiar with selected anatomical structures in the frog.

Materials

Dissecting set; frog board and clips or thread; beaker of frog Ringer's solution; medicine dropper; and a living frog (if the frog is not to be used in physiological experiments, a preserved specimen may be substituted).

Discussion

Frogs are vertebrates possessing many external and internal parts that are similar in structure and function to those of humans. As you study the frog, always try to answer the question: "How does this part or action of the frog relate to the human body?"

Procedure

1. To determine the sex of your specimen (before dissection), examine the inner digit (thumb), and compare with other specimens. The thumb of the male frog is thickened as a result of a glandular area in this region. During the mating season, the area is enlarged; after the breeding season, the thumb is somewhat reduced in size.
2. Pith the brain of the frog. This destroys the brain and as a result the frog does not feel pain.
3. To pith the frog, hold it in your left hand (as illustrated in Fig. 12-10). The fingers and thumb should hold the arms to the trunk. Place the index finger on top of the head and press the head downward. With the nail of the right index finger, locate the junction between the occipital bone and the vertebral column. Indent the skin at this point.
4. Insert a probe through the skin at this junction. As soon as the probe enters the cord between the bones, the frog should become flaccid. This condition is called spinal shock.

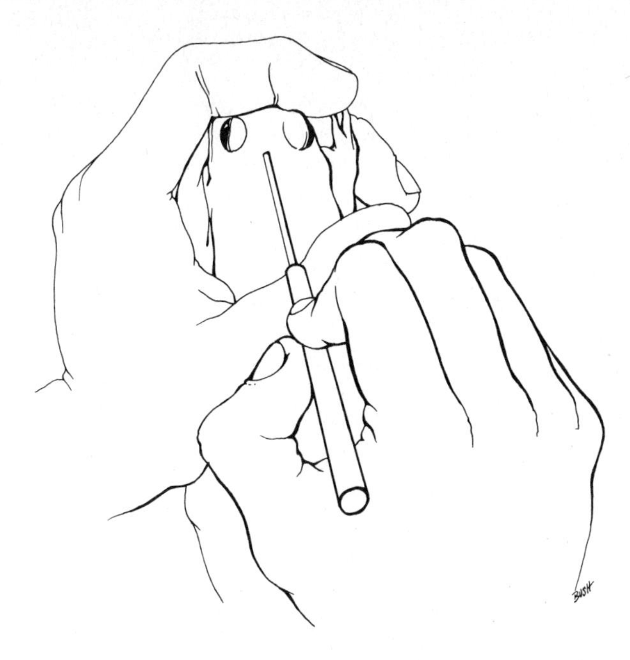

Figure 12-10. Pithing the Frog

5. When it is certain that the probe has entered the cord, turn and direct the probe forward through the foramen magnum inside the cranial cavity. Twist the probe until the brain is destroyed. It should be possible to feel bone on all sides of the probe if the probe is positioned correctly in the cranial cavity.
6. Check to make certain that the probe is not in the oral cavity by looking into the mouth.
7. The frog is now single-pithed and is called a spinal frog. If the probe were turned to destroy the cord as well as the brain, the frog would be called a double-pithed frog.
8. The frog will emerge from the period of spinal shock in approximately five minutes and regain spinal reflexes. To determine whether pithing is complete, check for the presence of the corneal reflex by touching the cornea. If the brain is **completely** destroyed, the frog will not raise the lower lid which closes the eye. If the corneal reflex is still present, repith the frog.
9. Tie the frog to a frog board with the ventral surface up and with the four legs extended. Do not tie the legs so tightly that circulation is interrupted.
10. Using scissors, make an incision in the skin from the abdomen up to the jaw, slightly to one side of the midline (see Fig. 12-11).
11. Now, cut through the body wall along the same line, from the abdomen through the shoulder girdle. Use a lifting action on the lower blade of the scissors to avoid damaging the viscera. Open the thoracic cavity, exposing the heart and lungs.
12. By making several lateral incisions, some of the skin and musculature can be pinned aside, thus more clearly exposing the internal organs.
13. Keep the skin and the internal organs moist with frog Ringer's solution. The skin should not be allowed to touch the internal organs since it contains secretions that are toxic to them.
14. Note the absence of a diaphragm. Occasionally the lungs may be distended with air. They can be gently pushed back into the pleural cavity or deflated by a snip with the scissors.
15. Locate these structures in the frog.
    a. Identify the **esophagus.** This is the short muscular tube which extends from the mouth cavity to the stomach.
    b. With forceps pinch the upper end (fundus) of the **stomach.** Watch closely for several seconds.
    c. Observe the **pyloric valve,** the thickened ring of muscle in the region where the stomach joins the small intestine. Slit open this area to observe the sphincter.
    d. The first portion of the **small intestine** is called the **duodenum.** The remainder is the **ileum.** Gently lift a loop of the intestine and note that it is suspended by means of a thin membrane (**mesentery**) from the dorsal part of the body wall. Nerves and blood vessels may be seen in the mesenteries.
    e. The **large intestine** is the expanded, short, straight tube connecting the small intestine to the cloaca.
    f. Locate the **cloaca,** the tubular region into which the products from the urogenital tract and fecal wastes from the digestive tract pass before leaving the body by way of the anus. (Humans have a cloaca only during the early fetal stages.)
    g. The **liver** is the large red-brown organ, divided into three lobes, near the anterior end of the body cavity. Lift the lobes of the liver and observe the thin-walled sac called the **gall bladder.** It will probably be filled with a green to yellow-green fluid called **bile.** The gall

bladder is connected to the duodenum by the **bile duct.**

h. Identify the dark red **spleen** lying in the mesentery near the large intestine.

i. The **pancreas** is a pale yellow, somewhat indistinct organ, lying in the mesentery which extends from the stomach to the duodenum. The **pancreatic duct** empties into the bile duct—thus the digestive secretions from the pancreas enter the duodenum through the same tube as does the bile.

j. Insert a straw (or medicine dropper) into the **glottis** (slitlike structure on the floor of the buccal cavity) and see if you can inflate the two **lungs.**

k. On the surface of the lungs observe the air pockets (**alveoli**) which show through as small, clear compartments.

l. The glottis leads into the small, cartilaginous **larynx.** Cut into this organ and observe the opening or passages into the lungs.

m. Locate the **trachea.** Since this organ consists of short lateral branches of the air passageway leading from the larynx to the two lungs, they are sometimes called the **bronchi.**

n. Push the abdominal viscera aside to expose the **kidneys** lying next to the dorsal body wall. They are long, flat, narrow, and reddish-brown.

o. Try to locate the ureters, the small ducts which carry urine from the kidneys to the cloaca and then to the **urinary bladder.**

p. The **urinary bladder** is attached to the ventral surface of the cloaca. Since it is usually collapsed, it may be difficult to see.

q. The **thyroid gland** lies across and at the sides of the larynx. Though difficult to locate, it can usually be found.

r. The **adrenal glands** are located on the ventral surfaces of the kidneys. They are long, narrow, yellow bands of tissue.

s. The **pituitary body** and **pineal gland** may be found only if the brain is removed and observed carefully. The location of these two glands is **optional** and should be attempted only when the other exercises are completed.

t. The **testes** are two small, pale yellow, oval bodies attached to the kidneys by mesenteries.

u. The **vas efferentia** are several tiny ducts which carry sperm from the testes to the kidneys. After passing through the kidneys, the sperm travel through the **ureters** (same as urine) into the **cloaca** and finally leave the body through the **anus.**

v. The male frog contains **oviducts** (egg tubes), which are coiled white tubes, lying on each side of the body cavity. Since they are nonfunctional in the male frog, they are said to be vestigial.

w. Locate the **ovaries** in the female frog. Depending on the season of the year, the female frog may have the abdominal cavity nearly filled with the paired ovaries and their contents, the undeveloped **ova** (eggs). Each ovary consists of several thin-walled lobes, filled with masses of small, globular developing ova, each with a dark as well as a light colored area.

x. In the female the **oviducts** are usually much larger than those of the male, and they are functional. They are not connected directly to the ovaries, but each has an opening, a funnel-shaped **ostium,** which connects directly with the body cavity near the esophagus. The enlarged posterior portions of the oviducts are sometimes called **uteri.** The eggs break through the ovarian walls, pass into the body cavity, and enter the openings of the oviducts. As the eggs move along the oviducts they receive a jellylike coating from glands that line these tubes. This coating enables the eggs to absorb water after they leave the body. As the eggs leave the body, they come in contact with sperm, are fertilized, and develop into young tadpoles.

y. The **fat bodies** are yellow or orange-colored structures found at the anterior end of both the ovaries and testes. These are stores of fat which may be used as needed for nourishment—especially during hibernation and the breeding season. While the fat bodies are not parts of the reproductive system, they and the gonads develop from the same kind of embryonic tissue.

KEY TO FIGURE 12-11

1. right and left systemic arches
2. right atrium
3. right lung
4. right lobe of liver
5. median lobe of liver
6. gall bladder
7. small intestine
8. left atrium
9. truncus arteriosus
10. left lung
11. ventricle
12. left lobe of liver
13. stomach
14. spleen
15. urinary bladder
16. large intestine

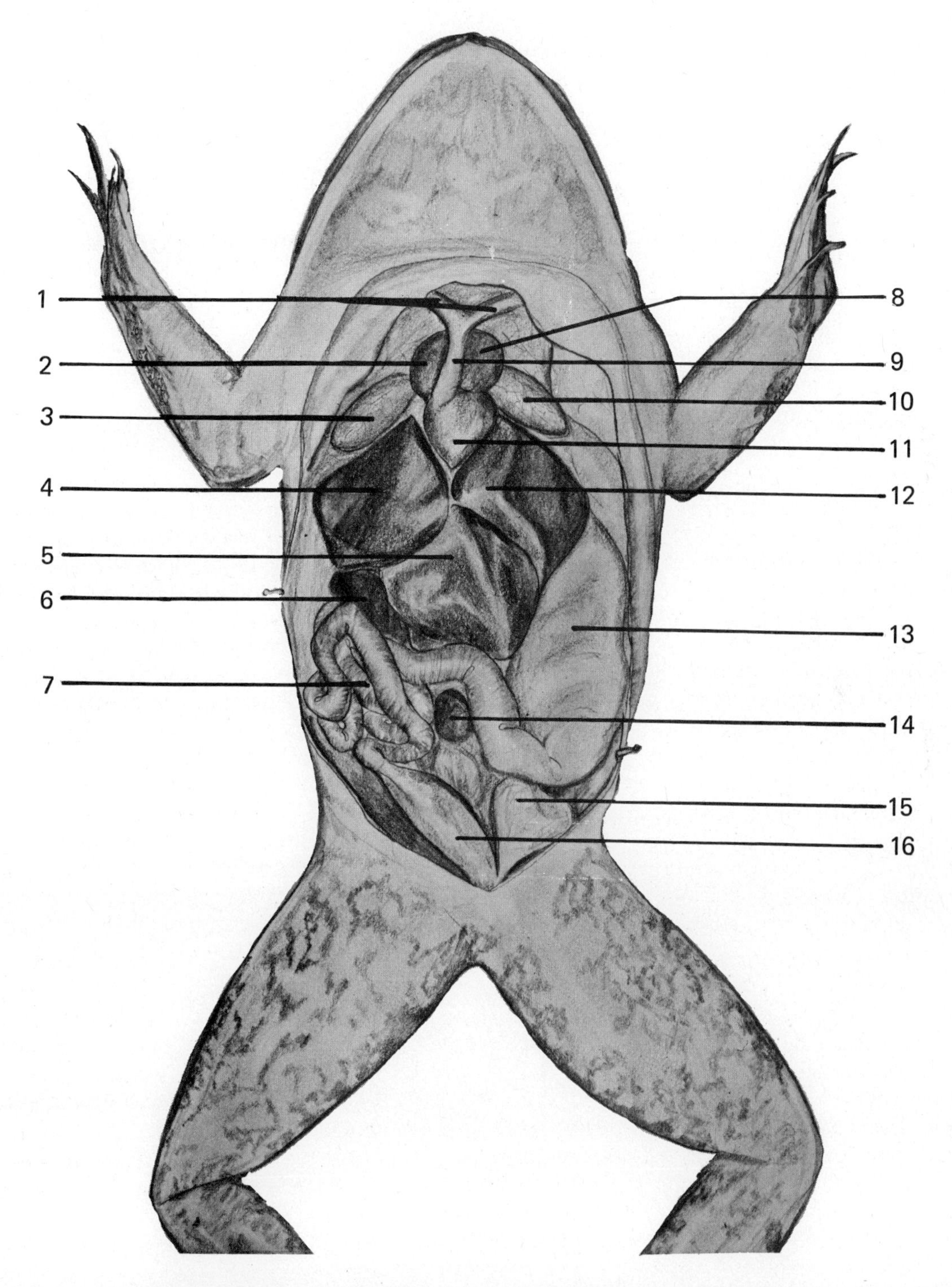

Figure 12-11. Ventral View of the Internal Organs of the Frog

Part 2. Observation of the Frog Heart Cycle

Objectives

To observe the parts of the frog heart and the sequence of contraction of each part.

Materials

Frog; dissecting set; frog board clips; beaker of frog Ringer's solution; and a medicine dropper.

Discussion

The frog heart will be used in this series of experiments to observe the action of the heart, and to demonstrate some of the properties of cardiac muscle. Although the frog has a three-chambered heart, it is similar in action to the human heart. The frog heart consists of the **sinus venosus,** a thin-walled, dorsal sac which receives blood from the **anterior** and **posterior caval veins**; the **right atrium,** into which the sinus venosus empties; the **left atrium,** which receives blood from the lungs; and the **ventricle,** which receives blood from both atria. The blood is then pumped out the **truncus arteriosus,** the large ventral artery which arises from the ventricle.

Procedure

1. Use the frog prepared and studied in Part 1, or prepare a fresh specimen using the same procedure as that of Part 1.
2. In the thoracic cavity, note the pericardium surrounding the heart. Wipe away any blood near the heart with filter paper or a paper towel.
3. Determine to which structures the pericardium is attached.
4. Observe the size of the heart.
5. Make an incision in the pericardium. Is any moisture present between the pericardial sac and the heart?
6. Note the size of the heart, after removing the pericardium, and compare it with the size before the pericardium was cut.
7. Moisten the heart with Ringer's solution.
8. Using Figure 12-14, 15, and 16 as guides, locate the **sinus venosus** (by lifting up the apex of the **ventricle**), the **right atrium,** the **left atrium,** and the **truncus arteriosus.**
9. Observe the sequence of contraction of the atria and the ventricle.
10. Determine the rate of the atria and the ventricle for 30 seconds. Multiply each figure by two to determine the heart rate per minute.

Part 3. Extrasystole

Objectives

To demonstrate the occurrence of extrasystole, the compensatory pause, and the absolute refractory period.

Materials

Frog board; stand; clips; thread; beaker of frog Ringer's solution; pin hook; medicine dropper; dissecting set; frog; and **either** kymograph, heart lever, induction stimulator, signal magnet, two double clamps, and platinum electrodes **or** physiograph, pin electrodes, stimulator output extension cable, transducer cable, and myograph B.

Discussion

The absolute refractory period, the period of time during which an excitable tissue will not respond to a stimulus, occurs with cardiac muscle tissue all during systole. The relative

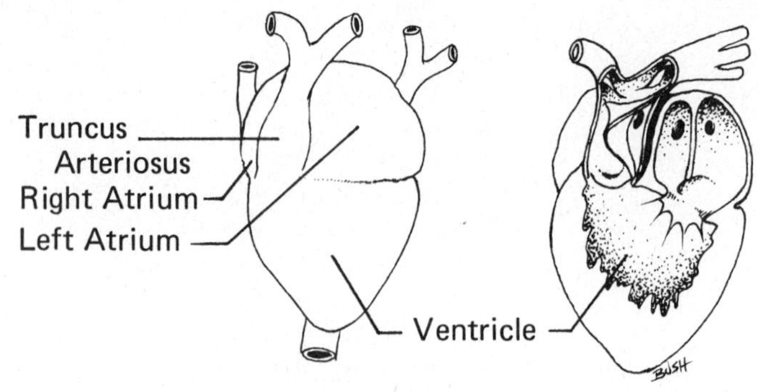

Figure 12-12. Ventral View (frog heart)

Figure 12-13. Longitudinal Section (frog heart)

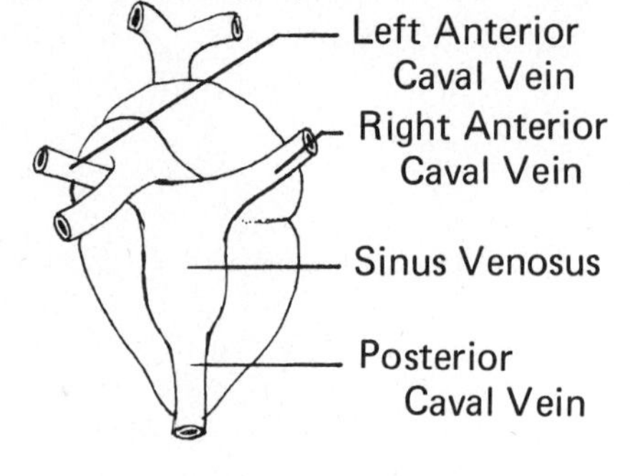

Figure 12-14. Dorsal View (frog heart)

refractory period, the period of time during which it takes a greater amount of stimulus than usual to stimulate the tissue, occurs during early diastole. During diastole, the heart responds to an adequate stimulus by giving an extrasystole, followed by a longer pause, a compensatory pause. In this experiment you will stimulate the heart at various times during the cardiac cycle, noting when the heart responds to a stimulus.

Procedure Using Kymograph

1. Connect the platinum electrodes and the signal magnet to the induction stimulator. Attach the platinum electrodes to the terminals marked STIM., and the signal magnet to the two terminals marked MAG. (refer to Fig. 12-15).
2. Prepare the kymograph. Place the frog heart preparation to the right of the kymograph with the writing facing the operator. Place the semicircular base of the stand toward the kymograph.
3. The heart lever is used for recording heart action. It consists of an aluminum stylus with an ink pen and an adapter, a yoke with a check nut to hold the stylus in any position, and a yoke handle (see Fig. 12-16).
4. Attach the yoke handle with a two-way clamp to the semicircular stand so that the check nut is up. Place the arm that protrudes from the adapter of the ink pen through the hole next to the check nut on the yoke (see Fig. 12-16). Tighten the check nut. Check to make sure that the aluminum stylus, the pen, and the adapter do not touch the yoke handle at any point in the excursion of the heart lever. If necessary, readjust the position of the arm in the check nut to achieve this goal.
5. Attach the signal magnet to the stand below the heart lever. Align the tip of the signal magnet pen immediately below that of the pen of the heart lever.
6. Use the same frog heart as in the preceding experiment. Pass a v-shaped pin hook (made from a straight pin) through the tip

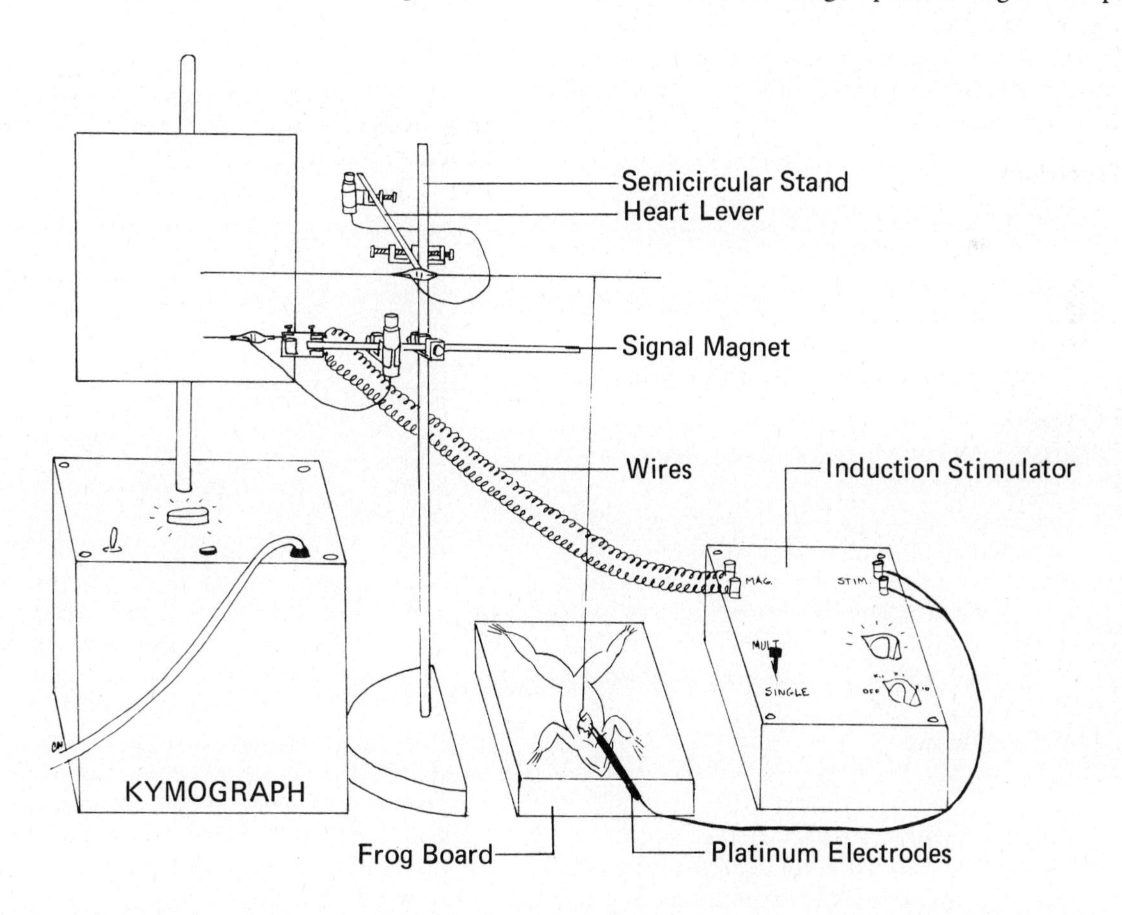

Figure 12-15. Diagram of Setup for Stimulation of Frog Heart Using Kymograph

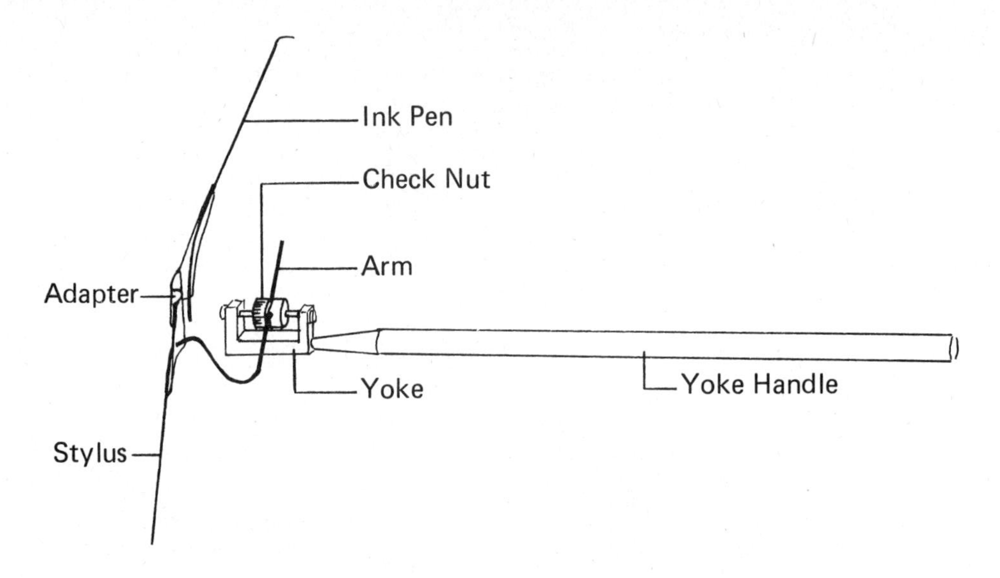

Figure 12-16. Heart Lever

of the apex of the ventricle. Do not puncture the ventricle, since a better record is obtained if the heart is full of blood. Tie a piece of **thread** to the pin hook.

7. Connect the thread from the pin hook to the aluminum stylus (short arm) of the heart lever with a small piece of colophonium cement. Adjust the length of the thread so that the ventricle is raised up from the body of the frog. The thread between the heart and the lever must be vertical.
8. Make sure that the frog heart lever is horizontal. The magnitude of the response on the record may be varied by adjusting the heart lever.
9. Line up the signal magnet so that the signal magnet pen writes immediately beneath the pen of the heart lever.
10. Use drum speed 7. Check that the equipment writes correctly on the drum.
11. Hold the stimulating electrodes so that they constantly touch the ventricle.
12. Record six to eight normal beats of the heart. The ink pen rises upon contraction of the ventricle and returns to the base line during diastole.
13. Using a weak, single stimulus (approximately 25 volts), stimulate the ventricle at 7 to 10 second intervals (make the circuit, **hold it for 7 to 10 seconds,** then break the circuit). Some of the stimuli should be administered during systole and others during diastole (both early and late). Continue the experiment until the heart record shows several extrasystoles.
14. Include the record under Results and Questions.

Procedure Using the Physiograph

1. Connect the myograph to the stand with the mounting rod and clamp. Attach the frog board to the base of the stand (see Fig. 12-17). See Exercise 7-F Part 1 for detailed operating instructions for the physiograph. Pens should be in the time and event channel and in the appropriate channel amplifier.
2. Attach the transducer cable to the myograph B and to the input connector on the channel amplifier of the DMP-4A and the Physiograph Four-A, or to the 9 pin connector on the transducer coupler of the DMP-4B.
3. Attach the stimulator output extension cable to the RF Isolated Output on the stimulator, red to red, black to black. Then, using the clip at the opposite end of the extension cable, attach the cable to the stand where the frog will be used.
4. Use the same frog as in the previous experiment. Double pith the frog (insert the probe into the spinal cord and turn it downward to destroy the cord). Then place the frog on its back on the frog board and fasten all four legs down.
5. Pass a v-shaped pin hook (made from a straight pin) through the tip of the apex of

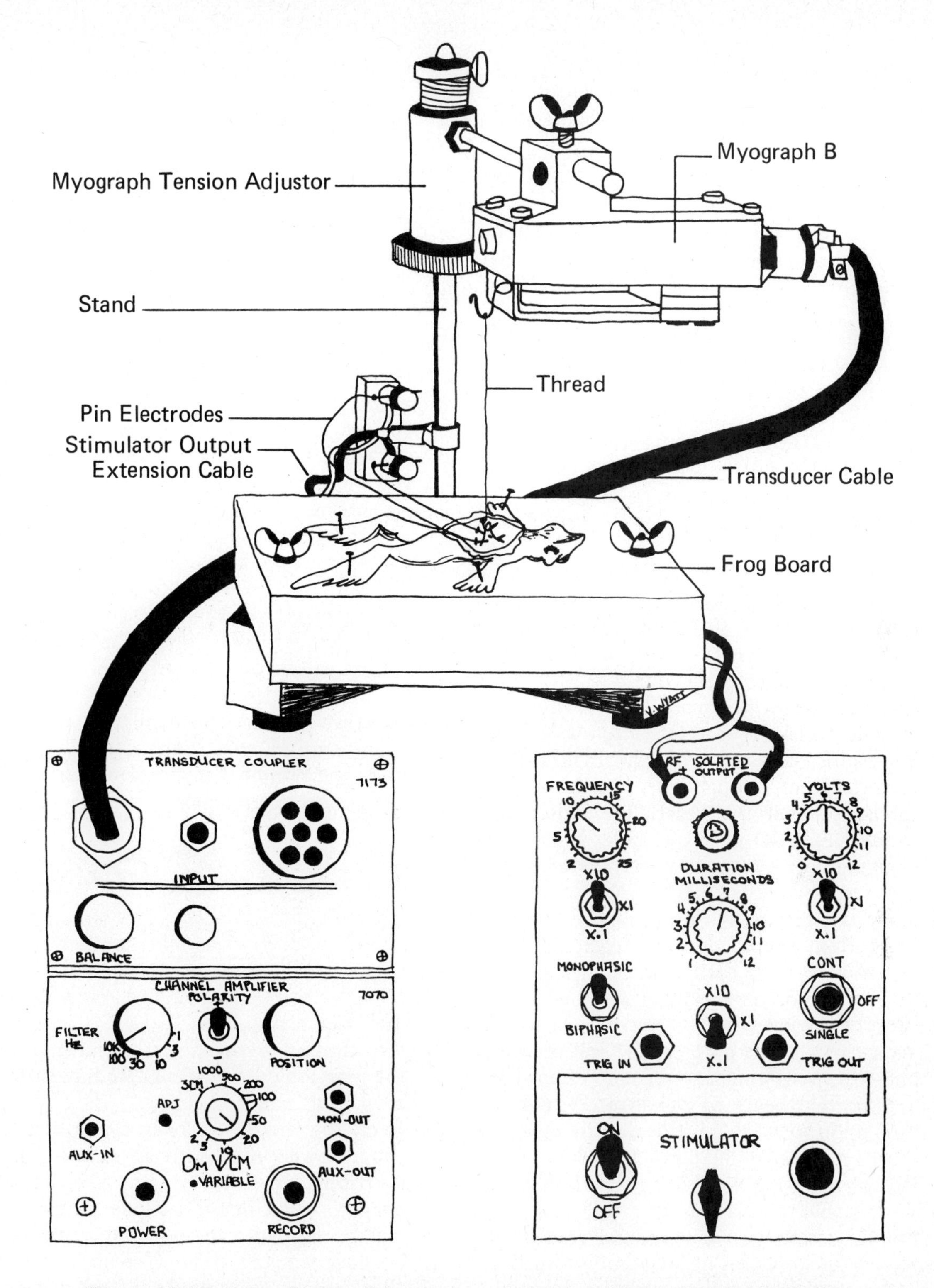

Figure 12-17. Setup for the Stimulation of the Frog Heart with the DMP-4B

the ventricle. Do not puncture the ventricle, since a better record is obtained if the heart is full of blood. Tie a piece of **thread** to the pin hook.

6. Tie the other end of the thread in a fixed loop (not a slip knot) and slip the loop over the myograph leaf spring hook. Adjust the vertical position of the myograph (or use the myograph tension adjustor) to maintain a constant slight tension on the thread

connected to the heart. The tension may need to be decreased later if the atrial contraction is visible on the record.

7. Attach a set of pin electrodes (blue and yellow) to the clip electrodes attached to the stand. Insert the pin electrodes into the ventricle of the heart.
8. Place the channel amplifier controls for the DMP-4A and the Physiograph Four-A in the following positions:
   a. ON-OFF to ON. Allow a five minute minimum warmup time.
   b. AMPLITUDE to 0 (fully counterclockwise).
   c. POSITION to center 0 position.
   d. RECORD-READY to READY.
9. With the Physiograph DMP-4B, set the channel amplifier controls as follows (see Exercise 7-F, Part 1, Step 9, for recalibration if necessary):
   a. Press the POWER switch ON.
   b. Press the RECORD switch OFF if it is glowing.
   c. Set the FILTER switch to 10K, the mV/CM switch to 100 mV/cm.
   d. Set the recording pen 1.5 cm below the center dark green line (or to any other desired baseline) with the POSITION control.
   e. Place the RECORD-READY switch in the RECORD position. Reset the baseline to 1.5 cm below the center dark green line with the BALANCE control.
   f. Adjust the amplitude to produce a deflection of the desired height (about 3/4 inch with each beat of the heart). The tension on the myograph may have to be readjusted.
10. Use a paper speed of 1 cm/sec and set the time marker at one per second. Record 6 to 8 normal beats of the heart. The pen rises upon systole of the ventricle, and returns to the baseline upon diastole.
11. Turn on the stimulator. Set the duration to 7, multiplier to X1. Using a weak, single stimulus (start with 0 volts and gradually increase the strength of the stimulus until a result occurs during diastole), stimulate the ventricle at 7 to 10 second intervals. Some of the stimuli should be administered during systole and others during diastole (both early and late). Continue the experiment until the heart record shows several extrasystoles and compensatory pauses.
12. Include the record, fully labeled, under Results and Questions.

Part 4 The Effect of Temperature on the Heart Rate

Objectives

To demonstrate the effect of temperature on the rate of the heart beat; to calculate the Q_{10} for the rate of the heart beat.

Materials

Dissecting set; beakers; frog Ringer's solution; ice; water baths of various temperatures; medicine dropper; thermometer; frog; frog board and clips; tripod; coarse salt; and wire gauze. For alternate procedure use stand; pin hook; thread; physiograph; myograph B; stimulator output extension cable; and pin electrodes.

Discussion

An increase in temperature speeds up the rate of chemical reactions. The rate of the heart beat will be determined at several temperatures in order to verify that an increase in temperature produces an increase in heart rate. The temperature coefficient or Q_{10} of a reaction is the rate of that reaction (rate of heart beat in this experiment) at one temperature divided by the rate of the same reaction at a temperature 10° C. below the first temperature. The Q_{10} is often 2 to 3 for biological reactions.

$$Q_{10} = \frac{\text{rate at a higher temperature}}{\text{rate at a temperature } 10^\circ \text{ lower}}$$

Procedure

1. Use the same frog heart preparation as in the previous experiment. Remove the pin hook from the frog heart.
2. Keep the frog board over the sink and use paper towels to absorb an excess Ringer's solution.
3. Ringer's solution of various temperatures will be needed, ranging from 2° to 32° C.
4. To cool Ringer's solution to approximately 2° C., place ice cubes, salt, and water in a large beaker. Place a small beaker containing Ringer's solution inside the large beaker.
5. Heat additional Ringer's solution. By a combination of very cold Ringer's solution and hot Ringer's solution prepare baths (as

needed) at the following centigrade temperatures: 2°, 7°, 12°, 17°, 22°, 27°, and 32° C. (Other temperatures may be used but they should be in a series with 5° intervals between them.)

6. Flood the heart with the coldest solution. This must be done for a **sufficient length of time to allow the heart to attain the temperature of the solution.**
7. After the temperature of the heart is stabilized (determine this by counting and checking for stability of the rate), count the heart rate for 1/2 minute and multiply by two.
8. Repeat the above procedure, increasing the temperature of the Ringer's solution 5° C. each time. Record the rate of the heart beat in the table under Results and Questions. **Do not use a temperature over 32° C.**
9. Calculate and record (under Results and Questions) the Q_{10} for at least three successive 10° C. increases in temperature, such as the increase between 2° and 12° C; 12° and 22° C; 22° and 32° C.

Alternate Procedure

1. Using the same frog heart preparation as in the previous experiment (using either the kymograph or physiograph), prepare the solutions and arrange the frog board over the sink as described in steps 1-7 above.
2. If using the kymograph, use drum speed 7. After the heart rate is stabilized at the lowest temperature Ringer's solution obtainable, record the heart rate at that temperature, starting at the bottom of the piece of kymograph paper, for one entire turn of the drum. Then lower the drum before beginning the next recording, stabilize the heart at a temperature 5° higher, and record. Repeat the above procedure, increasing the temperature of Ringer's solution 5° C. each time. Calculate the rate of the heart beat at each temperature by counting the number of beats in 15 cm and multiplying by 2. (At a drum speed of 7, the paper moves 0.5 cm/sec.)
3. The procedure is the same for using the physiograph (use paper speed .5 cm/sec). You may record the heart beat at each temperature on a separate entire sheet of paper, cut out the strips, and superimpose them on one sheet of paper.

Part 5. Heart Block

Objective

To demonstrate the occurrence of heart block.

Materials

Frog board and string; frog; beaker; frog Ringer's solution; medicine dropper; thread; and dissecting set.

Discussion

Pressure applied to the conducting system by means of a ligature in the atrioventricular groove decreases the rate of the heart beat, resulting in heart block. After release of the pressure, the ventricle may resume its normal beat. However, if the conducting system was permanently damaged by the ligature, the ventricle may beat at a much slower rate than the atria. This demonstrates that the automaticity of the ventricle is less than that of the remainder of the heart.

Procedure

1. Expose the heart of a frog or use the same frog as in the previous experiment. This experiment must be the last experiment in any series as the heart often will not resume beating.
2. Count the rate of the whole heart. Since the atria and the ventricle beat at the same rate, record this figure in the columns for the rate "before experiment" for both the atria and the ventricle.
3. Moisten a **thread** in Ringer's solution and pass it around the heart in the atrioventricular groove (see Fig. 12-18). Tie a loose, single knot so that pulling the ends will tighten the thread in the groove.

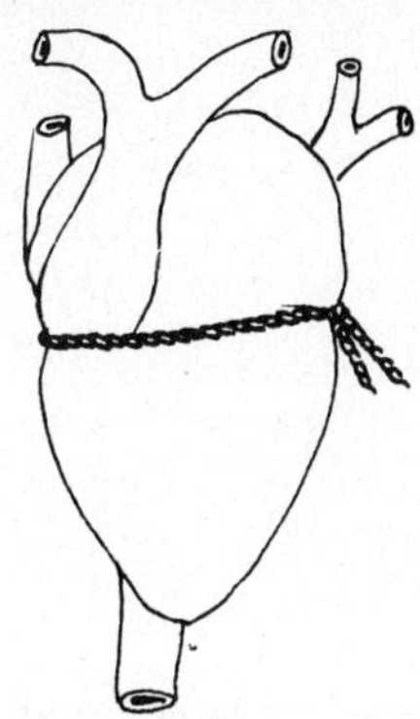

Figure 12-18. Ligature in Atrioventricular Groove

4. Tighten the thread slightly and count and tabulate the rates of both the atria and ventricle. Continue tightening the thread and tabulate the rates of both parts. (One student should count the rate of the atria, and, at the same time, another student should count the rate of the ventricles, for the same length of time.)
5. Continue tightening the thread until the ventricle stops beating. Count the contractions of the atria and record their rate and that of the ventricle (which is zero) in the row in the table headed "rate during ventricular arrest" under Results and Questions.
6. Cut or untie the thread in the atrioventricular groove. Now count the rate of both the atria and the ventricle and record these figures in the table. If the ventricle does not resume its beat, stimulate it. Record the rate of the beat of the ventricle if it resumes beating.

Part 6. Automaticity of the Heart

Objectives

To demonstrate the degree of automaticity possessed by the individual parts of the frog heart, and the gradual decrease in irritability from the sinus venosus to the apex.

Materials

Dissecting set; frog; beaker of frog Ringer's solution; watch glass; induction stimulator, and platinum electrodes.

Procedure

1. Use the same frog if the ventricle resumed a normal beat in Part 5, or prepare a fresh heart.
2. Determine the rate of heart beat when the heart is still within the body of the frog.
3. Remove the heart by first cutting through the arteries and veins near the heart. Lift up the heart and, to remove the sinus venosus, cut deep, dorsal to the heart.
4. Place the heart in a watch glass containing Ringer's solution.
5. After one minute count the rate of the entire heart.
6. Determine the location of the sinus venosus on the dorsal surface of the right atrium; then carefully separate the sinus venosus from the remainder of the heart.
7. After several minutes, count the rate of each of the two separated parts of the heart (the sinus venosus and the combined atria-ventricle).
8. Now separate the atria from the ventricle and determine the rate of each part after several minutes. If the ventricle does not initiate its own beat, stimulate it. Record in the chart (under Results and Questions) the initial rate (zero) and the rate after stimulation.
9. If the ventricle begins to beat, cut off the apex. The apex generally will not beat by itself. If the apex does not initiate its own beat, stimulate it and record the rate of the beat obtained.

Exercise 12-G Microcirculation in the Fish or Frog

Objective

To demonstrate circulation in the tail of a fish or the web of a frog.

Materials

Goldfish (or frog); microscope; balsa wood (or frog board and frog web board); thumb tacks; cotton; and a cover glass.

Discussion

The tail of a fish or the web of a frog can be used to demonstrate microcirculation, since the tail or web is thin enough for light to pass through, allowing the vessels to be observed with a microscope. The vessels in the web or tail can be distinguished on the basis of the rate of flow of blood, the pulsations, and the direction of flow of the blood.

Procedure A (Using Fish)

1. Cut a hole approximately 1/2 inch in diameter near one end of a piece of balsa wood approximately 3 inches wide, 6 inches long, and 1/4 inch thick.
2. Saturate with water two pieces of cotton, each slightly larger than the fish and from 1/4 to 1/2 inch thick, and place these near the hole in the board.

3. Lay the fish on the first cotton pad. (Handle the fish with care. Hold it firmly enough to prevent it from dropping, but do not crush it.) Orient the tail of the fish over the hole.
4. Lay the second wet cotton pad on top of the fish. (This pad has sufficient weight to hold the fish in position.)
5. Spread the tail over the hole in the balsa wood, and pin it in this position by applying a thumb tack at the margin of the tail on either side. Do not puncture the tail.
6. Place the fish so that the hole in the balsa wood is in position on the stage of the microscope for viewing with the low power objective. Cover the tail with a cover glass. Keep the tail moist with tap water.
7. Examine the vessels in the tail under low power of the microscope. Distinguish between an artery and a vein by noting the direction of blood flow (toward or away from a bifurcation).
8. Locate a capillary. The erythrocytes (nucleated) can be seen passing through the capillary in single file. Note the twisting movement of some of the RBC's in the capillaries.
9. Observe the difference in rate of blood flow in the three types of vessels.
10. Note in which type of blood vessels pulsations are observed.
11. The microcirculation may be observed for up to 30 minutes in this manner without damage to the fish. When you complete the experiment, release the fish into its container.

Procedure B (Using Frog)

1. Pith the frog. Place the frog (ventral surface up) on a frog board. Tie both forelegs and one hindleg loosely to the frog board, leaving the other hindleg free to extend over the web board.
2. Spread the web of the free leg over the hole in the web board and fasten the web to the web board with pins. Do not stretch the web too tightly.
3. Keep the frog and the web moist with tap water throughout the experiment. It may be necessary at intervals throughout the experiment to massage the hindleg to restore circulation. Select a light area of the web and examine this with the low power of the microscope, following steps 7-11 in the procedure outlined above with the fish.

Exercise 12-H Heart Sounds and Pulse Rate

Objectives

To listen to the heart sounds and to determine the effect of exercise on heart sounds; to determine the effect of body position and running upon heart rate.

Materials

Stethoscope and bicycle ergometer. For Alternate Procedure: physiograph and photoelectric pulse pick-up.

Discussion

As the heart beats, it makes certain characteristic sounds that can be heard through the stethoscope. These sounds are described as lub-dup. The first sound is caused by the contraction of the ventricle and the snapping shut of the atrioventricular (AV) valves; the second by the closing of the semilunar valves. These sounds furnish information concerning the condition of the heart valves. Heart murmurs are caused by defects in the heart valves.

The pulse reflects the rate of the heart beat. It can be felt at points where an artery lies near the surface over a bone; however, the radial artery is customarily used in taking the pulse.

Procedure

1. Place the stethoscope in the ears so that the top of the earpiece points inward and upward into the ear.
2. Hold the bell of the stethoscope firmly against the chest of the seated subject at the fifth intercostal space, about 3 inches to the left of the midline.
3. Listen to and describe the two heart sounds.
4. Have the subject run in place for 2 minutes. Again listen to the heart sounds. Describe any change in the sounds.
5. Practice taking your pulse at the radial artery by placing the tips of your fingers in the groove medial to the styloid process of

the radius (on the thumb side of the forearm). Determine the pulse rate for an entire minute each time.

6. Count your pulse for one full minute under each of the following conditions and record your values under Results and Questions:
   a. sitting
   b. standing
   c. while exercising (running in place or bicycling with a belt setting of 2 for females and 3 for males, pedaling 35 kph)
   d. immediately after exercise
   e. second minute following exercise
7. Obtain the resting, sitting pulse rate from each member of the class and average the two sexes separately.

Alternate Procedure Using the Physiograph

1. Perform steps 1-4 as outlined above.
2. Refer back to Exercise 7-F, Part 3, for detailed directions for the procedure and channel settings for use of the photoelectric pulse pickup for either model of the physiograph. Refer to Figure 7-17 for the connections on the Physiograph Four-A and the DMP-4A, and to Figure 12-19 for the DMP-4B.
3. Connect the photoelectric pulse pickup to the physiograph and to the ventral surface

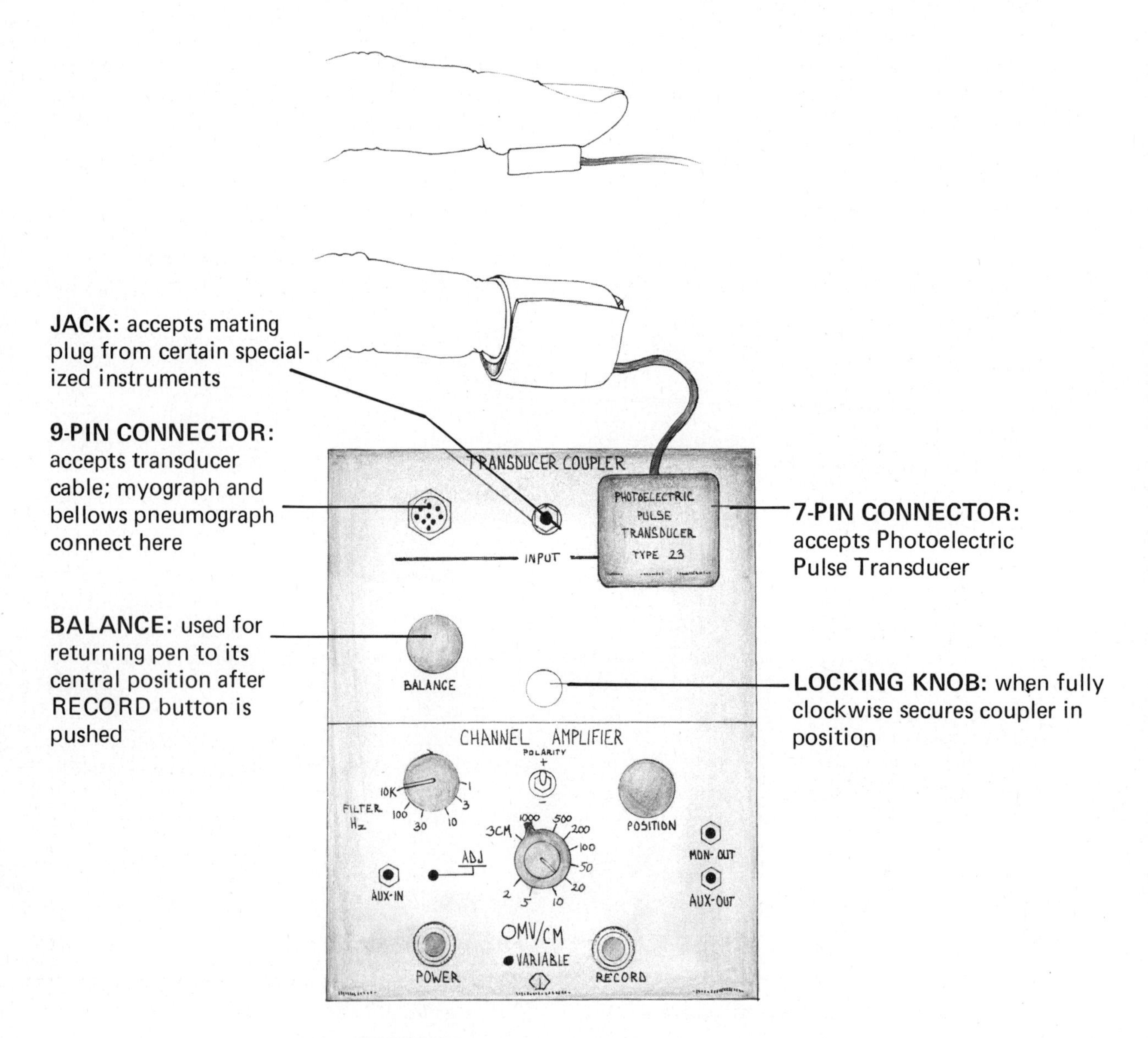

Figure 12-19. Photoelectric Pulse Pickup with DMP-4B

of a finger of the left hand so that the small (4 mm in diameter) photoconductor is next to the skin. Fit the foam rubber band over the photoelectric pulse pickup and wrap the band snugly around the finger. The pulse pickup is light sensitive and must be shielded with the black band.

4. The subject should stand by the bicycle. After the 1 minute warmup period for the 4A models and immediately with the 4B Physiograph model, place the RECORD-READY switch on RECORD. After the pen stabilizes its position, adjust the channel amplifier POSITION or BALANCE control to position the recording tracing in the center of the pen's movement.
5. Adjust the amplitude control (the mV/CM switch on the 4B model) for the desired sensitivity or amplitude. An amplitude of no more than 1 inch is desirable since the amplitude will increase after exercise. It may be necessary to reposition the pulse pickup on the finger if no record is obtained.
6. Use a paper speed of .5 cm/second. Record the pulse for 1 full minute while standing quietly.
7. Then sit on the bicycle, resting the left hand on the handle bar or on a pillow and record the pulse for another minute.
8. Exercise for one minute on the bicycle ergometer, recording the pulse continuously during the exercise. Do not grip the handle bar with the left hand. Use a belt setting of 2 for females, 3 for males, at 35 kph.
9. If a bicycle ergometer is not available, turn the channel amplifier back to READY and run in place for one minute, counting the pulse in the common carotid artery in the neck while running.
10. Record the pulse for 2 minutes on the physiograph immediately after exercise, while sitting quietly on the bicycle or on a laboratory chair.
11. Calculate the pulse rate during the last 30 seconds of each minute and multiply the rate by 2, to obtain the heart rate per minute. See the sample record below (Fig. 12-20) for method. At a paper speed of .5 cm/second, 5 seconds elapse between each two dark green vertical lines. Include your results under Results and Questions.
12. Obtain the before-exercise resting, sitting pulse rate from each member of the class and average the two sexes separately.

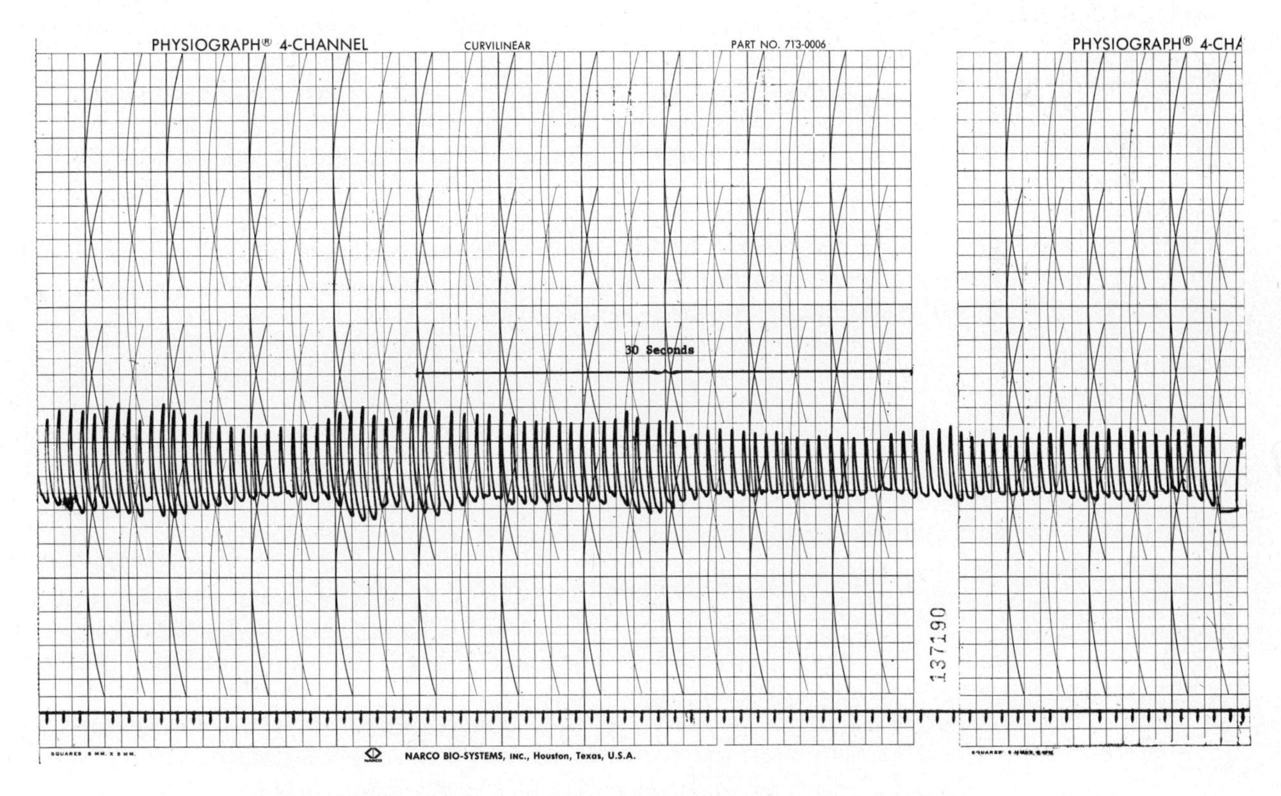

Figure 12-20. Record of Pulse with Photoelectric Pulse Pickup

Exercise 12-I
Blood Pressure

Objectives

To determine blood pressure and the effect of exercise upon blood pressure, heart rate, and respiratory rate.

Materials

Sphygmomanometer; stethoscope; bicycle ergometer (Alternate Procedure: physiograph; electrosphygmograph, transducer and impedance pneumograph couplers; photoelectric pulse pickup; cuff; **either** plate electrodes, adaptor cable AC-13, shielded input extension cable SC-13, **or** bellows pneumograph and transducer cable).

Discussion

Since direct measurement of blood pressure involves instruments which enter the blood vessel, the indirect method, which depends upon the amount of pressure that is required to close off the flow of blood through the brachial artery, is most commonly used. The brachial artery, located in the upper arm, lies close to the ventral surface of the humerus.

The sphygmomanometer consists of (1) a compression bag surrounded by an unyielding cuff for application of extra-arterial pressure; (2) a mercury manometer or pressure gauge by which the applied pressure is read; (3) a bulb for inflation of the cuff in order to create pressure in the system; (4) a valve to release the air from the cuff; and (5) attachments for connecting the cuff to the manometer and bulb. The manometer is used to measure the pressure within the cuff in millimeters of mercury.

Procedure

1. This experiment requires the collaboration of three students. One student should take blood pressure, one determine respiration rate, and the third (subject) should take his own pulse. (Pulse must be taken in the arm not used for the determination of blood pressure.)
2. The subject should sit in a quiet room for 5 minutes in a relaxed position with his right arm on the table. After this period of time, determine his blood pressure, respiration rate, and pulse. (Always take the pulse and respiration rate at the same time as the blood pressure reading.)
3. Wrap the cuff around the arm, approximately 1 inch above the antecubital space (elbow). The cuff should fit snugly. If a hook cuff is used, circle the arm so that the correct arrow is over the artery. Pull the tail end of the cuff snugly over the arm and hook the metal buttonhole over the metal button.
4. Place the stethoscope over the brachial artery just below the cuff. The stethoscope should not be in contact with the cuff. The stethoscope should be applied firmly throughout the experiment but not so tightly that the pressure compresses the brachial artery. No sound should be heard in the artery at this time. Verify this.
5. Pump air into the cuff until the mercury column or pressure reading on the gauge is 200 mm. If the pressure is high enough, the brachial artery is compressed, and no sound should be heard through the stethoscope.
6. Release the pressure slowly (the pressure should fall 2-3 mm per heart beat) by means of the valve on the bulb, while listening through the stethoscope for the first sound. This sound, which indicates that blood is again flowing through the artery, is heard with each beat of the heart and is taken as the **systolic pressure.** The height of the column of mercury or pressure on the gauge should be read at this point.
7. Continue reducing the pressure carefully. The sound becomes louder, then dull and muffled, and then ceases. The point of complete cessation of sound is taken as the **diastolic pressure.** If no cessation of sound occurs, the point of muffling should be taken as diastolic pressure, and should then be recorded as the point of muffled sounds. (On the average, dulling of sounds appears 5 to 10 mm above the point of cessation of sounds.)
   Diastolic sounds may persist all the way down to zero in certain clinical states, such as aortic insufficiency, hyperthyroidism, anemia, and in an occasional normal person.
8. Rapidly release the pressure once the diastolic pressure is determined.

9. With blood pressure readings taken in this manner a mean error of ± 8 mm Hg may be expected in individual readings of systolic and diastolic pressure (the systolic pressure tends to be too low and diastolic too high). Variations in auditory acuity are important causes of error. Errors are also likely to be obtained in subjects with unusually large or unusually small arms; here the error may exceed 30 mm Hg.
10. Determine systolic and diastolic pressures two additional times. (Allow 2 minutes between readings since, due to relaxation of arterial tone, the last readings may be lower than the previous ones.) Average these, if the determinations agree reasonably well, and record the value in the "Before Exercise" column in the table under Results and Questions.
11. Subtract diastolic pressure from systolic pressure to obtain pulse pressure.
12. The subject should exercise by running vigorously around the building (with blood pressure cuff attached if possible—consult instructor) or bicycle 2 minutes on the bicycle ergometer. Consult instructor for settings. **Immediately** upon the return of the subject to the room, repeat all measurements.
13. Again repeat all measurements 2 minutes after the return, then 4 minutes after the return, etc., until the rates return to normal. Try to be exact about the times. Record all results under Results and Questions.

Alternate Procedure with the Physiograph

1. The same experiments on the effect of exercise on circulation and respiration may be performed with the physiograph, using three channels to simultaneously record blood pressure, respiration rate, and heart rate. See Exercise 7-F, Part 1 for detailed directions for the operation of the physiograph. The bellows or impedance pneumograph and the photoelectric finger pulse pickup have been described previously (Exercises 11-E and 7-F, Part 3). The electrosphygmograph is a transducer/preamplifier for recording indirect systolic and diastolic blood pressures (range 0-300 mm Hg). The electrosphygmograph combines a pressure transducer and a preamplifier to produce single channel recordings of occluding cuff pressure and superimposed Korotkoff sounds. These sounds are the sounds heard as blood passes through a partially occluded blood vessel.
2. With the Physiograph Four-A and DMP-4A, set up the equipment as directed below in steps 3-11. For the DMP-4B, follow steps 12-19. The procedure is the same for all models for the remainder of the experiments. The subject should sit on the bicycle ergometer.
3. Open the bellows pneumograph valve by turning the screw one-half turn counterclockwise. Fasten the strap around the chest (the subject must wear a shirt) with the transducer plug facing upward. Tighten the strap as necessary so that the bellows pneumograph remains in position under the arms. See Figure 11-8 for the position of the pneumograph. Compress the bellows slightly and close the transducer valve by rotating the wing screw clockwise. Release the bellows after the valve is closed to produce a slight internal negative pressure.
4. Connect the bellows pneumograph to the input connector of the channel amplifier with the transducer cable. The cable should extend up over the subject's shoulder to support the transducer.
5. Connect the photoelectric pulse pickup to the ventral surface of a finger of the left hand (as described in detail in Exercise 7-F, Part 3) and to the physiograph.
6. Mount the electrosphygmograph on the stand using the mounting rod and clamp. Using the transducer cable, connect the upper surface of the electrosphygmograph to the input receptacle on the physiograph channel amplifier (see Fig. 12-21.) Wrap the cuff around the right arm of the subject. Check that the words "place distally over artery" on the inner surface of the cuff are on the middle of the ventral surface of the arm above the elbow (on the brachial artery).
7. Following the procedure illustrated by Figure 12-21, connect the plug on the cuff to the input marked MIC on the ESG. Connect the plastic tube on the bulb to the input marked BULB on the ESG. Turn the sound amplifier and the pressure amplifier on the ESG to 0 (fully counterclockwise).
8. Turn on all three physiograph channel

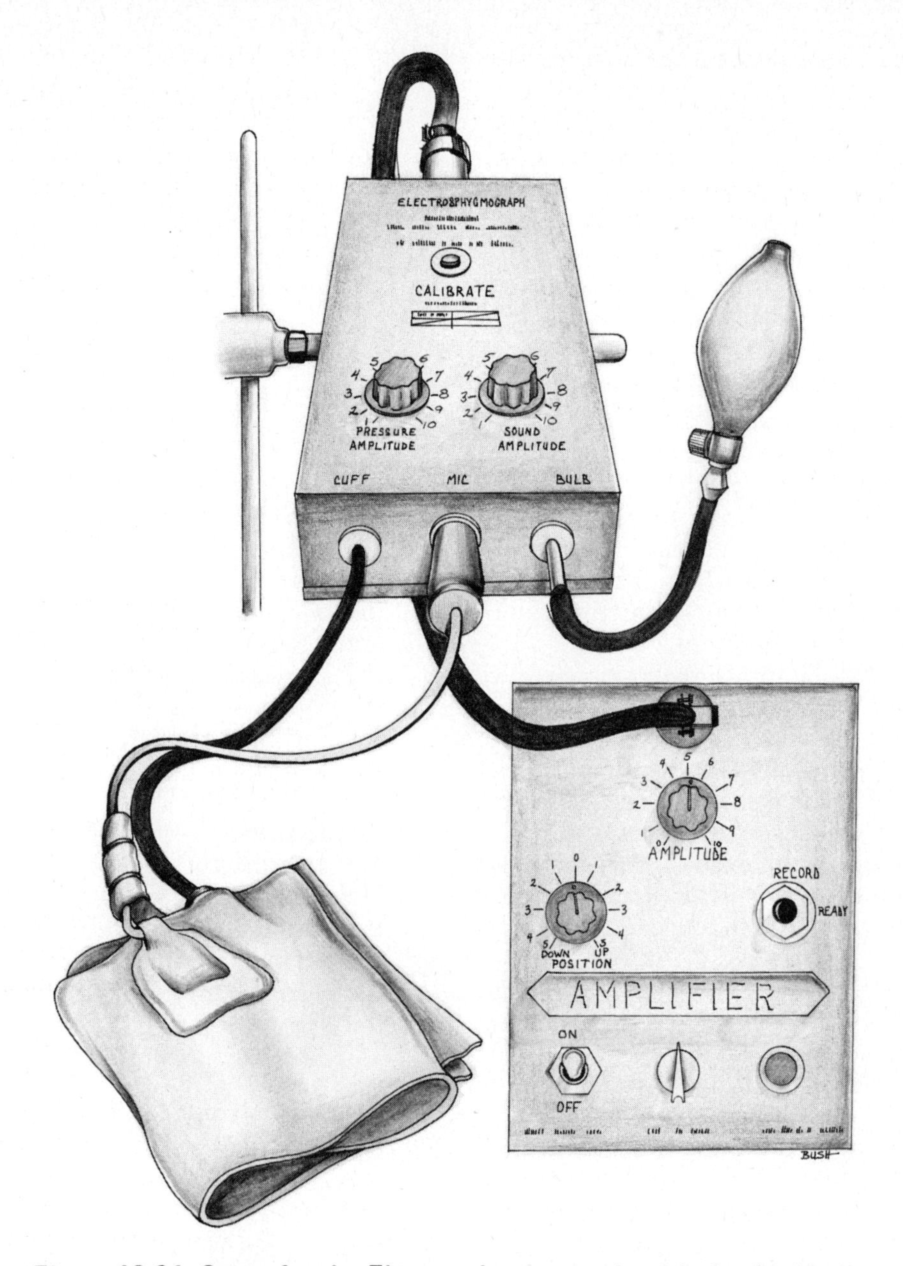

Figure 12-21. Setup for the Electrosphygmograph with the 4A Models

amplifiers.

a. ON-OFF to ON position.
b. POSITION to center 0 position.
c. RECORD-READY to READY position.
d. AMPLITUDE to 0 (nonfunctional with the ESG).

9. Set the paper speed to .1 cm/sec while testing the channels. Set the time marker to 1/sec. Place the RECORD-READY switch on RECORD on the channels connected to the photoelectric pulse pickup and the bellows pneumograph. After a one minute warmup adjust the amplitude so each record is approximately one inch in height. Adjust the position of the pen with the position control so that tracing is centered about the dark green line.

10. Allow 5 minutes for warmup of all 3 channels. Then turn the RECORD-READY switch on the channel amplifier for the ESG to RECORD. Set the paper speed to .1 cm/sec for calibration. With the pressure amplitude on the ESG on 0, adjust the channel amplifier position control to locate the pen 2.5 cm below the recording pen center line (five lines below center dark

line). Press the button marked 100 mg Hg Calibrate, and at the same time turn the ESG pressure amplitude control clockwise to position the pen on the center line (the heavy green line). With this calibration, five squares equals 100 mm Hg; therefore each square equals 20 mm and 1 cm equals 40 mm. Release the button and check that the pen returns to the zero pressure line 2.5 cm below the heavy green line. Repeat the above procedure, if necessary. At this point 0 mm Hg should be 2.5 cm below the center line, 100 mg Hg on the center line, 200 mg Hg 2.5 cm above the center line. Once the calibration has been established, it may be used throughout the experiment unless the channel amplifier position control or the ESG pressure amplitude control is reset.

11. Turn the sound amplitude control to 2 and inflate the cuff to 160 mm and then decrease the pressure until sound marks appear. Then adjust the sound amplitude so that the marks are at least 2 cm in amplitude. If the Korotkoff sounds are not clearly defined on the record, the cuff and microphone may require repositioning.
12. With the DMP-4B model, the electrosphymograph coupler and either two transducer couplers or one transducer coupler and one impedance pneumograph coupler are needed. Recalibrate the three channel amplifiers if necessary (see Exercise 7-F, Part 1, Step 9).
13. Place the photoelectric pulse pickup on the finger and connect the other end to the 7-hole input connector on the transducer coupler (see Fig. 12-19). If a second transducer coupler is available, attach the bellows pneumograph around the chest as directed in steps 3-4 and connect the transducer cable to the bellows pneumograph and to the 9 hole input on the second transducer coupler.
14. If a second transducer coupler is not available, record respirations with the impedance pneumograph. Apply a salt solution to two plate electrodes and attach them to the interspace between the fifth and sixth ribs near the sides of the chest, fastening them by means of a long rubber strap. Attach an additional plate electrode on the center of the subject's chest for grounding. Attach the long loose ends of the Adaptor Cable AC-13 to the subject, red on the left side, black on the right side, and green to the center plate electrode on the chest. Connect the short twisted ends of the Adaptor Cable AC-13 to the Shielded Input Extension Cable SC-13. Connect the red wire to the 1 position, the black to the 2 position, and the green to the G position.
15. Using Figure 12-23 as a guide, connect the cuff on the arm of the subject so that the microphone is over the brachial artery. Connect the cuff pneumatic tube to the ESG Coupler Connector CUFF, the hand pressure bulb to the ESG Coupler connector BULB, and the microphone plug to the ESG Coupler MIC jack.
16. Set the paper speed to .1 cm/sec while testing the channels. Press the POWER switch on for each channel. For the photoelectric pulse pickup, set the mV/CM switch on 20 mV/cm, set the pen to the center dark green line with the POSITION control, press the RECORD switch ON, and reposition the pen, if necessary, with the BALANCE control, to record symmetrically around the center dark green line. Increase or decrease the amplitude with the mV/CM switch as necessary to obtain a tracing approximately 1 inch high.
17. With the bellows pneumograph, press the POWER switch ON, set the mV/CM switch on 100 mV/cm, position the pen on the center dark green line with the POSITION control, press the RECORD switch ON and reposition the pen, if necessary, with the BALANCE control, to record symmetrically around the center dark green line. Increase or decrease the amplitude with the mV/CM switch as necessary to obtain a reading approximately 1-2 inches high. With the impedance pneumograph, press the POWER switch ON, set the FILTER switch to 30, the mV/CM switch to 10 mV/cm, and the DIRECT/CAPACITIVE switch to CAPACITIVE.
    Set the recording pen to the center dark green line using the POSITION control. Press the channel amplifier RECORD switch ON. Return the recording pen to the previously established baseline using the TRACE RESET button. A tracing of the breathing pattern should now be obtained. To increase the amplitude of the recording, turn the mV/CM switch to a lower number;

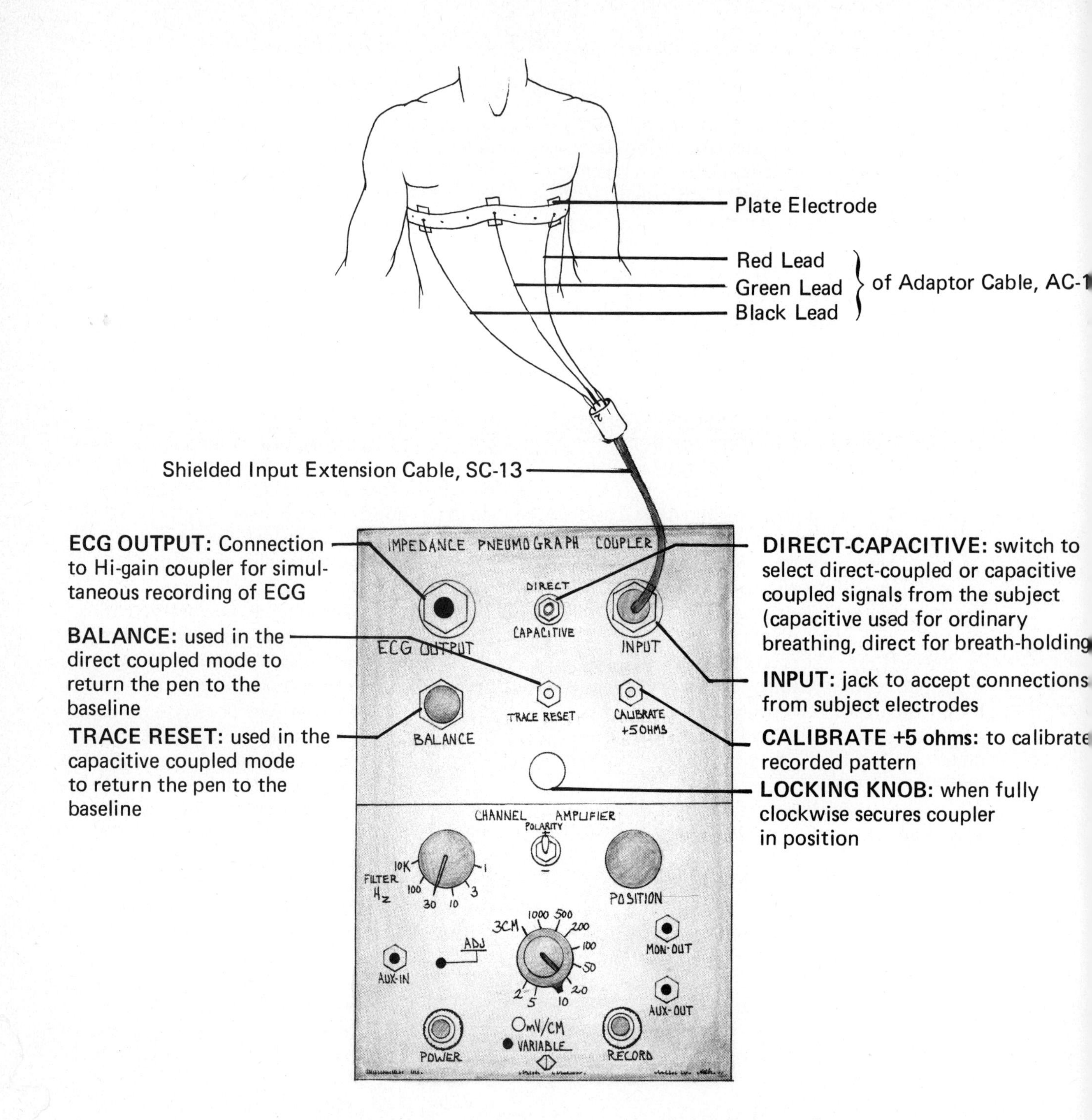

Figure 12-22. Impedance Pneumograph, with Physiograph DMP-4B

to decrease, turn it to a higher setting.

18. Initial settings for the channel amplifier for the electrosphygmograph:
    a. Press the POWER switch ON.
    b. Press the RECORD switch OFF if it is glowing, and the mV/CM switch to 2 mV/cm.
    c. Set the recording pen to a baseline 2.5 cm (five lines) below the center dark green line using the POSITION control.
    d. Set the ESG coupler PRESSURE switch to the 0-200 mm range.
    e. Press the channel amplifier RECORD switch ON.

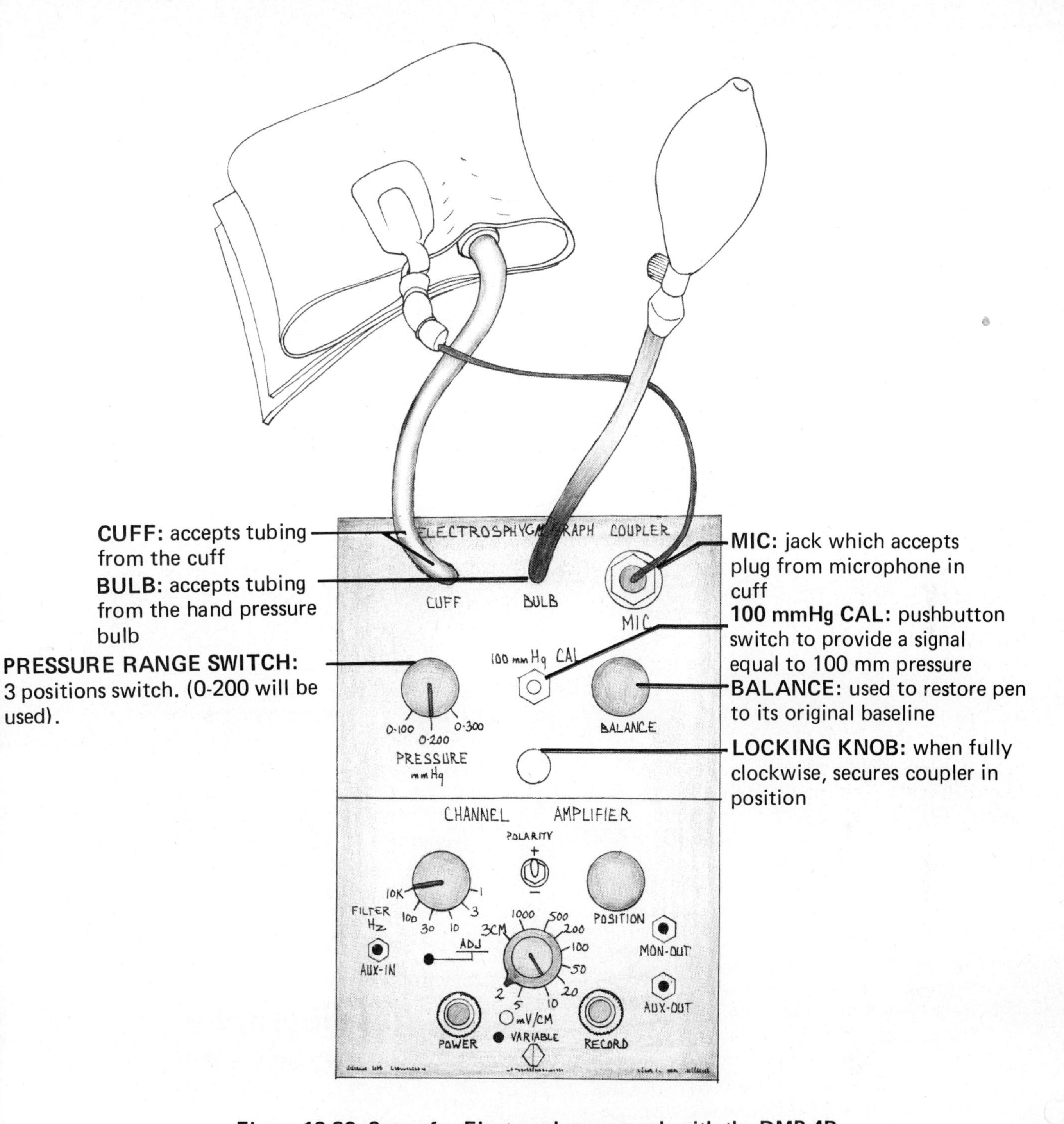

Figure 12-23. Setup for Electrosphygmograph with the DMP-4B

f. Adjust the ESG Coupler BALANCE control to set the recording pen to the original baseline 2.5 cm (5 lines) below the center dark green line. With this calibration, five squares equal 100 mm Hg; therefore, each square equals 20 mm and 1 cm equals 40 mm.

19. Inflate the cuff until sound marks appear. Then adjust the amplitude of the marks by varying the mV/CM switch until the marks are at least 2 cm in amplitude. If the Korotkoff sounds are not clearly defined on the record, the cuff and microphone may need repositioning. Always turn the RECORD switch off before making any changes in equipment.

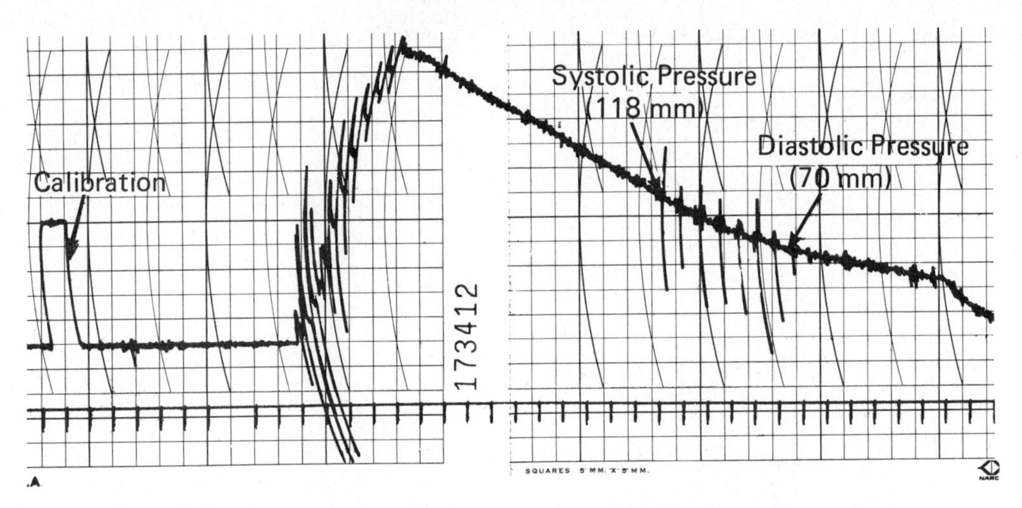

Figure 12-24. Tracing of Blood Pressure with Electrosphygmograph

20. Follow all the remaining steps with each procedure.
21. Position each pen so that it does not interfere with the tracing of the pen from another channel. The hand with the photoelectric pulse pickup should rest on a pillow on the handlebar or on the handlebar itself. The subject should not grip the handlebar with this hand. During readings of the blood pressure the arm with the cuff attached should hang down at the side. Turn the RECORD button off on the channel amplifier connected to the cuff while the subject is pedaling.
22. Turn the paper speed to .5 cm/sec. Turn all three channel amplifier RECORD-READY switches to RECORD.
23. Record the pulse, repiration rate, and blood pressure for 2 minutes while the subject is sitting on the bicycle. At the end of each minute of recording, take the subject's blood pressure. With the bulb, inflate the cuff above the systolic pressure (to at least 160 mm on the resting subject and to at least 200 mm after exercise). Slowly lower the cuff pressure (usually 5 mm/heart beat), slowly passing through the systolic and diastolic pressure ranges. Where the first vertical line intersects the diagonal descending line is the systolic pressure; where the last vertical line intersects the diagonal descending line is the diastolic pressure (see Fig. 12-24).
24. Have the subject pedal the bicycle ergometer for 2 minutes (belt setting of 2 for women, 3 for men, at 35 kph) continuing the recording of pulse and respiration during the exercise. Turn off the channel amplifier connected to the cuff while exercising. The subject should grip the handle bar with his right hand (the cuff should be on the right arm).
25. Immediately after the cessation of the exercise, the subject should drop his right arm to his side. Continue the recording of pulse and respiration and immediately take his blood pressure.
26. Continue the recordings for 6 minutes after exercise, taking the blood pressure at the beginning of each minute.
27. Calculate heart rate, respiration rate, and blood pressure each minute. Write these on your record (show where you made each calculation of rate per minute) and also enter these on the data chart under Results and Questions.

Exercise 12-J Electrocardiography with the Physiograph

Objectives

To record the electrocardiogram; to simultaneously record the electrocardiogram and respiratory pattern.

Materials

Physiograph; plate electrodes; strap; and **either** (with the 4A models) impedance pneumograph, hi-gain preamplifier, stand, clamp, impedance pneumograph plate electrode cable, impedance pneumograph to EKG coupling cable, two transducer cables; **or** (with the 4B models) hi-gain coupler, impedance pneumograph coupler, adaptor cable AC-13, impedance pneumograph/ECG preamplifier coupling cable, shielded input extension cable SC-13.

Discussion

The hi-gain preamplifier is used in recording a wide range of bioelectric potentials. It can be used to record EKG, EEG, EMG, and heart sounds. The impedance pneumograph can be used with the hi-gain preamplifier to simultaneously record respiration and EKG from the same set of electrodes.

With an instrument called an electrocardiograph, the electrical activity of the heart may be recorded from the exterior surface of the body. As the wave of excitation passes through the heart, electrical currents spread into the tissues surrounding the heart; a small proportion of these spread all the way to the surface of the body. If suitable electrodes are placed on the body on opposite sides of the heart, the electric potential generated by the heart can be recorded.

Electrodes are usually attached to the arms and legs of the subject and connected to the electrocardiograph. The right arm, left arm, left leg, and chest electrodes, combined in various combinations, are referred to as electrocardiograph leads. The right leg is included in the circuit as a neutral reference point. These electrocardiograph leads have certain designations which, by convention, always indicate certain electrode combinations. Only two chest leads will be used in this laboratory exercise since the EKG will be recorded in conjunction with breathing rate.

A normal electrocardiogram, showing three beats of the heart, is reproduced in Figure 12-25. Note that each beat is composed of several waves. These waves are due to changes in the electrical potential of the heart as the wave of depolarization (excitation wave) which precedes the contraction spreads from the SA node over the heart in a definite manner. When the wave of depolarization is moving towards the recording electrode, an upward deflection occurs; when the wave is receding from the same electrode, a downward deflection occurs. From an EKG it is possible to determine the magnitude, duration, and direction of the potential changes occurring within the heart.

The electrical event represented on the EKG is labeled on the normal record (see Fig. 12-25). The **P wave** is caused by the wave of excitation spreading from the SA node over the atria. The P wave immediately precedes the contraction of the atria. The duration of the P wave does not normally exceed .11 seconds.

The **QRS complex** is caused by the depolarization of the ventricles. Ventricular systole immediately follows this event. This is a complex wave with three distinct components. It consists of the downward deflection Q, the upward deflection R, and the negative deflection S. The normal range of duration of the QRS complex is .06 – .10 seconds.

The **T wave** is caused by the repolarization of the ventricles. (Atrial repolarization usually occurs during the QRS complex and therefore does not show up on the record.)

The distance between P and R indicates the delay in transmission of the impulse at the AV node. The P-R interval (sometimes called the P-Q interval) starts with the beginning of the P wave and ends with the beginning of the Q wave if one is present, or the R wave. The upper limits of normal for the P-R interval for cardiac rates above 70 beats per minute is .20 seconds. Any time longer than that indicates some delay in conduction.

The clinical value of the EKG lies in the fact that many functional abnormalities can be detected from these records, such as a defective valve, a defect in the conducting mechanism, a myocardial infarction, etc. If the tracing is definitely abnormal, the patient has heart disease. However, a patient may have a normal EKG and still have heart disease.

The following experimental setup may be used to obtain an electrocardiogram, and to measure both the heart rate (using the EKG

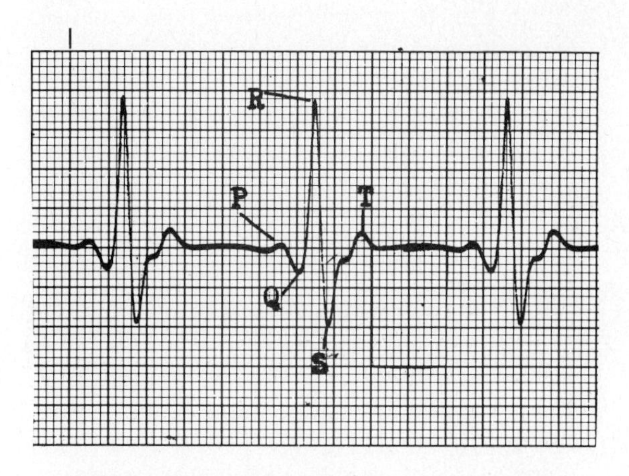

Figure 12-25. Normal Electrocardiogram Showing Three Beats of the Heart

record as an indicator of heart rate) and the respiration rate. Do not look for abnormalities in the EKG as it is not taken under standard conditions.

Procedure

1. See Exercise 7-F, Part 1, for detailed operating instructions for the physiograph and Exercise 11-E, Alternate Procedure Using the Impedance Pneumograph, for detailed directions for the setup for the impedance pneumograph and subject electrodes.
2. Apply a salt solution to three plate electrodes. Attach two of them to the interspace between the fifth and sixth ribs near the sides of the chest, fastening them by means of the long rubber strap. Attach an additional plate electrode on the center of the subject's chest for grounding. (See Figure 12-26 for the 4A models and Figure 12-27 for the 4B model.) Using either the impedance pneumograph plate electrode cable for the 4A models or the adaptor Cable, AC-13 for the 4B models, attach the long loose ends to the subject, red lead on the left side of the chest, black on the right, and green to the ground in the middle. Attach the short twisted ends (with the 4A models) to the front of the impedance pneumograph, the black electrode to the black input terminal, the red electrode to the red input terminal, the green to the ground terminal on the lower surface. With the 4B models connect the short twisted ends to the shielded input extension cable SC-13. Connect the red wire to the 1 position, the black to the 2 position, and the green to the G position.
3. With the 4A models, follow steps 4-12. For the 4B models, follow steps 13-20.
4. Connect the impedance pneumograph to the hi-gain preamplifier by means of the impedance pneumograph to EKG coupling cable (see Fig. 12-26). Run this cable from the jack on the impedance pneumograph labeled COUPLING TO EKG PREAMP on the lower end to the jack marked TRANSDUCER on the hi-gain preamplifier. Then connect the hi-gain preamplifier by means of a transducer cable to a channel amplifier. Connect the impedance pneumograph by means of a transducer cable to a different channel amplifier. Set the impedance pneumograph to CAPACITOR COUPLED OUTPUT, and the amplitude control fully counterclockwise.
5. If it is not necessary to obtain a tracing of the respiratory pattern of the subject, connect the wires from the plate electrodes directly to the hi-gain preamplifier (site labeled INPUT).
6. Set both physiograph channel amplifier controls as follows:
   a. ON-OFF to ON (allow 5 minutes warmup)
   b. POSITION at center 0 position
   c. RECORD-READY to READY
   d. AMPLITUDE—not functional with the preamplifier attached
7. Set the preamplifier controls on the hi-gain preamplifier as follows (one minute warmup):
   a. AMPLITUDE fully counterclockwise
   b. RECORD-CAL on CAL
   c. Time constant switch to EKG
8. After the warmup period, set the RECORD-READY switch on the physiograph channel connected to the hi-gain preamplifier to RECORD. Center the recording channel pen by adjusting the channel amplifier position knob. (The restore-trace button on the preamplifier can be pressed to accelerate the return of the pen to the base line position.) The amplitude of the recorded signal is controlled only by the preamplifier amplitude control.
9. Place the RECORD-CAL switch on the hi-gain preamplifier to RECORD. Adjust the amplitude control on the preamplifier to the desired amplitude. Return to CAL when not recording.
10. Place the channel amplifier connected to the impedance pneumograph on RECORD. Adjust the position of the pen to the centerline by turning the position knob. Adjust the height of the record by turning the amplitude knob on the impedance pneumograph slowly clockwise until the recorded patterns are of the desired magnitude.
11. In order to record from the subject, place the two channel amplifiers RECORD-READY switches onto READY, and the hi-gain preamplifier RECORD-CAL switch in the RECORD position. The restore trace pushbutton can be depressed to rapidly restore the pen to its recording base line.
12. If repositioning of the electrodes is re-

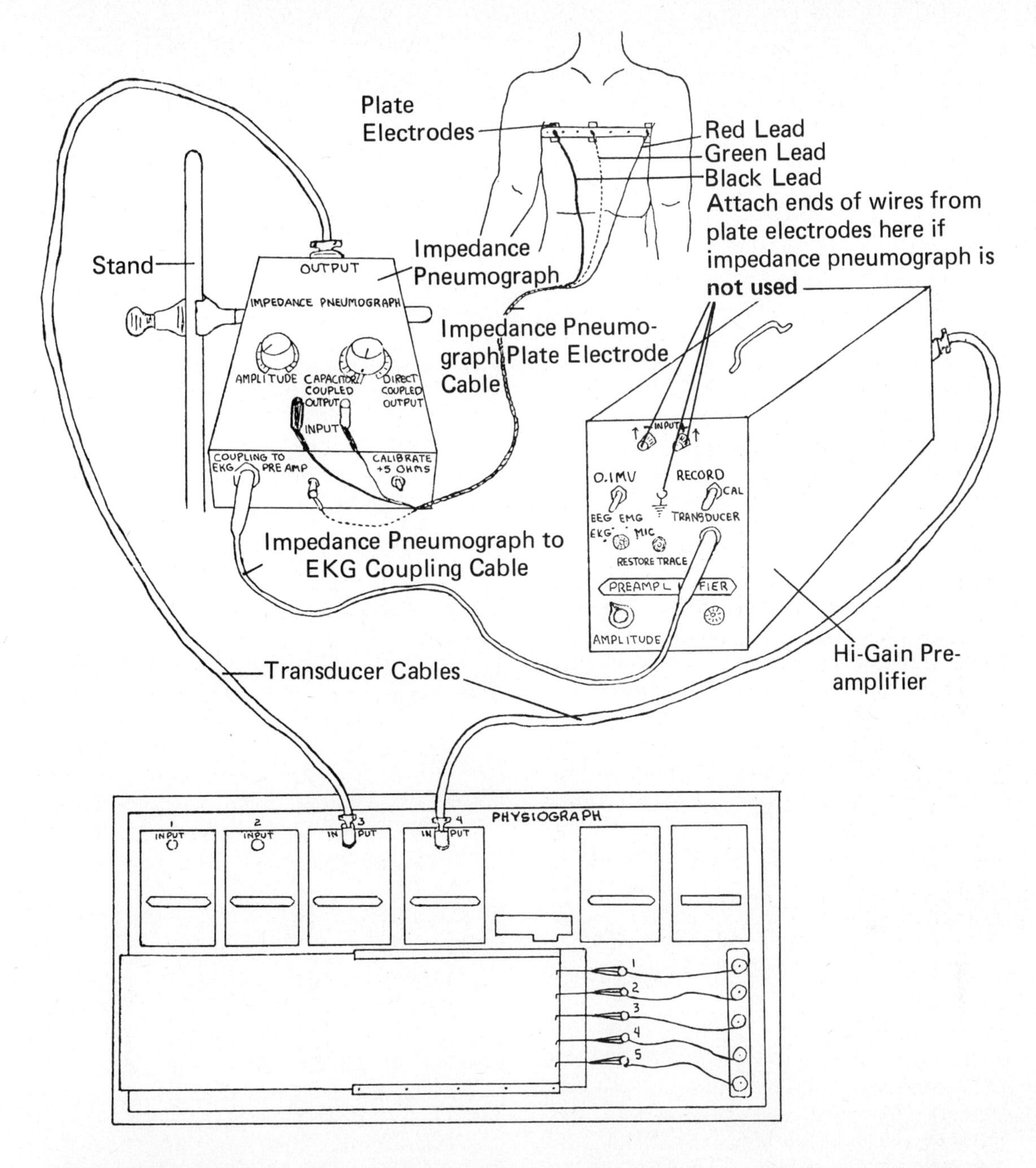

Figure 12-26. Setup for Simultaneously Recording Respiration and Heart Action with the Physiograph Four-A

quired at any time, put the channel amplifier switches on READY and the preamplifier switch on CAL.

13. For the 4B models, set up the equipment as in Figure 12-27. Insert the Shielded Input Extension Cable SC-13 to the INPUT jack of the impedance pneumograph coupler. Connect the Impedance Pneumograph/ECG Preamplifier Coupling Cable to the ECG OUTPUT on the impedance pneumograph. Insert the other end in the INPUT jack on the hi-gain coupler.
14. If it is not necessary to obtain a tracing of the respiratory pattern of the subject, connect the shielded input extension cable SC-13 directly to the INPUT jack of the hi-gain coupler (see Fig. 12-28).
15. If it is necessary to recalibrate the physiograph channel amplifiers, see Exercise 7-F, Part 1, Step 9.

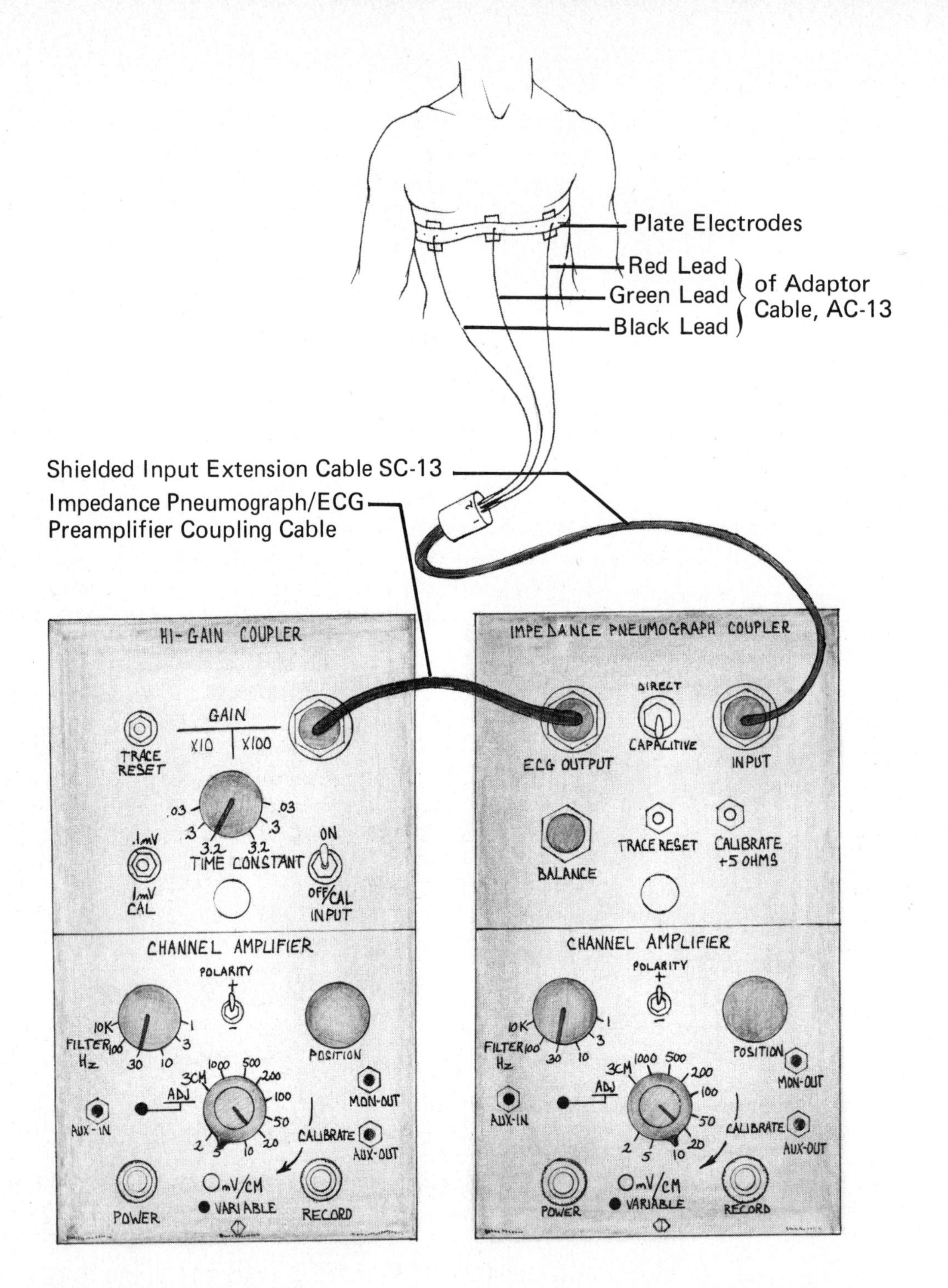

Figure 12-27. Equipment Setup for Simultaneously Recording EKG and Respiration with the Physiograph DMP-4B

16. Initial settings for the channel amplifier to be used with the hi-gain coupler: FILTER switch to 30, mV/CM switch to 5 mV/cm, the recording pen on the center dark green line (set with the POSITION control).
17. Initial settings for the hi-gain coupler: time constant switch to 3.2 and INPUT switch to OFF/CAL.
18. Initial settings for the channel amplifier to be used with the impedance pneumograph:

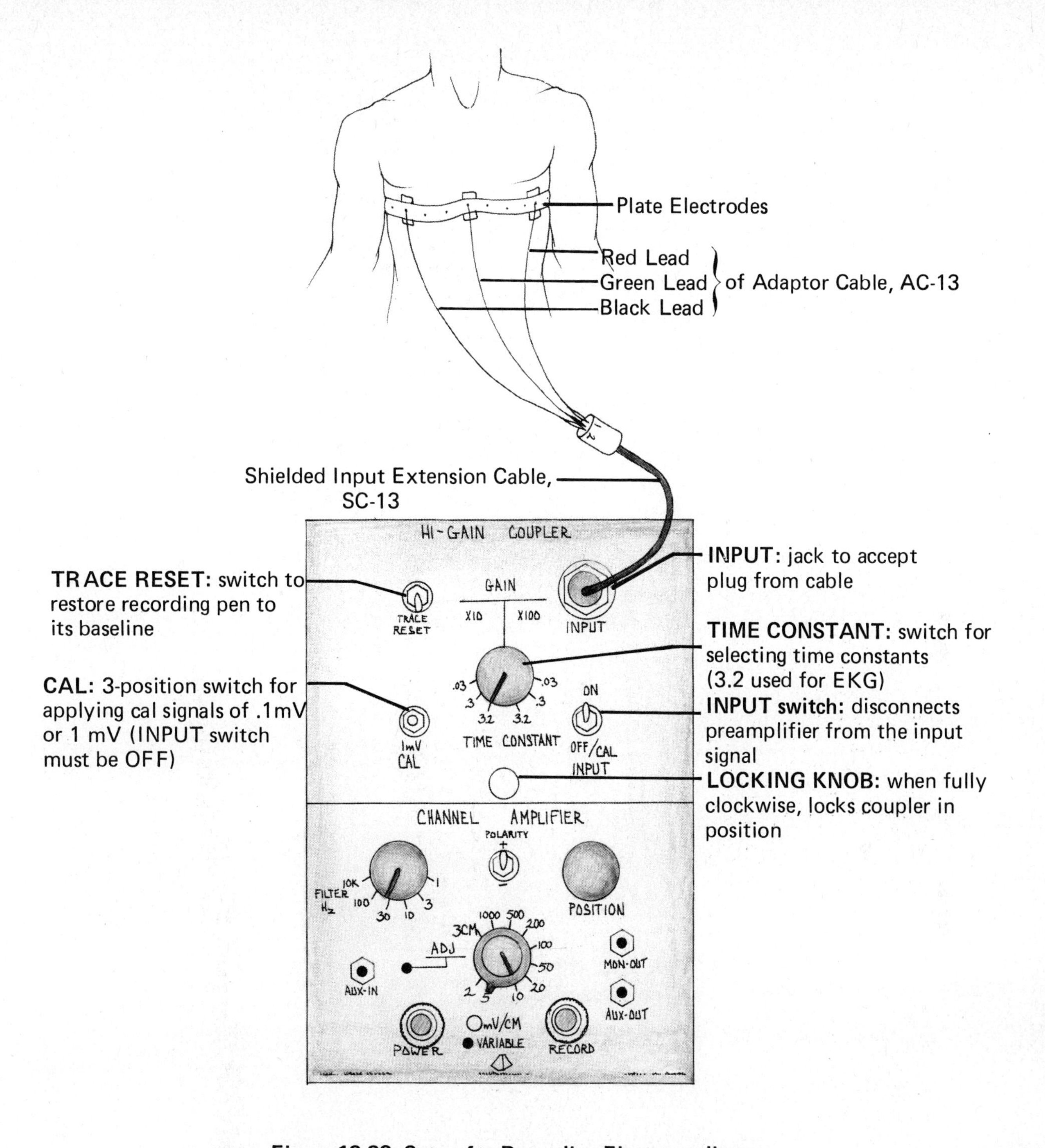

Figure 12-28. Setup for Recording Electrocardiogram

FILTER switch to 30, mV/CM switch to 10 mV/cm, the recording pen on the center dark green line (set with the POSITION control).

19. Initial settings for the impedance pneumograph coupler: DIRECT-CAPACITIVE switch on CAPACITIVE.

20. Set the hi-gain coupler INPUT switch to ON and press both channel amplifier RECORD switches on. A tracing of both the EKG and breathing should be obtained. To increase the amplitude of either recording, turn the appropriate mV/CM switch clockwise; to decrease the amplitude, turn the

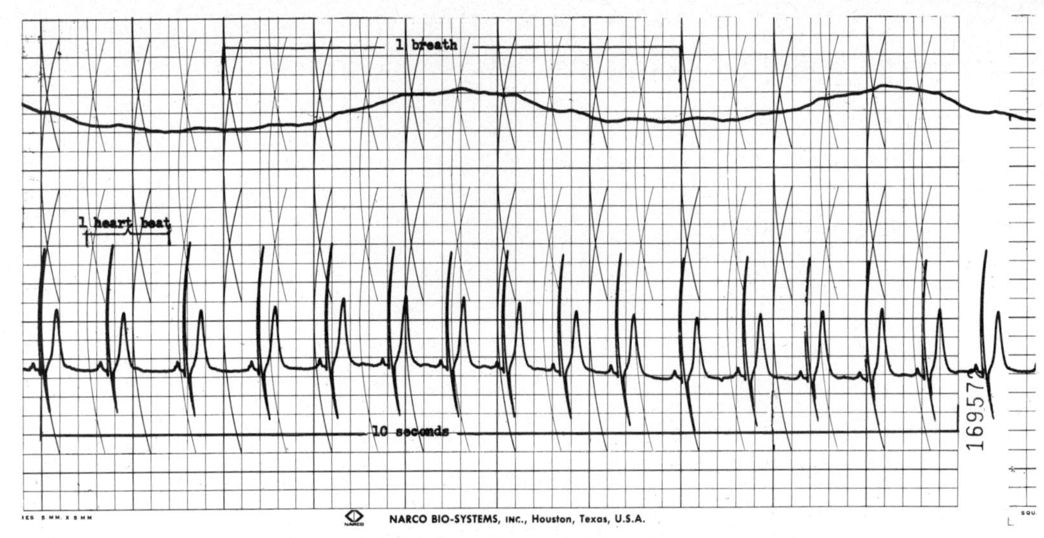

Figure 12-29. Tracing of EKG and Respiration

switch counterclockwise. With either record, return the pen to the baseline by pressing the TRACE RESET button on the coupler. If the recording is reversed, move the POLARITY switch to negative.

21. With either the 4A or 4B models, set the paper speed to 2.5 cm/sec. Set the time marker at 1 second.
22. Record the EKG and respirations for one complete piece of recording paper at paper speed 2.5 cm/sec. The record should be similar to Figure 12-29.

Self-Test—Circulatory System

Directions: See Chapter I, Self-Test, p. 9.

1. The right and left common iliac veins unite to form the **superior vena cava.**
2. The pulse is usually taken at the **radial** artery.
3. The blood vessel that carries blood from the fetus to the placenta is the **umbilical artery.**
4. Two fetal structures which allow blood to bypass the lungs are the **foramen ovale** and the **ductus arteriosus.**
5. The **aorta** has the highest blood pressure of any blood vessel.
6. The average pulse rate before birth is **160.**
7. The vagus nerve **accelerates** the heart.
8. Between heart beats the elasticity of the arterial walls **maintains** blood pressure.
9. Blood pressure is determined by **peripheral resistance** and **cardiac output.**
10. After a hemorrhage the pulse will be **slower.**
11. The vasomotor center, located in the **cerebellum,** regulates blood pressure.
12. Low blood pressure is called **hypotension.**

13. A hemocytometer is used for blood counts while a **sphygmomanometer** is used to measure blood pressure.

14. Lymph from the thoracic duct drains into the **left subclavian vein.**

15. An artery is a vessel which leads **from** the heart while veins lead **toward** it.

16. Small arteries are called **arterioles.**

17. The **endocardium** is a sac surrounding the heart.

18. The left atrium receives **deoxygenated** blood.

19. Contraction of the heart is called **systole,** and relaxation is called **diastole.**

20. The **atrioventricular** node is the pacemaker of the heart.

21. Constriction of a valve of the heart, making it difficult for blood to get through, is called **cardiac thrombosis.**

22. **Heart block** results from a defect in the conducting system.

23. The internal carotid artery, the vertebral artery, and the Circle of Willis are involved in supplying blood to **the brain.**

24. The **internal jugular** veins drain the brain.

25. The common iliac veins drain the **legs and pelvis.**

26. The brachial artery supplies the **leg.**

27. The **great saphenous** vein drains the legs.

28. The **celiac** artery supplies the stomach, liver, spleen, and pancreas.

29. The hepatic **vein** carries oxygen to the liver.

30. The portal vein carries the **end products of digestion** to the sinusoids of the liver.

31. The principal function of an erythrocyte is to **carry oxygen,** while that of a leukocyte is to **fight infection.**

32. The average number of RBC's per cubic millimeter is from 4½-5 million for **males.**

33. A symptom of leukemia is **a decrease** in leukocytes.

34. Red bone marrow is the source of **erythrocytes, leukocytes,** and **thrombocytes.**

35. **Neutrophils** are the most abundant type of leukocyte.

36. **Vitamin K** is necessary for blood clotting because it catalyzes synthesis of **prothrombin** in the liver.

37. Calcium is **essential** in the conversion of prothrombin to thrombin.

38. **Fibrin** is the solid clot and is derived from **fibrinogen.**

39. Clotting is **slowed** when heat is applied to the wound.

40. **Coagulants** include heparin and dicumarol.

41. Type A blood may be given safely to a patient with **either A or AB** blood.

42. **Antigens** are located in the plasma membrane of RBC's and **antibodies** are in the plasma.

43. Veins contain valves—arteries **do not.**

44. Arteries **always** carry oxygenated blood.

45. The Rh factor is present in **most** Caucasians in the United States.

46. Women are Rh negative **more often** than men.

47. Hemophilia is a sex-linked hereditary disease occuring **primarily in males.**

48. **Hypertension** is the term designating hardening of the arteries.

49. **Erythrocytes** are phagocytic.

50. When an individual lacks sufficient oxygen to burn fuel and remove wastes he develops an "**oxygen debt.**"

51. Blood is pumped to the lungs by the **left ventricle.**

52. **Endocarditis** is an inflammation of the lining and valves of the heart.

53. Congenital heart disease **appears during old age.**

54. Swelling, or edema, of the feet and legs is a common sympton of **heart failure.**

55. The blood protein which contains antibodies opposing many pathogens is **globulin.**

56. The **mitral (bicuspid) valve** is located between the right atrium and the right ventricle.

57. Rheumatic fever frequently affects the **heart.**

58. The products of the exchange that take place between the blood and tissues must **pass through the tissue fluid** that bathes the cell.

59. Receptors sensitive to changes in blood pressure are located in the **right atrium, carotid sinus,** and the **aortic arch.**

60. A slow, strong heart rate would result from nerve impulses arising mainly from the **thoracolumbar (sympathetic)** portion of the nervous system.

61. Blood is **acid** in reaction.

62. Formed elements account for approximately **60%** of the volume of blood.

63. Cholesterol **is** a normal physiological constituent of the blood.

64. White cell counts of **5,000–10,000** per cubic millimeter of blood are in the normal range.

65. The **hemocytoblast** is the common stem cell from which both erythrocytes and leukocytes are developed.

66. The usual effect of cold is to **constrict** superficial blood vessels and so help control hemorrhage.

67. The erythrocyte, in the mature stage, is **nucleated.**

68. Hemopoiesis **in the embryo** occurs in the liver, spleen, red bone marrow, and lymph nodes.

69. The **inorganic salts** of the plasma aid in CO_2 transport.

70. A solution which has the same osmotic pressure as the blood is an **isosmotic** solution.

71. Antibodies against the Rh factor are present in the blood of Rh negative individuals **who have been exposed to Rh positive blood.**

72. Blood serum is blood plasma minus **fibrinogen.**

73. Type AB blood may be called the universal **donor.**

74. An important function of the spleen is to produce **erythrocytes** in the adult.

75. The heart muscle is called the **endocardium.**

76. Cardiac muscle has **striated** myofibrils.

77. The heart and portions of the great vessels fill the **pericardial cavity.**

78. The membrane surrounding the heart is a **serous** membrane.

79. The blood in the pulmonary **artery** is fully oxygenated.

80. The heart beat is initiated by the **Bundle of His.**

81. **Systole** of the atria lasts longer than **diastole.**

82. Blood entering the left atrium of the heart comes directly from the **inferior and superior venae cavae.**

83. There is a slight delay in transmission of the impulse at the **atrioventricular** node.

84. Venous sinuses are found in the **skull** and **heart.**

85. The inner layer of tissue in the arterial wall is called the **endocardium.**

86. An **anastomosis** is a passageway or connection between two vessels.

87. The aorta arches to the **right.**

88. The coronary arteries are the **only** branches from the ascending aorta.

89. The **right and left common carotid arteries** divide into the internal and external carotid arteries.

90. There are **more** veins than arteries in the body.

91. The myofibrils in the walls of arteries are **striated.**

92. There are **two** umbilical arteries and **one** umbilical vein which provide for blood circulation between the placenta and the embryo.

93. A unique feature of the portal system is that blood from the digestive tract is detoured through the **kidney** instead of being returned directly to the inferior vena cava.

94. The umbilical vein carries **oxygenated** blood.

95. The prominent superficial vein in the neck is the **external carotid.**

96. The carotid sinus is located at the bifurcation of the **common carotid artery.**

97. The pulse pressure is the **sum of** the systolic and diastolic blood pressures.

98. The azygos vein drains blood from the intercostal veins into the **inferior vena cava.**

99. Systolic blood pressure is **higher** than diastolic blood pressure.

100. Hemorrhage from an artery results in a **steady** flow of blood.

101. Blood flows least rapidly through **veins.**

102. Blood pressure in **veins** normally exceeds that in capillaries

103. Blood pressure is taken at the **radial** artery.

104. The vessels which contribute the greatest resistance to blood flow are the **venules.**

105. Syncope or fainting results from **an increase** in blood pressure.

106. The formation of a clot in a branch of the coronary arteries is called a **cerebral vascular accident.**

107. Leukopenia is a **decrease** in the number of white blood corpuscles.

108. A reduction in erythrocytes or in the quantity of hemoglobin they contain is called **polycythemia.**

109. Dilated veins of the rectum produce a condition known as **hemorrhoids.**

110. Studies show that vigorous exercise **enlarges** the heart.

111. Vitamin **C** is needed for blood to clot.

112. The basilic vein drains the **brain.**

113. The pigment responsible for the color of red blood corpuscles is **myoglobin.**

114. **Leukocytes** are necessary for blood to clot.

115. Fluid which is located between the cells is called **intracellular** fluid.

116. **Leukocytosis** refers to the destruction of bacteria by leukocytes.

117. In making a complete circuit through the human body, blood passes through the heart **once.**

118. Hemoglobin carries **more** oxygen than carbon dioxide.

119. Leukocytes pass into the tissues through the walls of the **venules.**

120. In the blood clotting process, **thromboplastin** is formed from prothrombin.

121. During the beating of the heart, sounds are produced by the closing of the **semilunar valves.**

122. Muscle cells are present in the layer of an artery wall known as the tunica **intima.**

123. Capillaries may be present in the walls of the **larger arteries.**

124. Blood pressure increases when the ventricles are in **diastole.**

125. The spleen is concerned with the destruction of **lymphocytes.**

126. Vasodilation is controlled by a center in the **medulla oblongata.**

127. Approximately **10%** of the body weight is blood.

128. **Proteins are** the most abundant compound in plasma.

129. The condition erythroblastosis fetalis occurs when the mother is **Rh+** and the baby is **Rh–**.

130. Hemoglobin contains the mineral **copper.**

131. The coronary sinus opens into the **right atrium.**

132. The nutrient artery of the lungs is the **bronchial artery.**

133. The artery which brings oxygen to the liver is the **portal artery.**

134. The vagus nerve carries impulses that **stimulate** the heart.

135. Apoplexy is a term used for a **heart attack.**

136. The normal thrombocyte count per cubic millimeter is **25,000 – 50,000.**

137. A thrombus is a **moving** blood clot.

138. Localized death of cardiac tissue due to obstruction of a coronary artery is known as a **myocardial infarction.**

139. Elephantiasis is a condition caused by **an excessive production of lymph.**

140. Heparin and dicumarol may be given to the patient to **prolong** the clotting time.

141. **Bradycardia** is an abnormally slow heart beat.

142. In hypothermia, the heart rate would **increase.**

143. Essential hypertension is that which results from **a known cause.**

144. The thoracic duct receives **blood** from the digestive tract.

145. Lymph flow is aided by **valves** in lymphatic vessels.

146. Lymph glands produce large numbers of **lymphocytes.**

KEY

1. inferior vena cava
7. slows down
10. faster
11. medulla oblongata
17. pericardium
18. oxygenated
20. sino-atrial
21. stenosis
26. arm
29. artery
33. an increase
39. speeded up
40. anticoagulants
44. do not always
46. no more often
48. arteriosclerosis
49. leukocytes
51. right ventricle
53. appears at birth
56. tricuspid
60. cranio-sacral (parasympathetic)
61. alkaline (basic)
62. 45%
67. non-nucleated (anucleate)
73. recipient
74. leukocytes (lymphocytes)
75. myocardium
79. vein
80. sino-atrial node
81. diastole, systole
82. pulmonary veins
85. endothelium
87. left
91. nonstriated
93. liver
95. external jugular
97. difference between
98. superior vena cava
100. pulsating
101. capillaries
102. arteries
103. brachial
104. arterioles
105. a decrease
106. coronary thrombosis
108. anemia
111. K
112. arm
113. hemoglobin
114. thrombocytes
115. intercellular (tissue)
116. phagocytosis
117. twice
119. capillaries
120. thrombin
122. media
124. systole
125. erythrocytes
127. 7%
128. water is
129. Rh−, Rh+
130. iron
133. hepatic
134. depress
135. stroke (cerebral vascular accident)
136. 250,000 - 500,000
137. stationary
139. blockage of the lymphatic vessels
142. decrease
143. an unknown cause
144. lymph

Name ______________________

Results and Questions for Chapter 12

Exercise 12-A. Blood Tests

Part 1. Determination of the ABO and Rh Blood Types

1. Record the results of the entire class in the following tables.

Blood Group	*Number of Individuals*	*Percent*	*Theoretical Percent*
O			45
A			41
B			10
AB			4
Total			100

Blood Group	*Number of Individuals*	*Percent*	*Theoretical Percent*
Rh+			85
Rh-			15
Total			100

2. Explain any differences in the percentages obtained in your class and the theoretical percentages.

____

3. What was your ABO blood type?______Rh blood type?______________

4. Which agglutinins are contained in your blood plasma?______________

5. Which agglutinogens are contained in your blood cells?______________

6. List two possible reasons why the antisera should be placed on the slide before the blood.

   A. ______________________

   B. ______________________

7. The Anti-D antisera must contain the Rh antibodies. Since these do not occur normally, can you suggest a source of the antibodies?______________

   ____

8. The a and b antibodies, if present, would interfere with the determination of Rh blood type. How are they removed from the antiserum?______________

   ____

Part 2. Determination of the Hemoglobin Content of Blood

1. Results of determination of hemoglobin:

   a. Tallquist scale:

      (1) grams of hemoglobin ______

      (2) percent hemoglobin ______

      (3) range (suggestive anemia, actual anemia, or normal) ______

   b. Spencer Hb-meter:

      (1) grams of hemoglobin ______

      (2) percent hemoglobin ______

2. What value in grams is taken as 100% on both the Tallquist scale and the Spencer hemoglobinometer? ______

3. Why is it important not to use the first drop of blood? ______

4. Why is it important not to squeeze the finger? ______

5. Discuss sources of error in the use of the Tallquist scale. ______

6. Which method of determining hemoglobin used in this experiment is more accurate? ______

7. What is the normal hemoglobin value (in grams/100 ml blood):
   For men? ______ For women? ______

8. Define anemia: ______

9. What was the class average hemoglobin value: For men? ______ For women? ______

10. What causes this characteristic difference in hemoglobin value between the sexes? ______

11. What is the function of hemoglobin? ______

12. Where is hemoglobin produced in the human body? ______

Name ____________________

Part 3. Coagulation Time and Bleeding Time

1. Record your results in the following data chart.

Test	*Time*
coagulation time – slide	
coagulation time – capillary tube	
bleeding time	

2. Were your results within the normal range? ____________________

3. Why was coagulation time longer than bleeding time? ____________________

____

4. Name two diseases or conditions in which the coagulation time is lengthened. ____________________

____

5. List four functions performed by blood plasma. ____________________

____

____

6. In the following table, indicate where each of the following substances is found and the function each performs in the coagulation of blood.

Substance	*Where Found in Blood*	*Function*
Fibrinogen		
Prothrombin		
Thrombokinase		
Calcium		

7. How can the blood clotting time be shortened? ____________________

____

8. What is the normal range for a RBC count? ____________________

9. What is the normal range for a WBC count? ____________________

10. What is the life-span of a typical RBC? ____________________

11. Where are RBC's destroyed? ____________________

12. Where are RBC's produced? ______

13. What do the following changes in a blood count indicate:

a. A white blood cell count above $15,000/mm^3$ ______

______

b. A red blood cell count below $4000/mm^3$ ______

c. A decrease in a normal white blood cell count ______

______

d. An increase in a normal red blood cell count ______

______

Exercise 12-B. Structure of an Artery, Vein, and Lymph Vessel Valve

1. What is the function of each of the following in the wall of an artery?

a. Endothelium ______

b. Elastic connective tissue ______

c. Vasa vasorum ______

2. Which type of vessels contain valves? ______

3. Explain the cause of the difference in shape of the interior of the artery and the vein. ______

______

______

4. Which contains proportionally more muscle, a large or small artery? ______

5. Briefly describe, in the table below, the distinguishing characteristics of the following vessels:

Vessel	*Structure*	*Function*
Artery		
Capillary		
Vein		

Exercise 12-C. The Gross Anatomy of the Human Circulatory System

1. List the major organ(s) supplied by each of the following arteries:

a. Brachial ______

Name ______________________

b. Internal mammary ______________________

c. Celiac ______________________

d. Inferior mesenteric ______________________

e. Lumbar ______________________

f. Internal iliac ______________________

g. Anterior tibial ______________________

h. Interior phrenic ______________________

i. Renal ______________________

j. Hepatic ______________________

2. Trace the path of a molecule of glucose from the small intestine to the right arm and name all blood vessels and chambers of the heart passed through. ______________________

____

____

____

3. Trace the path of a drop of blood from the left fingers to the dorsum of the foot, following the same directions as for No. 2. ______________________

____

____

4. Describe the functions of blood as they pertain to each of the following organ systems:

System	*Functional Relationship of Blood*
Respiratory	
Digestive	
Excretory	
Endocrine	

5. How do the following organ systems aid the circulatory system? ______________________

a. Skeletal system ______________________

b. Muscular system ______________________

c. Nervous system ______________________

6. Blood that flows away from the heart is always carried by blood vessels called ________________.
Is this blood always oxygenated? ________________ Explain: ________________

____

7. Blood that flows toward the heart is carried by blood vessels called ________. Is this blood always oxygenated? ____________ Explain: ________________

____

8. What functions do the capillaries perform in the circulation of blood? ________________

____

9. Explain why the heart is considered a double pump. ________________

____

10. What is the function of the portal system? ________________

____

Exercise 12-D. The Dissection of the Sheep Heart

1. Describe the location of the parietal layer of the pericardium. ________________

____

2. Describe the location of the visceral layer of the pericardium. ________________

____

3. Name the major tissue in the myocardium. ________________

4. What is the function of the myocardium? ________________

5. What is the function of the pulmonary artery? ________________

6. What is the function of the blood vessels in the anterior longitudinal sulcus? ________________

____

7. Which ventricle has thicker walls? ________________

8. Which chamber makes up the apex of the heart? ________________

9. How many pouches are present in the pulmonary semilunar valve? ________________

10. Are chordae tendineae attached to these semilunar valve pouches? ________________

11. Does the semilunar valve open during systole or diastole of the right ventricle? ________________

____

____

12. Is there a valve visible at the entrance of the superior vena cava into the heart? ____________

____

13. What is the function of the coronary sinus? ____________

14. What is the function of the chordae tendineae? ____________

15. What causes the tricuspid valve to open? ____________

16. What causes the tricuspid valve to close? ____________

17. What is the function of the pulmonary valves? ____________

18. Are chordae tendineae attached to the bicuspid valve? ____________

19. Is a moderator band visible in the interior of the left ventricle? ____________

20. How many pouches are visible in the aortic semilunar valve? ____________

21. Name the three blood vessels branching from the aortic arch in humans. ____________

____

22. What is the internal layer of the artery wall called? ____________

23. In No. 22 name the tissue forming the lining of this layer. ____________

24. Name the tissue found in largest amounts in the tunica media of the aorta. ____________

25. Based on your observations of the **sheep heart**, what is the easiest way to distinguish between an artery and a vein?

____

26. What is the function of the ductus arteriosus? ____________

____

27. What is the function of the foramen ovale? ____________

____

____

28. What is the name of the structure containing cartilage in the walls dorsal to the aorta? ____________

____

29. Trace a drop of blood through the heart from the superior vena cava to the aorta. ____________

____

____

Exercise 12-E. The Dissection of the Circulatory System of the Fetal Pig

1. To what structures was the pericardium attached? ______

______

2. Name the branches of the celiac artery. ______

3. Name the remnant of the ductus arteriosus after birth. ______

4. Compare the branches of the aortic arch in man and in the fetal pig or cat. ______

______

5. Is there a common iliac artery in the pig? ______

6. Describe the base of the aorta in the fetal pig. ______

______

7. What is the special importance of the hemiazygos or azygos vein in the human? ______

______

8. Which type of blood (oxygenated or deoxygenated) is located in each of the following in the adult pig?

a. Pulmonary artery ______ d. Renal artery ______

b. Pulmonary vein ______ e. Hepatic vein ______

c. Aorta ______

9. Which type of blood (oxygenated, deoxygenated, or mixed) is located in each of the following in the fetal pig? (Study the coloring carefully in the model of the human fetal circulation.)

a. Aorta ______ f. Pulmonary artery ______

b. Ductus arteriosus ______ g. Superior vena cava ______

c. Ductus venosus ______ h. Umbilical artery ______

d. Inferior vena cava ______ i. Umbilical vein ______

e. Internal iliac artery ______

10. Trace the path of a drop of blood from the ankle of the human fetus to the right side of the fetal brain; name all blood vessels and chambers of the heart passed through. ______

______

______

______

11. Trace the path of a drop of blood from the fetal brachial artery to the stomach. ______________

12. List several causes of a "blue baby." ______________

13. From which blood vessels do the umbilical arteries arise? ______________

14. What is the function of the umbilical arteries? ______________

15. What is the function of the ductus venosus? ______________

Exercise 12-F. Physiology of the Frog Heart

Part 1. Frog Anatomy

1. What is the cloaca? ______________
2. What sex was your frog? ______________
3. In Step 15b, what happened when you pinched the fundus of the stomach? ______________ What is this called? ______________
4. What type of muscle is located in the wall of the stomach? ______________
5. What is the special term for the fold of peritoneum that connects the intestine to the dorsal body wall? ______________
6. Compare the trachea of the frog with that of the human. ______________
7. Trace the path of sperm in the male frog. ______________
8. What is the function of the ovaries? ______________
9. The bile duct carries bile to which organ? ______________

Part 2. Observation of the Frog Heart Cycle

1. Rate of contraction of the parts of the frog heart:

   a. Rate of atria ______________

   b. Rate of ventricle ______________

2. Should the rate of the parts of the frog heart agree?______Explain:____________________

____

3. List the parts of the frog heart in the order of their contraction.____________________

____

4. List the functions of the pericardium observed in this experiment.____________________

____

5. What is meant by systole?____

6. What is meant by diastole?____

7. Compare the structure of the frog heart with that of the human heart.____________________

____

Part 3. Extrasystole

1. Paste the kymograph or physiograph record showing extrasystole, completely labeled, in the space below. Label absolute refractory period, extrasystole, compensatory pause, systole, and diastole. Indicate on the record where the heart was stimulated during systole and where it was stimulated during diastole.

2. What is meant by the absolute refractory period of the heart?____________________

____

3. When does the absolute refractory period occur in the cardiac cycle?____________________

____

4. What results were obtained when the heart was stimulated during systole?____________________
____________________Explain these results:____________________

____

5. What results were obtained when the heart was stimulated during diastole?____________________
____________________Explain these results:____________________

____

____

Name ______________________

6. During which phase of the cardiac cycle (early or late diastole) was the greatest height of extrasystole obtained?________Explain:____

____

7. Define the compensatory pause.____

8. Can the heart be stimulated to produce a tetanic contraction?______________Explain:________ ____Of what possible

advantage to circulation is this fact? ____

Part 4. The Effect of Temperature on the Heart Beat

1. Record the rate of the heart beat at each temperature in the data chart. If the kymograph or physiograph was used, include your record. Count the heart rate for at least 15 seconds at each temperature. Calculate the Q_{10} as directed in the Discussion.

Temperature	*Rate/minute*	Q_{10}	*Force of Beat*

2. Why does an increase in temperature, within limits, cause an increase in the rate of physiological processes?____

____

____

3. Which would bring about a larger decrease in the rate of ventricle contraction: applying a cold probe to the sinus venosus or to the ventricle directly?________Explain your answer:__________

____

____

4. If you recorded your heart rates with either the physiograph or kymograph, determine at which temperature the beat was most forceful. (The force may be determined by measuring the height

of the contractions in millimeters on the record. Draw a base line connecting the bottom of each heart beat. Measure from there to the top of a typical beat at each temperature.) Record your values on the data chart.

5. At which temperature was the beat most forceful?____________________

Part 5. Heart Block

1. Record your results from the experiment on heart block in the following table:

	Atria	*Ventricle*
Rate before experiment		
Rate during gradual tightening		
Rate during ventricular arrest		
Rate after thread cut		

2. How did the pressure cause the production of heart block in this experiment?__________

3. What causes the production of heart block in humans?__________

4. What are the symptoms of heart block in humans?__________

5. As the thread was tightened, was there any change in the rate of beat of the atria?__________

   Should there have been any change?______Explain:__________

6. As the thread was tightened, was there any change in the rate of beat of the ventricles?______ Explain:__________

7. Did the beating of the ventricle (after arrest) have any relationship to that of the atria? ______Explain:__________

8. Trace the path of a message over the conducting system in the human heart.__________

Name ____________________

Part 6. Automaticity of the Heart

1. Record your results in the following table:

	Parts of the Heart	*Heart Rate/Minute*
	Entire heart before removal	
	Entire heart after removal	
Step 7	Separated sinus venosus	
Step 7	Combined atria-ventricle	
Step 8	Separated atria	
Step 8	Separated ventricle	
	Apex	

2. Was the rate of beat of the entire heart the same before and after removal from the body?____Explain any difference observed:____________________

____

3. Define automaticity:____________________

____

4. How was automaticity demonstrated in this experiment? ____________________

____

5. In this experiment which parts of the heart were demonstrated to possess automaticity?____

____

6. Arrange the parts of the heart in order of decreasing irritability and automaticity.____

____

7. Name the pacemaker of the human heart.____________________

8. Name the pacemaker of the frog heart.____________________

9. Is the heart dependent on nerves for contraction?______What is the function of the nervous supply to the heart?____________________

____

10. If the Bundle of His were destroyed, would you expect that the ventricles of the human heart would beat faster or slower than the atria?______ Explain:____________________

____

Exercise 12-G. Microcirculation in the Fish or Frog

1. Fill in the following table, comparing circulation in the three types of blood vessels based on your observations:

Circulation Indices	*Arteries*	*Veins*	*Capillaries*
Rate of movement			
Pulsations			
Diameter			

2. Of what advantage to the body is the slow rate of blood flow through the capillaries?________

3. What determines the rate of flow of blood in the different types of blood vessels?________

Exercise 12-H. Heart Sounds and Pulse Rate

1. Describe the heart sounds with the heart at rest.________
2. What is the cause of the first heart sound?________
3. What caused the second heart sound?________
4. Which sound may be called the systolic sound?________
5. Describe the change in the sounds after exercise.________
6. What causes the increase in force of the sounds?________
7. Define apical pulse.________
8. Define radial pulse.________
9. What is pulse deficit?________
   What causes this condition?________
10. Define pulse.________
11. What is the normal pulse rate of women?________ Of men? ________

Name ______________________

12. Record your heart rate in the following table. (Include your labeled record if the physiograph was used.)

Body Position and State	*Heart Rate*
Sitting	
Standing	
During exercise	
Immediately after vigorous exercise	
Two minutes following exercise	

13. What was the average pulse rate for women in the class (use the figures obtained while sitting at rest)? ______________________

14. What was the average pulse rate for men in the class? ______________________

15. What is the effect of physical training on pulse rate? ______________________

____

16. What was the effect of body position on the heart rate? ______________________

____

17. What was the effect of exercise on heart rate? ______________________

Exercise 12-I. Blood Pressure

1. If you did not use the physiograph record all measurements of blood pressure, heart rate, and respiration rate in the following table.

Time	*Systolic Pressure*	*Diastolic Pressure*	*Pulse Pressure*	*Pulse Rate*	*Resp. Rate*
Before exercise					
After exercise Immediately after					
Two minutes after					
Four minutes after					
Six minutes after					
____ minutes after					

2. If you used the physiograph, record your figures in the following table. Include your labeled record with this table.

Time	*Systolic Pressure*	*Diastolic Pressure*	*Pulse Pressure*	*Pulse Rate*	*Respiration Rate*
Before Exercise Minute 1					
Minute 2					
During Exercise Minute 1					
Minute 2					
After Exercise Minute 1					
Minute 2					
Minute 3					
Minute 4					
Minute 5					
Minute 6					

3. The sounds heard through the stethoscope in this experiment were not heart sounds. What caused the production of the sounds?__________

____

4. What are the usually accepted limits of normal blood pressure for your age and sex? __________

If your values were outside these limits, suggest possible reasons.__________

____

5. The following commonly occur during and immediately after exercise. Discuss the causes of each.

a. Increase in systolic pressure__________

____

b. Increase in heart rate__________

____

c. Increase in respiration rate__________

____

d. Decrease in diastolic pressure__________

____

Name ______________________

6. How rapidly did the circulatory and respiratory systems return to normal after exercise?______

____

7. How would the trained athlete differ from the untrained with respect to the time of return to normal?____

____

8. Define each of the following:

   a. Hypotension____

   b. Hypertension____

   c. Orthostatic hypotension___________________________________

   d. Basal pressure____

Exercise 12-J. Electrocardiography with the Physiograph

1. Fasten the record obtained in this experiment below. Label the EKG and the respiratory record. Label P, QRS, and T waves on the EKG.

2. Calculate the duration of one heart beat by measuring the distance between the beginning of two P waves in centimeters. Label on the record which beat was used. Divide the distance by the paper speed (2.5 cm/sec) to obtain the duration in seconds of one beat.

length of 1 beat in cm. ____________________

Duration of 1 beat ____________________

3. Calculate the heart rate per minute. Divide the duration of one heart beat into 60 to determine the heart rate per minute.

Heart rate per minute ____________________

4. Determine in a similar way the respiratory rate per minute.

length of 1 breath in cm. ____________________

Duration of one breath ____________________

Respiratory rate per minute ____________________

5. Determine the heart rate for a period of 10 seconds. Since the paper speed is 2.5 cm/sec, 25 cm of paper represents 10 seconds (see Fig. 12-29). Mark off 25 cm on the record (10 sections between dark green lines). Count the number of QRS waves in this 25 cm. Multiply your answer by 6 to obtain the heart rate per minute.

No. beats in 10 sec. ________ Heart rate per minute ________

6. Do the answers obtained in No. 3 and No. 5 agree? ________

What do you think is the most likely explanation for any difference in answers? ____________________

____

7. What causes the occurrence of each of the following?

a. P wave ____

b. QRS complex ____

c. T wave ____

8. Examine carefully your recording of heart action. Is the rate uniform throughout the record? ________ Can you correlate any variation in the rate with inspiration or expiration? ________

9. Compare the duration of one heart beat with the theoretical rate for humans. ____________________

____

10. Calculate the P-R interval. This interval starts with the beginning of one P wave and ends with the beginning of the Q wave if one is visible (use the R wave if there is no Q wave visible). Measure the distance in centimeters as accurately as possible. Divide by the paper speed. (Show which section of the record you used.)

Distance between P and Q ________ P-R interval ________

11. Compare the P-R interval obtained in this experiment with the theoretical results. ________

____

12. What is the cause of the elapsed period of time between the P wave and the QRS complex? ________

____

____

13. In which disease condition may the P-R interval be lengthened? ________

____

13

The Nervous System

Exercise 13-A Anatomy of the Nervous System

Objectives

To become familiar with the gross and microscopic anatomy of the nervous system.

Materials

Prepared slide of cross sections of the spinal cord and of a nerve; models of the brain; charts; and reference books.

Discussion

For convenience of study, the nervous system may be divided into a central (CNS) portion, a peripheral portion, and the autonomic (ANS) portion. The latter is usually subdivided into the sympathetic and the parasympathetic divisions. The central nervous system is composed of the brain and the spinal cord; the peripheral nervous system of the spinal and cranial nerves. The sympathetic division of the autonomic nervous system consists of two parallel lateral chains of ganglia which lie on each side of the anterior surface of the spinal column, and collateral ganglia which lie along the front of the abdominal aorta and the celiac artery. The parasympathetic division of the ANS consists of ganglia which lie in or near visceral effectors. Both divisions of the ANS include fibers which conduct impulses to the ganglia (preganglionic fibers), and others which conduct impulses away from the ganglia (postganglionic fibers). The ganglia of these divisions contain the cell bodies of the postganglionic neurons.

The nervous system serves to coordinate body activities by receiving, integrating, and transmitting messages. The structural units of the nervous system are neurons. (The structure of the neuron and medullated nerve fiber has been described in Exercise 3-F.)

Procedure

Part 1. Structure of a Nerve

1. Obtain a slide of a cross section through a nerve. Examine the slide with the low power objective.
2. Observe the connective tissue sheath called the epineurium surrounding the entire nerve. This sheath contains connective

tissue cells, collagenic fibers, adipose tissue, and nutrient blood vessels called the vasa nervorum.

3. Note that the nerve fibers comprising the nerve are arranged in bundles; each bundle (funiculus) is surrounded by connective tissue called the perineurium.
4. Locate the endoneurium, the connective tissue that surrounds each individual nerve fiber.
5. Make a diagram of the entire cross section of the nerve showing the preceding structures. Label the epineurium, perineurium, endoneurium, nerve fibers, vasa nervorum, and adipose tissue. It is not necessary to draw any cells in this diagram.
6. Now, examine the slide with the high power objective. Locate a nerve fiber that contains a nucleus in the neurilemma. Draw this fiber below the diagram of the nerve. Include on this drawing the axis cylinder, located in the center of the fiber; the thick myelin sheath; the thin neurilemma surrounding the fiber; and the nucleus of the neurilemma. (Compare with Figure 3-21.)

Part 2. The Spinal Cord

1. Obtain a slide of a cross section through the spinal cord. First, study this slide with the naked eye to observe the shape and size of the cord. Since the cord was mounted upside down on the slide (so that it will appear right side up when using the microscope), be certain to hold the slide so that the printing is at the bottom of the slide and upside down while examining it with the naked eye.
2. Diagram the shape of the entire cord and adjacent roots of the spinal nerve on a full sheet of paper. This drawing should include the shape of the central gray matter.
3. Now, study the slide under low power of the microscope. (Return the slide to the usual position for microscope viewing.)
4. Observe the shape of the central canal, the dorsal median septum, and the ventral median fissure. Determine whether the latter two touch the central gray matter. Include these on the diagram. Use your text and/or other references to aid you in this study.
5. Locate nerve cell bodies in the dorsal root ganglion and in the ventral horn of the gray matter.
6. Include the following labels on your diagram: dorsal root, dorsal root ganglion, ventral root, dorsal horn, ventral horn, dorsal, lateral and ventral funiculi, dorsal median septum, ventral median fissure, central canal, white matter, and gray matter.

Part 3. The Gross Structure of the Brain

1. Figures 13-1 and 13-2 of the human brain may be studied as self-tests. See SectionJ, p. xx (Illustrations as Self-Tests) for a suggested approach.
2. Identify the structures indicated on Figures 13-1 and 13-2 on models of the human brain, preserved human brain, sheep brain (see Exercise 13-B) and with similar illustrations on charts and in reference material.

Exercise 13-B The Dissection of the Sheep Brain

Objectives

To study the structure of the sheep brain and to compare it with the human brain.

Materials

Sheep brains (whole and sagittal sections); human brain; and a dissecting set.

Discussion

The structure of the sheep brain is very similar to that of the human brain. Note any differences as you study the brains. The brain is surrounded by three layers of **meninges.** The outermost, the **dura mater,** was removed in preparing the sheep brains. This layer may be seen surrounding the brain of the pig if the surrounding cranial bones are carefully cut away with scissors.

Procedure

1. Obtain a sagittal section of the sheep brain and an entire sheep brain and rinse both with tap water. Use the entire brain for steps 2-15.
2. The inner two layers of the meninges can be seen covering the brain. The **arachnoid,** the middle layer, lies between the **dura mater** and the **pia mater,** the innermost, vascular layer of the meninges. The

KEY TO FIGURE

1. corpus callosum
2. septum pellucidum
3. thalamus
4. mammillary body
5. hypothalamus
6. infundibulum
7. hypophysis (pituitary)
8. pons
9. medulla oblongata
10. fornix
11. splenium
12. pineal body
13. corpora quadrigemina
14. superior colliculus
15. inferior colliculus
16. cerebral aqueduct
17. cerebral peduncles
18. cerebellum
19. fourth ventricle

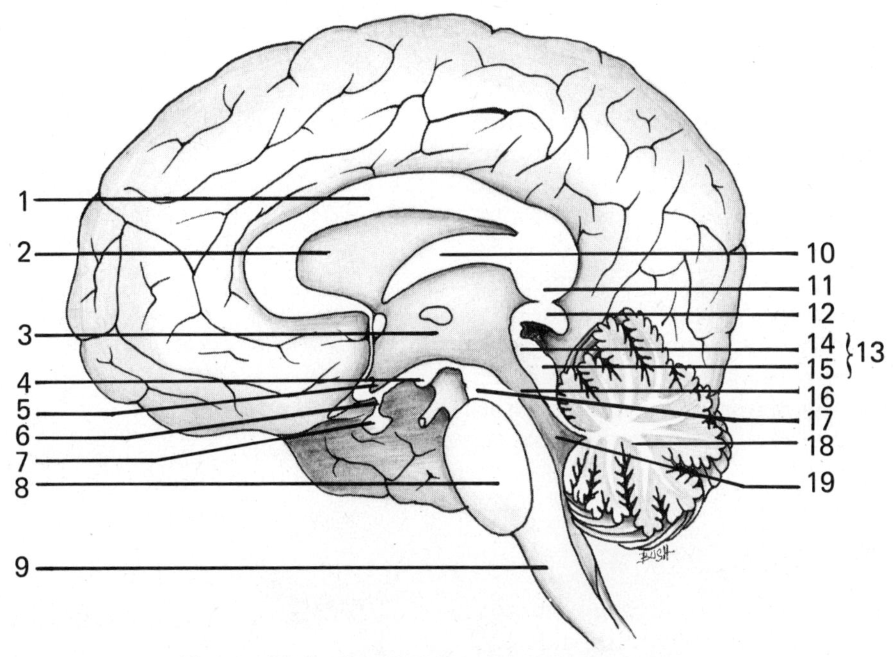

Figure 13-1. Sagittal Section of the Brain

KEY TO FIGURE

1. optic chiasma
2. hypothalamus
3. pons
4. medulla oblongata
5. olfactory bulb
6. olfactory tract
7. optic nerve
8. infundibulum
9. optic tract
10. mammillary body
11. oculomotor nerve
12. motor root
13. sensory root
14. trigeminal nerve
15. abducens nerve
16. vagus nerve
17. cerebellum

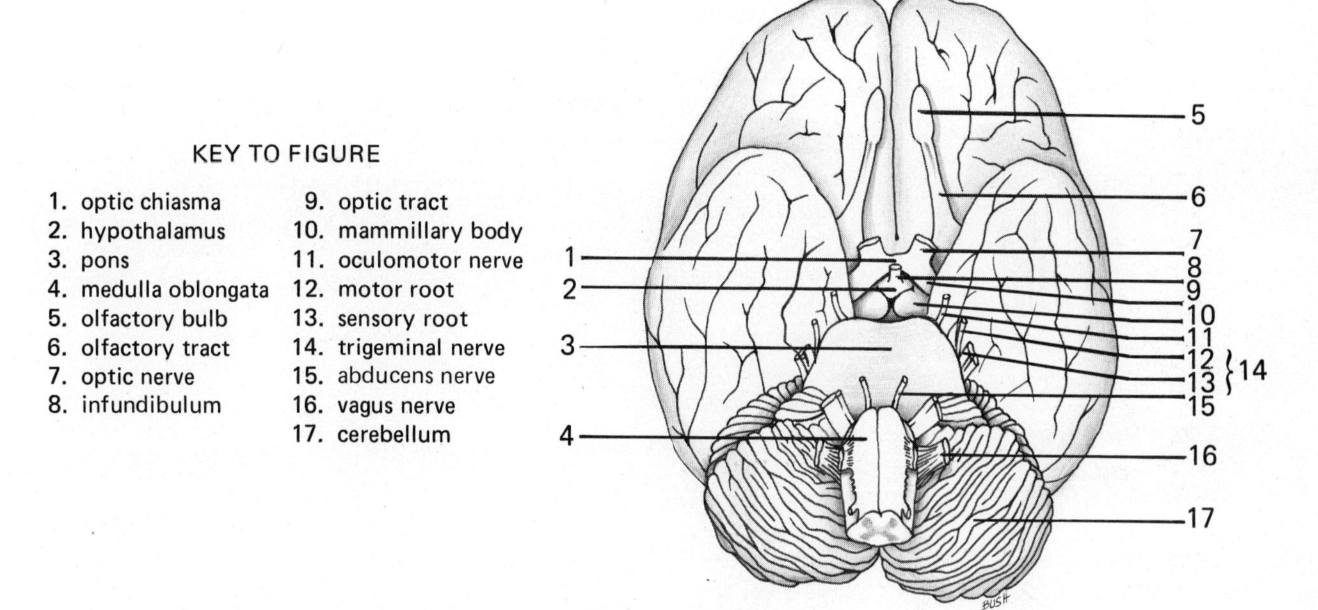

Figure 13-2. Inferior Surface of the Brain

arachnoid is most easily distinguished from the pia mater in the region overlying the grooves on the brain surface, since the pia mater dips into the grooves and the arachnoid does not.

3. Observe the anterior paired **cerebral hemispheres** and the posterior **cerebellum** on the dorsal surface of the brain. The cerebral hemispheres are separated from each other by the **longitudinal fissure**; the cerebellum is separated from the cerebral hemispheres by the **transverse fissure** (see Figure 13-3).
4. Spread the cerebral hemispheres apart and observe deep in the longitudinal fissure the thick transverse band of fibers, the **corpus callosum**, that connects the cerebral hemispheres.
5. The surface of each hemisphere is composed of numerous convolutions. The raised area of each convolution is called the gyrus, the depression a sulcus.
6. The roof of the midbrain (mesencephalon) can be seen by spreading the cerebral hemispheres and cerebellum apart (see Fig. 13-3). Four prominent round swellings, the **corpora quadrigemina**, form the roof of the midbrain. The larger, anterior pair is called the **superior colliculi**, the smaller, posterior pair the **inferior colliculi**. The **pineal body** can be seen between the superior colliculi. The **trochlear nerve** appears as a thin white strand, directed ventrally, slightly posterior to the inferior colliculi.
7. Posterior to the cerebral hemispheres is the **cerebellum**. The cerebellum is connected with the brain stem by three prominent fiber tracts or peduncles. Lift up the lateral edge of the cerebellum. The middle **cerebellar peduncle** can be seen connecting the cerebellum with the pons. Slightly posterior to this is the **inferior cerebellar peduncle** connecting the cerebellum to the medulla. Locate the **superior cerebellar peduncle**, which is composed of the fibers

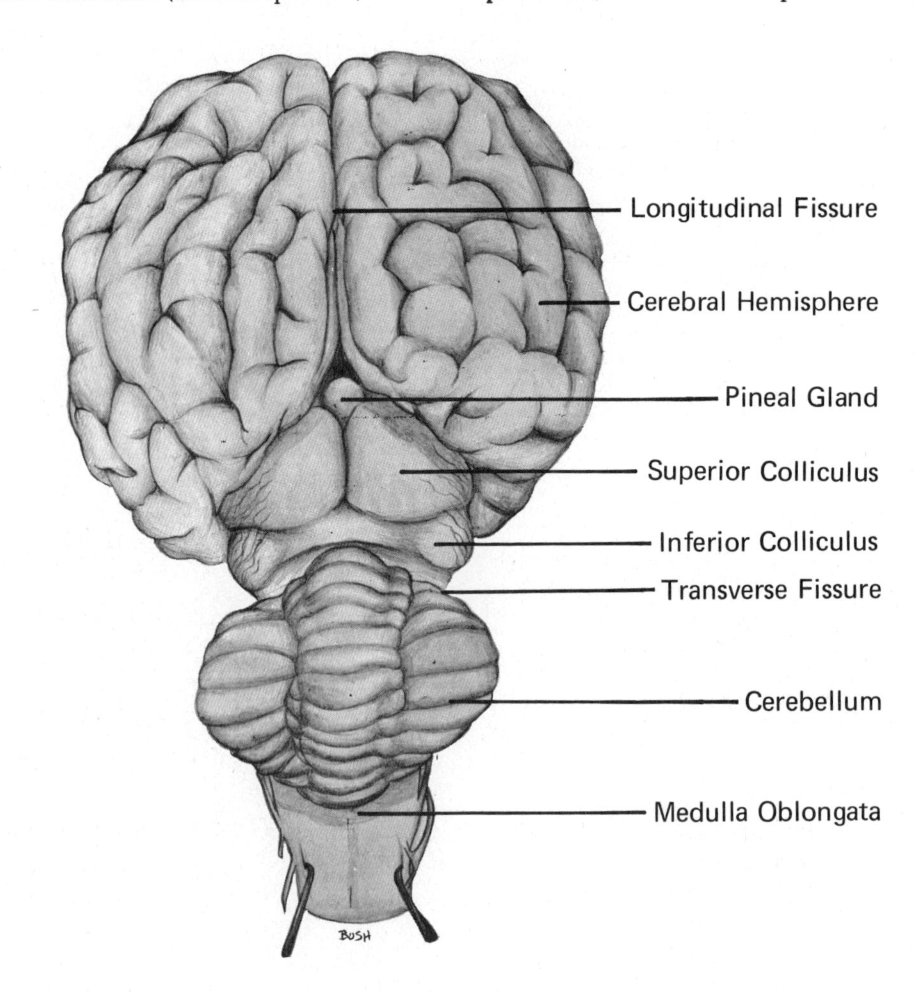

Figure 13-3. Dorsal View of Sheep Brain (with cerebellum separated from cerebral hemispheres)

connecting the midbrain to the cerebellum.

8. Observe the ventral surface of the sheep brain. A pair of **olfactory bulbs** can be seen beneath the cerebral hemispheres. These bulbs lie over the cribriform plate of the ethmoid and receive the olfactory neurons from the nose (see Fig. 13-4).
9. A white band, the **olfactory tract**, extends from each bulb along the ventral surface of the cerebral hemispheres.
10. The ventral surface of the diencephalon, the **hypothalamus**, is posterior to the olfactory tracts. The optic nerves undergo a partial crossing over (decussation) at the anterior border of the hypothalamus, forming the cross known as the optic **chiasma.**
11. The remainder of the hypothalamus is the oval area lying posterior to the optic chiasma, covered by the **pituitary gland** (hypophysis). Do not remove the gland. The **infundibulum** can be seen connecting the pituitary to the hypothalamus.
12. Posterior to the infundibulum is the rounded **mammillary body** (there are two in humans).
13. Observe the **cerebral peduncles** on the ventral surface of the midbrain. The large **oculomotor nerves,** covered by the pituitary gland, arise from the cerebral peduncles posterior to the mammillary body.
14. Posterior to the midbrain is the **pons.** This is composed primarily of white fibers, many of which run transversely across the pons out to the cerebellum.
15. The **medulla oblongata** is posterior to the pons. The longitudinal bands of tissue at each side of the **ventral median fissure** on the ventral surface of the medulla are known as the **pyramids.**
16. **To see the remaining** parts of the brain, use the sagittal section of the sheep brain. (Compare the specimen with Figure 13-5),

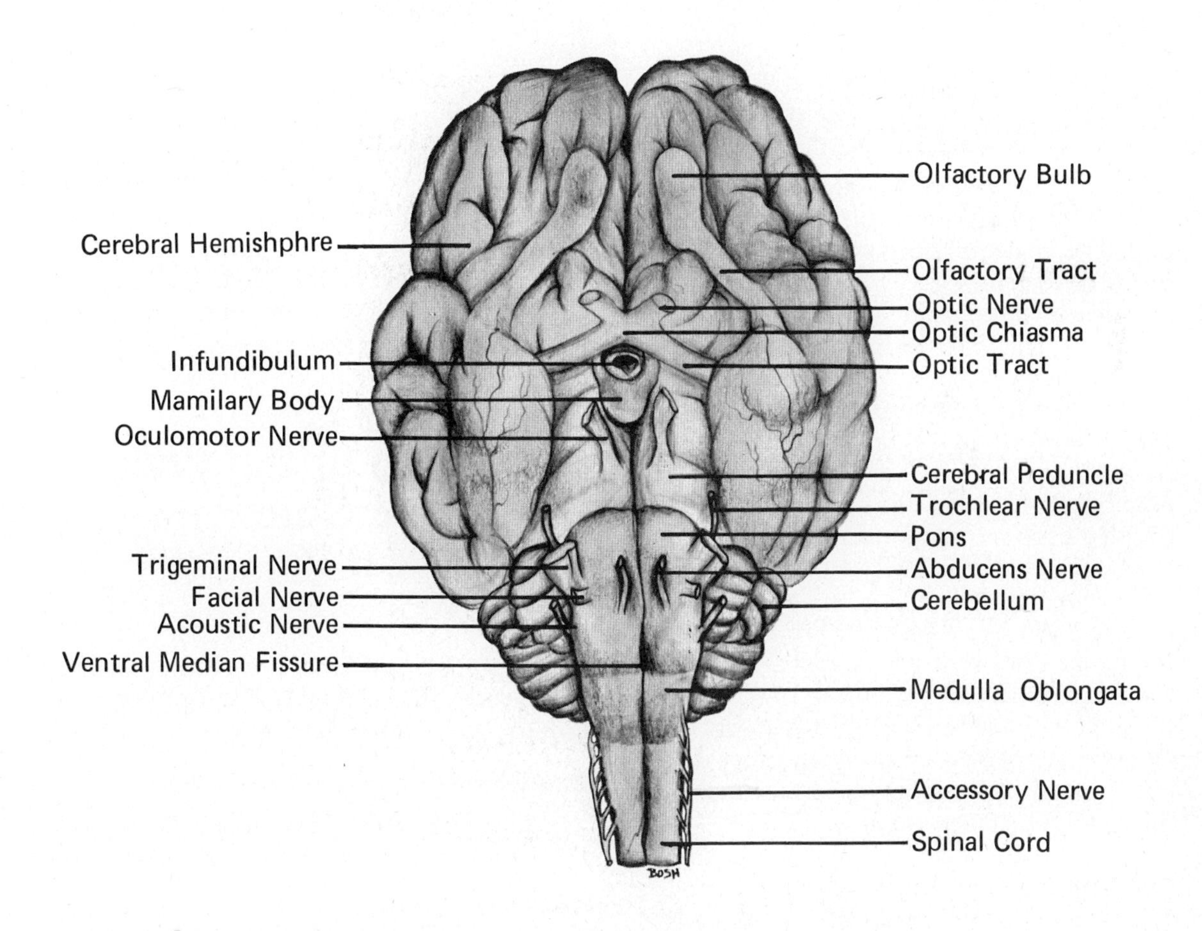

Figure 13-4. Ventral View of Sheep Brain

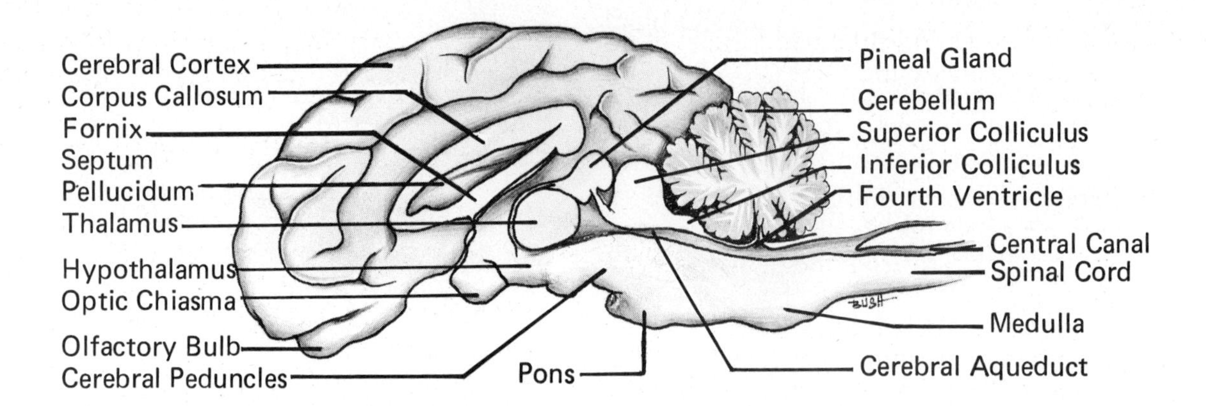

Figure 13-5. Sagittal Section through Sheep Brain

sagittal section of the sheep brain.) Relocate the **corpus callosum,** which consists of the white fibers connecting the two cerebral hemispheres.

17. A thin, vertical septum of tissue, the **septum pellucidum,** lies ventral to the corpus callosum. The **lateral ventricle** lies behind this septum and may be seen by breaking the septum.
18. The **fornix,** a band of white fibers, lies ventral to the septum.
19. The **third ventricle** and the **thalamus** lie ventral to the fornix. The narrow third ventricle, the walls of which are covered by a shiny layer of epithelium, is in the midline. The thalamus forms the lateral walls of the third ventricle.
20. The **massa intermedia** extends across the third ventricle, connecting the two sides of the thalamus. This structure appears as a dull circular area not covered by epithelium.
21. The **foramen of Monro,** the opening through which each lateral ventricle communicates with the third ventricle, lies in the depression anterior to the massa intermedia. Find this connection by passing a dull probe through it.
22. Relocate the **hypothalamus.** This lies ventral to the third ventricle.
23. Note the **pineal body** dorsal to the midbrain, near the **superior colliculus.**
24. Observe the narrow **cerebral aqueduct** leading through the midbrain, connecting the third and fourth ventricles.
25. The **fourth ventricle** lies above the pons and medulla, below the cerebellum.
26. The beginning of the spinal cord may be seen connected to the medulla. A canal known as the **central canal,** which is connected to the fourth ventricle, is present in the center of the cord.
27. Note the treelike arrangement of gray and white matter in the cerebellum. This arrangement is known as the **arbor vitae** (tree of life). The gray matter of the cerebellum is on the outside, the white is toward the center.
28. The outer layer of the cerebral hemispheres, the **cortex,** is also composed of gray matter. Make a shallow incision through the cortex in order to see the white matter located beneath.
29. Examine the human brain and locate as many of the preceding structures as possible. Note the paired mammillary bodies, the larger size of the cerebral hemispheres, and the deeper convolutions.

Exercise 13-C Dissection of the Brain and Spinal Cord of the Fetal Pig

Objectives

To study the structure of the brain and spinal cord of the fetal pig and to compare it with the human brain and spinal cord.

Materials

Fetal pig; dissecting board or pan; and dissecting set.

Procedure

1. The brain and spinal cord of the fetal pig are very similar in structure to the human

and sheep brain, although smaller. Since the cranial bones and vertebral column are not fully ossified, it is relatively easy to remove the brain and to expose the spinal cord of the fetal pig.

2. Make an incision through the skin of the fetal pig from the shoulders to the pelvis. Separate the skin from the muscles and pin it to the side.
3. Cut through the muscles on each side of the vertebral column from the midthoracic region to the pelvis to expose the vertebral column.
4. With fine, sharp scissors cut off the tops of the vertebral arches. Then insert the tip of the scissors in the vertebral cavity, and cut lengthwise through the bones of the vertebral column, just lateral to the center of the column. It may be easier to cut the bone if the other hand is placed under the abdominal region of the pig, arching the back of the pig. Remove the bone and expose the spinal cord.
5. Examine the **spinal cord**. The **dura mater** surrounds the cord. The **dorsal roots** of the spinal nerves should be visible emerging from the cord. The **spinal nerves** are visible as fine white threads running out to the muscles.
6. Place the pig on its back and examine the interior of the thoracic cavity. A white thread should be visible running lengthwise, parallel to each side of the vertebral column. This is the **sympathetic trunk**. The swellings on the trunk are the **ganglia** of this lateral chain of ganglia.
7. Remove the skin from the entire dorsal surface of the skull, from the eyes to the neck.
8. Chip away the skull bones, starting with the fontanels. Use scalpel and fingers until the brain is completely exposed.
9. Separate the spinal cord from the medulla at the foramen magnum, by making a cut through the spinal cord caudal to the medulla.
10. Carefully life out the brain from the cranial cavity, severing each cranial nerve and the stalk of the pituitary as far from the brain as possible. The **olfactory bulbs** usually tear off from the brain and may be visible in the floor of the cranial cavity. Identify the stalk of the pituitary in the floor of the cranial cavity.
11. Remove the dura mater; then follow steps 2-28 of the procedure for the sheep brain.

Exercise 13-D Reflex Action in the Normal, Spinal, and Double-Pithed Frog

Objectives

To demonstrate which reflexes require the presence of the brain, and which merely require the presence of the spinal cord.

Materials

Frog; dissecting set; beakers; tripod; wire gauze; stand; thermometer; clamp; frog board; and string.

Discussion

A reflex is an involuntary response to a stimulus. Reflexes are purposeful, predictable, and persistent. That is, they are useful (often protective); a given stimulus always produces the same response; they last throughout the lifetime of the animal. Some reflexes, such as the photopupil reflex, require the presence of the brain; others, such as the withdrawal of the leg if the toes are dipped in acid, require only the spinal cord.

Procedure

1. Make all the following tests and observations upon
   a. a normal frog,
   b. a spinal frog, and
   c. a double-pithed frog.
2. Upon completion of the tests, record all data in the chart under Results and Questions.
3. Normal frog:
   a. Place the frog on the deck. Note the position of the head and legs.
   b. Determine the rate of respiration by observing the closing and opening of the nostrils or by the raising and lowering of the floor of the mouth.
   c. Place the frog on its back and observe to see if it rights itself.
   d. Place the frog on the frog board, right side up, and secure it gently around the midtrunk region with string. Tip the frog board so the head is depressed.

Note whether the frog elevates his head.

e. Place the frog in a sink full of water. Can he stay afloat and swim? A spinal frog may make swimming movements.

f. Let the toes of the right foot of a suspended frog dip into a beaker of water at 50° C. Does he withdraw his foot?

g. Pinch the toes of the right hind leg. Does he withdraw his leg?

h. Test for the presence of the corneal reflex by observing if the frog blinks when the cornea is touched.

i. Observe the frog's response to a loud noise (drop a book on the table near the frog).

4. Spinal frog:
   a. Pith the frog.
   b. Wait until the frog emerges from the period of spinal shock, then perform all the tests in Step 3 of Procedure.
5. Double-pithed frog:
   a. To double-pith the frog, insert the probe into the occipitovertebral junction. Direct the probe downward into the spinal cavity and destroy the cord.
   b. Repeat all the tests in Step 3 of Procedure.

Exercise 13-E Demonstration of Individual Human Reflexes

Objectives

To determine the response obtained in various common reflexes.

Materials

Rubber reflex hammers and cotton.

Discussion

A variety of human reflexes will be demonstrated in this exercise. The first two reflexes (knee and ankle jerks) are stretch reflexes. Stretching the tendon (in this experiment caused by striking the tendon with a reflex hammer) stimulates stretch receptors. An impulse travels to the cord, then back (by way of the same nerve) to the muscle in which the impulse originated. The muscle contracts, thus resisting further stretch. This type of reflex is important in maintaining an upright position.

Procedure

Part 1. Patellar Reflex (knee jerk)

1. Have the subject sit on a table so that the leg hangs free.
2. Strike the patellar ligament just below the knee with the edge of the hammer. This should result in reflex contraction of the quadriceps femoris muscle, resulting in extension of the leg.
3. Have the subject clasp his hands behind his back and pull, just prior to and during the striking of the tendon. What effect does this have on the extent of the knee jerk?
4. Give the subject a column of figures to add and while he is doing this, tap the patellar ligament. What effect does this have on the extent of the knee jerk?

Part 2. Achilles Reflex (ankle jerk)

1. The subject should kneel on a chair, his feet hanging free over the edge of the chair.
2. Tap the tendon of Achilles. The gastrocnemius muscle should now contract, resulting in plantar flexion of the foot.

Part 3. Corneal Reflex

1. Touch the cornea with a piece of cotton.
2. What reflex action immediately followed this stimulus?

Part 4. Photopupil Reflex

1. The subject should close his eyes for two minutes while facing a bright light (the microscope lamp).
2. Have the subject open his eyes and note the large pupil, which should constrict shortly after the light strikes the eye.
3. After the initial constriction, a lesser dilation of the pupil occurs, which is then followed by another constriction.
4. See if you can observe the preceding theoretical results in the subject.

Part 5. Accommodation Reflex

1. Have the subject look at a distant object (20 or more feet away) in the hallway. (Do not look into bright light.)

2. Examine the size of the pupil.
3. Now have the subject look at a pencil held about 10 inches from the face (without changing the illumination) and observe the size of the pupil.

Part 6. Convergence Reflex

1. Observe the position of the eyeballs while the subject is looking at a distant object.
2. Now have the subject focus on the pencil held about 10 inches from the face.
3. Observe the change in the position of the eyeballs.

Exercise 13-F Reaction Time

Objective

To determine the length of human reaction time to sight, sound, and touch.

Materials

Two keys; kymograph; semicircular stand; clamp; signal magnet; wire; and two dry cells.

Discussion

The reaction time is the length of time required for a voluntary response to the stimulation of a sense organ. It varies with the particular sense organ stimulated, and is affected to some extent by practice and by the intensity of the stimulus.

Procedure

1. Use a drum speed of 4 throughout the experiment.
2. Connect the two dry cells, the signal magnet and the two keys in series (see Fig. 13-6).
3. The subject whose reaction time is to be measured operates key A. The investigator operates key B.
4. To determine the reaction time to **sight**, key B must be a silent key.
5. Start the drum rotating.
6. Have the subject close key A and, while the drum is rotating, watch for the deflection of the signal magnet.
7. The investigator should then close key B. This completes the circuit; the pen of the signal magnet should descend.
8. As soon as the subject sees the pen descend on the paper, he should open key A, breaking the circuit. This causes the pen to return to its original position.
9. Repeat the test at least three additional times and calculate the average reaction time to sight. Do this by measuring the length of the deflection of the signal magnet tracing and dividing by the drum speed. Record the reaction time in hundredths of seconds. Show all calculations.
10. Repeat this determination for the other students at the table.
11. To measure the reaction time to **sound**, key B must make an audible noise when closed.
12. The subject should close his eyes throughout the experiment.
13. Have the subject close key A. The investigator should then close key B.
14. As soon as the subject hears the click of key B, he should open key A, breaking the circuit.
15. Repeat this action at least three additional times and calculate the reaction time to sound.
16. Repeat this determination for the other students at the table.

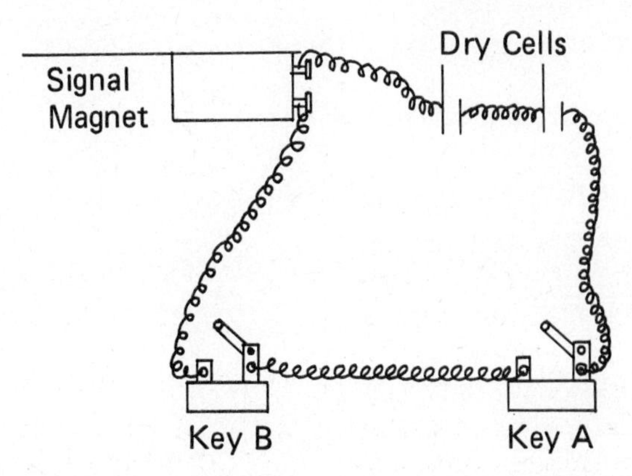

Figure 13-6. Diagram of Setup for Measuring Reaction Time

17. To determine the reaction time to **touch**, key B must again be silent. The subject must close his eyes throughout this experiment.
18. The subject should close key A.
19. The investigator should close key B with one hand, and touch the subject on the shoulder with the other hand at exactly the same time.
20. As soon as the subject feels the touch, he should open key A, breaking the circuit.
21. Repeat this action at least three times and calculate the reaction time to touch.
22. Repeat this determination for the other students at the table.
23. Record all results in the table under Results and Questions.

Self-Test — Nervous System

Directions: See Chapter 1, Self-Tests, p. 9.

1. Cerebral palsy is characterized by **loss of muscular control.**

2. Multiple sclerosis is a disease of the **autonomic nervous system.**

3. The rate of oxidation of alcohol in the body is **fixed.**

4. When a nerve is stimulated by a number of subliminal stimuli in a short space of time, **summation** may occur.

5. At the synapse impulses pass from the axon of one neuron to the **axon** of the next neuron.

6. A threshold stimulus results when energy is **just sufficient** to initiate the response of normally excitable tissue.

7. **White matter** of the spinal cord contains the long ascending or descending motor and sensory pathways.

8. The radial nerve is a branch of the **cervical plexus.**

9. The **phrenic** nerve is a branch of the plexus that is formed by the first four cervical nerves.

10. Depolarization is associated with excitation of a membrane, while **hyperpolorization** is associated with inhibition.

11. The **reflex arc** is the functional unit of the nervous system.

12. In the reflex arc impulses pass from the internuncial neuron to the **afferent** neuron.

13. The motor area of the cerebral cortex lies in the **post-central** gyrus.

14. The spinal cord extends to the base of the **first sacral** vertebra.

15. **Alpha waves** of the electroencephalogram are associated with alert, relaxed wakefulness.

16. Visual and other types of directed attention, or deep concentration, are associated with interruption of the **alpha rhythm.**

17. The **hypothalamus** is an important relay station for sensory pathways between the brain stem and cerebral cortex.

18. Cerebrospinal fluid has a composition resembling **lymph.**

19. Temperature regulation is under the control of the **thalamus.**

20. The aggregations of nerve cell bodies located deep in the brain and spinal cord may be called **nuclei.**

21. Alcohol **stimulates** the central nervous system.

22. Tobacco, when smoked, may act as a **stimulant** on the brain.

23. A **synapse** is the small gap between a nerve and muscle.

24. The surface of the brain contains raised parts called convolutions or **sulci.**

25. Walking is an example of an **acquired** reflex.

26. High blood pressure is a common cause of brain damage.

27. Alcohol first affects the **cerebrum.**

28. The cortex of the brain consists of **white** matter.

29. Inflammation of a membrane of the brain or spinal cord is called **meningitis.**

30. Injury to a speech center affecting speech and word understanding is called **aphasia.**

31. Neurosyphilis **may be dormant for several years** and then attack the brain.

32. **Cerebral palsy** may affect the use of arms and legs and may be due to a brain injury at birth.

33. Most of us could live longer without **food** than without **sleep.**

34. Motor nerves to one side of the body originate, for the most part, in the **same** side of the brain.

35. The relationship between the mind and the body is called **psychosomatic.**

36. The major control mechanisms for balance and posture are located in the **cerebellum.**

37. The cranial nerve conveying impulses concerned with equilibrium is the **acoustic.**

38. The cranial nerve conveying impulses to most of the viscera of the body is the **oculomotor** nerve.

39. The nerve carrying impulses causing us to shrug our shoulders and turn our head is the **accessory.**

40. The cranial nerve that would carry impulses warning you that the coffee was hot is the **trigeminal.**

41. The cranial nerve that carries impulses giving you information about decaying substances is the **acoustic.**

42. In general, the craniosacral division of the autonomic system is concerned with conservative and restorative processes while the **thoracolumbar** division is concerned primarily with processes involving energy exchange.

43. A lumbar puncture is done between **L5-S1** to avoid hitting the spinal cord which ends at L1-L2.

44. In a lumbar puncture cerebrospinal fluid may be removed from the **subarachnoid space.**

45. A red color of the cerebrospinal fluid may indicate bleeding somewhere in the **central nervous system.**

46. An amputee sometimes has sensations in the missing extremity because the nerve fiber may be stimulated at any point along its **course.**

47. When giving intramuscular injections, they should be given in the **upper, outer quadrant** of the buttocks.

48. It is important to give injections in the correct quadrant to avoid the **blood vessel** which courses through the buttocks.

49. A tic douloureux is a spasmodic neuralgia of the **glossopharyngeal** nerve.

50. A patient with an extremely dominant **sympathetic** system will probably be nervous, excitable, have cold hands and feet, and may perspire excessively.

51. It takes just a few seconds for the body to mobilize, because there are extensive synaptic connections and diffuse distribution of the **sympathetic** fibers.

52. A reflex is a **voluntary** response to a harmful stimulus.

53. The Nissl bodies in neuroplasm are related to **the metabolism of the cell.**

54. Axons carry nerve impulses **toward** the nerve cell body.

55. Gray matter of the brain and spinal cord is predominantly made up of **nonmyelinated** nerve cell processes.

56. The **myelin sheath** is responsible for regeneration of injured peripheral nerve tissues.

57. The **central nervous system** is composed of the brain and spinal cord.

58. The connective tissue cells of the central nervous system are called **neuroglial cells.**

59. The myelin sheaths of spinal and cranial nerves are broken into segments at sites known as **nodes of Ranvier.**

60. **Proprioceptors** are nerve endings that receive stimulation from muscles, tendons, and joints.

61. Interoceptors are nerve endings that receive stimulation from **skin surfaces.**

62. A nerve impulse may be transmitted **in any direction** in a nerve fiber, but follows a specific direction governed by the positions of synapses.

63. The outer surface of nerve cell membrane is electrically **negative** with respect to the interior of the cell at rest.

64. **White** matter of the brain consists of masses of nerve cell bodies.

65. The localization area for voluntary motion is found in the **frontal lobe** of the cerebrum.

66. The **great longitudinal fissure** divides the cerebrum into right and left hemispheres.

67. The **thalamus** is responsible for conscious motor activity and conscious sensations.

68. The normal control of vital mechanisms of the viscera is mediated by the **autonomic nervous system.**

69. A rapid heart rate, an increase in respiratory rate, and an increase in blood pressure would result from stimulation of the **parasympathetic** division of the autonomic nervous system.

70. Normal digestion and elimination are under the control of the craniosacral division of the autonomic nervous system.

71. The pupil of the eye dilates in response to stimulation of the **craniosacral** division of the autonomic nervous system.

72. The **dura mater** is the inner layer of the meninges at the surface of the spinal cord and brain.

73. The agent liberated at the terminals of all preganglionic neurons and the postganglionic neurons of the parasympathetic divisions is called **acetylcholine.**

74. The passage of a nerve impulse from one neuron to another is due to the presence of **acetylcholine** at the synapse.

75. The central nervous system of man is formed from the **ectoderm** of the early embryo.

76. The strength of an impulse traveling along a nerve fiber **is** related to the strength of the stimulus.

77. When an impulse travels along a nerve fiber, the speed of the impulse indicates that it is **the same** as electricity.

78. The region of the spinal cord which contains cell bodies of motor neurons is the **posterior** horn.

79. A **conditioned** reflex is the product of repeated similar stimuli.

80. The vagus nerve emerges from the **medulla.**

81. The control of the rate of the heart beat is a function of the **pons varolii.**

82. Various disorders of the nervous system can be detected by means of an **electroencephalogram.**

83. The **arachnoid** forms the floor of the subarachnoid space.

84. The rate of respiration is controlled by a center in the **cerebellum.**

85. The cerebellum has a cortex composed of **white matter.**

86. In the development of the nervous system, the central nervous system first appears as a **tube**.

87. The nerve that carries impulses resulting in the action of chewing is the **trigeminal**.

88. The nerve carrying impulses causing us to frown is the **glossopharyngeal**.

KEY

2. central nervous system
5. dendrite
8. brachial plexus
12. efferent
13. precentral
14. first lumbar
17. thalamus
19. hypothalamus
21. depresses
23. myoneural junction
24. gyri
28. gray
34. opposite
38. vagus
41. olfactory
43. L3-L4
48. sciatic nerve
49. trigeminal
52. an involuntary
54. away from
56. neurilemma
61. viscera
63. positive
64. gray
67. cerebrum
69. sympathetic
71. thoracolumbar
72. pia mater
76. is not
77. not the same
78. anterior (ventral)
81. medulla oblongata
83. pia mater
84. medulla oblongata
85. gray matter
88. facial

Name ____________________

Results and Questions for Chapter 13

Exercise 13-A. Anatomy of the Nervous System

Part 1. Structure of a Nerve

1. How is it possible to distinguish between an axon and a dendrite in the slide of the cross section of a nerve?

   ____

2. Define each of the following:

   a. Nerve ____________________

   b. Tract ____________________

   c. Ganglion ____________________

   d. Vasa nervorum ____________________

   e. Neuron ____________________

3. What is the function of the neurilemma? ____________________

   ____

4. What is the function of the myelin sheath? ____________________

   ____

Part 2. The Spinal Cord

1. What is the function of each of the following?

   a. Dorsal horn ____________________

   b. Lateral horn ____________________

   c. Ventral horn ____________________

   d. White matter ____________________

   e. Dorsal root ganglion ____________________

   f. Dorsal root ____________________

   g. Ventral root ____________________

2. Figure 13-7 is an outline of a cross section through the spinal cord. Complete the diagram by drawing the roots and the beginning of the spinal nerve. Add to the diagram a simple reflex arc and label its components.

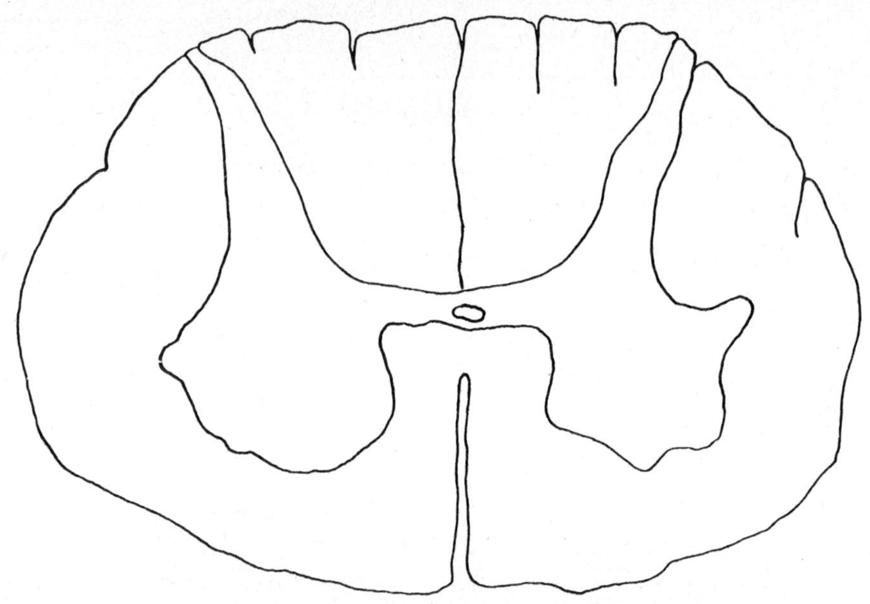

Figure 13-7. Cross Section of Spinal Cord

3. What is the function of each of the following tracts and where does each cross over?

Tracts	*Site of Crossing Over*	*Function*
Posterior columns		
Lateral spinothalamic tract		
Spinocerebellar tract		
Lateral corticospinal tract		

4. How many pairs of spinal nerves emerge from the cord?________________

5. What is the cauda equina?________________

6. Name the major plexuses of spinal nerves and list which spinal nerves form each plexus.

____

____

7. Classify the neuron whose cell body was visible in the dorsal root ganglion as to structure ________ and funtion.________________

8. Classify the neuron whose cell body was located in the ventral horn as to structure________ and function.________________

9. Where does the spinal cord terminate?________________

Part 3. The Gross Structure of the Brain

1. What is the function of each of the following?

   a. Cerebellum________________

Name ______________________________

b. Hypothalamus ______________________________

c. Thalamus ______________________________

d. Pons ______________________________

e. Superior colliculi ______________________________

2. Name the lobes of the cerebral cortex. ______________________________

3. Name the three parts of the forebrain. ______________________________

4. Which cranial nerves emerge from the medulla? ______________________________

5. Name one example of commissural fibers seen in the sagittal section of the brain.

____

6. Name the three layers of the meninges that surround the brain and the cord.

____

Exercises 13-B and C. The Dissection of the Sheep and/or the Pig Brain

1. Where is cerebrospinal fluid manufactured? ______________________________

2. Trace the path of cerebrospinal fluid through the brain. ______________________________

____

____

3. What is the advantage of having convolutions in the cortex? ______________________________

____

4. Describe the composition of white matter in the cord. ______________________________

____

5. Of what is gray matter composed? ______________________________

____

6. What is the function of cerebrospinal fluid? ______________________________

7. Where is a spinal tap usually made? ______________________________

8. What is the cause of hydrocephalus? ______________________________

9. Compare the structure of the dura mater of the brain with that of the cord.

____

Exercise 13-D. Reflex Action in the Normal, Spinal, and Double-Pithed Frog

1. Record all results in the chart below.

Reflex Activity	*Normal Frog*	*Single-Pithed Frog*	*Double-Pithed Frog*
a. Position of head and legs			
b. Rate of respiration			
c. Righting from back			
d. Elevation of head			
e. Swimming			
f. Withdrawal from pinch			
g. Withdrawal from hot water			
h. Corneal reflex			
i. Response to noise			

2. Which reflexes of those tested require the presence of the brain?____________

3. Which reflexes of those tested require only the presence of the spinal cord?____________

4. What evidence do you have that reflexes are purposeful?____________

5. What structures must be present for most protective reflexes to occur?____________

6. List one example, demonstrated in this exercise, of a flexor reflex.____________

7. What is the function of flexor reflexes?____________

Name ________________

8. Define a synapse. ________________

9. How do nerve impulses cross a synapse? ________________

____

10. What is the function of cholinesterase? ________________

____

Exercise 13-E. Demonstration of Individual Human Reflexes

1. Fill in the following chart concerning the reflexes demonstrated in this exercise.

Name of Reflex	*Results Obtained*	*Nerve*		*Muscle*
		Afferent	*Efferent*	*Involved*
Knee Jerk				
Ankle jerk				
Corneal reflex				
Photopupil reflex				
Accommodation reflex				
Convergence reflex				

2. What is the purpose of the photopupil reflex? ________________

3. What is the purpose of the convergence reflex? ________________

____

4. What effect did clenching the hands or adding a column of figures have on the extent of the knee jerk? ________________

5. What sensation was perceived when the cornea was touched? ________________
Why was the sensation of touch not felt? ________________

Exercise 13-F. Reaction Time

1. Tape your kymograph records for sight, sound, and touch below. Label each record.

2. Calculate the duration of your reaction time in seconds by measuring the length of each in centimeters and dividing by the drum speed (in centimeters per second). Record your results in the table below.

	1	*2*	*3*	*4*	*Average*
Sight					
Sound					
Touch					

3. Compare your results with the theoretical results. ______________________________

____

4. Why were there differences in the reaction times to the three types of stimulation?

____

5. Did practice have any effect on the reaction time? ______________________________

6. What might be some possible sources of error in this experiment? ______________________

____

____

14

Sense Organs

Introduction

Stimuli from both the external and internal environment are received by a variety of receptors. Specific stimuli are received by specific receptors which are, for the most part, modified histologically for their particular function. Receptors may be classified functionally, depending upon the particular kind of stimulus they respond to, as photoreceptors, mechanoreceptors, chemoreceptors, pressoreceptors, or osmoreceptors. A classification of receptors follows which also includes the type of stimulus received:

A. Exteroceptors

These are receptors which are sensitive to stimuli arising at the surface of the body as well as those coming from the external environment. Included are the following:

1. Rods and cones in the retina of the eye (photoreceptors).
2. Hair cells of the Organ of Corti in the inner ear.
3. Nasal membrane receptors in the superior concha (chemoreceptors).
4. Taste receptors in the tongue (chemoreceptors).
5. The following skin receptors:
   a. Meissner's corpuscles—touch (mechanoreceptors).
   b. Krause's end bulbs—cold (mechanoreceptors).
   c. Organ of Ruffini—warmth (mechanoreceptors).
   d. Pacinian corpuscle—deep pressure (mechanoreceptors).
   e. Bare (unencapsulated) nerve endings—pain (respond to a variety of stimuli).

B. Interoceptors

Interoceptors are receptor end organs which are located in and transmit impulses from the viscera. Stimulation of these receptors may give rise to sensations of discomfort or internal pain that may be the result of hunger, thirst, cramping pains accompanying disorders of the gastrointestinal tract, the sharp pain of acute appendicitis, etc. Other interoceptors, based on their functions, may be classified as follows:

1. **Pressoreceptors** are sensitive to changes in blood pressure. When stimulated, the following responses result:

a. **Right heart (Bainbridge) reflex**—Pressoreceptors in the superior and inferior venae cavae, as they enter the right atrium, when stimulated by increased pressure, convey impulses to the cardiac center in the medulla along vagus nerve fibers. This results in an accelerated heart beat.
b. **Aortic reflex**—Pressoreceptors in the aortic sinus (located in the aortic arch) are sensitive to changes in pressure. Increased pressure slows the heart rate (impulses to cardiac center via vagus nerve).
c. **Carotid sinus reflex**—Pressoreceptors in the carotid sinus, which is a slightly dilated area of the internal carotid artery at the point where the common carotid A. divides (bifurcates) into the internal and external carotid arteries, respond to fluctuations in pressure. Impulses are conveyed to the cardiac center of the medulla by the glossopharyngeal nerve. Increased pressure slows the heart rate.

2. **Osmoreceptors** are sensitive to changes in osmotic concentrations in the blood (extracellular fluid), especially as influenced by sodium and chloride ions. They are found in the hypothalamus and are important in regulating fluid balance within the body.

C. Proprioceptors

These are receptors which give information concerning movements and position of the body. They are located primarily in muscles, tendons, joints, and the nonauditory portion of the inner ear.

The action of many of these receptors will be demonstrated in the exercises that follow.

Exercise 14-A Gross Anatomy of the Human Eye

Objectives

To locate and identify the structures of the human eye.

Materials

Text; reference books; models; and wall charts.

Discussion

Before proceeding to other exercises concerning the eye,it is usually advantageous to become thoroughly familiar with the gross features of the eye. Much of this learning activity may be accomplished outside of the laboratory.

Procedure

1. Use Figures 14-1, 14-2, and 14-3 of the eye as self-tests. See Section J, p. xx (Illustrations as Self-Tests) for a suggested approach.
2. When you have become familiar with the indicated structures on the figures, locate them on the models of the eye and on wall charts.

Exercise 14-B Dissection of the Sheep Eye

Objectives

To locate and identify the structures of the mammalian eye by examining and dissecting a sheep eye.

Materials

Preserved sheep eye and dissection instruments.

Discussion

The size and structures of the sheep eye compare favorably with those of the human eye. This, coupled with their availability, makes them ideal for studying the anatomical structures of the eye.

Procedure

Part 1. External Aspect of the Sheep Eye

1. Note the fat on the surface of the eye. This cushions the eye from shock in its bony orbit.
2. Identify the following structures:
   a. The **sclera**, the tough external white coat.
   b. The **conjunctiva**, reflected over the anterior surface of the eye and attached to the eyeball a short distance from the edge of the **cornea**.

KEY TO FIGURE 14-1

1. posterior cavity (contains vitreous humor)
2. fovea centralis
3. optic disc
4. retinal blood vessels
5. optic nerve
6. nerve sheath
7. canals of Schlemm
8. posterior chamber
9. anterior chamber
10. anterior cavity
11. aqueous humor
12. pupil
13. cornea
14. crystalline lens
15. iris
16. conjunctiva
17. suspensory ligaments
18. ciliary body
19. sclera
20. choroid
21. retina

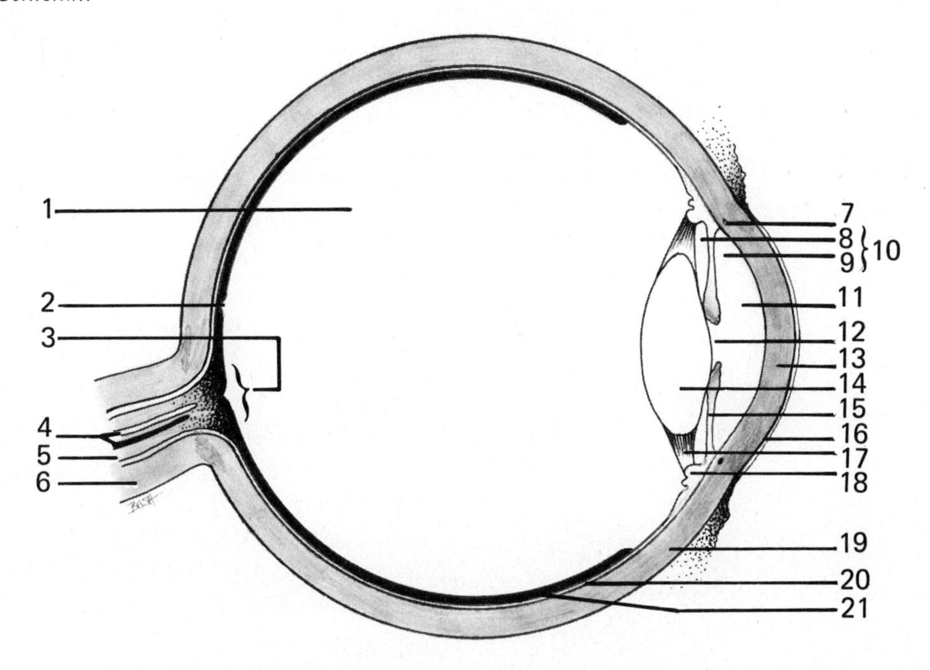

Figure 14-1. Midsagittal Section through the Eye

KEY TO FIGURE 14-2

1. superior oblique
2. medial rectus
3. lacrimal duct
4. lacrimal sac
5. lacrimal puncta
6. nasolacrimal duct
7. superior rectus
8. lacrimal gland
9. excretory ducts
10. lateral rectus
11. sclera
12. pupil
13. iris
14. inferior rectus
15. inferior oblique

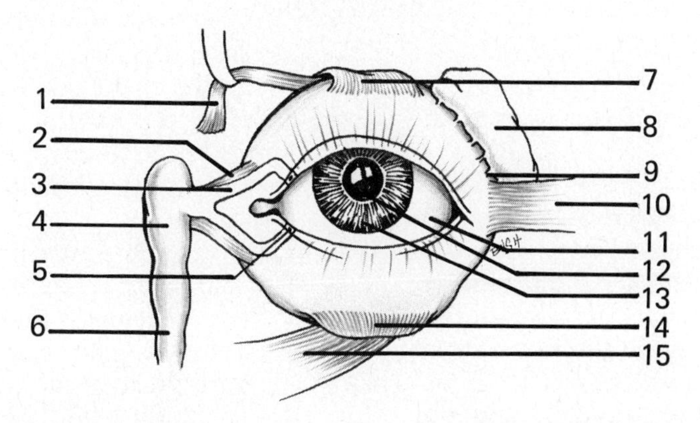

Figure 14-2. Anterior View of Eye Showing Lacrimal Apparatus

KEY TO FIGURE 14-3

1. levator palpebrae superioris
2. pulley
3. superior oblique
4. superior rectus
5. cornea
6. medial rectus
7. lateral rectus
8. sclera
9. inferior oblique
10. inferior rectus

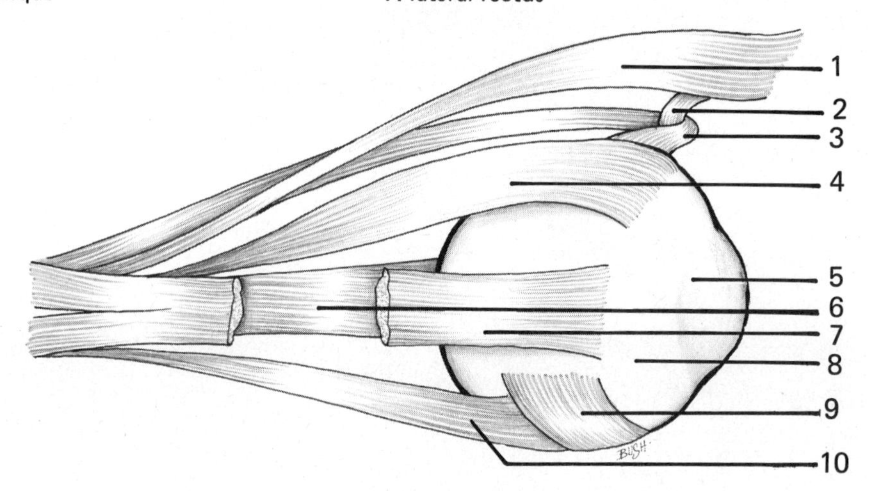

Figure 14-3. Extrinsic Muscles of the Right Eye

c. The **cornea,** the anterior, transparent (opaque in your specimen, due to action of the preservative), portion of the sclera.
d. The **optic nerve,** located on the posterior surface. The nerve has the appearance of a solid white core and is approximately 3 mm thick.

3. Carefully dissect away the connective tissue (fat, etc.) from the posterior surface of the eyeball to free the six **extrinsic eye muscles.** These resemble flattened straps.

Part 2. Internal Aspect of the Sheep Eye

1. Hold the eye so that the cornea is in an inferior position.
2. Making an incision into the eyeball about 1/2 cm from the edge of the cornea, cut completely around the eye (parallel to the cornea).
3. If the incision was done properly, it should now be possible to carefully loosen the vitreous humor (vitreous body) from the crystalline lens and to keep the vitreous humor in the posterior portion of the eyeball and the lens in the anterior third of the eye.
4. Examine the interior of the anterior part of the eye.
   a. Observe the **ciliary body,** the black structure which has the appearance of narrow, radial, folds.
   b. Locate the **suspensory ligaments,** the very delicate fibers connecting the ciliary body to the lens. They hold the lens in position.
   c. Free the lens from the ciliary body and remove it. Remnants of the suspensory ligaments can be seen attached to the lens.
   d. The **iris** is now visible anterior to the former position of the lens. This also appears black. Try to distinguish between the **circular** and **radial** fibers comprising the iris.
   e. Hold the lens up to the light. Does any light pass through? The lens in your specimen may be opaque due to the action of the preservative.
5. Examine the external surface of the anterior third of the eye. It is now relatively easy to distinguish the iris, **pupil,** and cornea.
6. Examine the posterior two-thirds of the eye and observe the following structures:
   a. The **vitreous humor** (in life, this substance is perfectly clear). Remove it from the eyeball.
   b. The **retina,** the white inner coat that was covered by the vitreous humor.

Determine the point at which the retina is attached dorsally.

c. The **choroid coat.** The retina covers this coat, and the two are easily separated. The iridescent appearance of the choroid is due to the presence of the **tapetum lucidum,** a special structure not present in the human eye. The function of the tapetum lucidum is to reflect some light back onto the retina. This reflecting device is found in vertebrates that live under conditions of low light intensity. This causes the animal's eyes to shine in the dark.

d. The **sclera,** the outer white coat.

Exercise 14-C Accommodation

Objective

To demonstrate accommodation.

Materials

Pins and 3 x 5 cards.

Discussion

The ability of the eye to focus on far objects and then to rapidly adjust and focus on objects near the eye is referred to as accommodation. Ability to accommodate depends upon changes in the shape of the crystalline lens.

Procedure

1. Face the window and hold your hand, with the fingers outspread, about 6 inches in front of your eyes.
2. Close your eyes for a moment, then open them and note which seems to be in focus—your fingers or some distant object such as a bush or tree or some other item. Record your observations under Results and Questions.
3. With a pin, prick two holes in a 3 x 5 card that are approximately 3 mm apart (a distance less than the diameter of the pupil).
4. Look through the pinholes and bring a pin into the line of vision so that it appears in the overlapping part of the luminous circles.
5. Now, look at a distant object in the back of the room and note whether the pin appears double or single.
6. Accommodate for the pin. Note whether the pin now appears double or single. Record your observations and answer the questions under Results and Questions.

Exercise 14-D Measuring the Near Point of Accommodation

Objectives

To measure the near point of vision, to determine if the value is more or less for the dominant eye; to determine which eye is the dominant one.

Materials

Pin; 3 x 5 card; and a meter stick.

Discussion

Since accommodation depends upon the resiliancy of the lens, the near point of accommodation will vary as the resiliency varies. The shape of the eyeball will also influence the measurement.

Procedure

1. Hold a pin, or a 3 x 5 card with a well-defined letter printed upon it, at arm's length.
2. While holding a meter stick just under the eye, gradually bring the pin (or card) closer to the eye by drawing it along the margin (or on the surface) of the meter stick.
3. When the pin appears to be double, or the letter on the card becomes distorted, the item is within the near point and your eye can no longer accommodate for it.
4. Move the pin (or card) away to the precise point where it appears single (or undistorted) again. This is your near point of vision.
5. Determine the near point for each eye in centimeters.
   Right eye____________________
   Left eye____________________
6. On the table on the following page locate the age group to which your near point corresponds for each eye and record in the spaces provided.
7. To determine your dominant eye, make a small hole (about this size ◯) in a piece of paper.
8. Place a coin at your feet on the floor.

9. Hold the paper with the hole in it at waist level.
10. Locate the coin through the hole in the paper and close your right eye. If the coin disappears you are right-eyed.

Table 14-1. Relationship between Age and Near Point of Vision

Age	Near Point	
	cm	*in.*
10	7.5	2.95
20	9.0	3.54
30	11.5	4.53
40	17.2	6.77
50	52.5	20.67
60	83.3	32.80

Right eye ____________________

Left eye ____________________

11. Now close your left eye. If the coin disappears you are left-eyed. Which eye was dominant? ________________

Exercise 14-E Snellen Chart Test for Visual Acuity

Objective

To measure visual acuity.

Materials

Snellen chart, illiterate "E" chart, and a 3 x 5 card.

Discussion

The Snellen method for measuring visual acuity is based upon the fact that certain sized letters can be clearly seen by eyes with normal acuity at specific distances. The first line (letter "E") is of such a height that it can be clearly seen by a normal-sighted individual standing 200 feet away from the chart. Another line on the chart contains letters of a size that should be clearly readable by the same individual at a distance of 20 feet. V = d/D where V stands for vision and d is the distance at which certain letters can be read by the subject. D is the distance at which the letters can be read by the normal eye. For example, if an individual reads a line of letters at a distance of 20 feet which the "normal" eye could read at a distance of 30 feet, his visual acuity = 20/30 (or two-thirds that of normal vision).

Procedure

1. The subject stands 20 feet from the Snellen chart; the observer should stand near the chart.
2. The subject should then cover one eye with a 3 x 5 card, but should not close the eye.
3. The observer indicates the lines to be read and the subject should read the chart as far down as possible with each eye separately, and with both eyes together (without glasses). Note: d/D can be read off the chart next to the last line that can be read successfully by the subject. Record your data in the table under Results and Questions.
4. If the subject usually wears glasses, the test should be repeated with the glasses on.
5. Repeat the above procedure with the illiterate "E" chart. Record your data in the table.

Exercise 14-F Test for Astigmatism

Objective

To test for the presence of astigmatism.

Materials

Astigmatism chart.

Discussion

Astigmatism is a condition in which the refracting surfaces of the cornea and crystalline lens are not perfect. A chart for the detection of this condition consists of a series of parallel lines arranged like the spokes of a wagon wheel. If the refracting surfaces are not distorted, all the spokes will be seen with equal distinctness and none will appear darker than others. If astigmatism is present, the horizontal spokes may be properlý focused and seen, while the vertical spokes are focused either in front of or behind the retina. They will, therefore, be more or less blurred. Of course, other spokes than the ones mentioned may be involved—it will vary with the location of the eye defect.

Astigmatism is corrected by means of a cylindrical lens. To test a lens (such as those in your glasses) designed to correct astigmatism, rotate the lens or glasses 90 degrees keeping the center of the lens fixed on some slender vertical or horizontal object (such as the window blinds). Any change in shape of the object indicates a cylindrical lens has been added to the glasses for the correction of astigmatism.

Procedure

1. Stand 10 feet away from a well-lighted astigmatism chart.
2. Cover the left eye and observe the lines on the chart. Note if some of the lines (e.g., vertical or horizontal) appear to be in sharper focus than others.
3. Move toward the chart, one step at a time, repeating your observations.
4. Repeat the preceding procedure with the right eye covered.
5. Record your data under Results and Questions.

Exercise 14-G Color Blindness

Objective

To determine whether the subject is colorblind.

Materials

Ishihara's color plates.

Procedure

1. Using sunlight whenever possible, hold the plates of the test book 75 cm (approximately 30 inches) from the subject.
2. Tilt the plates so that the plane of the paper is at right angles to the line of vision.
3. Each answer should be given without more than 3 seconds delay. The correct answer for each plate is given in the front of the test book.
4. Record how many errors the subject makes.

Exercise 14-H The Blind Spot

Objective

To demonstrate the occurrence of the blind spot.

Materials

None.

Procedure

1. Hold this page approximately centered, about 20 inches in front of the face, with the left eye closed.
2. Focus on the cross seen below. At this distance both the cross and the circle should be seen.
3. Gradually bring the paper closer until the circle cannot be seen (look steadily at the cross the entire time).
4. Bring the paper still closer to the face. Now the circle should reappear.
5. Repeat the above procedure with the right eye closed, but this time the opened eye should focus on the circle.

Exercise 14-I Anatomy of the Ear

Objectives

To become familiar with the structures of the ear and their functions.

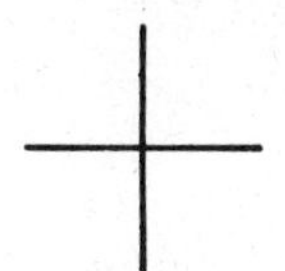

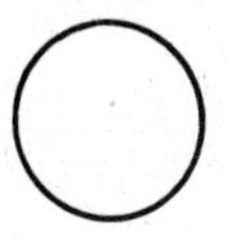

Materials

References; models; wall charts; and ear ossicles.

Discussion

Unlike light rays, which may travel through a vacuum, sound waves require some medium, such as air or water, in order to be conducted. In the ear, both air and water are involved in the transmission of sound waves to the auditory receptors located within the cochlea. A knowledge of the anatomy of the ear is essential if these principles, which relate to its physiology, are to be understood.

Procedure

1. Use Figures 14-4 and 14-5 of the ear as self-tests.
2. Compare the structures indicated on Figures 14-4 and 14-5 with those found on models of the ear, wall charts, and in references.
3. Using your text or other current references, determine the functions of the structures of the ear.
4. Demonstration: Examine the auditory ossicles (malleus, incus, and stapes) that have been placed under the stereoscopic microscope.
5. Obtain a slide of the **cochlea** labeled Internal Ear, Cochlea. Since the cochlea makes two and a half spiral turns around a pillar of spongy bone, the **modiolus**, the slide contains approximately five views of the canals of the cochlea. Selecting one portion that resembles some model of the cochlea (refer to Figs. 14-6 and 14-7), diagram a radial section through the canals and label the following structures: **basilar membrane, tectorial membrane, vestibular membrane, scala vestibuli, scala media (cochlear duct), scala tympani, hair cells of the Organ of Corti, supporting cells,** and **cochlear nerve.** Indicate the location of **perilymph** and **endolymph.**

Exercise 14-J Bone Conduction of Sound Waves

Objective

To demonstrate bone conduction.

KEY TO FIGURE 14-4

1. posterior semicircular canal
2. lateral semicircular canal
3. fenestra vestibuli (ovalis)
4. fenestra cochlea (rotunda)
5. superior semicircular canal
6. vestibule
7. cochlea
8. modiolus

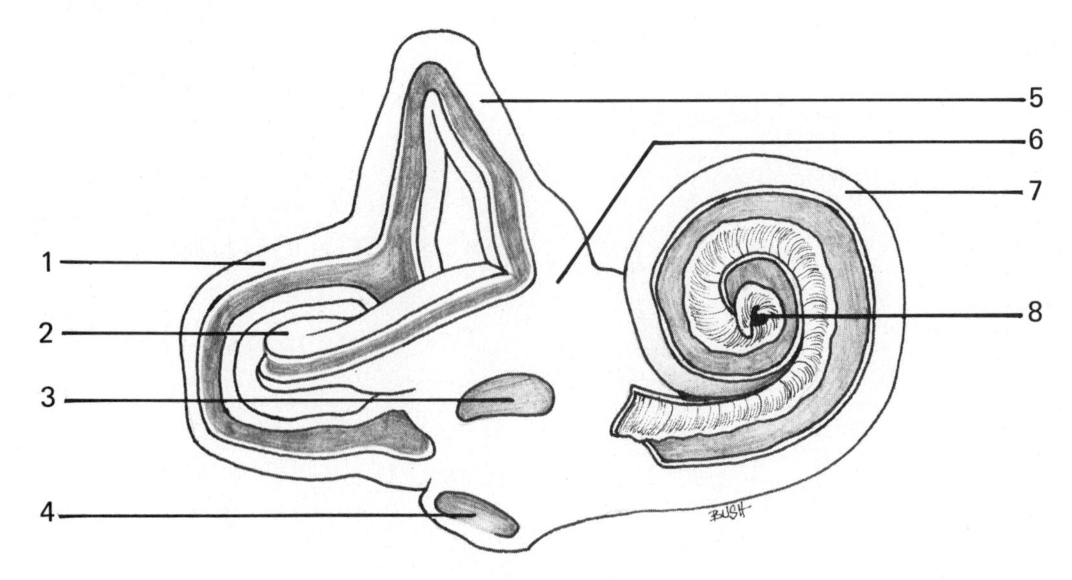

Figure 14-4. Osseous Labyrinth

KEY TO FIGURE 14-5

1. temporal bone
2. malleus
3. incus
4. tensor tympani (cut)
5. external auditory meatus
6. tympanic membrane
7. pinna
8. auditory nerve
9. vestibular nerve
10. cochlear nerve
11. semicircular canal
12. ampulla
13. utricle
14. saccule
15. modiolus
16. spiral lamina
17. cochlea
18. fenestra ovalis
19. stapes
20. fenestra rotunda
21. blood vessels
22. styloid process
23. mastoid process
24. eustachian tube

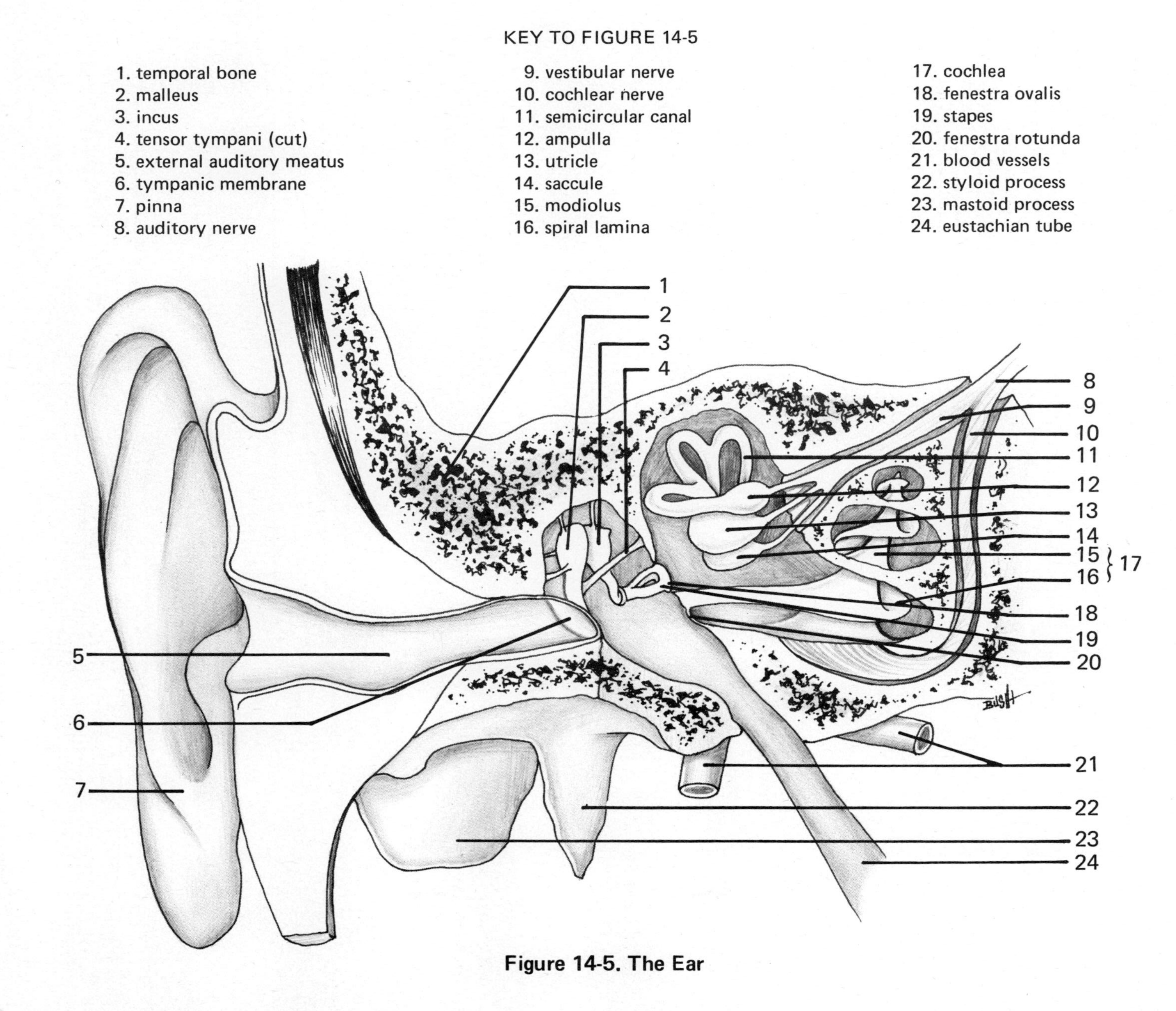

Figure 14-5. The Ear

KEY TO FIGURE 14-6

1. scala vestibuli
2. vestibular membrane
3. tectorial membrane
4. organ of Corti
5. scala tympani

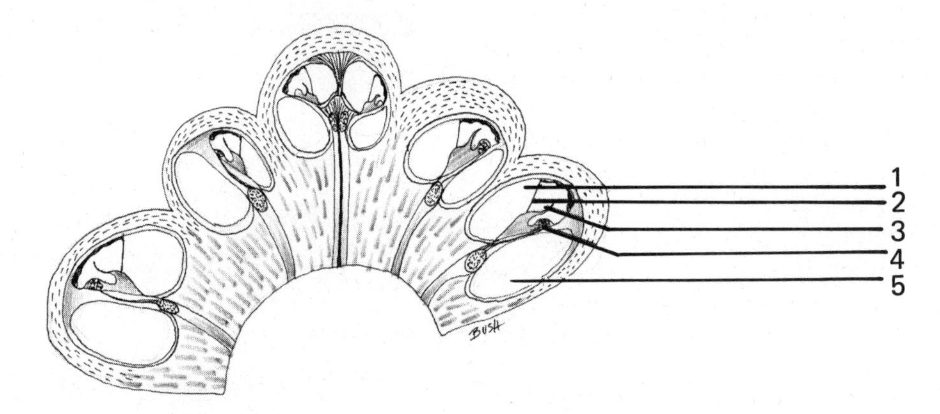

Figure 14-6. Axial Section through Entire Cochlea Showing Canals

KEY TO FIGURE 14-7

1. hair cells of organ of Corti
2. supporting cells
3. basilar membrane
4. scala tympani
5. scala vestibuli
6. vestibular membrane
7. cochlear duct
8. tectorial membrane
9. cochlear nerve

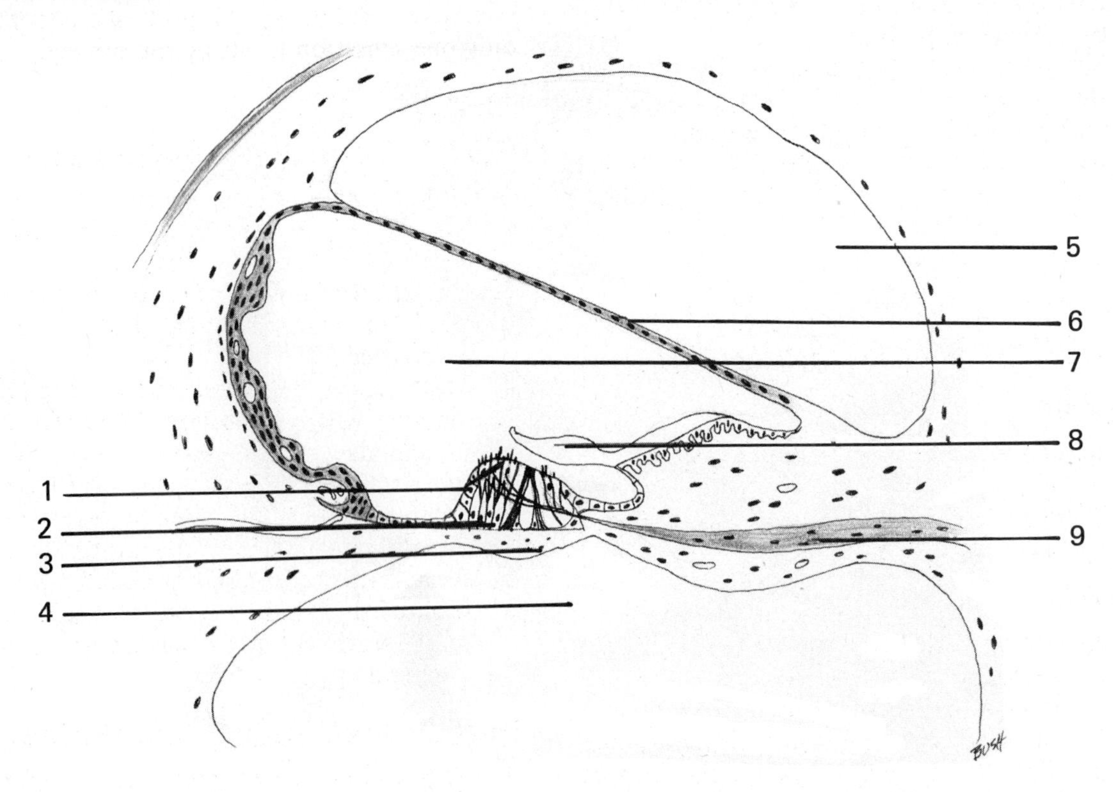

Figure 14-7. Radial Section through One Coil of the Cochlea

Materials

Tuning fork and 70% alcohol.

Procedure

1. To demonstrate bone conduction of sound waves, hold a ticking watch or the handle of a vibrating tuning fork (these may first be cleaned with 70% alcohol) between the teeth. If a tuning fork is used, start it vibrating by striking it against the palm of the hand (never against a hard surface).
2. With the watch or tuning fork in position, close first one ear, then both. Compare the intensity of the sound heard with both ears open, one ear closed, and both ears closed. Record your observations under Results and Questions.
3. Now perform the Rinné test for middle-ear deafness by holding the handle of a vibrating tuning fork against the mastoid process (the "bump" located just behind the ear). The tuning fork should be pointed back away from the ear but not touching the head (only the handle touches the head).
4. When the sound can no longer be heard, hold the tuning fork (by the handle) just outside of the external ear. In middle-ear deafness, no sound is heard by air conduction after an individual ceases to hear it by bone conduction.
5. Simulate middle-ear deafness, by plugging the ear to be tested with cotton (or close with a finger), then repeat the test.
6. Record your data and answer the questions under Results and Questions.

Exercise 14-K Localization of Sounds

Objective

To demonstrate localization of sounds.

Materials

Watch or tuning fork.

Procedure

1. To demonstrate localization of sounds have the subject close both eyes.
2. Now hold a ticking watch or vibrating tuning fork (or use some other sound-making device such as snapping fingers) approximately 6 inches: (a) in front of the head; (b) above the head; (c) behind the head; (d) on either side of the head. Record your observations under Results and Questions.

Exercise 14-L. Localization

Objectives

To determine whether the ability to localize a point of stimulation varies on different parts of the body; to determine whether there is a difference in the ability to discriminate between two points applied simultaneously on different parts of the body.

Materials

Pencil, ruler, and an esthesiometer.

Discussion

Both tactile localization and tactile discrimination are demonstrated in this exercise. Tactile localization involves the ability to tell more or less accurately which portion of skin is touched. In the brain there is a field which corresponds to the field of the periphery of the body. By tactile discrimination, we mean the power to discern two distinct sensations when the skin is stimulated at two points. When the stimulated points are close enough together, only one sensation is felt by the subject.

Procedure

Part 1. Localizing a Stimulus

1. Have the subject close his eyes.
2. Touch the palm of the subject's hand with a pencil.
3. The subject should then try to place the point of a pencil upon the precise spot stimulated.
4. Measure, in millimeters, the distance (if any) by which he missed the precise point of stimulation. Repeat the procedure two more times. Average the results of the three determinations and record the values in the table under Results and Questions (Part 1).
5. Repeat the foregoing procedure on the back of the hand, fingertip, ventral surface of the forearm, ventral surface of the upper arm, and upper back.

Part 2. Two-point Discrimination

1. The esthesiometer resembles a compass with two sharp points.

2. The eyes of the subject should be closed throughout this exercise.
3. To determine how close the two points of the esthesiometer can be and yet be felt as two points, the observer should touch the subject in a random manner with either one point or with two points of the esthesiometer to prevent guessing.
4. Measure (in millimeters) how close the two points of an esthesiometer can be and yet be felt as two points when applied to the skin on the ventral surface of the thumb. Make each determination three times. Average the results and record all values in the table under Results and Questions (Part 2).
5. Repeat the experiment on the palm and back of the hand, ventral surface of the forearm, ventral surface of the upper arm, back, and tip of the tongue.

Exercise 14-M Cutaneous Sensations

Objective

To demonstrate the punctiform distribution of cutaneous sense receptors.

Materials

Bristle; hot nails; cold nails; pins; rubber stamp grid; and a stamp pad.

Discussion

The cutaneous sensations include heat, cold, touch, and pain. Receptors for heat, cold, and touch are lacking in the internal organs. Pain receptors are quite general in distribution and this sensation can be elicited from most organs.

Procedure

1. Stamp a grid on the ventral surface of the forearm or, if too hairy, on the arm (on the biceps brachii).
2. Touch each square in the grid on your forearm in at least six different spots with a bristle to determine the location of the **touch** receptors. Apply just sufficient pressure to bend the bristle. Use the same pressure each time. Don't assume you are feeling a touch sensation merely because you see the bristle touching. Indicate the location of each receptor with a "T" in the appropriate location on the grid under Results and Questions.
3. To locate the **cold** receptors, use nails cooled in ice water. Dry the nails and then lightly touch just the tip of the nail to at least six different sites in each square of the grid. Change nails at frequent intervals. Mark the site of each cold receptor located by this procedure with a "C" on the grid under Results and Questions.
4. Using nails heated in water to 40° or 50° C., and dried, locate the **heat** receptors in accordance with the procedure of Step 3, and indicate their location on the grid with an "H."
5. To locate **pain** receptors, explore the same area of skin as before with a pin. The sensation perceived when a pain receptor is stimulated by slight pressure resembles a slight electric shock, not a definite pain. Indicate the sites of these receptors on the grid with the letter "P."
6. Relocate a touch receptor and stimulate the site with a cold nail. Record your observation under Results and Questions.

Exercise 14-N Pressure Sensation

Objective

To determine if sensitivity to pressure varies from one site to another in the body.

Materials

Bristle; a piece of glass plate (or other perfectly smooth, hard surface); and about 50 sheets of smooth paper.

Discussion

The receptors for this sensation are located immediately below the skin. A related sensation occurs when the urinary bladder or rectum is filled with urine or feces (sensation of fullness).

Procedure

1. Place a bristle (or hair) on a smooth, hard surface (e.g., a piece of glass) and cover it with from 10-50 sheets of paper (count them). Do this while the subject has his eyes closed or is looking elsewhere.
2. The subject should now attempt to deter-

mine the location of the bristle by lightly passing the tips of the fingers across the surface of the papers. Determine the maximum number of sheets of paper under which the subject could feel the bristle.

3. Repeat the test using the knuckle of the index finger, the proximal portion of the palm of the hand, the elbow, the chin, and the nose. Record your data under Results and Questions.

Exercise 14-O Receptor Adaptation

Objective

To demonstrate adaptation of receptors to stimuli.

Materials

Coins; paper cups; warm water; ice water; and tincture of iodine or perfume.

Discussion

Disappearance of a sensation is due to the fact that the receptors adapt to stimuli and thus initiate no more nerve impulses until a change takes place in the stimulus.

Procedure

Part 1. Touch Stimulation

1. Place an object (e.g., a coin) upon the skin of the ventral surface of the forearm. How many seconds does the touch sensation last?
2. Repeat the experiment at a new spot on the skin. After the sensation disappears, add two more coins of the same size on top of the first coin. If the sensation returned, how many seconds did it last? Record your data under Results and Questions.

Part 2. Temperature Stimulation

1. Immerse an index finger in warm water for 2 minutes. Now dip the other index finger into the same container of warm water. Note the difference between the sensations in each of the fingers.
2. Now immerse one index finger in warm water and the other in ice water. After 2 minutes immerse both fingers in the same container of cold tap water. Record your observations under Results and Questions.

Part 3. Chemical Stimulation

1. With one nostril held closed, sniff (breathe relatively deeply) a bottle containing tincture of iodine or perfume. Note the beginning time and the time it takes for the odor to disappear.
   Beginning time________Time required for odor to disappear________________
2. As soon as the odor disappears, release the other nostril and attempt to smell the tincture of iodine with it. If you can still detect the odor in the single nostril after 5 minutes, release the other nostril and smell the iodine with it. Note any difference in intensity. Record your observations under Results and Questions.

Exercise 14-P Referred Pain

Objective

To experience referred pain.

Materials

Enamel pan and ice water.

Discussion

Referred pain is a common sensory phenomenon in which stimulation in one part of the body gives rise to a sensation (often pain) that seems localized in a different or remote part of the body.

Procedure

1. Place your elbow in a tray of ice water deep enough to cover it.
2. Keep your elbow in position long enough to feel the following progression of sensations: first, a feeling of some discomfort in the region of the elbow; later, pain sensations that are felt elsewhere. Record your observations under Results and Questions.

Exercise 14-Q Location of Taste Buds

Objective

To determine where the greatest concentration of receptors (taste buds) for each of the four basic taste sensations is located.

Materials

5% sucrose; 1% acetic acid or 5% unsweetened lemon juice; 5% NaCl; 1/2% quinine sulfate; cotton-tipped applicators; and paper cups.

Procedure

1. An outline of the tongue appears under Results and Questions on page 344. As you obtain data from the activities below, record on this outline the site where each substance was tasted using the following symbols to represent each spot where a specific taste receptor was stimulated: x for sweet, o for salt, s for sour, and b for bitter.
2. Before doing a given test, thoroughly dry the tongue with a paper towel.
3. With a cotton-tipped applicator apply a small amount of 5% sucrose to the back, sides, and tip of the tongue. Record the results on your drawing.
4. Rinse your mouth with water and dry the tongue before using each of the remaining three solutions.
5. Repeat Step 3, in the following sequence, for each of the remaining solutions: 5% NaC1, 1% acetic acid, 1/2% quinine sulfate. Record your results on the drawing.

Self-Test—Sense Organs

Directions: See Chapter 1, Self-Test, p. 9.

1. A **detached retina** occurs when the tissue lining the eyeball is pulled loose in one area and causes blurred vision.

2. **Conjunctivitis** is an inflammation of the membrane which lines the eyelids and covers the cornea.

3. **Glaucoma** is a disease of the eye characterized by increased tension within the eyeball and progressive loss of vision.

4. If you look into an eye with an **ophthalmoscope** you may observe the retina, macula lutea, optic disc, and the retinal blood vessels.

5. "Watery eyes" are frequently associated with acute rhinitis because the inflammation of the nasal mucosa causes swelling which closes the **eustachian tube** and blocks tear drainage.

6. Patients with one eye bandaged have difficulty in feeding themselves because they have poor depth perception due to the loss of **chromatic** vision.

7. Normal vision is called **emmetropia.**

8. **Myopia**, or farsightedness, is caused when the eyeball is too short, or the refractive power too little, which causes the rays to focus behind the retina.

9. A patient with **nystagmus** will have involuntary oscillations, or "to and fro" movements, of the eyeball.

10. A cold reduces the sense of smell and causes food to be relatively tasteless. This is because many "tastes" are really **olfactory impressions.**

11. **Otitis media** is usually a complication of a severe cold and sore throat, because the infection ascends through the eustachian tube.

12. A **myringotomy** (an incision in the tympanic membrane) is sometimes performed in severe otitis media to allow pus to escape.

13. A hearing aid is effective in conduction deafness because the hearing aid amplifies the sound waves and transmits them through the **external and middle ears** to the inner ear.

14. Our ears become "stopped up" with sudden changes in altitude because the eustachian tubes are **usually closed**; thus changing air pressure causes bulging of the tympanic membrane.

15. Cutaneous sensations **are equally** distributed over the body.

16. The brushes of Ruffini receive sensations of **heat.**

17. Pressure sensations are received by **Pacinian corpuscles.**

18. The end bulbs of Krause receive **cold** stimuli.

19. Hearing and balance are localized in the **frontal** lobe of the cerebrum.

20. The conscious awareness of heat, cold, and pressure is localized in the **parietal** lobe of the cerebrum.

21. The perception of objects seen is localized in the **occipital** lobe of the cerebrum.

22. Adaptation to odors is a relatively **slow** process.

23. The olfactory cells are **modified hair cells.**

24. Taste buds to detect sweetness are most numerous on the **lateral parts** of the tongue.

25. Injury to the right optic tract behind the optic chiasma would destroy the nasal visual field sight of the **left** eye and the temporal visual field sight of the **right** eye.

26. The shape of the crystalline lens is regulated by the **ciliary muscle.**

27. A pigment that is thought to be associated with the cones is called **iodopsin.**

28. The brown pigment (melanin) which causes eye color is found in the **pupil.**

29. The **vitreous** humor of the eye circulates through the anterior and posterior chambers and is drained by the canal of Schlemm.

30. The ability of the eye to adapt for near vision and far vision is known as **accommodation.**

31. Astigmatism is due to irregularities in the **cornea.**

32. Tears aid in keeping bacteria out of the eye due to the presence of **lysozyme.**

33. Conduction deafness may be caused by cerumen in the ear canal.

34. The stapes is attached to the **fenestra cochlea** or **round window.**

35. Hair cells in contact with the **tectorial** membrane (in the inner ear) are stimulated when the membrane vibrates.

36. The receptors for hearing are located on the **organ of Corti** within the inner ear.

37. The utricle is generally associated with **static posture** or the orientation of the body in relation to gravity, while the semicircular canals are concerned with **dynamic equilibrium.**

38. The eustachian tube leads from the pharynx to the **internal** ear.

39. The **semicircular canals** contain hair cells which move when the head is moved.

40. **Ossicles** of the middle ear transmit sound waves to the inner ear.

41. The fluid in the **scala vestibuli** communicates with the round window.

42. The fibers of the cochlear nerve pass through the **modiolus** to reach the organ of Corti.

43. The organ of Corti is located on the **basilar membrane.**

44. Recovery from visual fatigue is aided by closing the eyes for **as little as ten minutes.**

45. The **cornea** is the clear transparent tissue covering a portion of the anterior surface of the eyeball.

46. The area where the optic nerve leaves the eye is **devoid of rods and cones** and is, therefore, insensitive to light stimuli.

47. The iris of the eye and ciliary body constitute the **extrinsic** muscles of the eye.

48. The white of the eye is the **choroid** portion of the eye.

49. The area of keenest vision of the eye is located at the **fovea centralis.**

50. The blind spot of the eye is located in the **fovea centralis.**

51. The right side of the cerebrum receives the nerve fibers from the **right side** of both eyes.

52. The **rods** are responsible for color vision.

53. Eye fatigue, from reading, is due to neuromuscular fatigue of **both extrinsic and intrinsic muscles.**

54. Vitamin A, frequently associated with night vision, is an essential part of **rhodopsin.**

55. Pink eye is a **very contagious** infection of the conjunctiva.

56. Appearance of a colored halo is an early symptom of the serious eye disease called **cataract.**

57. When the point of focus of your eyes are in front of the retina, you are **near-sighted.**

58. When treating crossed eyes (strabismus) the physician may sometimes **put a patch over the good eye** of the child.

59. A foot candle is the light given by a candle of **standard** size at a distance of **one foot.**

60. Visual purple, which aids in twilight vision, is not regenerated after being bleached by light if the person's diet is deficient in **vitamin C.**

61. Accommodation or the focusing of an object on the retina is brought about by the **change of shape in the lens.**

62. With advancing years the lens becomes firmer and the ability to alter its curvature decreases. This condition is called **hypermetropia.**

63. Flickering light would be most trying to the muscle of **the iris.**

64. In vision, accommodation is brought about by action of the **ciliary muscle.**

65. **Cataract** involves opaqueness of the lens.

66. The small air-filled chamber which contains the malleus, incus, and stapes is the **tympanum.**

67. Sound waves travel **at the same velocity** as light waves.

68. The parts of the ear concerned with balance include **otoliths** and **hair cells.**

69. The process of hearing involves the action of the **tensor tympani** and the **stapedius** muscles.

70. In the order of their occurrence from the outside inward, the bones of the middle ear are the **incus, malleus, and stapes.**

KEY

5. nasolacrimal duct
6. binocular
8. hyperopia
13. cranial bones
15. are unequally
19. temporal
22. rapid
24. tip
25. right-left
28. iris
29. aqueous
34. fenestra ovalis or oval window
38. middle
41. scala tympani
47. intrinsic
48. sclera
50. optic disc
52. cones
56. glaucoma
60. vitamin A
62. presbyopia
66. tympanic cavity
67. only a fraction as fast
70. malleus, incus, and stapes

Name ______________________

Results and Questions for Chapter 14

Exercise 14-A. Gross Anatomy of the Eye

1. Indicate, in the blanks provided, whether the following structures are part of the sclera, choroid, or retina.

   a. Rods and cones ______________ c. Cornea ______________

   b. Iris ______________ d. Ciliary body ______________

2. Identify the structures having the following functions (a structure may be used more than once since it may have several functions):

   a. Protects and gives form and shape to the eyeball. ______________

   b. An outer, transparent, structure which refracts light and, thus, aids in focusing light rays. ______________

   c. Essential in accommodation of the eye. The contraction of its muscle fibers results in change of shape of the lens. ______________

   d. Contains many blood vessels to provide for the nourishment of the retina. ______________

   e. Provides place of attachment of extrinsic eye muscles. ______________

   f. Contains a pigment which prevents internal reflection of light rays. ______________

   g. Contraction results in dilation of the pupil. ______________

   h. Contraction results in constriction of the pupil. ______________

   i. Contains the light receptors. ______________

3. Give the primary function of each of the following:

   a. Rods ______________________________

   b. Cones ______________________________

   c. Fovea centralis ______________________________

   d. Vitreous humor ______________________________

4. Answer the following questions:

   a. What is the location of the aqueous humor? ______________________________

   ____

   b. Which humor is more gelatinous? ______________________________

c. Which one of the three eye coats is complete anteriorly?__________

d. At what point is the retina attached dorsally?__________

e. Which cranial nerves supply the extrinsic eye muscles?__________

f. Describe the location of the optic nerve.__________

5. Name the intrinsic eye muscles:__________

6. Name the extrinsic eye muscles and, after each, give its function.

Muscle	Function

Exercise 14-C. Accommodation

1. When you opened your eyes with your fingers in front of them, which was in focus, your fingers or some distant object?__________

2. Did the pin appear to be double or single when you looked at a distant object (see No. 5 under Procedure) in the back of the room?__________

3. In the preceding question, where in the eye was the focus of the distant object?__________

4. Where in the eye was the focus of the pin?__________

5. When you accommodated for it, where was the focus of the pin?__________

6. Make two diagrams to account for the appearance of the pin in No. 5 and No. 6 of the Procedure.

Name________________________

Exercise 14-D. Near Point of Accommodation

1. Record the near point for each eye in centimeters. Right eye______Left eye______

2. Record the age group from Table 14-1 to which your near point corresponds.

   Right eye ______Left eye______

3. Did the near point of each of your eyes fall within an age range that is compatible with your present age?________________
   Discuss your answer:________________

   ____

4. Were you right-eyed or left-eyed?__________Based on your knowledge of which eye is dominant, explain any differences in the near point of your right eye when compared with your left eye.________________

   ____

   ____

5. List three changes that take place in the eyes during accommodation.

   ____

   ____

   ____

6. What property of the lens is essential for accommodation?________________

7. The property indicated in No. 6 is lost, to some extent, in what condition?

   ____

8. How would myopia affect the near point?________________

9. How would hyperopia affect the near point?________________

Exercise 14-E. Snellen Chart Test for Visual Acuity

1. Record your data for visual acuity in the following table:

Eyes	*Snellen (d/D) (without glasses)*	*Snellen (d/D) (with glasses)*	*Illiterate "E" Chart*
Right eye			
Left eye			
Both eyes			

2. What lens would be used to correct a nearsighted eye?________________

3. What lens would be used to correct a farsighted eye?

4. What lens could be used to make an emmetrope nearsighted?

5. What lens could be used to make an emmetrope farsighted?

6. If you wear glasses, does your prescription completely correct the abnormality? Explain:

7. What does a visual acuity of 20/15 indicate?

8. What does a visual acuity of 20/40 indicate?

9. Explain the cause of presbyopia and tell how it is corrected.

Exercise 14-F. Test for Astigmatism

1. In the spaces provided, indicate if astigmatism was detected, and whether it was the vertical or horizontal type.

   Right eye:

   Left eye:

2. What causes astigmatism?

3. How is the condition corrected?

Exercise 14-G. Color Blindness

1. How many errors were made in this test?

2. Were you color-blind? List any of your relatives who are also color-blind.

3. Give a precise explanation as to why color blindness occurs more frequently in males than in females.

Exercise 14-H. The Blind Spot

1. What is the blind spot?

2. Why is this spot blind?

Name ______________________

3. Draw three diagrams below which show the precise points of focus on the retina of the cross and circle that occurred in the eye in Steps 2, 3, and 4.

4. In the preceding, what was the point of focus of the circle when it was not visible?

____

5. Why are we unable to observe the blind spot when we see with both eyes at the same time?

____

Exercise 14-I. Anatomy of the Ear

Give the function of each of the following structures:

1. Tympanic membrane ______________________
2. Ear ossicles ______________________
3. Eustachian tube ______________________
4. Semicircular canals ______________________
5. Cochlear nerve ______________________
6. Vestibular nerve ______________________
7. Organ of Corti ______________________
8. Endolymph ______________________
9. Fenestra ovalis ______________________
10. Fenestra rotunda ______________________

Exercise 14-J. Bone Conduction of Sound Waves

1. Compare the intensity of the sound heard with:

   a. both ears open, ______________________

   b. one ear closed, and ______________________

c. both ears closed. ______

2. How was the sound received when both ears were closed? ______

______

3. Describe and discuss the results obtained in the Rinné test for middle-ear deafness.

______

4. How has the principle involved in the Rinné test for middle-ear deafness been used in some hearing aids? ______

5. Which is superior, air or bone conduction? ______

Exercise 14-K. Localization of Sounds

1. Results (indicate the precise location of the sound and the subject's localization of the sound):

Actual Location of Sound	Subject's Localization of Sound
a. in front of the head	______
b. above the head	______
c. behind the head	______
d. on either side of the head	______

2. Explain why it was not possible to localize the source of the sound each time. ______

______

______

Exercise 14-L. Localization

Part 1: Fill in the following table with the results obtained from the localization experiment. Record the number of millimeters by which the subject missed the precise point of stimulation.

Point of Stimulation	*mm*	*mm*	*mm*	*Avg. mm*
Palm of hand				
Back of hand				
Fingertip				
Back				
Ventral forearm				
Ventral upper arm				

Name ______________________

Part 2: Complete the following table with the results obtained while using the esthesiometer.

Point of Stimulation	*mm*	*mm*	*mm*	*Avg. mm*
Ventral thumb				
Palm of hand				
Back of hand				
Ventral forearm				
Ventral upper arm				
Back				
Tip of tongue				

2. In Part 1, was there a general pattern to the distribution of the ability to localize the precise point of stimulation?_______Explain:____

____

3. In Part 2, why was there a difference in ability to feel the points of the esthesiometer as two points in different parts of the body?____

____

4. Give a general conclusion concerning the distribution of touch receptors:________________

____

Exercise 14-M. Cutaneous Sensations

1. Record the sites of the receptors on the grid below:

Touch "T"	Cold "C"	Heat "H"	Pain "P"

2. Tabulate the total number of each type of receptor that was mapped in the skin area and record in the space provided below:

Touch __________ Cold __________ Heat __________ Pain ____________

3. What results did you obtain from stimulating a touch receptor site with a cold nail?__________

Explain your results:____

4. Are different kinds of receptors located at the same sites? ____________

5. How do your results compare with the theoretical distribution of cutaneous receptors?

____

6. What is meant by punctiform distribution of cutaneous receptors? ____________

____

7. Identify the cutaneous receptors localized in this experiment by their scientific names:

____

____

Exercise 14-N. Pressure Sensation

1. In the spaces provided record the total number of sheets of paper through which the bristle (hair) could be detected.

   a. Fingertips ____________ d. Elbow ____________

   b. Knuckle ____________ e. Chin ____________

   c. Palm of hand ____________ f. Nose ____________

2. What area of your body was most sensitive to pressure? ____________

3. Which part of your body was least sensitive to pressure? ____________

Exercise 14-O. Receptor Adaptation

Part 1. Touch Stimulation

1. Record, in seconds, the length of time the coin was felt on the ventral surface of the forearm.

____

2. When two coins were added to the first coin, did the touch sensation return? ____________

3. If the sensation did return, record, in seconds, the length of time it lasted (with the three coins).

____

4. Briefly explain your results: ____________

____

____

Part 2. Temperature Stimulation

1. Describe your sensations for each of the exercises involving water stimuli:

   a. Warm water (first one index finger, then the other). ____________

   ____

Name ____________________

b. One index finger in warm water, other in ice water. ____________________

____

c. One finger in warm water, one in ice water, then both in cold tap water. ____________________

____

____

2. Explain any observed differences in the preceding. ____________________

____

____

Part 3. Chemical Stimuli

1. Record, in seconds, the time required for the odor to disappear: ____________________
2. Did you observe any difference in the intensity of the odor with the newly opened nostril? ______ Describe the sensation: ____________________

   ____
3. Based on your results in all three parts, what are your conclusions concerning receptor adaptation? ____________________

   ____
4. Is it advantageous for receptors to adapt? ______ Explain: ____________________

   ____
5. Pain receptors show limited adaptation. Why is this advantageous? ____________________

   ____

Exercise 14-P. Referred Pain

1. Exactly where did you feel the referred pain? ____________________
2. Using the following information, give an explanation for the referred pain you felt: the ulnar nerve, supplying the ring and little fingers and the inner side of the hand, passes over the elbow joint.

   ____

   ____

Exercise 14-Q. Location of Taste Buds

1. Indicate on the diagram on the next page where each of the four solutions could be tasted most readily on the tongue.

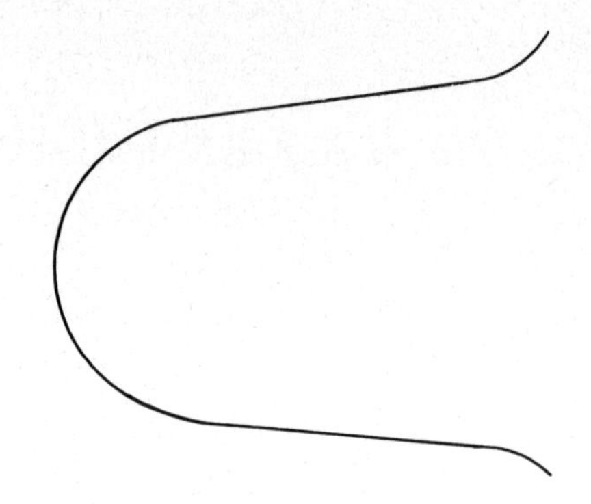

2. Name the nerves for taste. ______________________________

3. Which nerve carries messages of touch from the tongue? ______________________________

4. Which nerve is necessary for movement of the tongue? ______________________________

15

The Urinary System

Exercise 15-A Anatomy of the Human Urinary System

Objectives

To become familiar with the location, structure, and function of the organs in the urinary system.

Materials

Dissectable torso; anatomic charts; reference books; models; and slide of nephron.

Discussion

The urinary system consists of the kidneys, ureters, urinary bladder, and urethra. The ureters conduct urine from the kidneys to the bladder; the urethra conducts urine (and sperm in the male) from the bladder to the outside.

The primary function of the kidneys is to preserve homeostasis. This term refers to the maintenance of a constant internal environment. The kidneys maintain homeostasis through the excretion of various toxins and metabolic wastes as well as other substances present in the blood in surplus amounts (e.g., sugar eaten in excess or in the disorder known as diabetes mellitus). Fluid and electrolyte balance, as well as acid-base balance, depends primarily upon the proper functioning of the kidneys. Malfunction of these organs results in a failure of homeostasis and, unless corrected, such failure finally results in death.

The structural and functional unit of the kidney is the nephron; there are approximately one million of these units in each kidney. The blood is filtered in these units, and urine is produced as a result of selective transport (which is partially under the influence of hormones) of water and other substances from the uriniferous tubule of the nephron.

Procedure

1. Using your text, charts, and reference books for orientation, locate the following organs on the dissectable torso. Be prepared to describe the location of each organ in anatomic terms.
   a. Kidneys
   b. Ureters
   c. Urethra
   d. Urinary bladder
2. On the model of a longitudinal section through the kidney, locate the structures indicated on the keys to Figures 15-1 and

KEY TO FIGURE 15-1

1. adipose tissue
2. renal pelvis
3. major calyx
4. minor calyx
5. ureter
6. pyramid
7. calyces
8. renal medulla (the semicircle of pyramids)
9. renal papilla
10. renal cortex
11. renal column
12. renal capsule

KEY TO FIGURE 15-2

1. renal artery
2. renal vein
3. renal pelvis
4. ureter
5. calyces
6. arcuate artery and vein
7. interlobar artery and vein
8. interlobular artery and vein

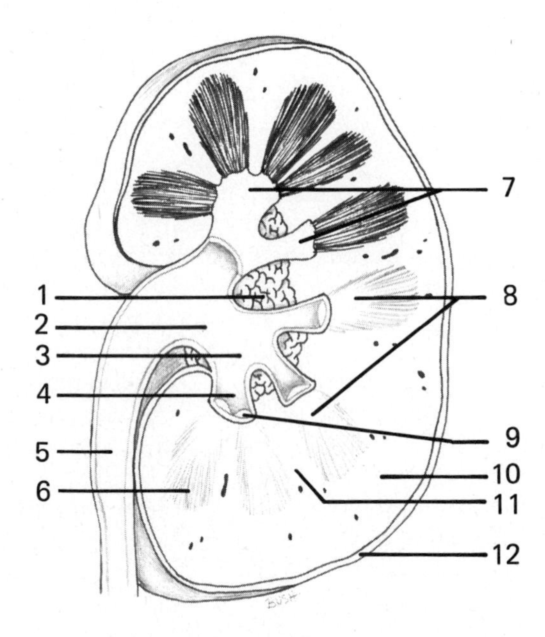

Figure 15-1. Longitudinal Section through the Kidney

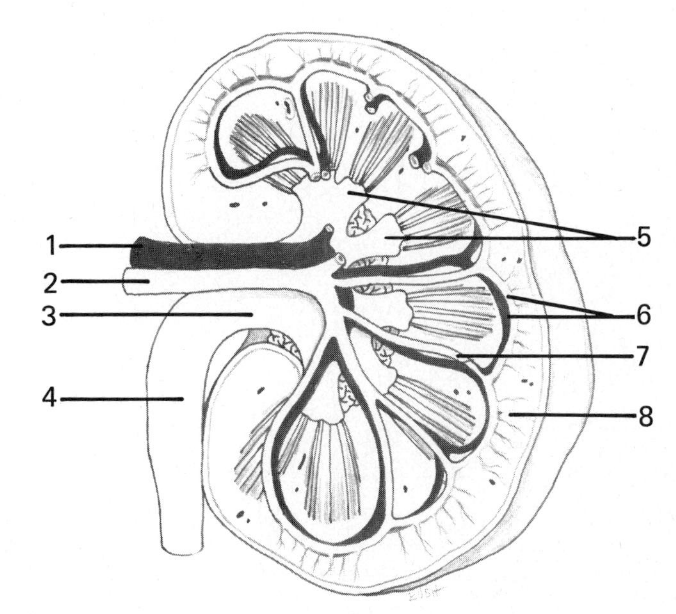

Figure 15-2. Renal Blood Supply

15-2. These figures may be used as self-tests. See Section J, p. xx (Illustrations as Self-Tests) for a suggested approach.

3. Examine a slide of the kidney showing nephrons. Compare a nephron observed on the slide with Figure 15-3, which depicts a single collecting tubule draining urine from two nephrons, and Figure 15-4, which is an enlargement of the renal corpuscle partially sectioned to show the tuft of capillaries contained within. Figures 15-3 and 15-4 may be used as self-tests.

Exercise 15-B Dissection of the Urinary System of the Fetal Pig

Objectives

To identify the urinary organs in the fetal pig and to compare them with the corresponding human structures.

Materials

Fetal pig; dissecting board or tray; and dissecting set.

Discussion

The organs in the urinary system of the fetal pig are very similar to those in the human. As you dissect the organs, be prepared to trace the path of urine from its site of production to the point at which it passes to the outside. The pig kidney will be sectioned in order to study its internal structure, since it provides a good example of a typical mammalian kidney.

KEY TO FIGURE 15-3

1. distal convoluted tubule
2. proximal convoluted tubule
3. efferent arteriole
4. renal corpuscle
5. afferent arteriole
6. Loop of Henle (ascending limb)
7. Loop of Henle (descending limb)
8. collecting tubule

KEY TO FIGURE 15-4

1. afferent arteriole
2. efferent arteriole
3. glomerulus
4. Bowman's capsule
5. uriniferous tubule (proximal convoluted tubule)

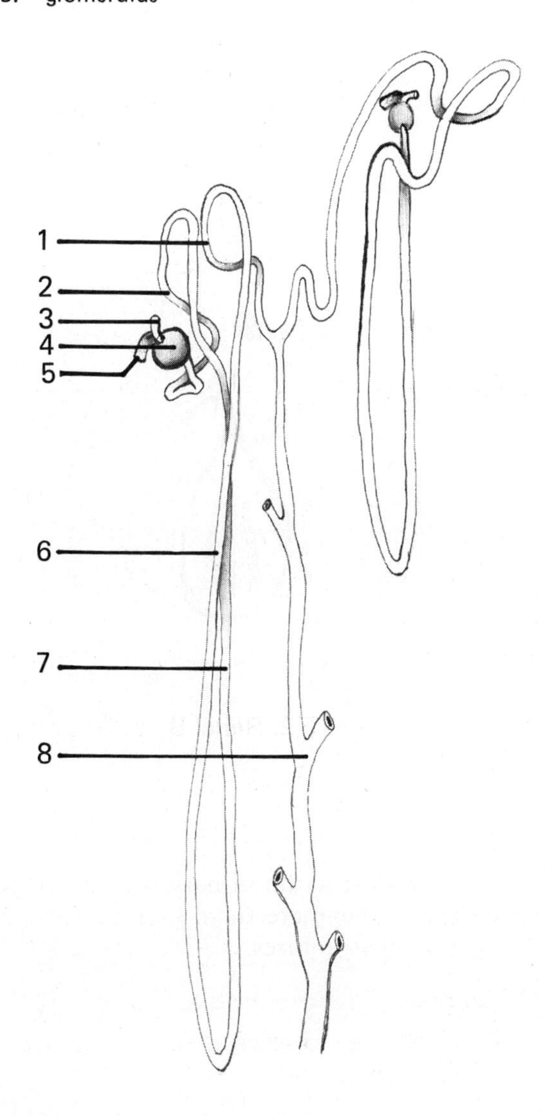

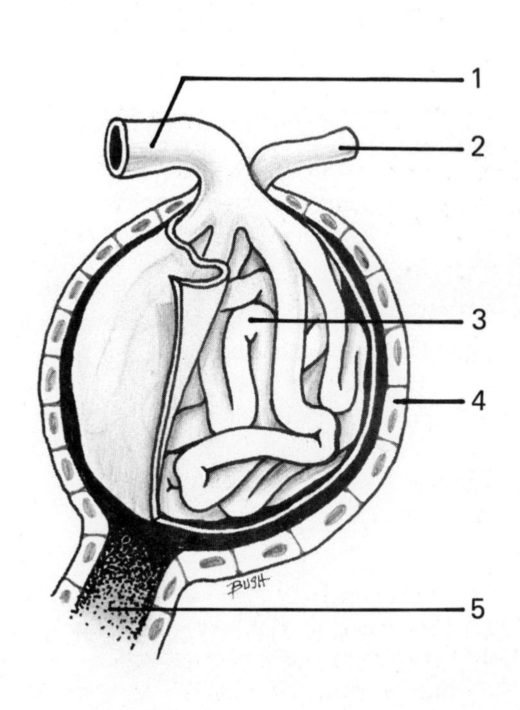

Procedure

1. Observe the paired kidneys, on the dorsal body wall of the pig. Use Figure 15-5 as a guide.
2. Remove the peritoneum, which covers only the ventral surface, from the left kidney. Since the kidneys are separated from the abdominal organs by a layer of peritoneum their location is described as being retroperitoneal.
3. Identify the **renal artery** and **renal vein,** which carry blood to and from the kidney.
4. Locate the **adrenal gland**, a narrow band immediately above and medial to each kidney.
5. Observe the **ureter**, the narrow, white, convoluted tube which drains the urine from each kidney. Trace the ureter from

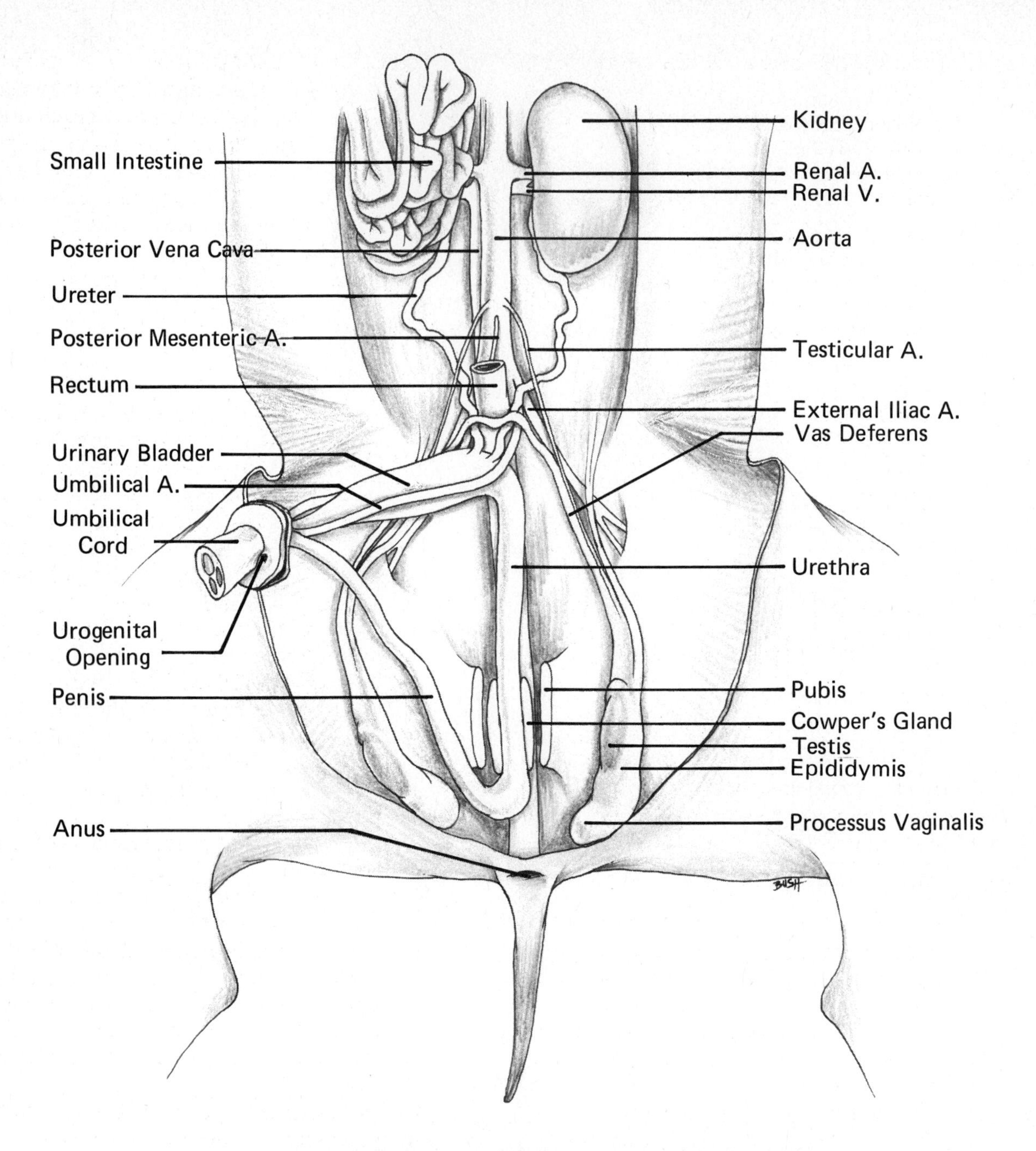

Figure 15.5. Ventral View of Urinary System (with intestine pushed to the right) of the Male Fetal Pig

the **hilum,** the opening on the medial border of each kidney, to the **urinary bladder,** freeing it from the peritoneum. The urinary bladder is attached to the reflected ventral strip of the abdominal wall.

6. Observe the **umbilical arteries** which lie lateral to the urinary bladder.
7. Locate the **urethra,** the duct which conducts urine from the posterior end of the bladder to the outside. The remainder of the urethra will be freed when the reproductive system is dissected.
8. Remove one kidney. Make a longitudinal (coronal) section through the kidney.
   a. The **renal capsule** is the thin layer of connective tissue around the outside of the kidney.

b. The **renal cortex** is the outer light brown layer of the kidney immediately beneath the capsule. This layer contains renal corpuscles.
c. The next layer of the kidney, the **renal medulla,** contains the **pyramids.**
d. Locate the **renal pelvis,** the funnel-shaped expansion of the ureter inside the **renal sinus,** which is the hollow interior of the kidney.
e. Identify the **papillae** of the pyramids which open into the **calyces** of the pelvis.
f. Trace the renal artery into the kidney. Locate the **interlobar, arcuate,** and **interlobular** arteries. The corresponding veins should also be identified.

9. Replace the kidney in the abdominal cavity of the fetal pig.

Exercise 15-C Urinalysis

Objectives

To determine some of the characteristics of urine; to perform tests for some abnormal constituents which may be present in urine; and to use the tests in determining unknowns in urine specimens.

Materials

Cylinder (or large diameter test tube); beakers; hydrometer; concentrated nitric acid; test tubes; test tube holder; Benedict's solution; funnel; filter paper; 10% acetic acid; albuminoscope; Combistix; Albutest Reagent tablets; Clinitest Reagent tablets; Tes-Tape; powdered sulfur; Ketostix; Acetest Reagent tablets; pHydrion paper; and pH meter.

Part 1. Specific Gravity, Appearance, and pH of Urine

Discussion

Specific gravity is an expression of the weight of a substance relative to the weight of an equal volume of water. Water, by definition, has a specific gravity of one. The urinometer is a float having a scale that indicates directly the specific gravity of the urine into which it is placed. It is necessary to correct specific gravity for temperature, since the hydrometer is calibrated to read correctly only at 15° C.

The color of normal urine varies from a light straw color to dark amber. Usually freshly voided urine is clear, but it may become cloudy upon standing. The presence of certain drugs, bile pigments or their products, and blood will affect the color. A cloudy appearance of freshly voided urine is usually caused by the presence of pus or phosphates.

Use Appendix 5 (The Determination of pH) for review of background in the determination of pH.

Procedure

1. Place a portion of the urine specimen in the cylinder or large diameter test tube.
2. By means of a urinometer (hydrometer) determine the specific gravity of the urine specimen. In making this determination make certain that the hydrometer floats freely. The reading should be to three decimal places.
3. Determine the temperature of the urine specimen. Add one to the last decimal (thousandth) place of the hydrometer reading for every three degrees above 15° C.; subtract one from the last decimal place for every three degrees below this figure. Record your figures under Results and Questions.
4. Note the color and transparency of the urine specimen.
5. If the urine is cloudy, add a drop of concentrated nitric acid to a 5 cc portion of the urine specimen. This will dissolve the phosphates of calcium and magnesium, which may cause cloudiness, but not pus. Therefore, this test may be used to distinguish between a phosphate precipitate and pyuria.
6. Determine the pH of the urine specimen with pHydrion paper (or a pH meter).
7. Record all data under Results and Questions.

Part 2. The Detection of Protein and Glucose in Urine

Discussion

This exercise will demonstrate several common tests for the presence of albumin and glucose in urine. Although a slight amount of albuminuria is normal in approximately 5% of the population, it also occurs abnormally in many renal diseases, in pregnancy, in patients with hypertension, etc. Glycosuria occurs in patients with diabetes mellitus, as well as in certain other diseases. Benedict's solution is commonly used to detect the presence of reducing sugars in urine. This solution, however, is not specific for glucose, since it will give a positive test with any reducing sugar. There are tests which are specific for the presence of glucose. For example, Tes-Tape contains an enzyme which converts glucose to gluconic acid, which then reacts with an indicator in the paper. This reaction occurs only with glucose.

Procedure

1. Heat Test
   a. If the urine is cloudy, it should be filtered to render it clear.
   b. Fill a test tube two-thirds full of urine and hold it slanting in the flame. Heat only the upper part of the tube.
   c. If the boiling causes a cloudiness in the tube, it may be due to albumin or to phosphates. Add a few drops of 10% acetic acid which will dissolve the phosphates. If the cloudiness persists, it was due to protein.
2. Heller's Cold Nitric Acid Test
   a. Add urine to the arm of the albuminoscope containing the black square.
   b. Let sufficient concentrated nitric acid flow down the side arm so that the interface between the two liquids is in the black square. Do not shake.
   c. If albumin is present a white ring will appear at the interface. A colored ring may appear due to the oxidation of any bile pigments present. (The latter test is called Gmelin's test for bile.)
3. Albutest Reagent Tablets (this reagent is also available in stick form as Albustix)
   a. Place the tablet, which contains bromophenol blue, on a clean surface and add one drop of urine to its upper surface.
   b. After the urine drop has been absorbed, add two drops of water, and allow this to penetrate before reading.
   c. Compare the color (in daylight or fluorescent light) on top of the tablet with the color chart provided. If negative, the original color of the tablet will not be materially changed at the completion of the test.
   d. If protein was present in the urine specimen, a blue-green spot will remain on the tablet surface after the addition of water. The amount of protein present is indicated by the intensity of the blue-green spot.
4. Combistix Reagent Strips
   a. Dip the test end of the strip in urine.
   b. The yellow tip is the indicator for protein. Compare this with the protein color chart on the side of the bottle.
   c. The red part of the strip provides a test for glucose. This should be compared with the glucose color guide 10 seconds after dipping. This is a qualitative test for glucose. Clinitest tablets provide a better quantitative test.
   d. The orange pH indicator is third from the top. Compare the color with the pH chart on the side of the bottle. Compare this result with that obtained in Part 1.
5. Benedict's Test
   a. To test for the presence of glucose, place 5 cc of Benedict's solution in a test tube.
   b. Add exactly 8 drops of urine and boil for 2 minutes (or about 5 minutes in a boiling water bath).
   c. A high concentration of phosphates may produce a white precipitate, whereas the precipitate of cuprous oxide in a positive Benedict's test is red.
   d. If the solution remains blue, the test is negative; green is recorded as +, yellow as ++, orange as +++, and red as ++++.
6. Clinitest Reagent Tablets
   a. Place ten drops of water and five drops of urine in a test tube.
   b. Add one Clinitest tablet. The concentrated sodium hydroxide in the tablet generates heat so that the liquid will boil in the test tube. The color of the solution is graded as in Benedict's test.
7. Tes-Tape
   a. Dip a 1½ inch strip of Tes-Tape in the urine specimen. Remove and wait one minute. Immediately compare the darkest area with the color chart.
   b. A yellow color indicates that the urine is sugar free. Positive colors are increasing intensities of green. If the tape indicates 1/2% or higher, wait one additional minute and make a final

comparison. Clinitest reagent tablets are usually used for quantitative tests.

c. To test the reliability of a roll of Tes-Tape, dip it into a 2% solution of glucose or a sample of Coca-Cola from a freshly opened bottle. Both will give a 4+ reading.

8. Record all data under Results and Questions.

Part 3. The Detection of Bile and Ketone Bodies in Urine

Discussion

Ordinarily, ingested fats are oxidized in the body and form ketones (acetone, etc.). Upon further oxidation they become carbon dioxide and water. If diabetes mellitus is present, and the patient is unable to utilize glucose, fat is then oxidized to produce energy. As a result, ketones are produced in greater quantities than usual, and usually are excreted into the urine.

Hay's test for bile salts involves the addition of sulfur to urine. Since bile salts lower the surface tension of the water in the urine, the presence of bile salts will cause the sulfur to sink.

Procedure

1. Hay's Test for Bile Salts in Urine
   a. Sprinkle a little finely powdered sulfur upon the surface of cool urine.
   b. If the sulfur sinks at once, bile salts are present in the amount of 0.01% or more.
   c. If the sulfur sinks only after gentle agitation, bile salts are present in the amount of 0.0025% or more.
   d. If the sulfur remains floating even after gentle shaking, bile salts are absent.
2. Ketostix
   a. Dip the test end of the strip in urine (do not touch the test end of the strip with the hand).
   b. After one minute, compare the color of the dipped end with the color chart.
   c. The color will remain unchanged, or appear cream colored from wetting, in a negative test.
   d. In a positive test, the color will vary from pink to purple, depending upon the amount of ketone bodies present.
3. Acetest Reagent Tablets
   a. Place a tablet on a clean surface.
   b. Put one drop of urine on the tablet, and take a reading after 30 seconds.
   c. Compare the color formed with the color chart.
   d. The color will remain unchanged or cream-colored in a negative test, but will vary from pink to purple, depending on the amount of ketone bodies present, in a positive test.
4. Record all data under Results and Questions.

Part 4. Analysis of Urine Specimens for Unknowns

Discussion

In chemistry, when the composition of a substance is not known, the specimen may be designated as an unknown. In this experiment, the unknowns will be urine specimens to which glucose and albumin may be added. The glucose and albumin may be added to any or all of three urine specimens, and in any combination. Each of the specimens will then be analyzed to determine which contain the additives.

If the preceding tests gave positive results for the presence of glucose and/or albumin in any student's urine, the instructor should be informed of this fact and the urine of some other student should then be used for this experiment.

Procedure

1. Number three clean test tubes (1, 2, and 3) near the top.
2. Add 10 cc of urine to each tube.
3. Take the tubes to the instructor to have glucose and albumin introduced into the specimens.
4. Test your specimens to determine in which tubes, if any, the glucose and albumin have been added. Record your results in the table provided under Results and Questions.

Self-Test—The Urinary System

Directions: See Chapter 1, Self-Test, p. 9.

1. The term glycosuria refers to the presence of **sugar** in the urine.
2. Inflammation of the urinary bladder is called **cystocele.**
3. Anuria means **scanty** urine production.
4. **Oliguria** refers to frequency of urination at night.
5. The general function of a diuretic drug is to **decrease** the excretion of urine.
6. The pyramids are located in the **medulla** of the kidney.
7. The renal pelvis opens directly into the **urinary bladder.**
8. The components of urine enter Bowman's capsule from the blood by the process of **osmosis.**
9. Urea is formed in the **kidneys.**
10. Sugar is normally **excreted** by the kidneys after its removal from the blood.
11. The formation of urea is accomplished by the breakdown of **amino acids.**
12. If the **pancreas** is malfunctioning, sugar might be found in the urine.
13. Renal blood vessels enter the kidney at the **hilus.**
14. The kidneys lie between the **periosteum** and the posterior body wall.
15. Urination is termed **micturition.**
16. The smooth muscle in the wall of the bladder is called the **detrusor** muscle.
17. **Stones** are materials (often albumin) which have taken the form of the tubules in which they are molded.
18. The kidneys excrete **only a small** amount of the water and other substances they receive from the blood.
19. The tuft of capillaries in Bowman's capsule is known as the **glomerulus.**
20. The efferent arteriole has a **larger** diameter than the afferent arteriole.
21. The urine contains a **higher** concentration of urea than the blood plasma.
22. The wastes of metabolic processes include **salts** that pass out of the body through the lungs.
23. The **pancreas** is superior in location to each kidney.
24. The glucose and amino acids are reabsorbed by the part of the nephron known as the **glomerulus.**

25. The connective tissue surrounding the outside of the kidney is called the **capsule.**

26. The urinary bladder serves to **concentrate** urine.

27. The function of the Loop of Henle is to reabsorb **water.**

28. The fundamental microscopic unit of structure of the kidney is the **calyx.**

29. Protein is present in blood, but absent from the **glomerular filtrate.**

30. The term albuminuria refers to the presence of **pus** in the urine.

31. Because of the size of the molecules, fats are usually **found** in urine.

32. The kidneys filter approximately **300** quarts of fluid each day.

33. The pressure of blood in the glomerulus is **higher** than in other systemic capillaries.

34. The renal **veins** rapidly subdivide to form arterioles.

KEY

2. cystitis
3. lack of
4. nocturia
5. increase
7. ureter
8. filtration
9. liver
10. reabsorbed
14. peritoneum
17. casts
20. smaller
22. gases
23. adrenal gland
24. proximal convoluted tubule
26. store
28. nephron
30. protein (albumin)
31. not found
32. 190
34. arteries

Name ______________________

Results and Questions for Chapter 15

Exercise 15-A. Anatomy of the Human Urinary System

1. What is the function of each of the following?

   a. Glomerulus ______________________

   b. Uriniferous tubule ______________________

   c. Renal pelvis ______________________

   d. Ureter ______________________

2. What is the function of the adipose tissue which surrounds the kidney? ______________________

   ____

3. Into which blood vessel does the renal vein empty? ______________________

4. What is the name given to the portion of the renal cortex which is located in between the pyramids? ______________________

5. What is the name given to the layer of the kidney which contains the pyramids? ______________________

   ____

6. What is the name given to the apex of a pyramid? ______________________

7. List the functions of a kidney. ______________________

   ____

   ____

8. What is the usual blood pressure in the glomerulus? ______________________

9. What is the advantage of this high blood pressure in the glomerulus? ______________________

   ____

   ____

10. In which body cavity are each of the following organs located?

    a. Kidneys ______________________ c. Ureters ______________________

    b. Urinary bladder ______________________

11. What is the trigone of the bladder? ______________________

12. Which subdivision of the autonomic nervous system prevents elimination of urine? ______________________

    ____

Exercise 15-B. Dissection of the Urinary System of the Fetal Pig

1. Trace the path of urine from its site of formation to the outside of the body.

2. Define the term retroperitoneal.

3. Which parts of the nephron are located in the cortex?

4. Which parts of the nephron are located in the medulla?

5. Name the structure located in the renal sinus.

6. Trace the path of blood from the renal artery to the glomerulus.

7. Trace the path of blood from the glomerulus to the renal vein.

8. Compare the internal structure of the pig kidney with that of the human kidney.

9. Compare the location of the adrenal glands in the pig with those in man.

Exercise 15-C. Urinalysis

Part 1. Specific Gravity, Appearance, and pH of Urine

1. Specific gravity:

   a. Temperature of urine specimen ____________

   b. Uncorrected specific gravity ____________

   c. Corrected specific gravity ____________

Name ______________________

2. What is the normal specific gravity range of urine? ______________________

3. If the result you obtained was outside the normal specific gravity range, suggest a possible reason. ______________________

____

4. Name two conditions of the body which result in an increase in the specific gravity of the urine.

____

____

5. Name two conditions of the body which result in a decrease in the specific gravity of the urine.

____

____

6. What is the function of the antidiuretic hormone (ADH)? ______________________

____

7. What is the usual relationship found between volume of urine voided and specific gravity (direct or indirect)? ______________________

8. What was the color of your urine specimen? ______________________

9. Name the pigment responsible for the normal color of urine. ______________________

10. As the volume of urine voided increases, what usually happens to the color of urine? ______________________

____

11. Describe the transparency of the urine specimen. ______________________

12. If your specimen was cloudy, what results were obtained when a drop of nitric acid was added to it? ______________________

13. What conclusion concerning the origin of the cloudiness of the urine (in No. 12) can be drawn?

____

____

____

14. All urine specimens become cloudy upon excessive standing at room temperature. Explain why.

____

____

15. What was the pH of the urine specimen? ______________________

16. What is the average pH of urine? ____________________

What is the normal range of pH? ____________________

17. Name the compound usually considered responsible for the acidity of urine. ____________________

____

18. Why do all urine specimens become alkaline upon standing? ____________________

____

19. Would you expect the following to result in an increase or a decrease in the pH of urine (check the appropriate space)? Explain each answer.

Diet and Body Conditions	*pH Increased*	*pH Decreased*	*Explanation*
Eating a large amount of protein			
Starvation			
Diabetes mellitus			
Vegetarian diet			
Renal tract infection			

Part 2. The Detection of Protein and Glucose in Urine

1. Fill in the following chart concerning the protein and glucose tests.

Protein and Glucose Tests	*Purpose of Test*	*Results*	*Explanation of Results*
Heat test			
Heller's cold nitric acid test			
Albutest			
Combistix			
Benedict's			
Clinitest			
Tes-tape			

2. Compare the value for pH obtained with Combistix with that obtained with pHydrion paper or the pH meter. ____________________
Which value is likely to be more accurate? __________ Why? ____________________

____

Name ______________________

3. How could you determine if the sugar resulting in a positive Benedict's test in a lactating woman was lactose or glucose? ______________________

____

4. Name the disease which is the most likely cause of a high amount of sugar in the urine. ______________________

____

5. What is the most likely explanation of the appearance of a trace of sugar in the urine? ______________________

____

6. Which term is used for the presence of each of the following in urine?

a. Pus ______________________ c. Glucose ______________________

b. Protein ______________________ d. Ketone bodies ______________________

Part 3. The Detection of Bile and Ketone Bodies in Urine

1. Record all results in the following table.

Tests	*Purpose of Test*	*Results*	*Explanation of Results*
Gmelin's test			
Hay's test			
Ketostix			
Acetest			

2. In which disease would you expect Ketostix to be positive?

3. Compare the composition of the blood in the glomerulus, the glomerular filtrate, and urine by filling in the following chart. Place + in the blank if the substance is normally present, – if it is usually absent.

	Blood	*Glomerular Filtrate*	*Urine*
Water			
Protein			
Glucose			
Sodium chloride			
Ketone bodies			
Erythrocytes			

Part 4. Analysis of Urine Specimens for Unknowns

In the table below, place a + sign to indicate the presence of the substance (positive) and a – sign to indicate the absence of the substance (negative). In the space provided, write the name of the test(s) that helped you obtain your results.

Tube No.	*Glucose*	*Protein*	*Tests Used*
1			
2			
3			

16

The Endocrine System

Exercise 16-A Endocrine Glands

Objectives

To become familiar with the locations and functions of endocrine glands.

Materials

Dissectable torso; brain model showing hypophysis and pineal gland; anatomical charts; text and other references; fetal pig; and embalmed cat.

Discussion

The secretions of endocrine glands (containing hormones) diffuse into the blood which transports them to the structures acted upon. In contrast to these ductless glands, exocrine glands have ducts which convey the glandular secretions to some surface either inside or outside of the body. For example, the secretions of the salivary glands are carried by ducts to the buccal cavity.

A compound, to be classed as a hormone, must meet the following criteria: (1) it is produced by a particular endocrine gland; (2) removal of the gland, or the part of the gland producing the compound, results in the failure of some other organ to develop or function normally; (3) in the absence of the endocrine gland, artificial replacement of the hormone in the body restores the target organ to its normal size and functional capacity. Chemically, hormones are either steroids, proteins, or derived from proteins.

In the autonomic nervous system, impulses traveling over preganglionic fibers stimulate the secretion of acetylcholine at the synapses with ganglion cells. The compound is also released at postganglionic parasympathetic terminals and most of the postganglionic sympathetic fibers that supply smooth muscle and sweat glands in the skin. All such fibers are known as cholinergic fibers; these fibers inhibit the action of certain target organs such as the heart and the sphincters of the gastrointestinal tract, and stimulate the action of others, such as the muscle in the gastrointestinal tract, bronchioles, urinary bladder, and gall bladder. Sympathin, which is the same as norepinephrine (noradrenalin), is released at the postganglionic sympathetic terminals (with the exceptions noted above). These fibers are known as adrenergic fibers; they are excitatory in certain areas such as the heart, blood vessels in the abdominal viscera, and sphincters, and inhibitory in the gastrointestinal tract and urinary bladder. While these fibers are not endocrine glands, the compounds secreted are chemical coordinators which act in a manner similar to hormones. The endocrine and nervous systems are the two main integrating mechanisms of the body. The normal functioning of these two systems results in the parts of the body working together as an integrated whole.

Procedure

1. Using your text, references, and anatomical charts as resource materials, locate the glands on the following page on the dissectable torso and the model of the brain:
   a. Hypophysis (pituitary)
   b. Ovaries
   c. Pancreas
   d. Parathyroids
   e. Pineal gland
   f. Suprarenals (adrenals)
   g. Testes
   h. Thymus
   i. Thyroid
2. Locate the following glands in the fetal pig and, in the space provided, give a brief, precise anatomical description of their location:
   a. Hypophysis

   b. Ovaries

   c. Pancreas

   d. Parathyroids

   e. Suprarenals (adrenals)

   f. Testes

   g. Thymus

   h. Thyroid

3. An embalmed cat will be dissected to show each of the glands listed under Step 2. The location of each will be pointed out by the instructor. Following the demonstration, compare the glands of the fetal pig with those of the cat. Note similarities and differences in the sizes and locations of the glands. Use care to avoid damaging the glands in the cat.
4. Identify the glands indicated in Figure 16-1 on the dissectable torso, in wall charts, and in illustrations in other reference material. Figure 16-1 may be used as a self-test. See Section J, p. xx (Illustrations as Self-Tests) for a suggested approach.
5. Answer the questions and complete the tables under Results and Questions.

Exercise 16-B Action of Hormones

Part 1. The Effect of Epinephrine (Adrenalin) on the Frog Heart

Objective

To determine the effect of epinephrine on the frog heart.

Materials

Frog board; stand; clamp; clips; beaker of frog Ringer's solution; dissecting set; pin hook; medicine dropper; 1:10,000 epinephrine (adrenalin) solution; syringe with very small diameter needle; live frog; and **either** kymograph and heart lever **or** physiograph, myograph B, pin electrodes, and stimulator output extension cable.

1. Expose the heart of a pithed frog and pass a pin hook through the apex. Attach the heart lever or myograph, by means of a thread, to the heart. (The procedure for pithing the frog and preparing the equipment and the heart is described in Exercise 12-G, Part 3.
2. Keep the heart moist with Ringer's solution until ready to apply epinephrine.
3. Obtain a tracing of normal heart action.
4. Apply a small amount of epinephrine (adrenalin) solution to the surface of the heart with a medicine dropper. Continue recording the contractions.
5. By comparing the records obtained before and after the administration of the hormone, note the effect of epinephrine on the rate and force of the heart beat. (If no response was obtained, inject 0.1 cc of the epinephrine solution directly into an atrium, and obtain a record of the results.)

KEY TO FIGURE 16-1

1. pineal gland
2. thyroid gland
3. thymus gland
4. adrenal gland
5. testis
6. hypophysis (pituitary gland)
7. parathyroid glands
8. pancreas
9. ovary

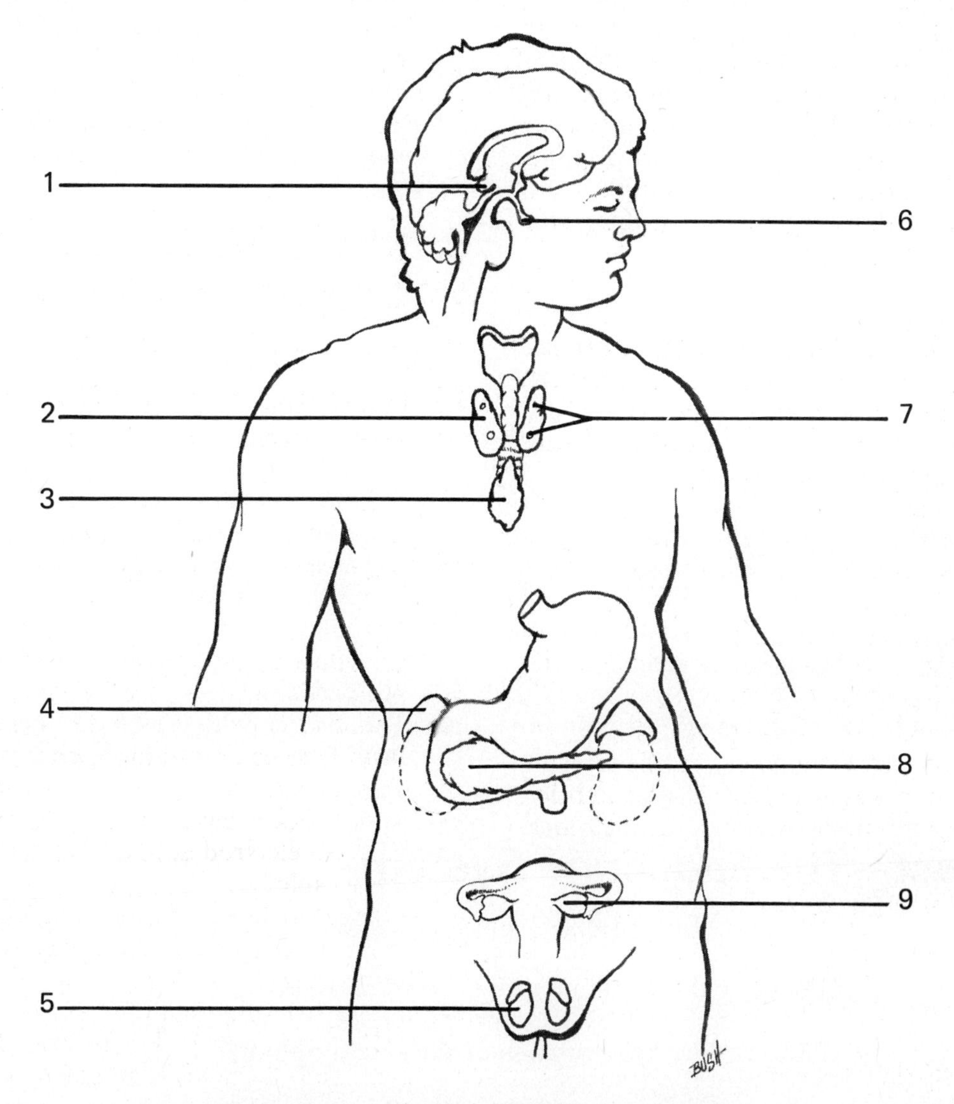

Figure 16-1. Location of Endocrine Glands

Part 2. The Effect of Acetylcholine on the Frog Heart

Objective

To determine the effect of acetylcholine on the frog heart.

Materials

Frog setup used in Part 1; medicine dropper; and 1% acetylcholine chloride (or acetylcholine bromide).

Procedure

1. Thoroughly wash the exposed frog heart used in Part 1 with Ringer's solution. If it was necessary to inject epinephrine into an atrium in the preceding experiment, use a second frog (the recovery period in this case will be too prolonged for the time alloted).
2. Obtain a tracing of normal heart action.
3. Apply a few drops of 1% acetylcholine chloride (or acetylcholine bromide) solu-

tion to the surface of the heart while you continue to record the heart action. If the heart fails to stop, apply a few more drops of the acetylcholine solution.

Part 3. The Effects of Insulin on a Fish

Objective

To observe the effects of insulin on a fish.

Materials

Commercial insulin; two 500 cc beakers; glucose; and 1½ to 2 inch fish ("sunfish" or "bluegill" of genus *Lepomis* are preferred; goldfish may also be used, but the results are more variable).

Discussion

A fish is a convenient animal in which to observe the effects of insulin and, since the action of this hormone on the fish involves the same mechanism as that in humans, the results obtained in the following experiment should provide insights concerning its administration to humans.

Insulin aids in maintaining the physiological balance between glycogen stored in the liver and muscles and the glucose in the blood. When insufficient insulin is produced by the Islets of Langerhans (found in the pancreas), as is the case in diabetes mellitus, injections of insulin are usually administered. Care must be exercised to avoid giving too much insulin, since this results in such a decrease in the level of blood glucose that the patient goes into shock. If the nervous system is to function properly, the blood glucose level must be adequate.

Procedure

1. Place a small (1½ to 2 inch in length) fish in a beaker containing 200 cc of water to which 10 drops of commercial insulin have been added.
2. Observe the fish for the following symptoms that occur as a result of the diffusion of the insulin through the gill membranes of the fish into its circulatory system: extreme irritability, convulsive motions, and finally coma (state of insensibility).
3. When the fish goes into a coma, place it in a second beaker containing 200 cc of water to which a half teaspoon of glucose has been added.
4. The fish should return to normal, and it may then be placed back in the aquarium.

Self-Test—The Endocrine System

Directions: See Chapter 1, Self-Test, p. 9.

1. Hypofunction of the thyroid gland in infancy is called **myxedema.**
2. Diabetes insipidus is caused by hypofunction of the **anterior pituitary.**
3. Tetany is caused by hyposecretion of the **thyroid.**
4. Insulin is not given by mouth because it is **digested by enzymes in the alimentary canal.**
5. The somatotropic hormone is produced by the **anterior pituitary.**
6. Goiter is a condition characterized by excessive growth of the **thymus.**
7. **ADH** stimulates the production of the adrenal cortical hormones.
8. A disease resulting from the failure of the nephrons to reabsorb water is called **diabetes mellitus.**
9. Secretions of the **adrenal** glands control calcium metabolism.

10. An exophthalmic goiter is a result of **hyperthyroidism.**

11. The **pineal** gland is located in the sella turcica.

12. **Hypertrophy** means an increase in the size of a structure.

13. True dwarfism results from a lack of the **parathyroid** hormone.

14. Protruding eyeballs may indicate **hyperthyroidism.**

15. Steroids are formed by the **adrenal cortex.**

16. As a person grows older, the thymus **remains the same size.**

17. Adrenalin **decreases** the consumption of oxygen.

18. The PBI test is a measure of **thyroid** function.

19. The hormone that causes the gall bladder to contract is **secretin.**

20. A symptom of diabetes mellitus is **alkalosis.**

21. ACTH is produced by the **adrenal cortex.**

22. Iodine is needed for the production of **parathormone.**

23. Acromegaly is caused by **a lack of thyroxin.**

24. Increased secretion of ADH may result in **diuresis.**

25. **The lactogenic hormone** causes the ejection of milk from the mammary glands.

26. Oxytocin is stored by the **posterior pituitary.**

27. The hormone controlling male secondary sex characteristics is **FSH.**

28. The adrenal **medulla** is essential for life.

29. Epinephrine stimulates **glycogenesis.**

30. A lack of **parathormone** can cause muscle twitches.

31. Dwarfism, gigantism, and acromegaly result from a defect in the **thyroid.**

32. The interstitial cells produce **ICSH.**

33. The hormones of the adrenal medulla have the same function as the **parasympathetic** nervous system.

34. The absence of insulin results in the formation of **ketone bodies.**

35. **Cretinism** is more common where there is a lack of iodine in the soil.

KEY

1. cretinism
2. posterior pituitary (or hypothalamus)
3. parathyroids
6. thyroid
7. ACTH
8. diabetes insipidus
9. parathyroid
11. pituitary
13. somatotropic
16. decreases in size
17. increases
19. cholecystokinin
20. acidosis
21. anterior pituitary
22. thyroxin
23. an excess of STH
24. edema
25. oxytocin
27. testosterone
28. cortex
29. glycogenolysis
31. pituitary
32. testosterone
33. sympathetic

Name ____________________

Results and Questions for Chapter 16

Exercise 16-A. Endocrine Glands

1. Define an endocrine gland. ____________________

2. Which endocrine glands are considered to be mixed glands? ____________________

3. Which endocrine gland is embedded in the dorsal surface of the thyroid gland? ____________________

____

4. Fill in the following chart concerning function of hormones.

Function of Hormone	*Hormone*
Absorption of glucose into cell	
Regulates BMR	
Regulates thyroid gland	
Stimulates spermatogenesis	
Regulates calcium balance	
Regulates reabsorption of water in kidney	
Stimulates contraction of pregnant uterus	
Stimulates ovulation	
Regulates mental development	
Regulates sodium and potassium balance	
Increases heart rate and blood pressure	

5. The table on the following page contains a list of symptoms resulting from endocrine disorders. In the spaces provided, write the name of the specific part of the gland (wherever possible), the name of the hormone involved, and the disease, and place a check mark under hyposecretion or hypersecretion

6. A ductless gland has been found to secrete a substance which influences growth and development, as well as metabolic rate. How would you determine whether the active substance in the secretion was a hormone? ____________________

____

7. How would you determine if more than one hormone was involved in Question 6? ____________________

____

Symptoms of Endocrine Dysfunction	*Gland (or Part of Gland)*	*Hormone*	*Name of Disease*	*Hyposecretion*	*Hypersecretion*
Failure of physical, sexual, and mental development					
Muscle spasms due to increased irritability of nerves and muscles					
Precocious sexual development					
Continued growth of bones in children					
Shock, convulsions, marked decrease in blood sugar					
Copious excretion of urine					
Loss of vigor, falling hair, puffy skin, lowered BMR					
Development of secondary sexual characteristics of male in adult female					
Obesity, sexual infantilism, often subnormal intelligence	both anterior and posterior pituitary	all hormones	Frolich's syndrome		
Coma, marked increase in blood sugar					
Demineralization of bones, with spontaneous fractures					
Bulging eyeballs, goiter, loss of weight, nervousness, rapid pulse					
Lower jaw, hands, and feet become larger in adults					

8. On the basis of your study of the endocrine system, which gland was most likely to be producing the secretion in Question 6? ______________________________

____

Name ______________________

9. Name three symptoms of Cushing's disease. ______________________

____

10. Name three symptoms of Addison's disease. ______________________

____

11. Name three hormones produced by the ovary. ______________________

____

12. Which endocrine gland produces vasopressin? ______________________

13. Which gland is called the master gland? ______________________

14. What is the function of the pineal gland? ______________________

____

15. Which hormone is sometimes called an emergency hormone? ______________________

16. Which two glands are called "glands of childhood"? ______________________

17. What is the common name of the somatotropic hormone? ______________________

Exercise 16-B. Action of Hormones

Part 1. The Effect of Epinephrine (Adrenalin) on the Frog Heart

1. Paste your recording below. This record should show both normal heart action and the effect of adrenalin. Label the point at which the adrenalin was applied.

2. Briefly describe the effect of adrenalin on the rate and force of the heart beat. ______________________

____

____

3. Which gland secretes adrenalin? ______________________

4. What are the symptoms of a pheochromocytoma? ______________________

____

____

Part 2. The Effect of Acetylcholine on the Frog Heart

1. Paste your recording below. This record should show both normal heart action and the effect of acetylcholine.

2. Briefly describe the effect of acetylcholine on the rate of the heart beat.________________

____

3. What is meant by vagal inhibition?____

   Relate this to the results obtained in this experiment.________________________

____

4. What is meant by a cholinergic nerve?____

5. Classify the vagus nerve (adrenergic or cholinergic).________________________

Part 3. The Effects of Insulin on a Fish

1. Briefly describe what happened to the blood glucose level of the fish as a result of diffusion of insulin into the blood stream.____

____

2. Explain what happened to cause the fish to return to normal when it was placed in the glucose solution.____

3. What is the purpose of a glucose-tolerance test?________________________________

4. What is insulin shock?____

5. What first aid treatment is recommended for a person in insulin shock?________________

____

6. How may a diabetic avoid insulin shock?____

17

The Reproductive System

Exercise 17-A The Human Male Reproductive System

Objectives

To become familiar with the location and function of the structures of the male reproductive system.

Materials

Dissectable torso; anatomic charts; references; and slides of the testis and epididymis.

Discussion

The male reproductive system consists of the two male gonads (testes), the ducts which conduct the sperm to the outside, and the accessory structures (glands and penis). Spermatogenesis begins at puberty under the influence of hormones, and continues throughout the life of the male.

Procedure

1. Examine a slide of a cross section through the human testis under both low and high power of the microscope. In order to locate the testis on the slide, compare it with Figure 17-1 (since some slides contain the epididymis on the same slide). The seminiferous tubules (one of which is illustrated in Figure 17-1), should fill the field.
2. Examine the shape of the seminiferous tubules. These are approximately circular, since each is a cross section through a long tubule (see Fig. 17-4).
3. Draw a portion of the field that includes three seminiferous tubules with the interstitial cells in between.
4. Study the wall of one seminiferous tubule. The cells here are in various stages of meiosis. The cells near the outside of the tubule are called **spermatogonia.** These cells begin the process of spermatogenesis (sperm production) by dividing by mitosis several times and enlarging to form **primary spermatocytes,** located closer to the center of the tubule. The primary spermatocytes undergo meiosis and form **secondary spermatocytes,** and finally **spermatids.** (Study Figure 17-6 which compares spermatogenesis and oogenesis.) The spermatids are the cells closest to the lumen of the tubule. These mature into **spermatozoa** (sperm)

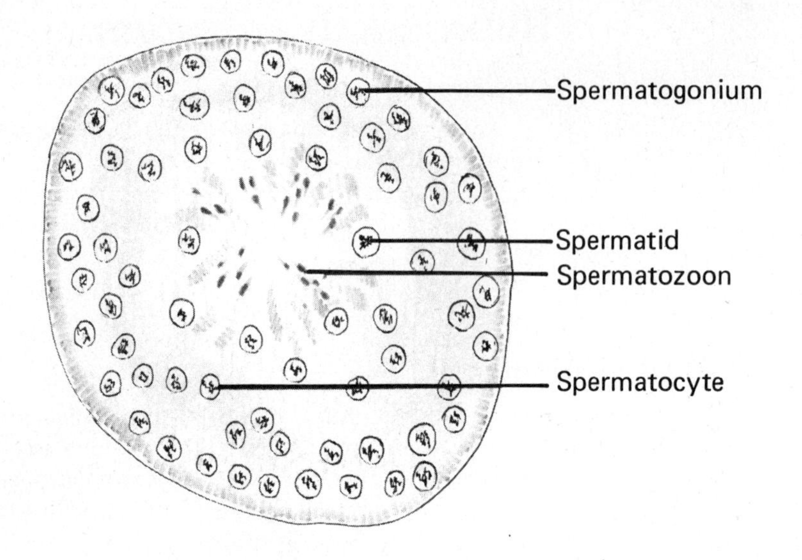

Figure 17-1. Cross Section through a Seminiferous Tubule

Figure 17-2. Cross Section through a Seminiferous Tubule (enlarged)

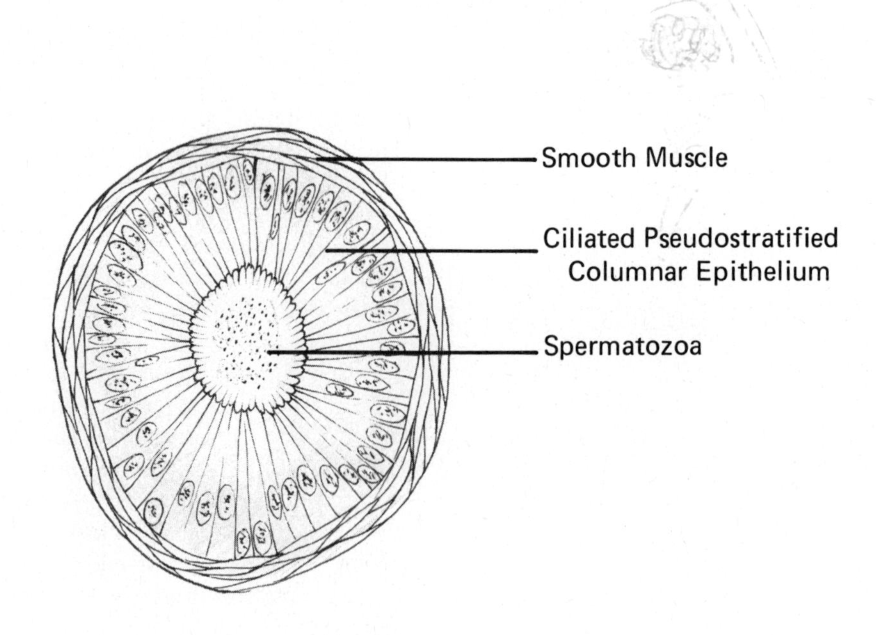

Figure 17-3. Cross Section through the Epididymis

located in the vicinity of the spermatids. Figure 17-2 shows an enlargement of part of the wall of the seminiferous tubule.

5. Draw a strip of cells from the outside to the center of one seminiferous tubule on your original drawing of the seminiferous tubule.
6. Study the slide of the epididymis (or use the portion of the epididymis on the slide of the testis). Identify each of the parts listed in Figure 17-3. Note the sperm in the lumen of the tubule. Compare the structure of the wall of the epididymis with the wall of the seminiferous tubule.
7. The testis, the epididymis, and the beginning of the vas deferens are located in the scrotum. These structures and some of their parts are indicated on Figure 17-4 which may be used as a self-test. See Section J, p. xx (Illustrations as Self-Tests) for a suggested approach.
8. Using your text or other references, locate the structures on the key to Figure 17-5 on the dissectable torso and compare them with Figure 17-5 which may be used as a self-test.

KEY TO FIGURE 17-4

1. vas efferens
2. tunica vaginalis
3. tunica albuginea
4. seminiferous tubule
5. septum
6. epididymis
7. vas deferens
8. rete testis

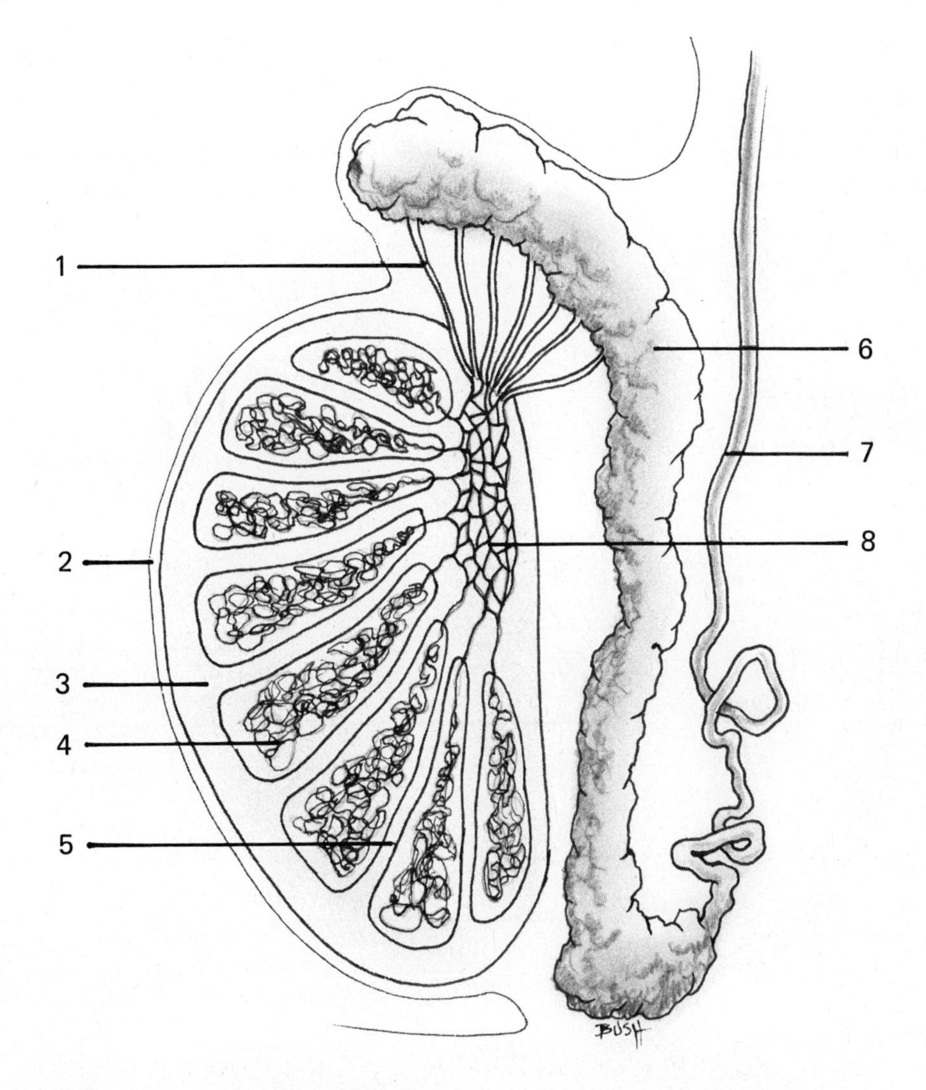

Figure 17-4. Vertical Section through the Testis and Adjacent Structures

KEY TO FIGURE 17-5

1. vas deferens
2. pubis
3. prostate gland
4. corpus cavernosum urethrae
5. urethra
6. penis
7. prepuce
8. glans penis
9. ureter
10. sacrum
11. urinary bladder
12. seminal vesicle
13. ejaculatory duct
14. rectum
15. Cowper's gland
16. external anal sphincter
17. anus
18. tunica vaginalis
19. testis
20. scrotum

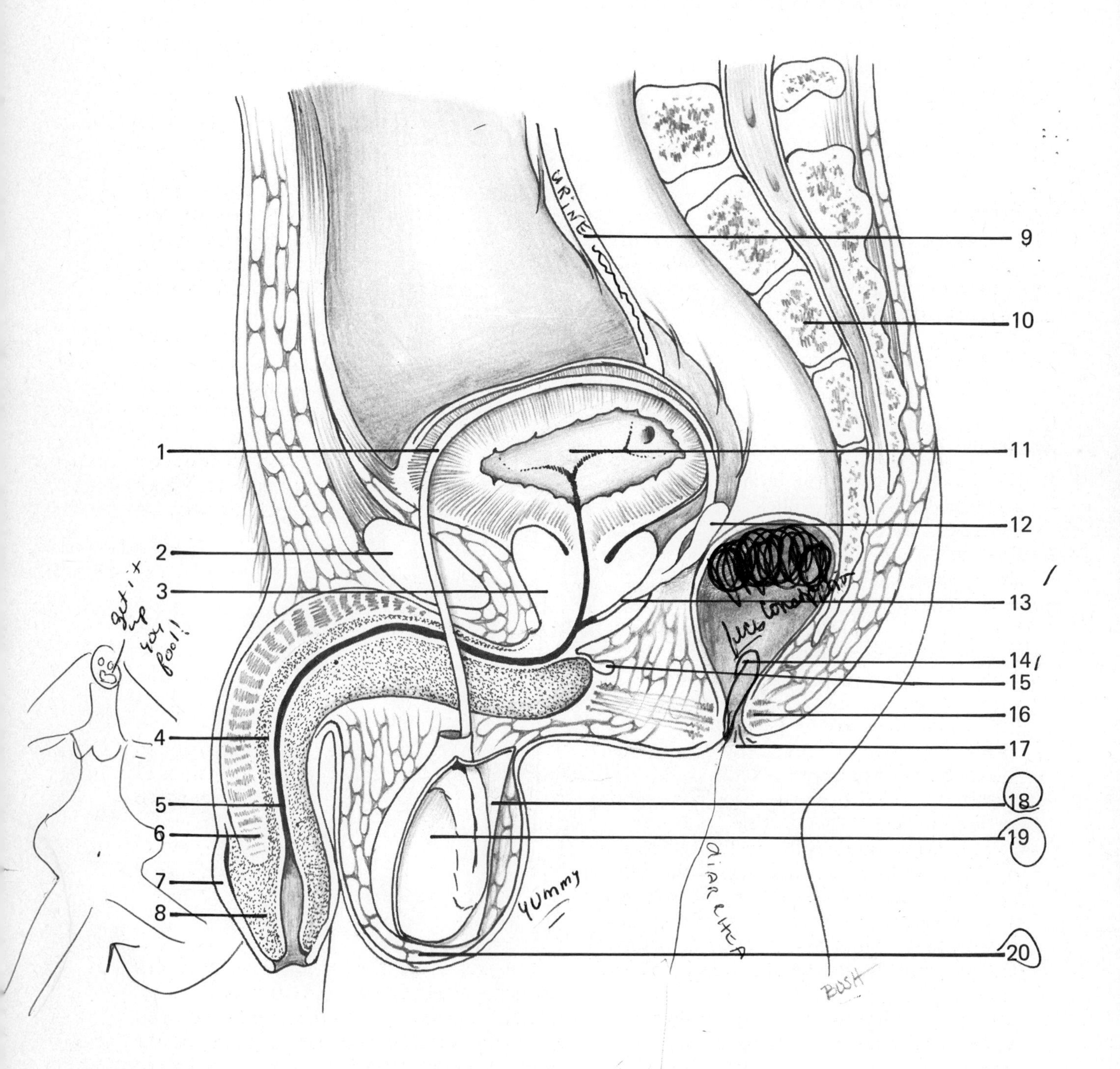

Figure 17-5. Parasagittal Section through Male Pelvis

Exercise 17-B The Human Female Reproductive System

Objectives

To become familiar with the location and function of the structures in the female reproductive system.

Materials

Dissectable torso; anatomic charts; references; and slides of the ovary.

Discussion

The female reproductive system consists of the female gonads (ovaries), the Fallopian tubes, uterus, vagina, and accessory structures. The process of oogenesis (egg production), which actually begins before birth, ceases at the time of climacteric, usually between ages 45 and 50.

Procedure

1. Obtain two slides of the human ovary, one showing the Graafian follicle and the other containing the corpus luteum.
2. First study the slide of the ovary showing the Graafian follicle. This slide shows various stages in the development of the follicle. Locate a **primordial follicle** under high power. This structure contains a large immature ovum with a conspicuous nucleus, surrounded by a single layer of follicle cells. Mark off a sheet of drawing paper into fourths. Draw the primordial follicle in the first area.
3. Returning to low power, locate an **immature follicle.** The follicle cells have divided many times and secreted follicular fluid. The immature ovum (which may not be visible, depending on how the follicle was sectioned) is located at one end of the follicle, surrounded by follicle cells. Draw this structure in the second area on your paper.
4. On the same slide locate and draw a **mature (Graafian) follicle** in the third area of your paper. This is a very large follicle filled with clear fluid.
5. On the second slide of the ovary, locate and draw the **corpus luteum.** This structure appears after ovulation. The wall of the follicle collapses and is thrown into folds. The cavity assumes an irregular stellate shaped appearance. The cells become filled with a yellow fluid. Draw this structure in the fourth area of your drawing paper.
6. The stages in oogenesis are similar to those of spermatogenesis (study Fig. 17-6). The oogonia multiply before the birth of a female child. At the time of birth, the primordial follicles contain the primary oocytes. These mature at various times during the reproductive years of the woman. Approximately 20 primordial follicles begin to develop each menstrual month; however, usually only one matures into a Graafian follicle and releases an egg.
7. Compare the appearance of structures on your slides with Figure 17-7, which may be used as a self-test.
8. Using a model, wall chart, your text, or other references, identify each of the structures indicated in the key below and compare with Figure 17-8, which may be used as a self-test.
9. Using your text or other references, locate the structures listed in the key to Figure 17-9 on the dissectable torso and compare with Figure 17-9.

Exercise 17-C The Endometrial (Menstrual) Cycle

Objectives

To identify the phases of the endometrial cycle; to learn the origins and the functions of the hormones involved in the endometrial cycle.

Materials

Textbook and/or other references.

Discussion

Menarche, the beginning of a functional menstrual cycle, usually starts at ages 12 to 15. Menstruation is the periodic disintegration of the highly vascular endometrial lining of the uterus. The endometrial tissue is discharged along with blood through the vagina about

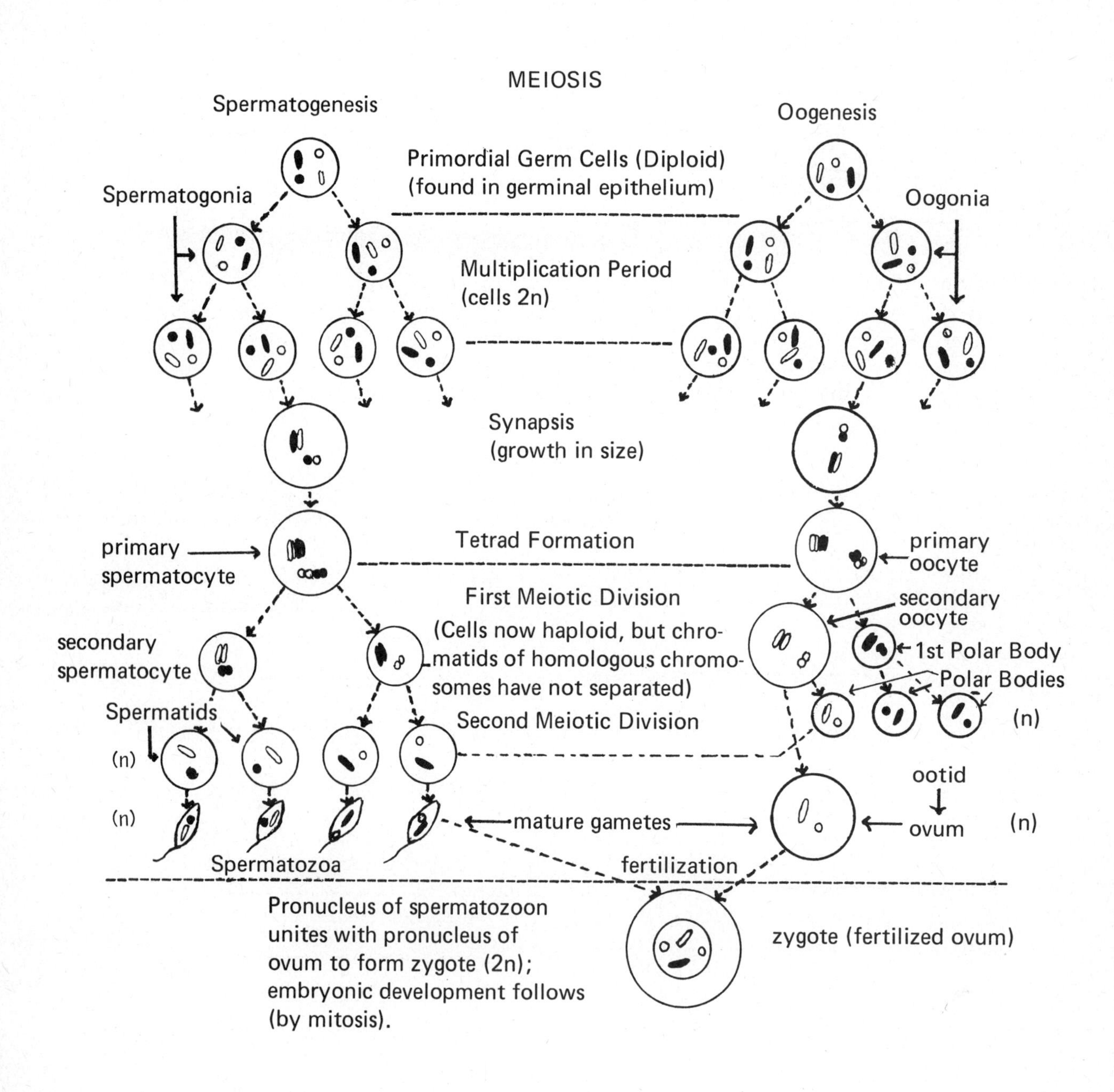

Figure 17-6. Comparison of Spermatogenesis and Oogenesis

KEY TO FIGURE 17-7

1. primordial follicles
2. blood vessels
3. immature follicle
4. follicular fluid
5. corpus luteum
6. Graafian follicle
7. immature ovum

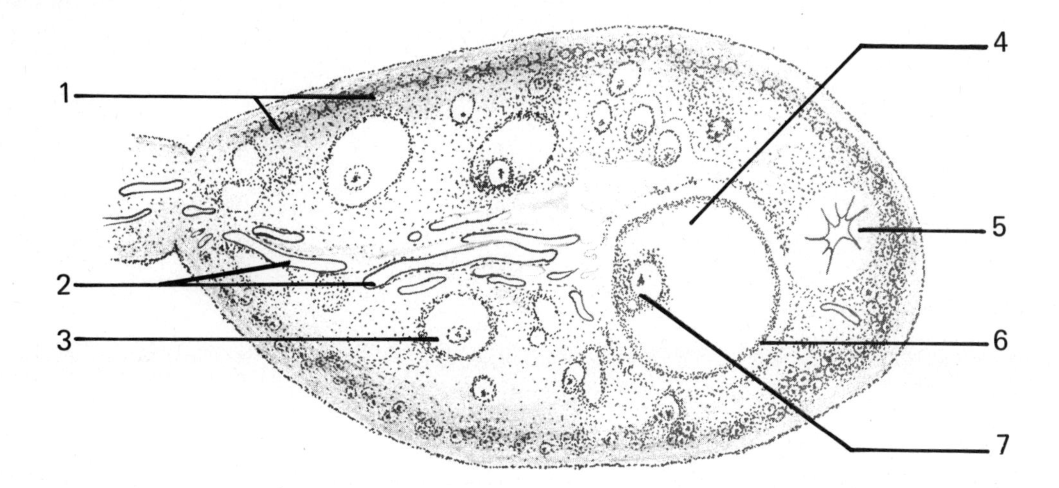

Figure 17-7. Diagramatic Section through Human Ovary

KEY TO FIGURE 17-8

1. fundus
2. ovary
3. ovarian ligament
4. body (uterine) cavity
5. myometrium
6. Fallopian tube
7. fimbriated opening
8. ovary (sectioned)
9. internal os
10. cervical canal
11. external os (mouth of uterus)
12. cervix
13. fornix
14. vagina

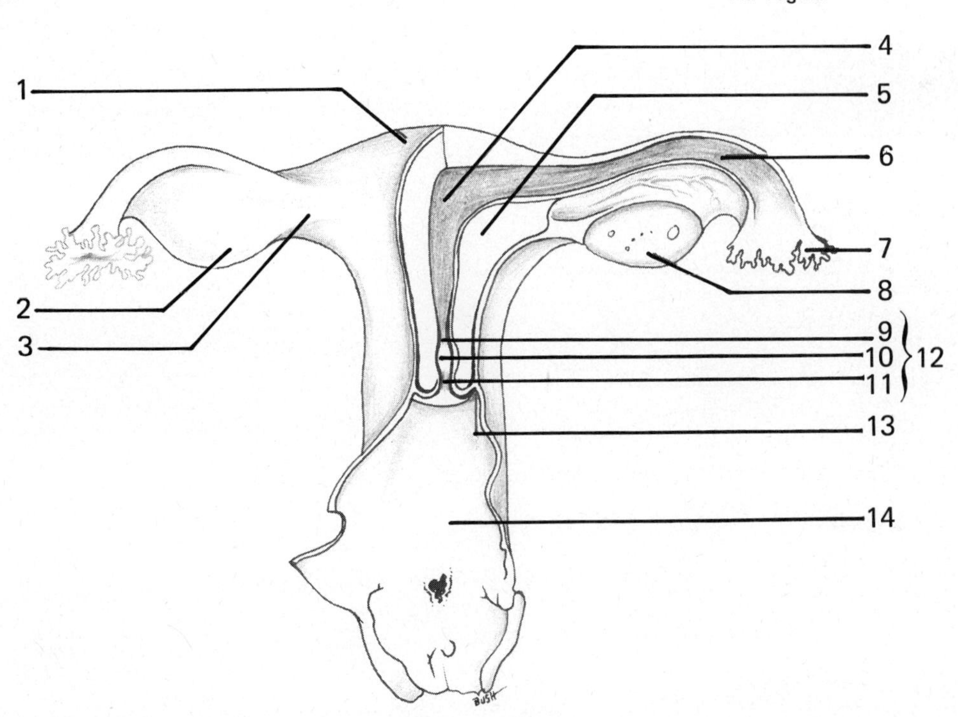

Figure 17-8. Coronal Section through the Female Reproductive Organs

KEY TO FIGURE 17-9

1. fimbriated opening
2. ovary
3. Fallopian tube
4. fundus
5. body cavity
6. urinary bladder
7. urethra
8. labium minus
9. labium majus
10. sacrum
11. rectum
12. cervix
13. rectouterine excavation
14. external os
15. fornix
16. vagina
17. external anal sphincter
18. anus

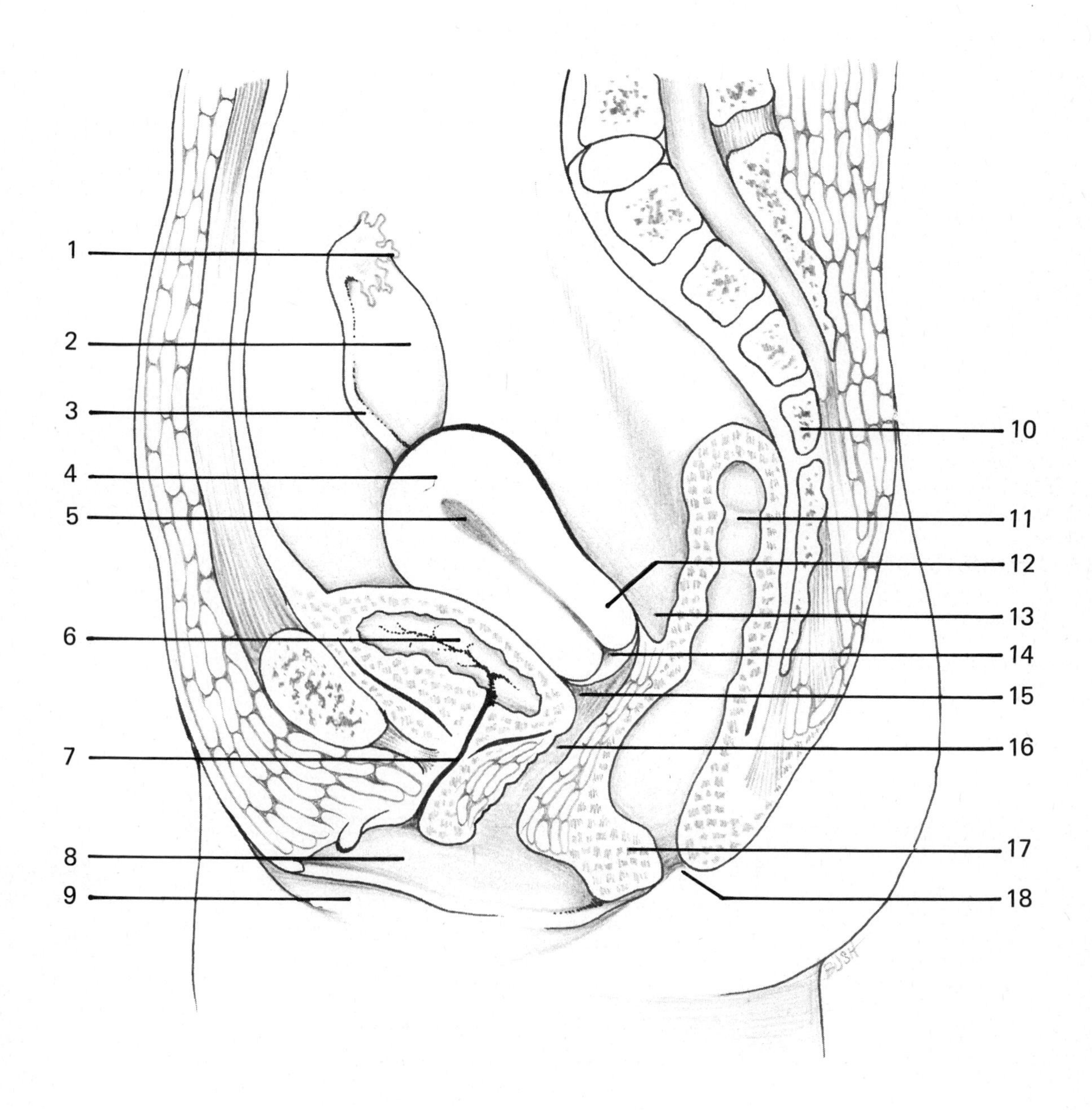

Figure 17-9. Median Sagittal Section through Female Reproductive Organs

KEY TO FIGURE 17-10

1. prepuce of clitoris
2. external urethral orifice
3. labium minus
4. labium majus
5. mons pubis
6. clitoris
7. vestibule
8. vaginal orifice
9. hymen
10. perineum
11. anus

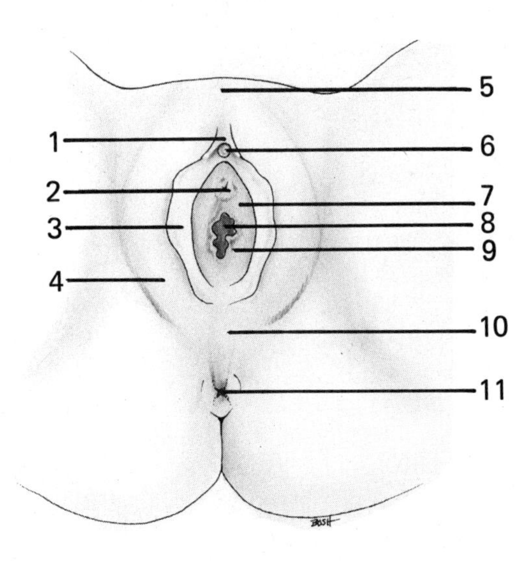

Figure 17-10. Female External Genital Organs (perineal view)

KEY TO FIGURE 17-11

1. clavicle
2. pectoralis major
3. adipose tissue
4. skin
5. intercostal muscles
6. rib
7. tubuli lactiferi (milk ducts)
8. nipple
9. areola
10. glandular tissue

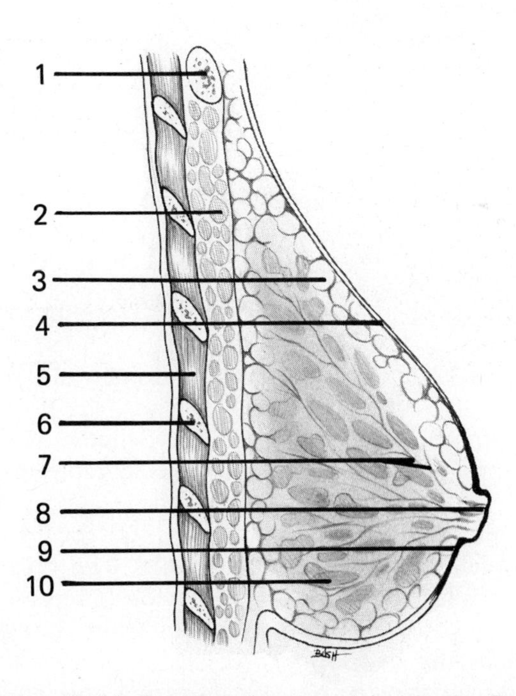

Figure 17-11. Sagittal Section through the Breast

every 28 days. The period of discharge lasts an average of four to five days. Normally, menstruation does not occur during pregnancy, and it stops permanently at menopause (which usually takes place between 45 and 50 years of age).

In the endometrial cycle (see Schème 17-1, the preovulatory phase (the time before ovulation) is often days 6 to 13. However, this phase may be much longer. The time of ovulation depends on the duration of the preovulatory phase. For example, in a 32 day cycle the preovulatory phase would probably last until cycle day 17 or 18 and ovulation would then take place on the 18th or 19th cycle day (instead of day 14 or 15). The length of the postovulatory or premenstrual phase, on the other hand, is fairly constant and usually lasts 14 cycle days. For example, in the 28 day cycle illustrated in the diagram (Scheme 17-1), the postovulatory phase is from days 15 to 28. Thus, any differences in the length of the total menstrual cycle are due primarily to differences in the duration of the preovulatory, rather than the postovulatory, phasȩ.

During the preovulatory phase a follicle containing an immature ovum (egg) begins to develop in response to the presence of follicle stimulating hormone (FSH), a secretion of the anterior lobe of the pituitary gland. The follicle soon begins to secrete the hormone estrogen which stimulates the initial formation of the endometrium. After ovulation (during the postovulatory phase), the anterior pituitary secretes luteinizing hormone (LH) which causes the follicle to thicken into the corpus luteum. This structure produces the hormone progesterone (and estrogens) which continues the development of the endometrium and maintains it for about seven or eight months in the event of pregnancy. An ovum will die if it is not fertilized within about 24 hours after ovulation. If fertilization does not occur, the pituitary gland (a few days after ovulation) does not provide enough hormones (e.g., LH) to maintain the corpus luteum which, in turn, ceases to provide hormones for maintaining the endometrium. The result is menses.

Procedure

1. Using your text and/or some other references, define each of the following:
   a. Menses ______________________________
   ____
   b. Preovulation ______________________________
   ____
   c. Premenstrual ______________________________
   ____
   d. Postovulation ______________________________
   ____
   e. Endometrium ______________________________
   ____
   f. Graafian follicle ______________________________
   ____
   g. Corpus luteum ______________________________
   ____
   h. Anterior pituitary ______________________________
   ____
   i. Ovulation ______________________________
   ____

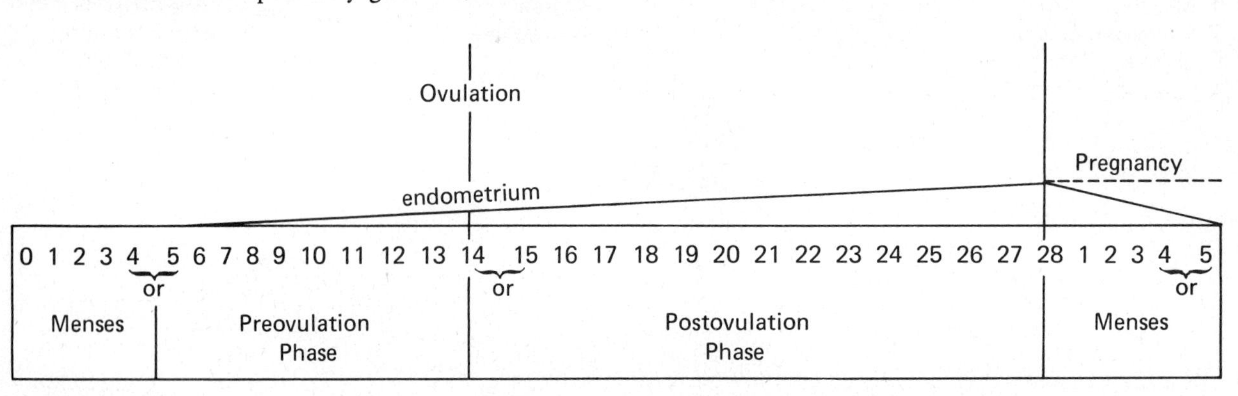

Scheme 17-1. Diagram of the Endometrial Cycle

In Scheme 17-1, the slanted line between days 5-28 represents the buildup of the endometrium. The slanted line between days 1-5 represents the disintegration of the endometrium. The dotted line with the word pregnancy on it represents the maintenance of the endometrium during pregnancy.

2. Determine which hormones are secreted during the preovulation phase and record them on the above diagram.

3. Determine which hormones are secreted during the postovulation phase and record them on Scheme 17-1.

Exercise 17-D The Placenta

Objectives

To identify and learn the structures of the placenta.

Materials

Model of placenta, preserved fetus with umbilical cord and placenta; text and/or other references.

Discussion

The placenta serves as an organ of nutrition, respiration, and excretion for the developing individual. It also serves as a temporary endocrine gland. The chorionic portion (see Fig. 17-12) of the placenta secretes chorionic gonadotropin, a hormone which maintains the functioning of the corpus luteum. The placenta also produces estrogens and progesterone.

Procedure

1. Use Figure 17-12 showing a section of the placenta, as a self-test.
2. Compare Figure 17-12 with the model of the placenta, the preserved fetus, and with drawings of this organ in your text and/or other references. As you study the placenta, note the following:
   a. The **trophoblast,** which develops an outer and an inner layer.
   b. The **placental villi,** which grow from the inner layer of the trophoblast.
   c. The **chorion,** which includes the trophoblast and its villi; it is the outer fetal membrane.
   d. The **umbilical arteries,** which lead into capillaries in the chorionic villi, and which carry waste materials from the embryo.
   e. The **marginal sinuses** and **intervillous spaces,** where the mother's blood surrounds the villi.
   f. The **maternal arterioles** and **venules,** which carry blood to and from the marginal sinuses and intervillous spaces.
   g. The **umbilical vein,** which carries

KEY TO FIGURE 17-12

1. placental septum
2. intervillous space
3. chorion
4. amnion
5. trophoblast
6. maternal arterioles
7. maternal venules
8. villus
9. marginal sinus
10. umbilical arteries
11. umbilical vein
12. umbilical cord

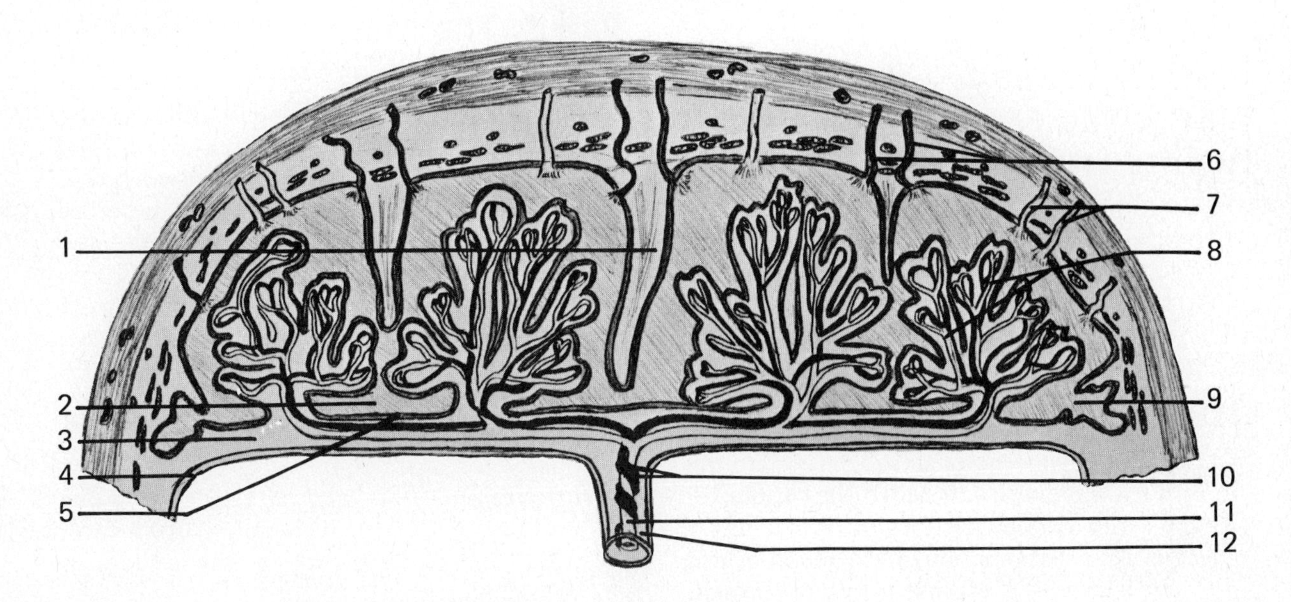

Figure 17-12. Section of Placenta Showing Scheme of Circulation

oxygenated blood and nutrients from the placenta to the embryo.

h. The **umbilical cord**, which contains the umbilical blood vessels. Note that it does not contain any nerves.

i. The **amnion**, which is a fetal membrane that rests against the chorion. The amnion is filled with amnionic fluid in which the embryo floats.

Exercise 17-E Dissection of the Reproductive System of the Fetal Pig

Objectives

To identify the reproductive organs of the fetal pig and to compare them with the corresponding human structures.

Materials

Fetal pig; dissecting set; and dissecting pan.

Discussion

The reproductive organs of the fetal pig are similar to those of the human. Note, however, the differences in the structure of the uterus and the lack of an ejaculatory duct in the male. Although you will dissect the reproductive system of only one sex, you are responsible for that of both sexes. Therefore, carefully study the reproductive structures on a fetal pig of the opposite sex.

Procedure

Part 1. The Female Reproductive System

1. Use Figure 17-13 as a guide in your dissection. Identify the **ovaries**, a pair of small light-colored oval bodies located posterior to the kidneys.
2. The **uterine tubes (Fallopian tubes)** are very small, highly convoluted tubes lying on the dorsal surface of the ovaries. The expanded end of the Fallopian tube, which partially covers the ovary and picks up the eggs from the ovary, is called the **ostium.**
3. Trace the Fallopian tubes until reaching a larger tube immediately next to each ovary. These tubes, the **uterine horns** or **cornua,** are the beginning of the uterus. The eggs are carried through the Fallopian tubes to the uterine horns where they develop. The fetuses tend to be equally spaced throughout the two horns.
4. The two horns unite in the midline to form the **body of the uterus** which lies dorsal to the urethra. The **broad ligament** can be seen running laterally from the body of the uterus to the uterine horns.
5. To dissect the rest of the female reproductive system, the pelvic cavity must be exposed. Remove the skin from the ventral pelvis and cut through the pelvic muscles and the pubic symphysis in the midventral line. Cut with care since the urethra lies immediately beneath the pubis.
6. Locate the **urethra,** the tube carrying urine from the urinary bladder.
7. Dorsal to the urethra, identify the **vagina,** the tube leading from the posterior end of the uterus.
8. Separate the urethra from the vagina. Toward the posterior end, the vagina and urethra unite to form a common passage called the **urogenital sinus (vulva)** which opens to the outside. An external **genital papilla** is located on the external surface at the opening of the urogenital sinus.
9. The lateral boundaries of the urogenital sinus are folds called the **labia.** These unite ventrally to form the genital papilla.
10. Locate the **rectum,** the continuation of the large intestine, dorsal to the vagina.

Part 2. The Male Reproductive System

1. Locate the **scrotum,** the sac visible under the skin ventral to the anus. Early in fetal development the testes are located below the kidneys; however, they migrate before birth through the inguinal canal into the scrotum.
2. Locate the **inguinal canals,** two openings in the abdominal wall, by tracing the internal spermatic (testicular) arteries posteriorly until they pass through the canals.
3. To expose the remainder of the male reproductive organs, cut through the skin in the midventral line ventral to the pubic symphysis. Cut through the scrotum carefully, to avoid damaging the structures contained within.

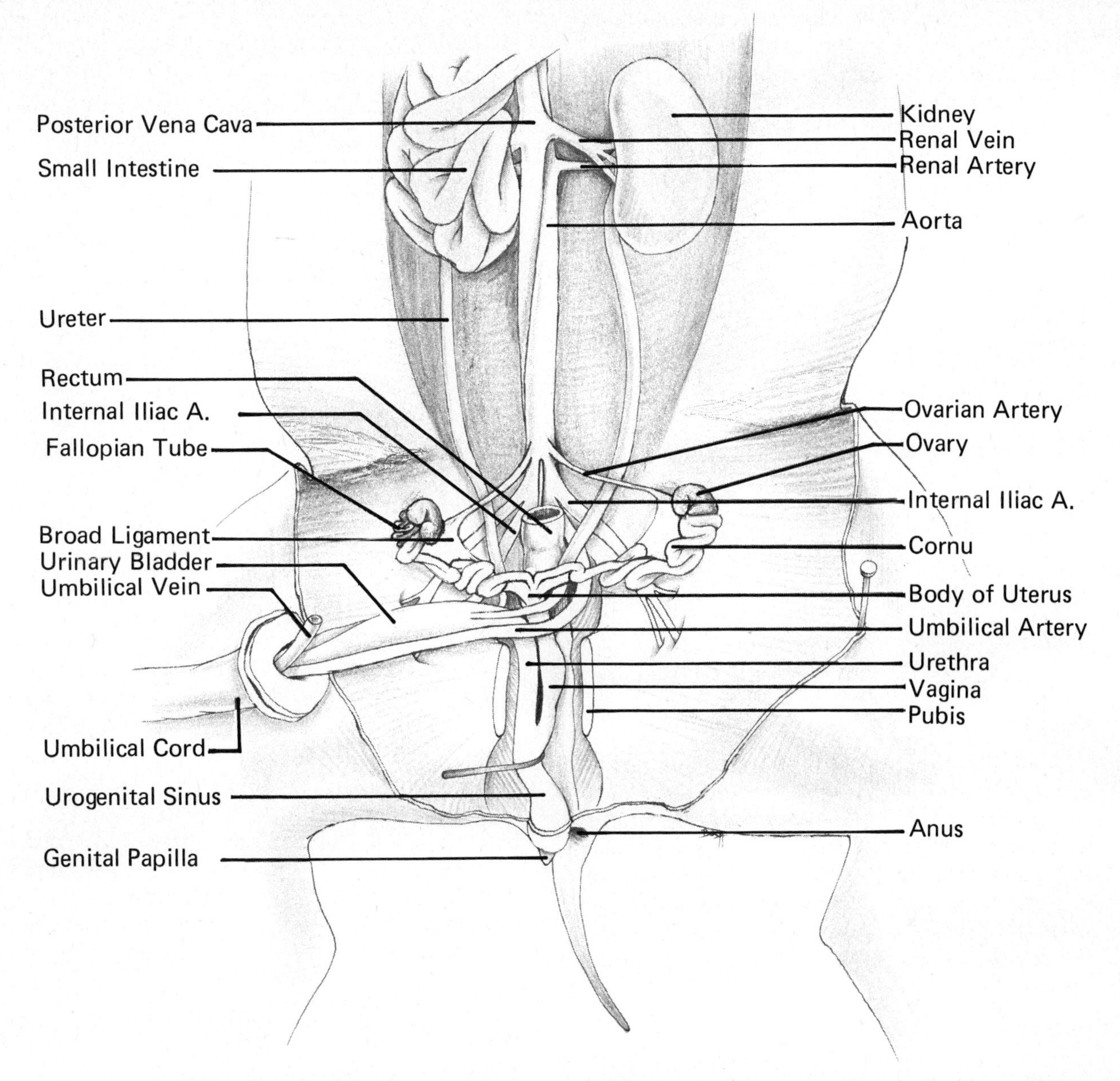

Figure 17-13. Reproductive Organs of the Female Fetal Pig (with intestine and urethra pulled to the right and right ovary turned to left)

4. Using Figure 17-14 as a guide, locate both the left and right **processus vaginalis,** large white sacs on each side which contain the testes; these may not completely fill the processus vaginalis at this stage of development. This structure is an evagination of the peritoneum that precedes the descent of the testes and surrounds them.
5. Pass a probe from the abdominal cavity through the inguinal canal and note that this emerges inside the processus vaginalis.
6. Cut open one of the sacs to expose the **testis.** The **epididymis** should be located along the medial side of the testis. This begins at the cranial end of the testis and extends to its caudal end. Identify the **gubernaculum**, the band of tissue which extends from the posterior end of the epididymis to the scrotal wall. This helps pull the testis posteriorly.
7. The **vas deferens** carries the sperm from the epididymis through the inguinal canal to

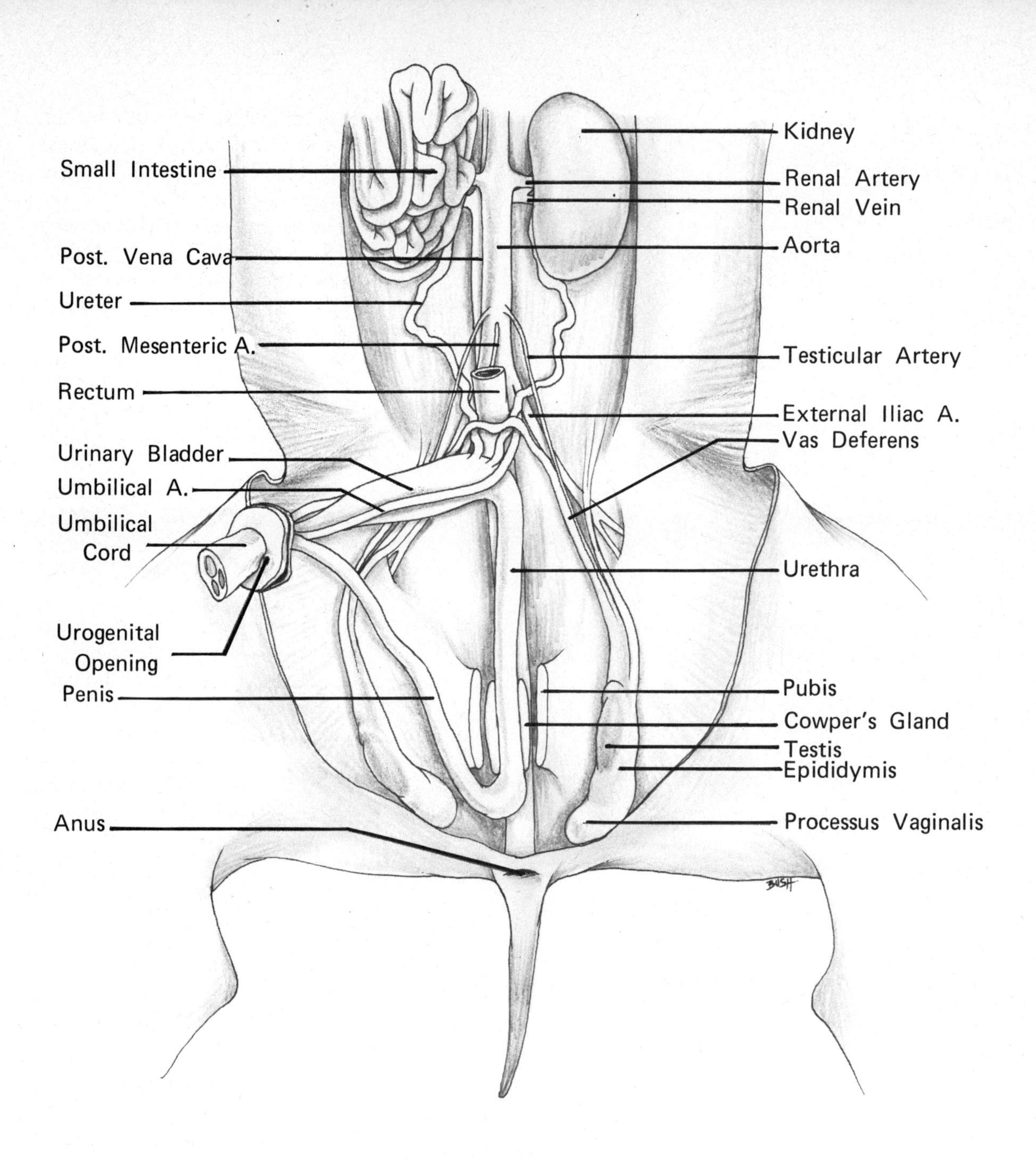

Figure 17-14. Ventral View of Urinary System and Reproductive Organs of the Male Fetal Pig (with intestine pushed to the right)

empty into the urethra. Trace the vas deferens through the inguinal canal to the urethra, noting how it loops over the ureter and enters the dorsal surface of the urethra.

8. Locate the **penis**, the long muscular tube lying just under the skin immediately posterior to the umbilical cord and the urogenital opening in the midventral strip of the abdominal wall. Remove the overlying skin so that the penis is exposed.
9. Now move the penis to one side of the midventral line and cut through the midventral portion of the pelvic muscles and the pubic symphysis. Spread the legs apart to expose the pelvic cavity. The **urethra** should now be visible emerging from the bladder.
10. Separate the **rectum** from the urethra and trace both tubes to the outside.
11. Identify the large pair of bulbourethral or Cowper's glands, each of which is located at one side of the urethra near the anus.

Exercise 17-F Embryology

Objective

To study various stages of development before birth.

Materials

Slides of starfish embryology and a series of preserved human fetuses.

Discussion

The stages in early embryology will be studied in the starfish. The stages in human embryology are similar, although not identical.

Procedure

1. Obtain three slides of starfish embryology (cleavage stages, blastula, and gastrula). Compare all structures observed with Figure 17-15.
2. After fertilization occurs, the fertilized egg divides repeatedly by mitosis. These cells normally remain attached together. The slide entitled "cleavage stages" contains the fertilized egg and the early cleavage stages. Locate the **zygote**, the 2, 4, 8, and 16 cell stages. Observe the **fertilization membrane** surrounding the fertilized egg and the early stages of cleavage. Observe the size of the fertilized egg and the size of the young embryo.
3. After a number of cleavage divisions, a solid ball of cells (the **morula**) is formed. Locate this structure on the slide.
4. The cells in the solid ball soon move to the outside, forming a hollow ball or blastula. On the slide entitled "starfish blastula," locate a sectioned **blastula** showing the cavity in the center.
5. Soon after the formation of the blastula, the ball becomes indented at one end. This invagination produces a two-layered embryo. Locate this stage (**gastrula**) on the slide entitled "starfish gastrula."
6. Study the stages of human development exemplified by specimens present in the laboratory. Note the age at which the nails and hair develop. Study Figures 17-16 through 17-19, which show stages in human development.
7. Examine Figure 17-20 which shows the position of a full term fetus in the uterus.

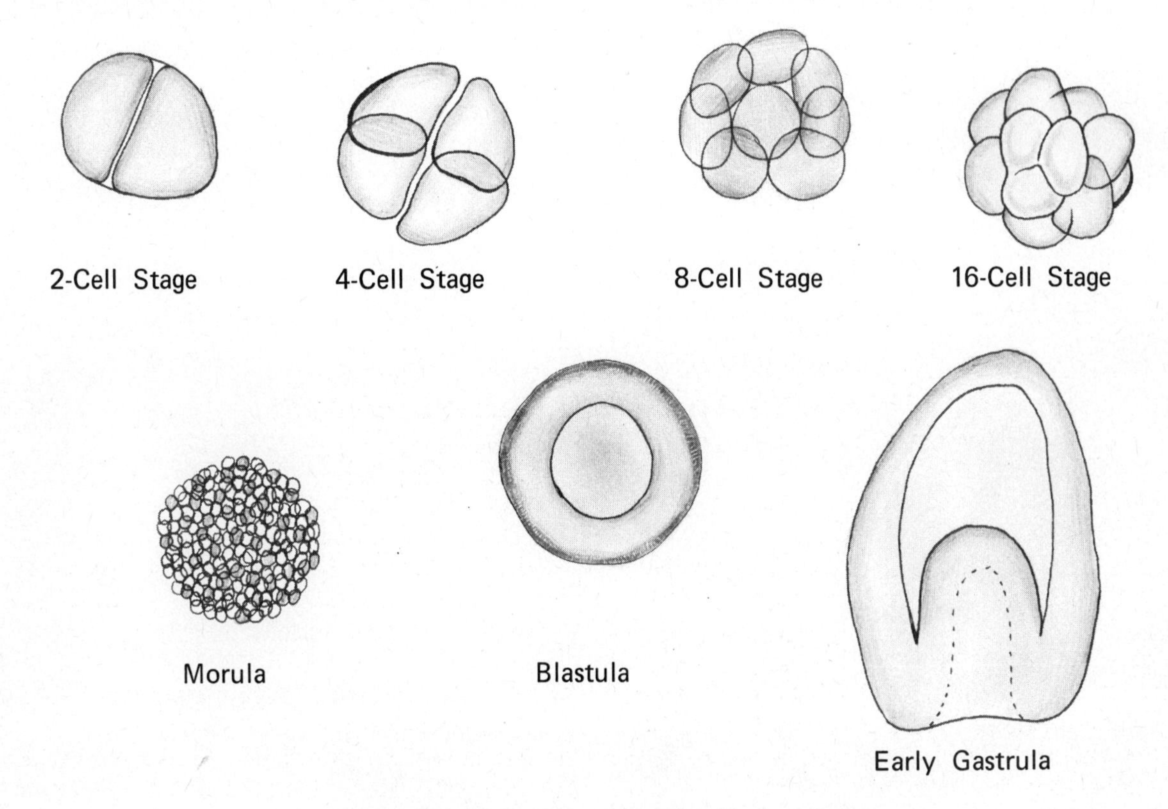

Figure 17-15. Stages of Starfish Embryology

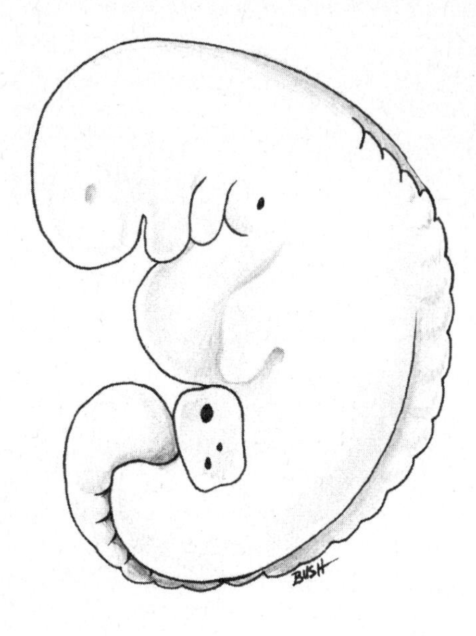

Figure 17-16. Twenty-six Day Old Embryo

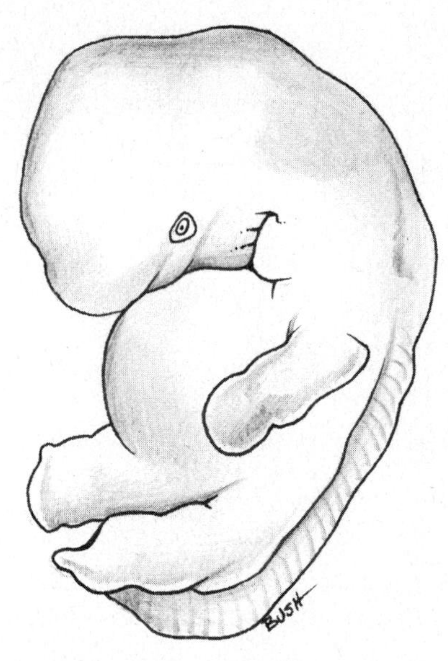

Figure 17-17. Thirty-four Day Old Embryo

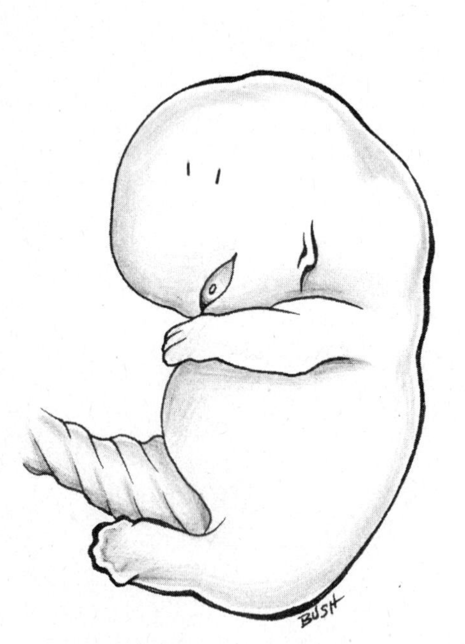

Figure 17-18. Forty-three Day Old Embryo

Figure 17-19. Fifty-six Day Old Embryo

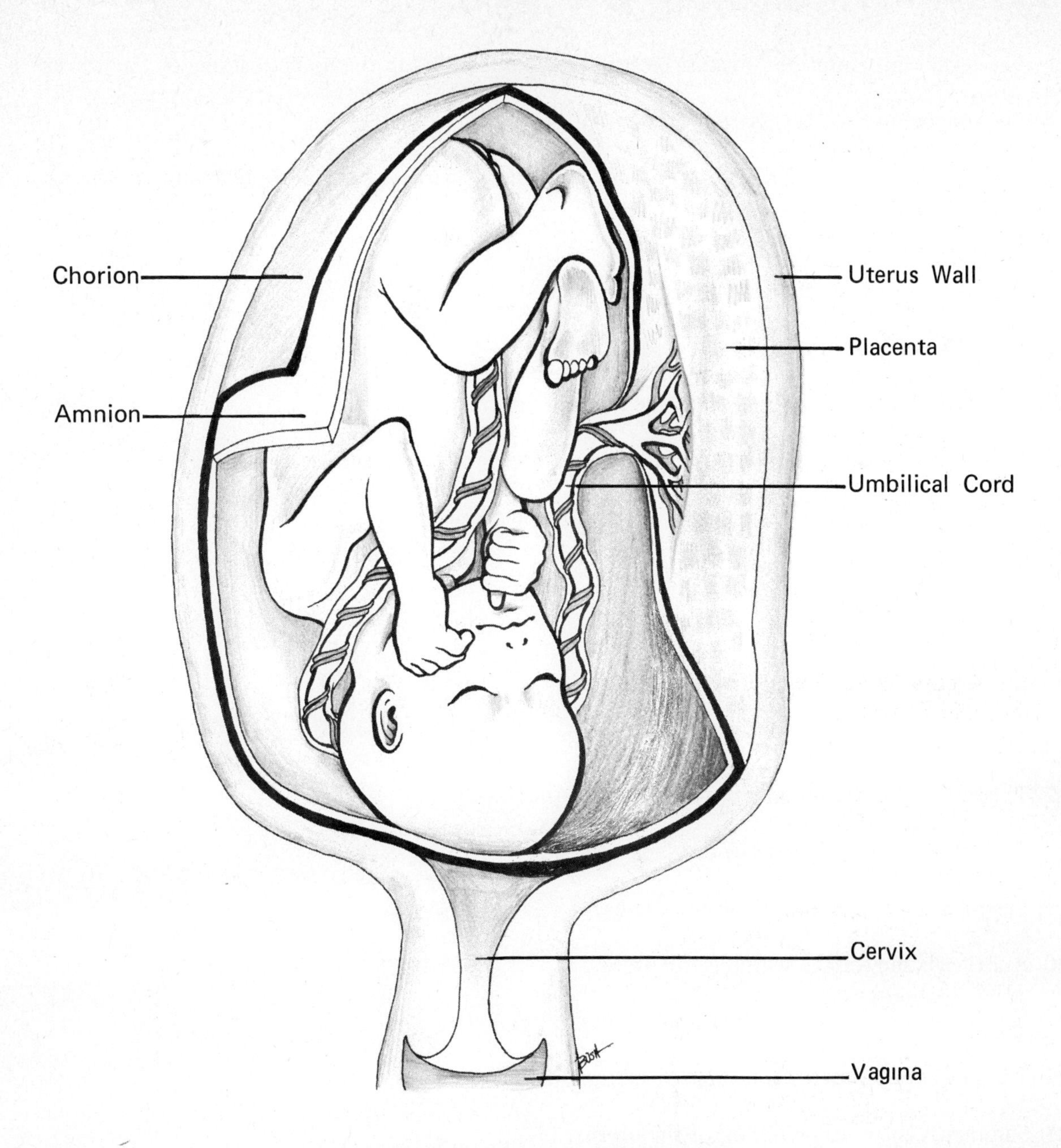

Figure 17-20. Full Term Human Fetus

Self-Test—The Reproductive System

Directions: See Chapter 1, Self-Test, p. 9.

1. The sperm cell on its way to the exterior of the male body follows this course: **seminiferous tubule, vas deferens, epididymis, urethra.**

2. An ovum on its way to the exterior of the female body follows this course: **ovary, oviduct, uterus, vagina.**

3. A zygote is a **follicle.**

4. The term "**implantation**" means: blastula being imbedded in the mucous lining of the uterus.

5. The **menopause** is the cessation of menstruation.

6. Unless fertilization occurs, the human ovum usually **disintegrates** before reaching the uterus.

7. The vagina normally has an **acid** pH.

8. Spermatozoa are formed in the **epididymis.**

9. **Androgens** are produced by the interstitial cells.

10. A surgically sterilized female **may** menstruate normally.

11. Ovulation involves the rupture of the **corpus luteum** with the release of the egg.

12. The upper rounded part of the uterus is called the **fundus.**

13. The female homologue of the penis is the **clitoris.**

14. The seminal vesicles produce a secretion giving an **acid** reaction to the semen.

15. Sperm may retain motility even after a period of **24 hours.**

16. Testosterone is produced by the **follicle** cells.

17. The mucosa of the pregnant uterus is called the **decidua.**

18. **Fraternal** twins are always of the same sex.

19. Fertilization usually occurs in the **vagina.**

20. Another name for the womb is the **uterus.**

21. Dysmenorrhea means **menstrual pain.**

22. The placenta serves for **transport** of material between mother and child.

23. Salpingitis is an inflammation of the **uterus.**

24. A **mastectomy** is surgical removal of the uterus.

25. **Amenorrhea** means the absence of menstruation.

26. **Rectocele** is the protrusion of the rectum into the vagina.

27. Cleavage of the zygote begins in the **uterus.**

28. The notochord is the precursor of the **spinal cord.**

29. The first division of the fertilized ovum results in **two identical cells,** each containing the same amount of male and female chromosome elements.

30. The skin is formed from the **entoderm.**

31. Ovulation usually occurs **five** days after the onset of menstruation, in a 28 day menstrual month.

32. Progesterone is secreted by **both the corpus luteum and the placenta.**

33. The **areola** is the pigmented area around the nipple.

34. The cul-de-sac of Douglas is the pocket between the **rectum** and the uterus.

35. The hormone responsible for ovulation is **FSH.**

36. Circumcision is the surgical excision of the **prepuce or foreskin.**

37. The testes descend through the **inguinal canal** before birth.

38. The umbilical **artery** contains deoxygenated blood.

39. The **chorion** is the inner of the two fetal membranes.

40. The cervix is the lower part of the **Fallopian tube.**

41. Pregnancy is not possible after a hysterectomy, **but ovulation still occurs.**

42. The normal position of the uterus is with the fundus tilted **forward over the urinary bladder.**

43. Sperm cells are produced in the **vasa efferentia.**

44. A mature sperm cell contains the **haploid** number of chromosomes as a result of **meiosis.**

45. The **diploid** number of chromosomes results from the fertilization of an ovum by a sperm cell.

46. Sperm cells are stored in the **epididymis and the vas deferens.**

47. Failure of the male gonads to descend into the scrotum is known as **a hernia.**

48. The common passageway for both sperm and urine is the **ureter.**

49. The erectile tissue of the ventral portion of the penis is called the **corpus cavernosum.**

50. The enzyme **hyaluronidase,** if present in sufficient amounts, apparently enables sperm to penetrate the corona radiata during fertilization.

51. An ovum dies within approximately **24 hours** after ovulation unless fertilized.

52. Some sperm cells have a life expectancy of approximately **72 hours.**

53. Fraternal twins result when **a single ovum is** fertilized by two sperm cells.

54. The secretions of the **bulbourethral (Cowper's)** glands lubricate the urethra in males during sexual excitation.

55. Another name for the birth canal is the **uterine cavity.**

56. Progesterone is produced in the nonpregnant body by **only the corpus luteum.**

57. Ovulation cannot occur with a low blood level of **luteinizing hormone (LH).**

58. Contraceptive pills contain **ovarian hormones or related substances.**

59. Contraceptive pills prevent **ovulation.**

60. In the nonpregnant body, estrogens are produced by the **follicle only.**

KEY

1. seminiferous tubule, epididymis, vas deferens, urethra
3. fertilized egg
8. seminiferous tubules
11. Graafian follicle
14. alkaline
16. interstitial cells
18. identical
19. Fallopian tubes
22. exchange
23. Fallopian tubes
24. hysterectomy
27. Fallopian tube
28. vertebral column
30. ectoderm
31. 14
35. LH
39. amnion
40. uterus
43. seminiferous tubules
47. cryptorchidism
48. urethra
49. corpus urethrae (corpus spongiosum)
53. two ova are
55. vagina
60. corpus luteum and follicle

Name ______________________

Results and Questions for Chapter 17

Exercise 17-A. The Human Male Reproductive System

1. Was there any evidence of spermatogenesis in the wall of the epididymis?__________Should there be?__________

2. What is the function of each of the following?

   a. Epididymis __________

   b. Vas deferens __________

   c. Prostate gland __________

   d. Cowper's gland __________

   e. Urethra __________

   f. Seminal vesicle __________

   g. Seminiferous tubules __________

   h. Interstitial cells __________

3. Define cryptorchidism. __________

   What is the significance of this condition? __________

4. Trace the path of sperm from the site of production to the outside. __________

   ____

5. Name the three male accessory reproductive glands. __________

   ____

6. Define semen. __________

7. What is the spermatic cord? __________

8. How many sperm are required to fertilize one ovum? __________

9. Why would enlargement of the prostate gland interfere with the passage of urine? __________

   ____

10. List three defects that might cause sterility. __________

    ____

11. Where are the Cells of Sertoli located? __________

    What is their function? __________

12. What structure in the male is homologous to the labia majora?________________

Exercise 17-B. The Human Female Reproductive System

1. Describe the location of each of the following:

   a. Broad ligament________________

   b. Round ligament________________

   c. Ovarian ligament________________

2. What is the endometrium?________________

3. Trace the path of the egg from the site of production to the outside.________________

   ____

4. What type of muscle is located in the wall of the uterus?________________

   Name the muscular layer.________________

5. What is meant by salpingitis?________________

6. Would pregnancy be possible in case of a bilateral salpingectomy?________________

   Explain.________________

   Would menstruation be possible in this case?________ Explain.________________

   ____

7. Define the hymen.________________

8. Name the two parts of the uterus.________________

9. Name the hormone produced by the ovarian follicle.________________

   Name those produced by the corpus luteum.________________

10. What is a retroverted uterus?________________

11. How many mature Graafian follicles are produced by the ovary at one time?________________

12. What is the mons pubis?________________

13. Define the perineum.________________

    What is the obstetrical perineum?________________

14. What is the function of Bartholin's glands?________________

15. What is the name of the erectile tissue located near the anterior junction of the labia minora?

    ____

Name ______________________

16. What is an oophorectomy? ______________________

Exercise 17-C. The Endometrial (Menstrual) Cycle

1. What is the name of the hormone secreted by the Graafian follicle? ______________________

2. What is the function of this hormone? ______________________

3. In which phase (see diagram in Procedure) is this hormone active? ______________________

4. What is the name of the hormone that stimulates the growth of the follicle? ______________________

____

5. Which gland produces the hormone in Question 4? ______________________

6. Which gland produces the luteinizing hormone? ______________________

7. In which phase of the menstrual cycle is this hormone active? ______________________

8. What is the function of the luteinizing hormone? ______________________

9. Name the hormones secreted by the corpus luteum. ______________________

10. On which day of a 40 day menstrual cycle does ovulation usually occur? ______________________

11. On which day of a 28 day menstrual cycle does ovulation usually occur? ______________________

12. How long does an ovum survive if it is not fertilized? ______________________

13. Lack of which hormone causes the endometrium to break down? ______________________

14. Failure of the fertilized ovum to become embedded in the endometrium results in cessation of the secretion of ________; this in turn results in the disintegration of the endometrium, a process called ______________________ or ______________________

15. Fill in the following chart concerning the hormones in the menstrual cycle.

Hormone	*Gland Secreting Hormone*	*Stimulus for Production of Hormone*	*Function of Hormone in Menstrual Cycle*
a. FSH			
b. Estrogen			
c. LH			
d. Progest-erone			

Exercise 17-D. The Placenta

1. Name the two human fetal membranes. ______________________

2. Define the decidua basalis. ______________________

3. What is the origin of the placenta?

4. What is the function of the placenta?

5. What are the major components of the afterbirth?

6. What is the function of the chorion?

7. What is the function of the amnion?

Exercise 17-E. Dissection of the Reproductive System of the Fetal Pig

1. List any differences noted between the reproductive organs of the human and the fetal pig.

2. What is the function of the urogenital sinus in the female pig?

3. Compare the place of development of the fetus in the human with that in the pig.

4. Trace the path of sperm from the site of production to the outside in the pig.

Exercise 17-F. Embryology

1. Compare the blastula stage of the starfish with that of humans.

2. Compare the gastrula stage in the starfish with that of humans.

3. When does implantation of the fertilized egg occur?

4. What is the vernix caseosa?

5. What is the average duration of pregnancy?

6. Name the three germ layers in the developing embryo and two structures produced from each layer.

Germ Layers	*Structures Produced*

18

Genetics

Exercise 18-A Directions for Working Genetics Problems

Objectives

To become familiar with basic principles of heredity; to learn how to work selected genetics problems.

Materials

Text and/or other references containing information on elementary genetics.

Discussion

Only a few genetic principles will be discussed in this chapter. It is assumed that the instructor will supplement this discussion with lecture material and appropriate reading assignments. See Appendix 4 for definitions of the terms used in this chapter.

The chromosomes in the nuclei of the cells of higher plants and animals occur in matched pairs, which are referred to as **homologous** chromosomes. One member of a homologous pair of chromosomes is derived from the male parent, while the other **member of the pair is derived from the female parent.** A typical human **cell** contains 23 pairs of chromosomes. During **meiosis** (see Fig. 17-6 in Chapter 17), which occurs during the formation of the ovum and sperm (**gametes**), the pairs separate, and each gamete then contains one member of each of the 23 pairs of chromosomes. The latter cells are said to contain the **haploid** (n) number of chromosomes, which is an indication that they have half the species number of chromosomes. The species number of chromosomes in humans is 46 (23 pairs) and is referred to as the **diploid** (2n) number of chromosomes. When a sperm fertilizes an ovum, the species (or diploid) number of chromosomes is reestablished in the **zygote** (fertilized ovum). With the exception of the sex chromosomes (to be discussed later), the two members of a pair of homologous chromosomes are indistinguishable.

Genes, the fundamental units of heredity, are arranged in a linear sequence on the chromosomes. The number of different genes in humans has been estimated to be from 10,000 to 50,000. Genes are composed primarily of protein and a nucleic acid called **deoxyribonucleic acid** (DNA). The DNA transmits the genetic information, since it contains in coded form all the information required to build new cells. Each DNA molecule is composed of two strands of nucleotides wound in a double helix. Each nucleotide consists of one phosphate molecule, one sugar (deoxyribose), and one base (either the purine adenine or guanine, or the pyrimidine thymine or cytosine). Weak hydrogen bonds join the two strands together at the bases. Each base in one strand pairs with its complementary base in the second strand. (Adenine pairs only with thymine, and guanine only with cytosine.) This structure (see Fig. 18-1) may be compared with a ladder. The rungs of the ladder are the bases, and the backbone or sides the alternating sugar and phosphate groups. Since the sugar and phosphate remain the same in each nucleotide and only the bases change, each gene must be determined by the changing sequence in which the four bases are arranged. This may be compared with a four letter alphabet which spells out the genetic information.

When a cell divides, the DNA of each chromosome produces an exact copy of itself

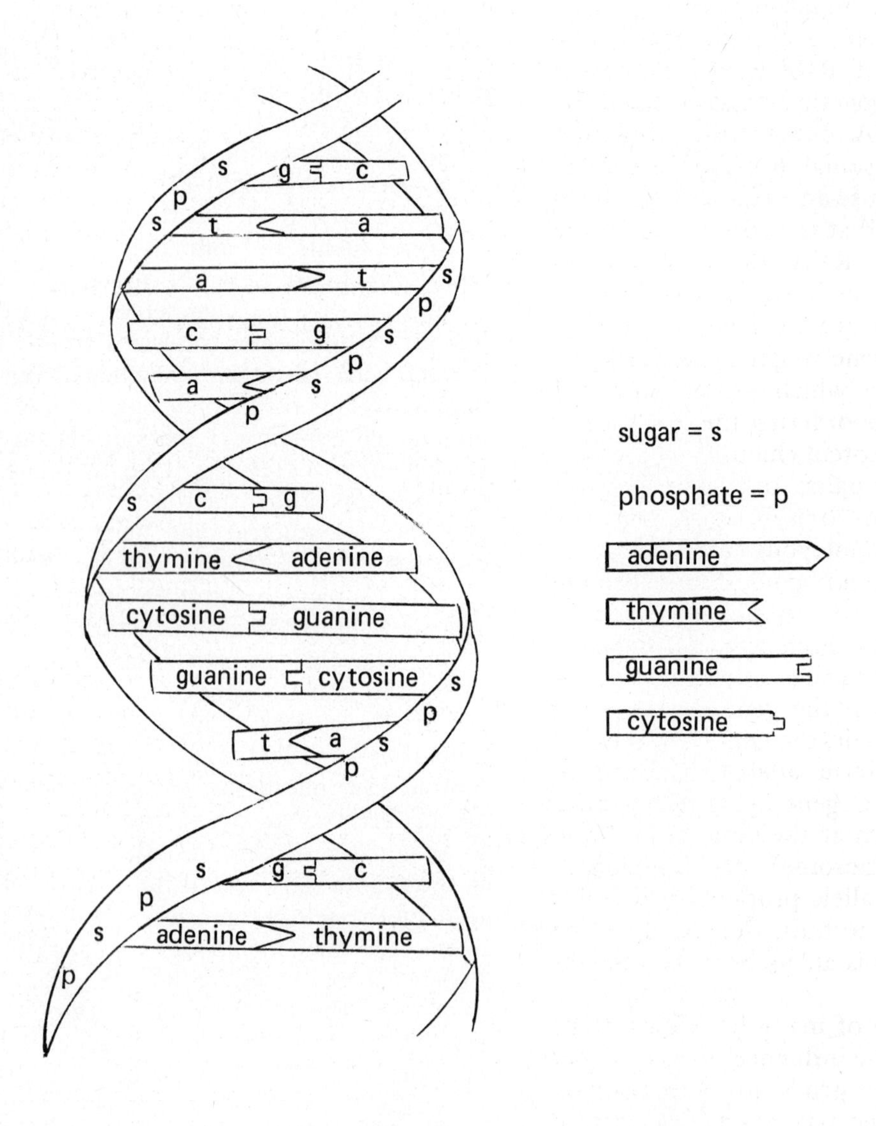

Figure 18-1. The Structure of DNA

(replication). The two strands that compose the DNA molecule separate and each strand duplicates its former member from chemical building blocks available in the cell. As a result of this replication of the DNA, each daughter cell (at the end of mitosis) contains the same genetic material as the parent cell.

The specialized nature of a cell depends upon the type and quantity of protein in the cell. DNA functions in protein synthesis by controlling the formation of a second nucleic acid, RNA (**ribonucleic acid**); RNA then serves as a pattern for protein synthesis. The structure of RNA is much like that of DNA, but it consists of only a single strand, the sugar is ribose, and the base uracil replaces thymine. Several types of RNA are produced in the nucleus. One type of RNA called messenger RNA carries the genetic message from its storage site in DNA out to the ribosomes (which contain ribosomal RNA) in the cytoplasm. Here proteins are synthesized from amino acids deposited at the correct site on the ribosome by transfer RNA. Every three bases in a particular sequence in DNA form the code for the placement of one particular amino acid in a protein. So a gene might be defined as a segment of DNA in which are encoded the complete instructions ordering the sequence of amino acids in one protein chain.

Genes occur in pairs, one gene on one chromosome, the other gene on the corresponding or homologous chromosome. If the two genes of a homologous pair of chromosomes are for the same expression of a trait, the condition is called **homozygous**. However, if one of the genes provides a different expression of the trait than the gene on the other chromosome of the pair, the condition is called **heterozygous**. The term **allele** is applied to alternate forms of a gene for a given trait. Alleles are genes born at the same **locus** (location on the chromosome) on homologous chromosomes. Each allele produces a different expression of the same trait. For example, the gene for brown eyes is an allele of the gene for blue eyes.

The transmission of many hereditary traits seems to be under the influence of a single pair of genes. If the two genes are heterozygous, usually only one gene will exert an effect. A gene is called **dominant** if it is fully manifested in the heterozygous state. If the gene is manifested only in the homozygous state, meaning that two identical genes are needed to produce the effect, the gene is called **recessive**. Occasionally the heterozygote is intermediate between the homozygous dominant and recessive. This **lack of dominance** occurs, for example, in some types of flower coloration and cattle coat color.

The **genotype** is the genetic makeup of an individual. It is an indication of the genes present, whether they are expressed or not. Usually symbols, such as letters of the alphabet, are used to express the genotype. The capital form of the letter is used as a symbol for a dominant trait, and the lower case form of the same letter is used to symbolize the recessive gene for the same trait. The **phenotype** is that which can be observed about an individual, such as hair color, eye color, or blood type, and it is usually described with words rather than symbols. As a result of the interraction between genes and the environment, the phenotype is potentially variable.

Part 1. Working Genetics Problems Involving a Single Pair of Genes (Monohybrid Cross)

Discussion

The method of working simple genetics problems involving a monohybrid cross will be illustrated by using a characteristic of pea plants, which were extensively studied by Gregor Mendel (sometimes referred to as the father of genetics). The method employed is readily applicable to human traits.

In pea plants, height is determined by a single pair of genes. The gene for tallness is dominant over that for dwarfness. If two plants heterozygous for tallness are crossed, approximately 3/4 of the offspring would be expected to be tall and 1/4 dwarf. Use the following procedure in working all problems of this type.

Procedure

1. Select and record symbols for each trait. In this problem, the symbol T = tall (since tallness is dominant) and t = dwarf (the recessive trait) may be used.
2. Write the genotype of each parent. In this problem, each parent would have the genotype Tt. Remember to use two genes for each trait since each cell in the gonads,

prior to meiosis, contains the diploid number of chromosomes.

3. Determine the number of different gametes each parent can produce by using the following rule: the number of different gametes = 2^n, where n = the number of heterozygous pairs. Therefore, a parent of genotype Tt could produce two different types of gametes (2^1), while a parent of genotype TtYy could produce four different types (2^2), and one of genotype TtYyAAbb could also produce only four, since the latter parent has only two heterozygous pairs of genes.
4. Write down all of the possible different gametes that could be produced. Each gamete must have one gene of each pair of genes; it can never have both members of one pair of genes and none of the other pair. For example, a plant with genotype TT would produce only one type of gamete, T. In a dihybrid cross (a cross involving two pairs of genes), a plant with genotype TTYY would produce only a single type of gamete, TY. It would not produce a gamete containing TT (both genes for tallness) or YY (both genes for the second characteristic). Be sure to consider all possible combinations. The number of gametes arrived at should agree with the theoretical number in Step 3. In our example of a cross between two parents having the genotype Tt, each parent can produce two kinds of gametes (in equal numbers): T and t.
5. Make a checkerboard (Punnett square) with the different gametes from one parent on one side, and the gametes from the other parent on the top, and complete the squares as shown below.

	T	t
T	TT	Tt
t	Tt	tt

6. Determine the phenotype and genotype of each of the resulting offspring. In this example, 3/4 of the offspring would be expected to be tall and 1/4 dwarf. The phenotypic ratio is, therefore, 3:1 (3 tall: 1 dwarf). The genotypic ratios of the offspring would be 1:2:1 as follows:

   1/4 TT
   1/2 Tt
   1/4 tt

7. Check to make sure that each new individual has two genes for each trait, and that the total number of genes in each offspring is the same as that in the original parents.

Part 2. Working Problems Involving Two Pairs of Genes (Dihybrid Cross)

Discussion

A dihybrid cross involves two different traits, each due to a pair of genes born on a different pair of chromosomes. In pea plants, the gene for yellow seed color is dominant over that for green seed color. If a pea plant heterozygous for tallness and seed color is inbred with an identical plant, the problem can be solved by following the same general procedure as that for a monohybrid cross.

Procedure

1. Symbols: T = tall, t = dwarf, Y = yellow, y = green.
2. Genotypes of parents: TtYy
3. The number of different gametes for each parent equals four, since TtYy contains two heterozygous pairs, and $2^2 = 4$.
4. Each parent, since the two have identical genotypes, would form these four gametes in equal numbers: TY, Ty, tY, and ty.
5. Checkerboard:

	TY	Ty	tY	ty
TY	TTYY	TTYy	TtYY	TtYy
Ty	TTYy	TTyy	TtYy	Ttyy
tY	TtYY	TtYy	ttYY	ttYy
ty	TtYy	Ttyy	ttYy	ttyy

6. Phenotypic ratios of the offspring

   9/16 tall with yellow seeds
   3/16 tall with green seeds
   3/16 dwarf with yellow seeds
   1/16 dwarf with green seeds

These ratios are obtained in a typical dihybrid cross.

Part 3. Working Problems Involving a Lack of Dominance

Discussion

In crosses involving a lack of dominance, both genes of the heterozygote express themselves in the offspring. Problems are worked in the same way as those in Part 1, but the heterozygous progeny show the interaction of both genes. For example, in snapdragons, if a plant producing red flowers (RR) is crossed with a plant producing white flowers (rr), the resulting plants produce only pink (Rr) flowers. If two plants producing pink flowers are crossed, approximately 1/4 of the offspring would have red flowers, 1/2 pink, and 1/4 white.

Procedure

1. Symbols: RR = red, Rr = pink, rr = white
2. Parental genotypes: Rr x Rr
3. Types of gametes: R, r and R, r
4. Place the gametes on the Punnet square; then complete the rest of the squares as follows:

	R	r
R	RR	Rr
r	Rr	rr

5. Determine the phenotypic ratio: 1:2:1 (1 red: 2 pink: 1 white).
6. Determine the genotypic ratio: 1:2:1 (1/4 RR: 1/2 Rr: 1/4 rr).

Part 4. Working Sex-linked Problems

Discussion

Of the 23 pairs of chromosomes in a human, one pair is called the **sex chromosomes** and the other 22 pairs are called **autosomes.** In the human female, the sex chromosomes look alike and are called X chromosomes, whereas in the male, the two chromosomes differ from each other. The male has one X chromosome and a smaller chromosome called the Y chromosome. During meiosis these two chromosomes separate. Therefore, the male produces two types of sperm, half containing the X chromosome and half the Y. All the ova of the female contain the X chromosome. Therefore, it is the male that determines the sex of the child.

The X chromosome, since it is larger than the Y, contains many genes not on the Y chromosome. These genes, located **only** on the X chromosome, are called **sex-linked** genes. Hemophilia and color blindness are both inherited as sex-linked recessives. Using color blindness as an example, if a woman has either one or two genes for normal vision, she will have normal vision. A woman having two genes for color blindness will be color-blind. However, since a man does not have a gene for color vision on the Y chromosome, he will be color-blind if he has only one gene for color blindness.

Procedure

Cross a woman who is heterozygous for color blindness with a male having a normal color vision.

1. The genotype for the woman (who has normal color vision but is a carrier of a gene for color blindness) may be written as follows: X^CX^c, wherein C = normal color vision, c = color-blind, and the X's represent the X chromosome where the genes for this trait are borne.
2. The genotype of a male, normal for color vision, may be written as follows: X^CY (a color-blind male would have the genotype: X^cY). Therefore, the problem could be set up as follows:

	♀		♂
Parents:	X^CX^c	x	X^CY
Gametes:	X^C, X^c		X^C, Y

The gametes are then placed on the Punnet square, and the rest of the squares are completed in the following manner:

	X^C	Y
X^C	X^CX^C	X^CY
X^c	X^CX^c	X^cY

3. From the preceding it is possible to answer the following questions:

a. From such a cross, what are the chances of a color-blind male being produced? (Answer: 1/2 or 50/50)
b. What are the chances of a color-blind female being produced? (Answer: none)

A key thing to remember is that half of all the sons produced as a result of such a union might be expected to be color-blind.

Part 5. Working Problems Involving Multiple Alleles

Discussion

The preceding problems assume that each gene exists in only two alternative forms; however, there are many genes that have more than two forms. These are referred to as multiple alleles. This is exemplified by the three different genes, borne at the same locus, involved in the determination of the ABO blood types. No individual will have more than two of these genes, since he has only two chromosomes and no more than one gene can occupy a given locus at a time.

The Rh factor also involves more than a single pair of genes (e.g., C, D, and E), although for elementary problem solving, only a single pair of genes may be considered wherein Rh+ is dominant to Rh−.

The MN blood type series is due to a pair of genes, neither of which shows dominance over the other. While this series is not important as far as transfusions are concerned (no antibodies are produced in response to introduction of the antigens), it may be of significance (along with the ABO and Rh types) in establishing possible paternity cases. In such cases it may be possible to prove that a certain man could not have been the father of the child in question, but it is not possible to prove that a particular man is the father of the child (only that he might have been).

Procedure

1. The phenotypes and genotypes that follow are necessary for working genetics problems involving blood types:

Phenotype	Genotype
O	OO
A	AO, AA
B	BO, BB
AB	AB
Rh+	++, +−
Rh−	−−
M	MM
N	NN
MN	MN

2. In the above genotypes, genes A and B (responsible for producing antigens A and B) are both dominant over O, but are not dominant over each other. Rh+ is dominant over Rh−. M and N show no dominance over each other.
3. Once the parents' genotypes are established, problems are solved in the same manner as that indicated for monohybrid and dihybrid crosses.

Part 6. Probability

Discussion

The solution of genetics problems involves the use of probability. We are asking what the chances are that a particular event will occur. Based on the genotypes of two heterozygous brown-eyed parents (see monohybrid crosses), it is possible to predict that there is one chance in four (1/4) that such parents may produce a blue-eyed child. It is not possible to say they will produce such a child.

Probability is defined as:

$$p = \frac{\text{number of favorable cases}}{\text{total number of cases}}$$

If an event is certain to happen, its probability is 1; if the event is impossible, its probability is 0. Other probabilities are expressed as fractions between 0 and 1. For example, the probability of drawing a king from a deck of cards is 4/52 or 1/13, since there are four kings in the deck. The probability of throwing heads in one toss of a coin is 1/2.

Procedure

1. To determine probability, ask what is the number of favorable cases or chances of a particular event occurring. The answer to this question in the case of the kings in the deck (mentioned in the discussion) was four.
2. Also, ask what is the total number of cases. The total cases in the deck of cards was 52.
3. Put the results of steps 1 and 2 in the form of a fraction. In the case under consideration, the resulting fraction was 4/52 which reduces to 1/13. Therefore, the probability

that a king would be drawn from a shuffled deck of cards is 1 in 13 or 1/13.

4. Another point to consider in such problems is that the probability that one or the other of two (or more) mutually exclusive events will occur is the **sum** of the separate probabilities. For example, the probability of drawing either an ace or a king from a standard deck of cards is $4/52 + 4/52 = 8/52$ or $2/13$.
5. The probability that both of two independent events will occur together is the **product** of the two separate probabilities. This can be extended to any number of different events. For example, the probability of drawing first an ace (then replacing it) and then a king from a standard deck of cards is $1/13 \times 1/13 = 1/169$.

Exercise 18-B Genetics Problems

Objective

To learn how to solve some of the basic types of genetics problems.

Materials

None.

Procedure

The following are practice problems involving the principles found in Exercise 18-A. On a separate piece of paper try to solve these problems without reference to the key at the end of the exercise. After completing the practice problems, solve those assigned by the instructor under Results and Questions.

1. How many different types of gametes can be produced by an individual of genotype AabbCcDDEe?
2. The gene for brown eyes is dominant over that for blue eyes. A woman with blue eyes marries a brown-eyed man whose mother had blue eyes. What proportion of the children would be expected to have blue eyes?
3. Curly hair shows lack of dominance over straight hair, the heterozygote possessing wavy hair. If two people with wavy hair marry, what hair types would you expect to find among their offspring?
4. Two children in a family have blood type O. Their mother has type B blood, their father type A. What is the genotype of the mother? What is the genotype of the father?
5. In humans the gene for farsightedness is usually inherited as a dominant. A man who is heterozygous for farsightedness marries a woman with normal vision. What fraction of their children would be expected to be farsighted?
6. The gene for astigmatism is inherited as a dominant. Two people heterozygous for astigmatism marry. What is the probability that they will have two normal children in a row?
7. What is the probability of having two boys in a row?
8. Two brown-eyed parents have a blue-eyed child. What is the probability that their next child will be blue-eyed?
9. Two normal parents are each heterozygous for a recessive gene for feeblemindedness. What is the probability that they will have four feeble minded children in a row?
10. In humans the gene for the ability to taste phenylthiocarbamide (PTC) is dominant over that for the inability to taste PTC. The gene for brown eyes is dominant over that for blue eyes. What proportion of the offspring of two parents, each heterozygous for both traits, would be blue-eyed tasters?
11. Red-green color blindness in humans results from a sex-linked recessive gene. Two normal parents (the mother is blue-eyed, the father heterozygous for brown eyes) produce a color-blind son.
    a. What are the genotypes of the parents?
    b. What proportion of their sons would you expect to be color-blind?
    c. What is the probability that two daughters in a row will be color-blind?
12. A woman carrying the genes for both color blindness and hemophilia (she inherited both of these sex-linked recessive traits from her father) marries a man who is color-blind, but not hemophilic.
    a. What is the probability that her first two children will be normal boys?
    b. What is the probability of having a hemophilic daughter?
13. In each of the following situations, state whether it is possible or impossible for these parents to have had the child indicated, and give a reason for your answer.

	Mother	Father	Child	Possible or Impossible?	Reason
a.	A, MN, Rh+	O, M, Rh+	AB, MN, Rh–	__________	__________
b.	AB	A	O	__________	__________

14. In humans, normal pigmentation is due to a dominant gene, albinism to its recessive allele. These are not sex-linked. A normal man marries an albino woman. Their first child is an albino. Use this information to solve the following problems:
    a. What is the genotype of the father?
    b. What is the probability that their next child will be an albino?

KEY TO PRACTICE PROBLEMS

1. $2^3 = 8$
2. 1/2
3. 1/4 straight hair
   1/2 wavy hair
   1/4 curly hair
4. mother's genotype = BO
   father's genotype = AO
5. 1/2
6. 1/16
7. 1/4
8. 1/4
9. 1/256
10. 3/16
11. a. $bbX^CX^c \times BbX^CY$
    b. 1/2
    c. 0
12. a. 1/16
    b. 0
13. a. Impossible; no B from parents
    b. impossible; no O from mother
14. a. Aa (A = normal, a = albino)
    b. 1/2

Exercise 18-C Variability in Man

Objective

To observe genetically determined variability in man.

Materials

PTC impregnated papers.

Discussion

Because of the almost infinite possible combinations of traits in humans, every individual, with the exception of identical twins, is unique. Even identical twins are not absolutely alike due to the modification of genes influenced by their environment. A study of the traits in this exercise exemplifies, in a limited way, human variability.

Part 1. Hereditary Basis for Tasting Phenylthiocarbamide

Procedure

1. Place a strip of phenylthiocarbamide (PTC) test paper on the tongue. If you experience no taste sensation, chew the paper.
2. Determine your genotype and record it under Results and Questions. Note: the ability to taste PTC is dominant to inability to taste the chemical (about 70% of subjects report a strong bitter taste, while approximately 30% report no taste). Use T for taster and t for non-taster. If you are a taster, you will not know (unless a study is made of your family) whether you are homozygous or heterozygous for the trait, and you should, therefore, record it as T_.

Part 2. Hereditary Basis for Pigmented Iris

Discussion

Pigmentation for eye color is located in the front part of the iris. A person with blue eyes has no pigment in the front of the eyes. The eyes become increasingly dark with increasing amounts of pigment in the front of the iris (passing through hazel or green or gray to brown or dark brown). Geneticists feel certain that there are several genes involved in determining eye color; however, in order to simplify, we shall group all types with a pigmented iris together.

Procedure

1. Determine your genotype for eye color and record it under Results and Questions.
2. A blue-eyed individual is homozygous recessive (bb) for nonpigmented iris. A person with a pigmented iris may be either homozygous (BB) or heterozygous (Bb) for the trait.

Part 3. Hereditary Basis for Middigital Hair

Discussion

The complete absence of hair from the middle segments of the fingers is a form of genetic variability in man.

Procedure

1. Determine your genotype for this trait and record it under Results and Questions.
2. Absence of middigital hair is due to a recessive trait (dd). Presence of hair on one or more middle segments may be due to multiple alleles, each of which is dominant to the recessive gene. Group all of the multiple alleles with the symbol D.

Part 4. Hereditary Basis for Ear Lobes

Procedure

1. Determine your genotype for ear lobes and record it under Results and Questions.
2. Free hanging ear lobes are dominant (E_) to adherent (ee) lobes (the latter has no gap between the lobe and the cheek).

Part 5. Hereditary Basis for Bent Little Finger

Discussion

The distal (terminal) segment of the little fingers may be straight or turned inward (toward the ring finger).

Procedure

1. Determine your genotype for the terminal segment of your little finger and record it under Results and Questions.
2. Bent is dominant (F_) to straight (ff) terminal segments.

Part 6. Hereditary Basis for Blood Groups

Procedure

1. Use the results obtained in Exercise 12-A, Part 1 wherein you determined your ABO and Rh blood type.
2. Determine your genotypes for ABO and Rh blood types by referring to the genotypes listed under Exercise 18-A, Part 4, and record them under Results and Questions.

Part 7. Hereditary Basis for Singing Voice

Procedure

1. Determine the genotype of your singing voice and record it under Results and Questions.
2. Singing voice is due to one pair of alleles which may be symbolized by V^1 and V^2. The expression of the genes is different in males and females. The genotype of bass and soprano is V^1V^1; of baritone and mezzo-soprano V^1V^2; of tenor and alto V^2V^2.

Part 8. Hereditary Basis for Double-Jointed Thumbs

Procedure

1. Determine your genotype for double-jointed thumb (actually loose ligaments which permit the thumbs to be thrown out of joint) and record it under Results and Questions.
2. Ability to throw the thumb out of joint is usually dominant (J_) to its allele (jj) which determines tight ligaments.

Part 9. Hereditary Basis for the Long Palmar Muscle

Procedure

1. Determine your genotype for the presence or absence of the long palmar muscle and its tendon.
2. Clench and flex your fist. You will observe either two or three tendons running from your forearm to the center of the heel of the palm. If three tendons are present, the prominent middle one is the tendon of the long palmar muscle. If only two tendons are present, then the long palmar muscle and its tendon are absent.
3. The presence of the muscle in both arms is determined by recessive alleles (pp). The presence of the muscle in only one arm is probably an expression of the heterozygous condition (Pp). The absence of the muscle in both arms is probably an expression of the homozygous condition (PP).

Part 10. Hereditary Basis for Hair Color

Discussion

Hair color seems to be determined by several factors; however, by considering only the larger categories, hair color may be analyzed in terms of two gene pairs. The brown pigment, melanin, provides either a brown, black, or blond hair color (if red is not present). The darker shades are dominant. Red hair color is determined by a homozygous recessive condition of another pair of alleles.

Procedure

1. As accurately as possible, determine your genotype for hair color and record it under Results and Questions.
2. Symbolize dark hair color with (D_) and blond with (dd). Red pigmentation is symbolized (rr) and lack of pigment is (R_). A person with genotype ddrr would have sandy red hair; one with ddRR or ddRr would have blond hair; one with DDrr or Ddrr would have red hair or dark hair with tints of red; one with DDRR or DdRR or DdRr or DDRr would have dark hair.

Self-Test—Genetics

Directions: See Chapter 1, Self-Test, p. 9.

1. Intellectual performance of some mentally deficient children has been improved by administering **thyroxin.**
2. If a person comes from a lineage showing epilepsy, **it is important** that the hereditary background of the proposed marriage mate be studied.
3. Color blindness is a sex-linked trait and is inherited **only by males.**
4. **All** color blindness is inherited.
5. **All** inherited conditions are apparent at birth.
6. A child with both parents showing allergy **is likely** to exhibit allergy early in life.
7. A frightening experience to an expectant mother **will** cause her child to become nervous.
8. A person may genetically transmit hypersensitivity, or allergy, to his offspring **without himself showing the condition.**
9. A person homozygous for allergy is most likely to show the condition **before age 16.**
10. Every person **has** an inherited disorder.
11. Consanguineous marriages **always result** in inferior offspring.
12. An appreciation of fine art is **inherited.**
13. Sex-linked disorders occur **only in males.**
14. A person who inherits a predisposition to a disorder **is certain to have** that disorder.
15. Hereditary factors result from the action of **genes found in chromosomes** of cells.
16. A certain organic disorder first appeared at the age of 50. This disorder **could not possibly** have been hereditary.

17. An organic disorder was present at birth. Therefore, this disorder **must have been** hereditary.

18. Consanguineous marriages tend to produce a greater tendency toward a **heterozygous** condition.

19. The only hereditary bridge between parents and offspring is the **blood.**

20. A person with a pair of genes which produce brown eyes **is always heterozygous** for eye color.

21. If a certain trait appears only when the person is homozygous for that trait, the gene is **recessive.**

22. If a person has one pair of genes for a trait, **only one** of these genes can be transmitted to an offspring.

23. A person with a pair of genes which produce blonde hair is **homozygous** for hair color.

24. A person who inherits a condition resulting in pernicious anemia actually inherits a deficiency **of the stomach.**

25. The constitutional factor resulting in hardening of the arteries **appears to have a genetic basis.**

26. A high level of intellect is probably due to a fortunate **chance combination** of genes.

27. **All** feeblemindedness is inherited.

28. Parents of average intelligence **may have** children of superior intelligence.

29. Parents of average intelligence **may have** children with inherited feeblemindedness.

30. Parents of superior intelligence **may have** children of average intelligence.

31. Mature sex cells are called **gametes.**

32. The chromosome number of a female is **44 + XX**; of a male **44 + XY.**

33. A recessive sex-linked characteristic in a man might be transmitted to some of his **son's daughters.**

34. So-called acquired characteristics of living organisms are **never** inherited.

KEY

3. by both males and females
4. not all
5. not all
7. will not
10. does not have
11. do not always result
12. acquired
13. in males and females
14. may not have
16. could possibly
17. could have been
18. homozygous
19. gene
20. may be heterozygous or homozygous
27. not all
33. daughter's sons

Name________________________

Results and Questions for Chapter 18

Exercise 18-B. Genetics Problems

In solving the following problems, show all symbols used and designate their meaning, give genotypes of the parents, and show all work (e.g., Punnet squares). Write answers in the spaces provided to the right.

1. How many different types of gametes can be produced by an individual of genotype aaBbCcDdEe?

   ____

2. The gene for brown hair is dominant over that for blond hair. If two people heterozygous for brown hair married, what types of hair color would you expect to find in their offspring. List the proportions of each type.

   ____

3. Curly hair shows a lack of dominance over straight hair, the heterozygote possessing wavy hair. If a person with curly hair marries a person with straight hair, what hair types would you expect to find among their offspring?

   ____

4. A mother is blood type O, her child O. She claims a man of type AB is the father. Could he be?________ Explain:________________________

   ____

5. A man of blood type A marries a woman of blood type O. Their first four children are all of type A blood.

   a. Can you be certain whether the man is homozygous or heterozygous?________________

   b. Which is more likely?________________________

   c. If the fifth child has blood type O will that settle the question?________________

   Explain:________________________

6. A trait is always transmitted directly from a father to all his sons and from these to all their sons. If is likely that the gene for this trait is on which chromosome?________________

7. In humans the gene for farsightedness is usually inherited as a dominant. A man and a woman (both heterozygous for farsightedness) marry. What fraction of their children would be expected to be farsighted?

   ____

8. The gene for astigmatism is inherited as a dominant. Two people heterozygous for astigmatism marry. What is the probability that their first child will be a girl with astigmatism?

____

9. What is the probability of a husband and wife having three girls in a row?

____

10. Two brown-eyed parents have a blue-eyed child. What is the probability that their next child will be brown-eyed?

____

11. Two normal parents are each heterozygous for a recessive gene for feeblemindedness.

   a. What is the probability that they will have four normal children in a row?

   ____

   b. If this family had one child which was feebleminded, what is the probability that the next child will be normal?

12. In humans the gene for ability to taste phenylthiocarbamide is dominant over that for inability to taste PTC. The gene for brown eyes is dominant over that for blue eyes. What proportion of the offspring of two parents, each heterozygous for both traits, would be brown-eyed nontasters?

____

13. Red-green color blindness is due to a sex-linked recessive gene. Two normal parents (the mother is blue-eyed, the father heterozygous for brown eyes) produce a color-blind son.

   a. What proportion of all sons born would be expected to be blue-eyed and color-blind?

   ____

   b. What proportion of their daughters would you expect to be color-blind?

   ____

Name________________________

c. What is the probability that two sons in a row in this family will be blue-eyed and color-blind?

____

d. What is the probability that two children in a row will be blue-eyed and color-blind?

____

14. A woman carrying the genes for both color blindness and hemophilia (she inherited both these traits from her father) marries a man who is color-blind but not hemophilic.

a. What is the probability that the first child will be either a color-blind boy or a color-blind girl?

____

b. What proportion of the sons would be expected to be both color-blind and hemophilic?

____

15. In each of the following situations, state whether it is possible or impossible for these parents to have produced this child and give the reason for your answer.

Mother	*Father*	*Child*	*Possible or Impossible?*	*Reason*
a. AB, M, Rh–	A, MN, Rh–	B, M, Rh+		
b. A	A	O		

16. In humans, normal pigmentation is due to a dominant gene, albinism to its recessive allele. These are not sex-linked. A normal man marries an albino woman. Their first child is an albino. Use this information to solve the following problems:

a. What is the genotype of the mother?

____

b. What is the genotype of the child?

____

c. What is the probability of these parents having two albino children in a row?

____

Exercise 18-C. Variability in Man

1. Write your genotype for ability or inability to taste PTC. ______

2. Write your genotype for pigmented or nonpigmented iris. ______

3. What is your genotype for middigital hair? ______

4. What is your genotype for ear lobes? ______

5. What is your genotype for bent little finger? ______

6. Write the genotypes for your blood types (ABO and Rh). ______

7. Write the genotype for your singing voice. ______

8. What is your genotype for double-jointed thumbs? ______

9. What is your genotype for the long palmar muscle? ______

10. What is your genotype for hair color? ______

11. In the preceding (Nos. 1-10), which genotypes are you certain are correct? ______

    ______

12. In the cases where there is a possibility you are either homozygous or heterozygous, how could you determine your exact genotype? ______

    ______

Appendix 1 Formulas Used in Preparation of Solutions

Benedict's Solution

Copper sulfate (pure crystallized)	17.3 g
Sodium or potassium citrate	173.0 g
Sodium carbonate (crystallized)	200.0 g
(or 100 gm. anhydrous sodium carbonate)	
Distilled water, to make	1000.0 cc

Dissolve the citrate and carbonate in 700 cc of water, with the aid of heat, and filter. Dissolve the copper sulfate in 100 cc of water and pour slowly into the first solution, stirring constantly. Cool, and make up to 1 liter. The reagent keeps indefinitely.

Blood Anticoagulant Solution

Dissolve 12 g of sodium oxalate in 100 cc of 0.9% sodium chloride. To this add 4000 cc of freshly drawn blood; mix thoroughly.

Buffer Solution for Use with Wright's Stain

A solution of pH 6.4 is made by dissolving:

6.63 g monobasic potassium phosphate and
3.20 g dibasic sodium phosphate in
1 liter of distilled water.

Hayem's Solution

Mercuric chloride	0.5 g
Sodium sulfate	5.0 g
Sodium chloride	1.0 g
Distilled water	200.0 cc

Lugol's Iodine Solution

Iodine	1.0 g
Potassium iodide	2.0 g
Distilled water	300.0 cc

Millon's Solution

Dissolve 100 g of mercury in 200 g of concentrated nitric acid; dilute the solution with two volumes of water.

Ringer's Solution

Sodium chloride	7.000 g
Potassium chloride	0.075 g
Calcium chloride	0.100 g
Sodium bicarbonate	0.100 g
Distilled water, to make	1.000 liter

The sodium bicarbonate must be completely dissolved before the calcium chloride is added.

Turk's Solution

Distilled water	100.0 cc
Glacial acetic acid	1.0 cc
1% aqueous solution of gentian violet	1.0 cc

This solution must be filtered frequently to remove yeasts and molds.

Appendix 2 Weights and Measures

Linear Measurements

1 meter (m)	= 100 centimeters (cm) = 1000 millimeters (mm) = 39.37 inches (in) = 3.2808 feet = 1.09 yards
1 inch (in)	= 2.540 cm
1 millimeter (mm)	= 0.001 m = 0.1 cm = 1/25 in (approximately)
1 micron (μ)	= 0.000001 m = 0.001 mm = 10,000 Angstrom units (Å) = 1/25,000 in
1 millimicron (mμ)	= 0.000000001 m = 0.000001 mm = 10 Å = 1/25,000,000 in

Volumetric Measurements

1 liter (1)	= 1000 milliliters (ml) = 1000 cubic centimeters (cc) = 0.9 dry quart = 1.0567 liquid quarts (U.S. liquid)
1 fluid ounce (fl oz)	= 8 fluid drams (fl dr) = 29.573 cc
1 minim (min.)	= 1/60 fluid dram (fl dr) = 1/480 fluid ounce (fl oz)

Weight

1 kilogram (kg)	= 1000 grams (g) = 2.20462 pounds (avoirdupois) = 35.27 ounces
1 ounce (oz)	= 28.3495 g
1 gram (g)	= 1000 milligrams (mg) = 15 grains = 0.03527 ounce (avoirdupois)
1 milligram (mg)	= 1000 micrograms = 0.001 gram = 1/60 grain
1 microgram (mcg), (μg) or gamma	= 0.001 mg = 1/28,000,000 ounce (approximately)
1 grain (gr)	= 0.0648 g
1 dram (dr) (Apothecary)	= 60 grains = 3 scruples (scr)

Pressure

1 atmosphere (atm)	= 14.696 lb per sq in = 2116.3 lb per sq ft
1 atmosphere	= 29.92 in mercury (Hg) = 760 mm of mercury

Appendix 3 Respiration and Metabolism

Appendix 3-A. Dubois Body Surface Chart*

Directions

To find body surface of a patient, locate the height in inches (or centimeters) on Scale I and the weight in pounds (or kilograms) on Scale II and place a straight edge (ruler) between these two points which will intersect Scale III at the patient's surface area.

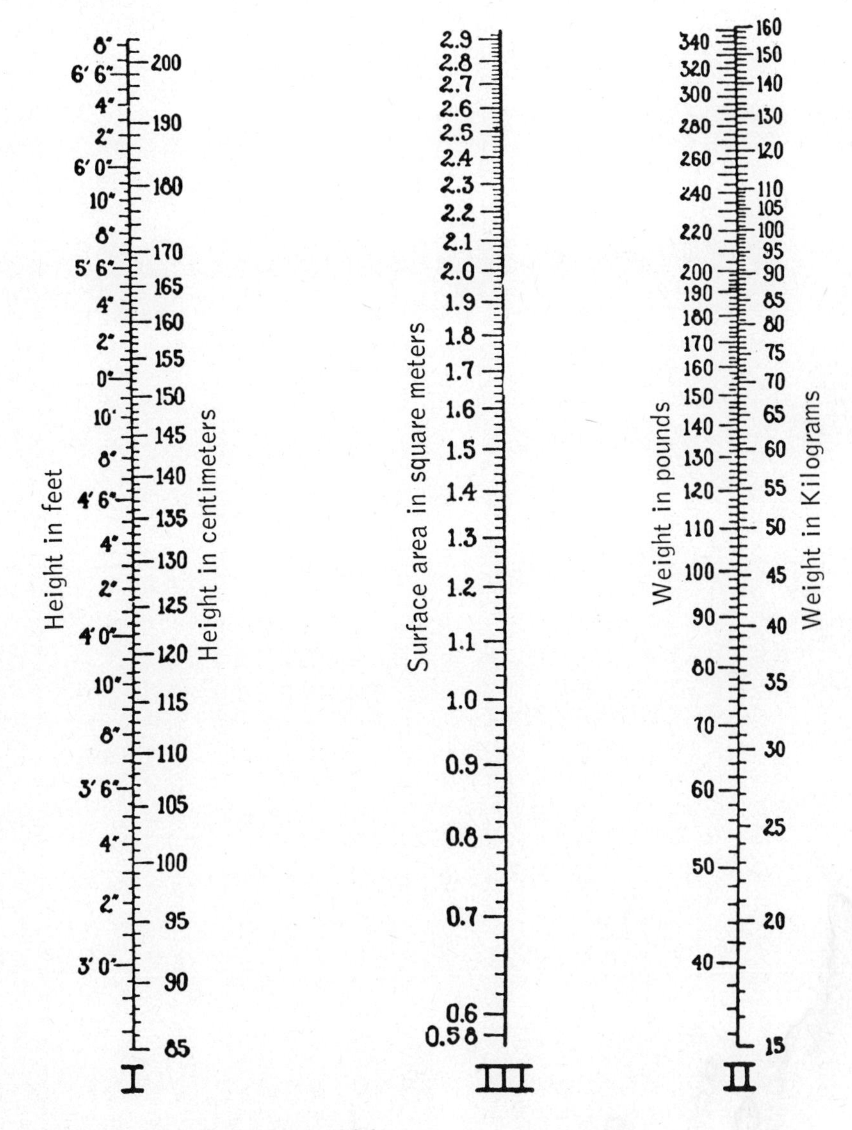

* As prepared by Boothby and Sandlford of the Mayo Clinic

Appendix 3-B. Predicted Vital Capacity—Males*

Height in Centimeters

Age	146	148	150	152	154	156	158	160	162	164	166	168	170	172	174	176	178	180	182	184	186	188	190	192	194
16	3765	3820	3870	3920	3975	4025	4075	4130	4180	4230	4285	4335	4385	4440	4490	4540	4590	4645	4695	4745	4800	4850	4900	4955	5005
18	3740	3790	3840	3890	3940	3995	4045	4095	4145	4200	4250	4300	4350	4405	4455	4505	4555	4610	4660	4710	4760	4815	4865	4915	4965
20	3710	3760	3810	3860	3910	3960	4015	4065	4115	4165	4215	4265	4320	4370	4420	4470	4520	4570	4625	4675	4725	4775	4825	4875	4930
22	3680	3730	3780	3830	3880	3930	3980	4030	4080	4135	4185	4235	4285	4335	4385	4435	4485	4535	4585	4635	4685	4735	4790	4840	4890
24	3635	3685	3735	3785	3835	3885	3935	3985	4035	4085	4135	4185	4235	4285	4330	4380	4430	4480	4530	4580	4630	4680	4730	4780	4830
26	3605	3655	3705	3755	3805	3855	3905	3955	4000	4050	4100	4150	4200	4250	4300	4350	4395	4445	4495	4545	4595	4645	4695	4740	4790
28	3575	3625	3675	3725	3775	3820	3870	3920	3970	4020	4070	4115	4165	4215	4265	4310	4360	4410	4460	4510	4555	4605	4655	4705	4755
30	3550	3595	3645	3695	3740	3790	3840	3890	3935	3985	4035	4080	4130	4180	4230	4275	4325	4375	4425	4470	4520	4570	4615	4665	4715
32	3520	3565	3615	3665	3710	3760	3810	3855	3905	3950	4000	4050	4095	4145	4195	4240	4290	4340	4385	4435	4485	4530	4580	4625	4675
34	3475	3525	3570	3620	3665	3715	3760	3810	3855	3905	3950	4000	4045	4095	4140	4190	4225	4285	4330	4380	4425	4475	4520	4570	4615
36	3445	3495	3540	3585	3635	3680	3730	3775	3825	3870	3920	3965	4010	4060	4105	4155	4200	4250	4295	4340	4390	4435	4485	4530	4580
38	3415	3465	3510	3555	3605	3650	3695	3745	3790	3840	3885	3930	3980	4025	4070	4120	4165	4210	4260	4305	4350	4400	4445	4495	4540
40	3385	3435	3480	3525	3575	3620	3665	3710	3760	3805	3850	3900	3945	3990	4035	4085	4130	4175	4220	4270	4315	4360	4410	4455	4500
42	3360	3405	3450	3495	3540	3590	3635	3680	3725	3770	3820	3865	3910	3955	4000	4050	4095	4140	4185	4230	4280	4325	4370	4415	4460
44	3315	3360	3405	3450	3495	3540	3585	3630	3675	3725	3770	3815	3860	3905	3950	3995	4040	4085	4130	4175	4220	4270	4315	4360	4405
46	3285	3330	3375	3420	3465	3510	3555	3600	3645	3690	3735	3780	3825	3870	3915	3960	4005	4050	4095	4140	4185	4230	4275	4320	4365
48	3255	3300	3345	3390	3435	3580	3626	3570	3615	3655	3700	3745	3790	3835	3880	3925	3970	4015	4060	4105	4150	4190	4235	4280	4325
50	3210	3255	3300	3345	3390	3430	3475	3520	3565	3610	3650	3695	3740	3785	3830	3870	3915	3960	4005	4050	4090	4135	4180	4225	4270
52	3185	3225	3270	3315	3355	3400	3445	3590	3530	3575	3620	3660	3705	3750	3795	3835	3880	3925	3970	4010	4055	4100	4140	4185	4230
54	3155	3195	3240	3285	3325	3370	3415	3455	3500	3540	3585	3630	3670	3715	3760	3800	3845	3890	3930	3975	4020	4060	4105	4145	4190
56	3125	3165	3210	3255	3295	3340	3380	3425	3465	3510	3550	3595	3640	3680	3725	3765	3810	3850	3895	3940	3980	4025	4065	4110	4150
58	3080	3125	3165	3210	3250	3290	3335	3375	3420	3460	3500	3545	3585	3630	3670	3715	3755	3800	3840	3880	3925	3965	4010	4050	4095
60	3050	3095	3135	3175	3220	3260	3300	3345	3385	3430	3470	3500	3555	3595	3635	3680	3720	3760	3805	3845	3885	3930	3970	4015	4055
62	3020	3060	3110	3150	3190	3230	3270	3310	3350	3390	3440	3480	3520	3560	3600	3640	3680	3730	3770	3810	3850	3890	3930	3970	4020
64	2990	3030	3080	3120	3160	3200	3240	3280	3320	3360	3400	3440	3490	3530	3570	3610	3650	3690	3730	3770	3810	3850	3900	3940	3980
66	2950	2990	3030	3070	3110	3150	3190	3230	3270	3310	3350	3390	3430	3470	3510	3550	3600	3640	3680	3720	3760	3800	3840	3880	3920
68	2920	2960	3000	3040	3080	3120	3160	3200	3240	3280	3320	3360	3400	3440	3480	3520	3560	3600	3640	3680	3720	3760	3800	3840	3880
70	2890	2930	2970	3010	3050	3090	3130	3170	3210	3250	3290	3330	3370	3410	3450	3480	3520	3560	3600	3640	3680	3720	3760	3800	3840
72	2860	2900	2940	2980	3020	3060	3100	3140	3180	3210	3250	3290	3330	3370	3410	3450	3490	3530	3570	3610	3650	3680	3720	3760	3800
74	2820	2860	2900	2930	2970	3010	3050	3090	3130	3170	3200	3240	3280	3320	3360	3400	3440	3470	3510	3550	3590	3630	3670	3710	3740

*Baldwin, E. DeF., Cournand, A., and Richards, D. W., Jr., Pulmonary Insufficiency 1, Physiologic Classification, Clinical Methods of Analysis, Standard Values in Normal Subjects/Medicine, 27:243, 1958.

Appendix 3-C. Predicted Vital Capacity—Females*

	Height in Centimeters																							
Age	146	148	150	152	154	156	158	160	162	164	166	168	170	172	174	176	178	180	182	184	186	188	190	192	194
16	2950	2990	3030	3070	3110	3150	3190	3230	3270	3310	3350	3390	3430	3470	3510	3550	3590	3630	3670	3715	3755	3800	3840	3880	3920
17	2935	2975	3015	3055	3095	3135	3175	3215	3255	3295	3335	3375	3415	3455	3495	3535	3575	3615	3655	3695	3740	3780	3820	3860	3900
18	2920	2960	3000	3040	3080	3120	3160	3200	3240	3280	3320	3360	3400	3440	3480	3520	3560	3600	3640	3680	3720	3760	3800	3840	3880
20	2890	2930	2970	3010	3050	3090	3130	3170	3210	3250	3290	3330	3370	3410	3450	3490	3525	3565	3605	3645	3695	3720	3760	3800	3840
22	2860	2900	2940	2980	3020	3060	3095	3135	3175	3215	3255	3290	3330	3370	3410	3450	3490	3530	3570	3610	3650	3685	3725	3765	3800
24	2830	2870	2910	2950	2985	3025	3065	3100	3140	3180	3220	3260	3300	3335	3375	3415	3455	3490	3530	3570	3610	3650	3685	3725	3765
26	2800	2840	2880	2920	2960	3000	3035	3070	3110	3150	3190	3230	3265	3300	3340	3380	3420	3455	3495	3530	3570	3610	3650	3685	3725
28	2775	2810	2850	2890	2930	2965	3000	3040	3070	3115	3155	3190	3230	3270	3305	3345	3380	3420	3460	3495	3535	3570	3610	3650	3685
30	2745	2780	2820	2860	2895	2935	2970	3010	3045	3085	3120	3160	3195	3235	3270	3310	3345	3385	3420	3460	3495	3535	3570	3610	3645
32	2715	2750	2790	2825	2865	2900	2940	2975	3015	3050	3090	3125	3160	3200	3235	3275	3310	3350	3385	3425	3460	3495	3535	3570	3610
34	2685	2725	2760	2795	2835	2870	2910	2945	2980	3020	3055	3090	3130	3165	3200	3240	3275	3310	3350	3385	3425	3460	3495	3535	3570
36	2655	2695	2730	2765	2805	2840	2875	2910	2950	2985	3020	3060	3095	3130	3165	3205	3240	3275	3310	3350	3385	3420	3460	3495	3530
38	2630	2665	2700	2735	2770	2810	2845	2880	2915	2950	2990	3025	3060	3095	3130	3170	3205	3240	3275	3310	3350	3385	3420	3455	3490
40	2600	2635	2670	2705	2740	2775	2810	2850	2885	2920	2955	2990	3025	3060	3095	3135	3170	3205	3240	3275	3310	3345	3380	3420	3455
42	2570	2605	2640	2675	2710	2745	2780	2815	2850	2885	2920	2955	2990	3025	3060	3100	3135	3170	3205	3240	3275	3310	3345	3380	3415
44	2540	2575	2610	2645	2680	2715	2750	2785	2820	2855	2890	2925	2960	2995	3030	3060	3095	3130	3165	3200	3235	3270	3305	3340	3375
46	2510	2545	2580	2615	2650	2685	2715	2750	2785	2820	2855	2890	2925	2960	2995	3030	3060	3095	3130	3165	3200	3235	3270	3305	3340
48	2480	2515	2550	2585	2620	2650	2685	2715	2750	2785	2820	2855	2890	2925	2960	2995	3030	3060	3095	3130	3160	3195	3230	3265	3300
50	2455	2485	2520	2555	2590	2625	2655	2690	2720	2755	2785	2820	2855	2890	2925	2955	2990	3025	3060	3090	3125	3155	3190	3225	3260
52	2425	2455	2490	2525	2555	2590	2625	2655	2690	2720	2755	2790	2820	2855	2890	2925	2955	2990	3020	3055	3090	3125	3155	3190	3220
54	2395	2425	2460	2495	2530	2560	2590	2625	2655	2690	2720	2755	2790	2820	2855	2885	2920	2950	2985	3020	3050	3085	3115	3150	3180
56	2365	2400	2430	2460	2495	2525	2560	2590	2625	2655	2690	2720	2755	2790	2820	2855	2885	2920	2950	2980	3015	3045	3080	3110	3145
58	2335	2370	2400	2430	2460	2495	2525	2560	2590	2625	2655	2690	2720	2750	2785	2815	2850	2880	2920	2945	2975	3010	3040	3075	3105
60	2305	2340	2370	2400	2430	2460	2495	2525	2560	2590	2625	2655	2685	2720	2750	2780	2810	2845	2875	2915	2940	2970	3000	3035	3065
62	2280	2310	2340	2370	2405	2435	2465	2495	2525	2560	2590	2620	2655	2685	2715	2745	2775	2810	2840	2870	2900	2935	2965	2995	3025
64	2250	2280	2310	2340	2370	2400	2430	2465	2495	2525	2555	2585	2620	2650	2680	2710	2740	2770	2805	2835	2865	2895	2925	2955	2990
66	2220	2250	2280	2310	2340	2370	2400	2430	2460	2495	2525	2555	2585	2615	2645	2675	2705	2735	2765	2800	2825	2860	2890	2920	2950
68	2190	2220	2250	2280	2310	2340	2370	2400	2430	2460	2490	2520	2550	2580	2610	2640	2670	2700	2730	2760	2795	2820	2850	2280	2910
70	2160	2190	2220	2250	2280	2310	2340	2370	2400	2425	2455	2485	2515	2545	2575	2605	2635	2665	2695	2725	2755	2780	2810	2840	2870
72	2130	2160	2190	2220	2250	2280	2310	2335	2365	2395	2425	2455	2480	2510	2540	2570	2600	2630	2660	2685	2715	2745	2775	2805	2830
74	2100	2130	2160	2190	2220	2245	2275	2305	2335	2360	2390	2420	2450	2475	2505	2535	2565	2590	2620	2650	2680	2710	2740	2765	2795

*Baldwin, E. Def., Cournand, A., and Richards, D. W., Jr., Pulmonary Insufficiency 1, Physiologic Classification, Clinical Methods of Analysis, Standard Values in Normal Subjects/Medicine, 27:243, 1958.

Appendix 4 Definitions for Genetics

1. Acquired character: a change from the normal type brought about by environmental influence.
2. Allele: one of two normally alternate forms of a gene, one being the normal type, the other the mutant type. When three or more alleles exist for any given locus (site) in the chromosome, they form a multiple allele series. Such cases are less frequent than those with only two alleles at a locus.
3. Autosomes: chromosomes other than sex chromosomes.
4. Backcross: a cross of an F_1 hybrid to either of its parents.
5. Carrier: an individual heterozygous for a recessive gene that is not expressed. In appearance it is indistinguishable from the homozygous dominant.
6. Chromosomes: a deep-staining, rodlike body in the nucleus of cells, visible at cell division. Chromosomes contain the genes, the hereditary determiners. All species have characteristic chromosome number.
7. Cumulative effect: the action of two alleles of a gene giving a more pronounced effect than one in the heterozygous condition. Probably one allele supplies insufficient enzyme for the reaction conditioned by two. The hybrids are distinguishable from parents. The effect is sometimes referred to as incomplete dominance.
8. Dihybrid cross: a cross of two pairs of genes; one pair for one trait and the other pair for a different trait.
9. Diploid (2n) (the species number of chromosomes): the chromosome number just twice the haploid gametic number (n). At meiosis homologous chromosomes form pairs; hence the name, meaning double or twofold (see Haploid).
10. Dizygotic twins: fraternal twins, produced by the fertilization of two separate eggs by two different spermatozoa.
11. Dominance: complete suppression of the expression of one allele in a chromosome by another allele at the same locus in the homologous chromosome.
12. Egg: the female germ cell; in animals, an ovum.
13. Eugenics: the science concerned with improvement of the heredity of the human race.
14. F_1: abbreviation for first filial generation, usually the hybrid between two homozygous types.

15. F_2: abbreviation for the second filial generation, usually produced by intercrossing two F_1's or by self-fertilizing an F_1, as is possible in plants.

16. Fertilization: the union of a male gamete (sperm) with a female gamete (ovum) to form a zygote which develops into an embryo.

17. Fraternal twins: see Dizygotic twins.

18. Gamete: a mature reproductive male or female germ cell, sperm, or egg.

19. Gene: the discrete particulate hereditary determiner located in the chromosome in linear order; the "element" of Mendel and "factor" of early genetic terminology. The gene is composed mostly of DNA. It was discovered by Mendel and named by Johannsen.

20. Genetics: the science of heredity of the similarities and differences among organisms.

21. Genotype: the genetic constitution or gene makeup of an organism.

22. Germ cell: a mature sex cell; a gamete, capable of uniting in fertilization with another gamete to reproduce the organism.

23. Germ plasm: a special kind of protoplasm transmitted from generation to generation. The chromosomes and genes constitute the germ plasm; this is the physical basis of heredity.

24. Gonad: a gland, such as an ovary or testis, in which gametes (sex cells) are produced.

25. Haploid: having half the number of chromosomes (gametic number) found in diploid organisms (half the species number of chromosomes).

26. Hemizygous: having but one gene (or chromosome) present, with no homologous allele. Genes on the X chromosome and Y chromosome of human males are hemizygous.

27. Heredity: transmission of traits from parent to offspring.

28. Heritability: the extent to which a given trait is determined by inheritance.

29. Heterozygous: the chromosome of one parent provides one form of allele for a trait, while the homologous chromosome (from the other parent) contains a different allele (a different form of the gene) at the same locus for the same trait.

30. Homologous chromosomes: a pair of chromosomes which are similar in size, shape, structure, and function. One of the pair is provided by the male parent, the other by the female parent. In humans, the male spermatozoon contains 23 chromosomes which are homologous to those chromosomes in the ovum (except for the X chromosome which pairs with the Y chromosome or another X chromosome).

31. Homozygous: the homologous chromosomes of one parent contain an allele for a trait that is exactly like the allele contained at the same locus on the homologous chromosome provided by the other parent.

32. Hybrid: a cross of unlike organisms (i.e., organisms with different traits – usually they are of the same species, but crosses have been made between closely related species, the offspring of which are usually sterile).

33. Identical twins: see monozygotic twins.

34. Inbreeding: mating of related types; **consanguineous** matings (i.e., the mating of relatives).

35. Incomplete dominance: lack of dominance. See Cumulative effect.

36. Lethal gene (lethal mutation): a gene which renders the gamete nonviable (gametic lethal) or one which prevents the complete development of a zygote (zygotic lethal). Zygotic lethals may produce their effects at any time from the fertilized egg on.

37. Linkage: the association of two or more characters in inheritance caused by two genes associated in the same chromosome. The nearer the genes to each other, the closer the linkage.

38. Linkage group: a group of characters all showing linkage with others in the same group. The genes for these characters are arranged linearly in one chromosome, the genes close together showing close linkage.

39. Locus (pl. loci): the physical location of a gene in the chromosome. Genes, normally, have two (sometimes more) alleles which may occur at a given locus (only one allele may occupy the locus at any given time).

40. Meiosis: a special type of cell division found in gamete production. It consists of two divisions, one of which is reductional. Homologous chromosomes pair and assort at random to produce gametes with the haploid (n) number of chromosomes.

41. Mendel's law: the principle that hereditary characters are determined by discrete particles (genes) that segregate at random in gamete formation (see Linkage).

42. Mitosis: cell division in which chromosomes are all divided equally in a manner so that daughter cells have exactly the same number and composition of genes as the original chromosomes in the original cell.

43. Monohybrid: a cross involving a single pair of genes for a given trait. (Also known as a monogenic cross.)

44. Monozygotic twins: twins that have developed from a single fertilized ovum (egg) that divided to form two zygotes with identical genotypes; identical twins.

45. Multiple alleles: where three or more alleles may occur at a single locus on a chromosome. All relate to the same trait (they may not occupy the same locus at the same time).

46. Mutable gene: an unstable gene that mutates frequently.

47. Mutagen: any substance, chemical or physical, that causes genes to mutate.

48. Mutant: a variant from the normal gene that is inherited in a Mendelian manner.

49. Mutation: a change that is inherited (see Mutant).

50. n, 2n: the gametic (haploid or n) and zygotic (diploid or 2n) chromosome numbers, respectively.

51. Outbreeding: mating of unrelated individuals or of individuals not closely related; the opposite of inbreeding.

52. Ovum: see Egg.

53. P: abbreviation for parent.

54. Parthenogenesis: reproduction without fertilization.

55. Pedigree: (1) the genotype of an individual; (2) a chart showing the ancestral history of an individual.

56. Phenotype: the appearance of an individual. The appearance is produced by the genotype in combination with the environment.

57. Predisposition: a condition of special susceptibility to a disease.

58. Probability: the chance or likelihood of a given possible event.

59. Progeny: offspring; individuals resulting from a mating.

60. Punnett square (after R. C. Punnett): the conventional checkerboard with male gametes on one axis and female gametes on the other. Zygotes are obtained by multiplying all male gametes by all female gametes.

61. Recessive: action of a recessive allele is suppressed by the dominant allele for the same trait.

62. Sex chromosome: a chromosome usually designated X or Y, particularly concerned with sex

determination; e.g., in *Drosophila* (fruit flies) and mammals including man, the female is XX (has two X chromosomes) and the male is XY.

63. Sex linkage: an unusual type of inheritance caused by genes located in the X chromosome. The Y chromosome is practically devoid of genes and those present are not homologous with those located on the X chromosome. The Y chromosome is not a carrier of alleles in sex-linked problems.

64. Sex-promoting alleles: certain alleles concerned with sex determination. Females are always heterozygous, males homozygous or hemizygous.

65. Sib mating: the mating of sisters and brothers (s - sister; i - with; b - brother; thus, sib).

66. Syndrome: a group of symptoms that occur together and characterize a disease.

67. Test cross: a cross of the heterozygote to an individual containing only recessive alleles of the gene in question.

Appendix 5 The Determination of pH

The concentration of hydrogen ion, H^+, in the environment (e.g., in food, water, solutions, etc.) and in living organisms (both plant and animal) is of great physiological importance. This ion is universally present in aqueous systems and has marked influence on the functions of living systems. It affects the solubility of all other elements as well as the availability and absorption of ions, the stability of proteins and enzyme systems, and numerous other functions of plants and animals.

The concentration of this ion is so important that a special scale (the pH scale) has been established to simplify its expression. The pH is the negative logarithm of the hydrogen ion concentration in a solution.

$$\text{Therefore, } pH = -\log (H^+)$$

In pure water the concentration of H^+ is equal to the concentration of OH^- (hydroxyl ion) and each is equal to 1×10^{-7} moles/liter. Since pH is defined as the negative logarithm of the hydrogen ion concentration, then in pure water, at 25°C., $pH = -\log (1 \times 10^{-7}) = \log 10^7 = 7.0$.

The pH scale of commonly used solutions extends from 0 to 14, with a pH of 7 representing neutrality. Pure water (lacking minerals and carbon dioxide) would have a pH of 7. As the pH goes up, the hydrogen ion concentration decreases, and the solution becomes more alkaline (basic). As the pH of a solution goes down, the hydrogen ion concentration increases, and the solution becomes more acidic. Since the pH scale is a logarithmic scale, a pH of 6 (10^{-6} moles/liter) would indicate an acid solution with a concentration of hydrogen ions ten times greater than that at a pH of 7.0 (10^{-7} moles/liter). The hydrogen ion concentration would be a hundred times greater at pH 6 (10^{-6} moles/liter) than at pH 8 (10^{-8} moles/liter).

Thus we see that acidity and alkalinity can be expressed in terms of pH. A pH less than 7.0 indicates an acid solution, while a pH greater than 7.0 indicates an alkaline solution.

Methods for Determining Hydrogen Ion Concentration (pH)

Taste

In general, acids have a sour taste while alkaline substances have a more or less bitter taste. For example, a soda cracker may have a pH as high as 8.5 (alkaline) and lemon may have a pH as low as 2.2 (acid). It is well to remember, however, that sourness may be masked by sugar. Thus a ripe apple may taste sweet but have a pH as low as 2.9.

Litmus Paper

Litmus paper is paper impregnated with the organic dye, litmus, obtained from several species of lichens. This paper is ordinarily provided in short strips and one end of the strip is immersed in the solution to be tested. Blue litmus paper turns red in an acid medium, while red litmus paper turns blue in an alkaline medium.

Hydrion Paper

This paper comes in a special dispenser from which strips of from one to two inches at a time may be removed. A portion of this is placed in the sample to be tested and then compared with a color scale. This is a quantitative measurement (the foregoing methods are qualitative) and is accurate to within a single unit of the pH scale.

Indicators

These are dyes which are used to test the pH of a solution. As the hydrogen ion content of the solution changes, rearrangement of the indicator molecule occurs, which results in a change in color. Indicators may also provide quantitative measurements of pH, though not as accurately as Hydrion papers.

pH Meters

These are instruments designed for rapid and accurate pH measurements of solutions, pastes, and semisolids. They are accurate, when properly standardized, to 0.10 pH unit, and determinations may be made in the 0-14 pH range.